SHORT-WAVELENGTH MAGNETIC RECORDING:
NEW METHODS AND ANALYSES

SHORT-WAVELENGTH MAGNETIC RECORDING:

NEW METHODS AND ANALYSES

JAAP J. M. RUIGROK

*Philips Research Laboratories,
Eindhoven, The Netherlands*

ELSEVIER
ADVANCED
TECHNOLOGY

Published by
ELSEVIER ADVANCED TECHNOLOGY
Mayfield House, 256 Banbury Road, Oxford OX2 7DH, UK

Distributed in the USA and Canada by
ELSEVIER SCIENCE PUBLISHING CO., INC.
655 Avenue of the Americas, New York, NY 10010, USA

WITH 32 TABLES AND 213 ILLUSTRATIONS

© 1990 ELSEVIER SCIENCE PUBLISHERS LTD

British Library Cataloguing in Publication Data

Ruigrok, J. J. M.
Short-wavelength magnetic recording.
1. Magnetic recording
I. Title
621.389′32

ISBN 0-946395-56-X

Library of Congress CIP data applied for

To my parents, Ada and our children:
Imke, Elmer, Anouk, Jorrit and Renske

Printed in Northern Ireland by The Universities Press (Belfast) Ltd.

Contents

Chapter 3. An alternative expression for the read flux in magnetic recording theory   53

Chapter 4 Further considerations concerning the alternative reciprocity theorem   63

Chapter 5 Sensitivity functions, gap loss functions and the efficiency   77

Chapter 6 Special and limiting cases of the general flux expression

Chapter 8 Video-head parameters: model and measurements

Chapter 9 Probe-type heads: models for efficiency and auxiliary-pole effect 375

Chapter 10 Cross measurements in magnetic recording: a new way of determining head performance

Chapter 11 Analytical description of thin-film yoke magnetoresistive heads

Chapter 13 Bandpass heads 499

Glossary of special definitions, symbols and notations (*)

Definitions

static $d/dt = 0$

quasi static $\varepsilon\ d/dt = 0$ and $\mu\ d/dt \neq 0$

retardation free   finite wave travel time neglected

dispersion   frequency dependent material property assumed

instantaneously
   reacting   no after-effects

time invariant   non-parametric material properties

Symbol(**)	Description
A	area
$\boldsymbol{A}$	tensor related to gap irregularity (5.84)
a	superscript denoting 'writing' state
a	subscript for auxiliary pole (chapter 9)
a	accuracy (chapter 5), auxiliary pole effect (chapter 9), head-to-tape distance (chapter 12), distance between gap centres (chapter 13)
a_d	transition width limited by self-demagnetization (a7.9)
a_t	transition width
ac	alternating current
$\boldsymbol{B}$	magnetic induction (p. 20)
B	bandwidth
$\boldsymbol{B}^b$	magnetic induction in 'reading' state on arbitrary surface (p. 59)
$\mathscr{B}^b$	magnetic induction in 'reading' state on fictitious equipotential surface (p. 60)

(*) This glossary lists the major special symbols, definitions and notations used in this thesis. Some symbols and notations, that are used only locally and explained there, have been left out of this list.
(**) The symbols printed in the actual text in roman (X) are scalars, in bold face (**X**) are vectors and the letters in bold face sanserif (**X**) are tensors or matrices.

BPH	bandpass head (chapter 13)
BYMRH	barberpole yoke magnetoresistive head (chapter 11)
b	superscript denoting 'writing' state (p. 59)
C	cross accuracy figure (10.3)
C or c	constant
Ci	cosine integral, see (5.44)
c	subscript for coercive, characteristic
$\boldsymbol{D}$	electric displacement (p. 20), head-to-backlayer distance
DL	distance loss, see (2.64)
DL	double layer
d or dem	subscript for demagnetization
d	head-to-tape distance, harmonic distortion (11.18)
d_4	head- (or smear layer-) to-tape distance
dc	direct current
dg	subscript for deep gap (p. 103)
$\boldsymbol{E}$	electric field
E	Young's modulus of elasticity (p. 290)
E_f	free energy density (p. 175)
EMM	exponentially-decaying magnetization model (section 7.2.2.3)
e	natural number (2.71828)
$\boldsymbol{e}_x, \boldsymbol{e}_y, \boldsymbol{e}_z$	unit vectors in x, y and z directions
eff	subscript for effective
F_R	remagnetization factor (2.46)
FFT	fast Fourier transform (p. 109)
FSF	flux sensitivity function (p. 80)
f	frequency, function
f_c	correction function (section 5.7.4)
$f(k; y)$	function of y and x Fourier transformed from x to k domain
G	generalized gap length of collection of gaps (13.5c)
GDL	generalized distance loss factor (a7.2)
GGL	gradual gap loss factor (sections 7.4.1, 5.7 and appendix 7.1)
GIL	gap-irregularity loss factor (section 5.9)
GLF	gap loss function (section 5.3)
GSL	gap-smear loss factor (2.65)
g	gap length, generalized gap length (5.50, 13.5d), 'shape function' of collection of gaps (Fig. 13.7)
gK	subscript, gradual Karlqvist (section 5.7.4)
$\boldsymbol{H}$	magnetic field
H_{an}	anisotropy field (p. 293)
H_{an}^{i}	induced anisotropy field (p. 110)
H_{appl}	applied field (field minus demagnetizing field of sample)

H_c	coercivity field of major loop (fig. 7.4.1)
H_{cr}	critical field of particle (p. 184)
H_f	free space field near head-facing tape surface (2.33)
H_m	maximum field in the tape (p. 205, 217)
H_{sw}	switching field of particle (p. 185)
H_{thr}	coercivity field of minor loop (fig. 7.4.1)
h	subscript for harmonic, head, hull curve
hifi	high fidelity
$\boldsymbol{h}^a$	normalized (divided by I^a) magnetic field in writing state (when h is Fourier transformed to the ω domain, I^a is too)
h	Planck's constant ($6.62620\ 10^{-34}$ Js), gap height, main pole (probe) height
h	reduced constant of Planck, i.e. $h/2\pi$
I	current
I_0, I_1	modified Bessel functions of the 1st kind and of order 0 and 1
Im	imaginary part of
ISL	ideal single layer ($\mu_2 = 1$) (p. 172)
i	subscript for ideal, intersection, intrinsic
iK	subscript, ideal Karlqvist
int	subscript, internal
J_0	ordinary Bessel function of the first kind and zero order
j	$\sqrt{-1}$
$\boldsymbol{j}$	current density
K	anisotropy energy density (p. 367)
K_0 or K_1	modified Bessel functions of the 2nd kind and of order 0 and 1
$\boldsymbol{K}, \boldsymbol{\chi}$	magnetic susceptibility tensor (p. 56, 174)
k	angular wavenumber, Boltzman constant ($1.38062\ 10^{-23}$ J/K)
L	subscript for load
L	inductance (p. 266)
LPL	low permeability loss factor (p. 226)
$\boldsymbol{M}$	magnetization
ME	metal evaporated tape
MP	metal powder tape
MS	subscript for magnetostrictive
MIG	metal in gap (chapter 12)
MRE	magnetoresistive element (chapter 11)
$\boldsymbol{M}_i$	induced part of magnetization (p. 56)
M_p	remanent magnetization of minor loop (p. 222)
M_r	remanent magnetization of major loop (p. 222)
m	subscript for main pole (chapter 9)
max	subscript, maximum
min	subscript, minimum

N	number of turns, demagnetization ratio, noise power
N_g	number of gaps (chapter 13)
n	complex refraction index (p. 544)
$\boldsymbol{n}$	unit vector normal to a surface
n	subscript for normalized (reduced)
$\mathbb{O}$	order of magnitude
opt	subscript for optical, optimum
P	(complex) permeance, playback figure (10.2)
PH	probe head (section 9.3-4, appendix 9.1)
PHRY	probe head with return yoke (section 9.5-6, appendix 9.2)
PHRYP	planary probe head with return yoke (section 9.7-8, appendix 9.3)
P_{ab}	permeance of head circuit at the 'source terminals' a and b
p	subscript for plane, playback
p	probability-density function (p. 117)
$p_{1,2}$	thin film or metal layer thickness (p. 127, 443, 468)
Q	quality factor (8.16)
R	reluctance, resistance, recording figure (10.1), remagnetization parameter (2.32)
R	superscript, reading
R_r	equivalent noise resistance (voltage noise) (p. 542)
R_n	equivalent noise resistance (current noise) (p. 542)
Re	real part of
RLx	reproduction loss for longitudinal recording
RLy	reproduction loss for perpendicular recording
$\boldsymbol{r}$	distance vector in space (x, y, z)
r	subscript for relative, recording
r0	subscript for rotational and zero frequency
rel	subscript for relative
res	subscript for resonance (often undamped)
$\boldsymbol{S}$	squareness vector (section 6.4.7)
S	squareness ($\equiv M_r/M_s$) (Fig. 7.4.I), shape function (13.5a, 13.7)
S^*	coercivity squareness (Fig. 7.4.I)
SEM	scanning electron microscope
Si	sine integral (5.43)
SL	single layer
s or sat	subscript for saturation
$\text{sign}(x)$	sign ($+$ or $-$) of x
$\text{sinc}(x)$	$\sin(\pi x)/\pi x$
str	subscript for stray
T	main pole (probe) thickness (chapter 9)

$\boldsymbol{T}$	orthonormal-transformation matrix (section 6.4.4)
T	superscript: transposed (p. 179)
TFH	thin film head (p. 126, chapter 11)
t	time
t_2	backlayer thickness (fig. 2.1)
t_3	coating thickness (fig. 2.1)
t_x	writing depth for longitudinal recording (p. 152)
t_y	writing depth for perpendicular recording (p. 152)
t_5	smear-layer thickness (fig. 2.1)
UMM	uniform magnetization model (sections 6.3.1, 7.2.2.1)
V	voltage
V_p	permanently magnetized domain (fig. 3.1)
V_i	domain with induced reversible magnetization (fig. 3.1)
VSM	vibrating-sample magnetometer (p. 223)
$(V)SF$	(voltage) sensitivity function (p. 79)
v	velocity
v_m	velocity of light in a medium (a13.11)
W	track width
W, w	overall width and mean (internal) peak width (p. 510)
x, y, z	coordinates in cartesian reference frame, subscripts for longitudinal, perpendicular and transversal, respectively
x', y', z'	coordinates in cartesian reference frame of particle (chapter 6)
YMRH	yoke magnetoresistive head (chapter 11)
Z	impedance, complex reluctance (chapter 8)
Z_{ab}	reluctance of head circuit at the 'source terminals' a and b
α	empirical write-distance loss coefficient, Gilbert's damping constant (appendix 8.1), winding filling factor (chapter 9), homogeneity coefficient (chapter 11)
α_2	tape parameter (2.37)
α_3	tape parameter (2.34)
α'_{32}	tape parameter (2.39)
α_{32}	tape parameter (2.35)
$\alpha_{1,2,3}$	direction cosines
β	anisotropy constant, $\beta \equiv \sqrt{\mu_x/\mu_y}$
Γ	reflection coefficient (11.6, a13.10)
γ	Euler's constant ($= 0.57721.....$), see (5.44), gyromagnetic ratio (p. 369)
δ	Dirac delta function, skin depth (p. 351)
Δ	difference operator
Δg	mean width of gap correction function (p. 105, roughly equal to the difference between generalized and optical gap

	length), difference between local gap length and average gap length (section 5.7)
ε	permitivity (p. 543)
$\eta_{(H)}$	efficiency ($\equiv$ field efficiency) (8.17, 9.2)
η_Φ	flux efficiency (8.20)
θ	angle, polar angle of major axis of particle with respect to x axis (Fig. 6.13)
θ'	angle between magnetization and major axis of particle (Fig. 6.12)
θ''	angle between field and major axis direction of particle (p. 184)
$\varkappa$	absorption index for light (p. 544)
Λ	triangle function (5.62)
λ	wavelength, magnetostriction coefficient (p. 367)
λ_c	characteristic length (11.4c, p. 49)
λ_0	gap-null wavelength
λ_m	wavelength of light in a medium (a13.11)
μ	relative (unless otherwise mentioned) magnetic permeability
μ'	real part of (relative) magnetic permeability
μ''	minus imaginary part of (relative) magnetic permeability
μ_0	magnetic permeability of vacuum ($4\pi\,10^{-7}$ Vs/Am)
μ_r	relative permeability (subscript r often omitted)
$\mu_{1,2,\text{etc}}$	relative permeability parameter, $\mu \equiv \sqrt{\mu_x \mu_y}$, of layer 1, 2, etc.
μ_{32}	tape parameter (2.36)
$\mathbf{v}$	unit vector normal to a surface, (Figs. 3.2 and 4.1),
ν	volumetric packing density
Π	rectangle function (5.60), product
π	pi ($= 3.1415926535...$)
ϱ	resistivity
Σ	summation
σ	standard deviation (p. 105), stress (positive if tensile) (p. 367), conductivity
σ^2	variance (p. 105)
σ_c	compressive stress (positive if compressive)
Φ	magnetic flux
Φ^R	read flux encompassed by filamentary reading coil (3.18, 4.17)
Φ_{ix}	longitudinal internal flux (6.14)
Φ_{iy}	perpendicular internal flux (6.15)
φ	angle, phase angle, azimuthal angle concerning particle orientation (with respect to the tape's $z = 0$ plane (Fig. 6.13))

φ'	azimuthal angle concerning magnetization direction in particle (with respect to the particle's $z' = 0$ plane (Fig. 6.12))
$\boldsymbol{\chi}$ or $\boldsymbol{K}$	magnetic susceptibility tensor (p. 56, 174)
Ψ	magnetic scalar potential (3.5)
ψ	angle, rotation angle of minor axes of particle (6.13)
ψ^{a}	normalized (divided by I^{a}) magnetic scalar potential of head in 'writing' state (when ψ^{a} is Fourier transformed to the ω domain, I^{a} is too) (3.16, 4.17)
ω	angular frequency
$\sim$	proportional to
∇	differential vector operator (nabla)
$\nabla\varphi$	gradient
$\nabla \cdot \boldsymbol{A}$	divergence
$\nabla \times \boldsymbol{A}$	curl
∇^{2}	Laplacian
$|\;|$	modulus (of complex or real number), length (of vector)
∞	infinity
$\hat{}$	complex number or time-independent quantity
$*$	convolution, if superscript then complex conjugate
$\times$	product, vector product
$\cdot$	scalar product, if above character then d/dt
1,2,3,4,5,6	subscript for free space, backlayer, coating, free space between tape and gapsmear, gap smear and infinite-permeability half-space, respectively (fig. 2.1)
$\equiv$	similar or per definition equal to
$\simeq$ or $\approx$	approximately equal to
$\lesssim$	less than or approximately equal to
$\gtrsim$	greater than or approximately equal to
$<$	less than
$>$	greater than
$\leqslant$	less than or equal to
$\geqslant$	greater than or equal to
$\leqslant$ or $\ll$	$\lesssim 0.1$ times
$\geqslant$ or $\gg$	$\gtrsim 10$ times
$\gg$	not $\gtrsim 10$ times
$\lll$	$\lesssim 100$ times
$\ggg$	$\gtrsim 100$ times
[001]	crystallographic directions
(001)	crystallographic plane with normal [001]
(001)	subscript for components in (001) plane
$\parallel$	parallel, $z_1 \parallel z_2 \equiv z_1 z_2/(z_1 + z_2)$
$\perp$	perpendicular
0	subscript for nominal, zero frequency

Curriculum vitae

Jaap J.M. Ruigrok was born in Leiden, the Netherlands, in 1951. He received the B.Sc. degree in electronic and radio engineering from the H.T.S. voor Radiotechniek en Elektronika in Haarlem in 1971 and in electrical engineering from the H.T.S. in The Hague in 1973. During the H.T.S. period he fullfilled his military service and worked for some periods at the astronomical observatory in Leiden and the radio telescope in Westerbork. He continued his study at the Delft University of Technology, where he received the M.Sc. degree (cum laude) in 1979. In this period he published some articles about photolithography and bubble dynamics.

He continued his investigations on the dynamics of bubbles in parabolic potential wells at the California Institute of Technology. Since the end of 1979 he is with the magnetics division of Philips Research Laboratories in Eindhoven. Till 1981 his work focussed on thin-film magnetoresistive heads for audio applications and from 1982 on the design of metal-in-gap ferrite heads, amorphous heads for high-coercive video tape, magnetic heads using superconductive films and squids, and heads for magneto-optic recording. This book reflects several results of this work, for which he recently received the doctor's degree at the University of Twente. His scientific interest covers both electronics and physics.

Acknowledgements

The investigations described in this book have been performed at Philips Research Laboratories in Eindhoven, the Netherlands. The author is therefore grateful to the Board of Directors of this Laboratory and to Dr. P.F. Bongers and Dr. U. Enz for permission to publish results of this research in this book and for twice allowing me to remain free from other business for several months in order to write the manuscript. I also with to thank Jelto Smits for encouraging me to publish parts of my work over the past ten years.

Furthermore I wish to thank the following individuals of the Philips Research Laboratories: Dick Tjaden for the helpful discussions concerning mathematical problems (in chapter 2 and 9 and appendix 7.2), and Bert de With and Eelco Visser for discussions concerning mechanical stress. Gratefully acknowledged is the contribution of Dirk Quak of the Delft University of Technology to the proof of the alternative reciprocity theorem in chapter 3. Prof. A.T. de Hoop is acknowledged for checking the analysis in this chapter.

I am very grateful for the enthusiastic support given by Ronald van Rijn who carried out numerical calculations for obtaining the numerous plots in this book and also carried out, together with Huub Maas, many of the experiments. The hardware and software of the rotating-tape measuring equipment used in several recording experiments described in this book are due to Huub Maas and Ronald van Rijn respectively. Thanks are due to Jean-Paul Morel, Jan Nooijen, Ad Sars, and Gerben Snijders for their contributions to the hardware and software of the helical-scan measuring machine used for almost all measurements in chapter 10. Part of the experiments in chapter 11 were carried out by Henk Landman. The VSM measurements in chapter 7 were carried out by Jan Bernards and the ellipsometry measurements used in appendix 13.3, by Jan Willem Martens.

Special gratitude goes to Prof. Jan Fluitman for carefully reading great parts of the manuscript and checking numerous analytical calculations in chapters 2, 4, 5 and 6. My colleague Steven Luitjens is greatfully acknowledged for reading and commenting upon especially the chapters 7, 10, 12 and 13, and my colleague Victor Zieren for reading and commenting upon chapter 2 to 6, 9, 10 and 12. Ernst Huijer of the Videq Head Laboratory is acknowledged for reading and commenting upon chapter 8. Also acknowledged is the reading and comments concerning other parts of the manuscript by Nol Broese van Groenou, Peter Sillen and Ulrich Enz. Sillen, Enz and Broese van Groenou and also Jean-Paul Morel of the Videq Head Laboratory and Dipl. Ing. Binder Krieglstein of Philips Videowerke Wien and members of the Head Workshop contributed to the development and realization of the MIG heads described in chapter 12. The section about MIG head technology, 12.2.5, and the section about the materials applied in the heads, 12.2.4, are due to Peter Sillen, assisted by Anita Bode. Peter Sillen is also responsible for the accurate manufacturing of the bandpass head described in chapter 13.

Thanks are also due to Hans Verbunt and the editorial staff of Philips Technical Review for the kind permission to incorporate parts of ref. [1.15] in the introduction, to Jean-Paul Morel for kind permission to publish the measuring data used in section 12.1.7 and Dr. W. Tuma of Philips Videowerke Wien for kind permission of publishing the experimental data used in section 10.6.

In the realization of the manuscript many persons and firms were involved. The drawings are from the graphical design group Delta-H and from Henny Alblas. Several corrections in the drawings are from Helmut Gassen. Type-setting is carried out by Pecasse Intercontinental B.V. in Maastricht, of which we want to thank especially the type-setters Mr. M. Woisch, Mr. P.G.M. Mommers and the corrector Mr. B.W. van der Woude. Ronald van Rijn, sometimes assisted by his wife Roze-Marie, proofread various versions of the manuscript. The english corrections are from K. Gilbert, G. Luton, D. Crowder, G. Bateman and my colleague Marc Johnson. Theo Schoenmakers of the 'Audio Visuele Dienst' was involved in the layout of the final manuscript.

Finally many thanks are due to my wife for her patience in enduring part of this work at home.

Foreword

It is, indeed, a great honour to provide a few words of introduction for Dr Jaap J. M. Ruigrok's book *Short-Wavelength Magnetic Recording.* The scope and mathematical rigour of this book can only be compared with W. K. Westmijze's 1953 landmark *Studies in Magnetic Recording.* It is difficult to believe that there now can remain unanswered any major questions about the reading (or reproducing) process in magnetic recording.

Moreover, Ruigrok is to be congratulated upon his easy writing style, which renders most of the mathematical treatments readily understandable as physical propositions. A careful study of this book cannot help but provide the reader with the most profound insights into the limits of short-wavelength recording—even my earlier published error, of a factor of 2 (p. 244), is noted and corrected!

JOHN C. MALLINSON
Center for Magnetic Recording Research
University of California, San Diego, USA

Chapter 1

Introduction

In this introductory chapter we will give a survey of the contributions made by this book to magnetic recording theory, models and measuring methods. The work is mainly inspired by practical needs concerning the development and design of heads. Firstly, a short historical overview of magnetic recording and the basic principles of this type of recording are given.

1.1 History

In 1888 the mechanical engineer Oberlin Smith published an article entitled 'Some possible forms of phonograph'. In this paper, published in the most widely read technical journal of that time, 'The Electrical World' [1.1], Smith described an original method for storing sound signals by leading small steel particles on a cotton or silk thread through the centre of a coil, carrying a current proportional to the sound signal. This invention was stimulated by another invention, namely that of the telephone by Bell in 1876. Playback was obtained by leading the magnetized thread through the coil again and listening to the induced signal via Bell's telephone (see Fig. 1.1a, b and c).

The first working apparatus was constructed by the Dane Valdemar Poulsen, who applied for a patent in 1898, followed by an improved and much more developed American patent and article in 1900 [1.2]. The first public demonstration followed in the same year at the World Exhibition in Paris. The apparatus consisted of a non-magnetic brass cylinder on the circumference of which a spiral-shaped groove was made. A steel wire with a diameter of 0.25 mm was put into this groove. By means of an electromagnet, signals were recorded onto and read back from the steel wire. Poulsen broke with the, for that time, general idea that a rod could only be magnetized over its full length, and applied this to a steel wire [1.3]. If the steel of the wire were hard enough, he

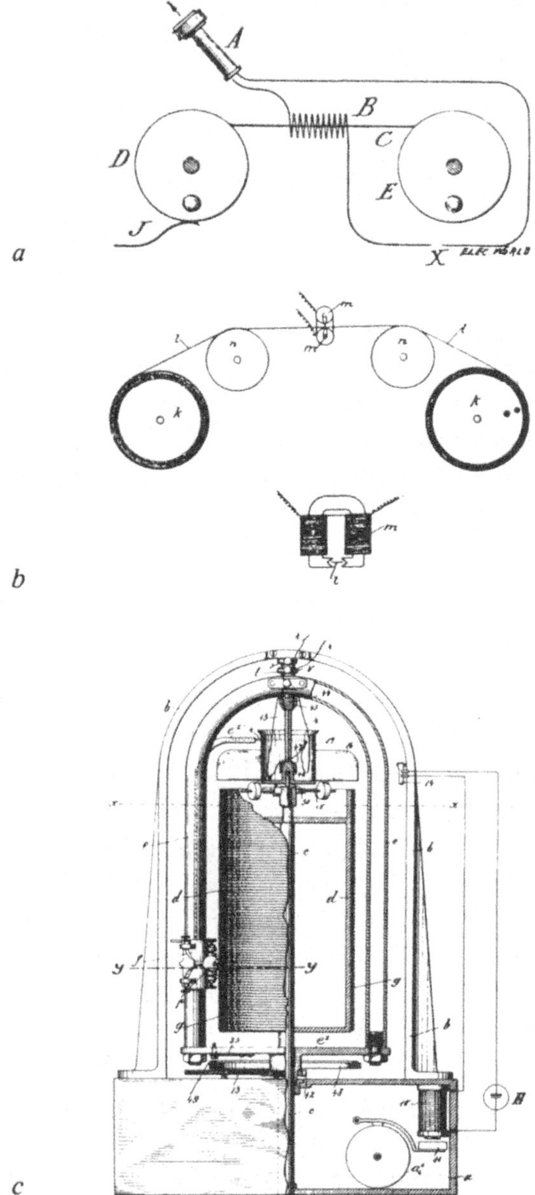

Figure 1.1. a) Original drawing of the oldest known proposal for magnetic recording (1888). The recording medium used is a wire. A dc source is located at X during recording.

The author refers to 'playback' in the following terms: '... it may be possible to insert at X ... some intensifying apparatus, such as a battery, but which has not yet been thought out'. So here electronics literally was the missing link.

b) Diagram from the original Danish patent taken out by V. Poulsen (1898). The recording medium here is a tape.

c) In his first American patent (1900) Poulsen proposed the use of a stationary drum to which a helix of steel wire was fixed as the recording medium. The write and read head was mounted on a carrier shaped like an inverted U. It could move vertically along the carrier, which could rotate around the drum.

supposed it had to be possible to magnetize it locally. The mechanically highest possible velocity of about 2 m/s of the circumference of the cylinder just sufficed to record speech. Yet the maximum playing time was half an hour. As a consequence the work and time necessary to put in a new wire with its length of a few kilometers were considerable. This drawback and the low signal level caused a rapid fading of interest in Poulsen's telegraphone.

In 1907 Poulsen proposed a combination of dc pre-magnetization and dc bias by which the quality of the recorded signal improved significantly. A further improvement was made by the Americans Carlson and Carpenter, who proposed ac pre-magnetization in a patent application. In fact this method is still the basic principle of today's analog recording. Their proposal was restricted to the use of steel wire, signals of about 1 kHz and an alternating signal of 10 kHz.

In a German patent of 1940, Von Braunmühl and Weber proposed ac biasing, in combination with the use of tape coated with a magnetic powder. By this method the noise level was reduced by more than a factor of five! The development of this particulate tape, made of paper or synthetic material, had already been started in 1927, especially by Pfleumer in Germany.

The first recorder using such tape was the magnetophone, developed by the Algemeine Elektricitaets Gesellschaft (AEG). The tape for this magnetophone was developed by IG Farbenindustrie AG, now known as the Badische Anilin- & Soda-Fabrik AG (BASF). The main advantage of this recorder was the price of 0.15 dollar per minute recording time compared to 1 dollar for the steel wire. Also the volume of the tapes on the reel was much smaller.

From that time all efforts were focused on improvements of tape and head, which led to a tremendous increase in sensitivity and bandwidth. These improvements are still going on.

The compact cassette for audio was introduced in 1963 by the Dutch company Philips and is a good example of a standard which has been accepted by many countries all over the world.

Ampex developed the video recorder for professional use in the fifties. In 1972 Philips was the first to introduce this recorder for consumer applications, the Video-Cassette Recording (VCR) system. Four years later, in 1976, the Japanese company JVC introduced the Video Home System (VHS).

Nowadays magnetic recording is applied in many consumer and professional products like video recorders (V2000, VHS, BETAMAX, 8mm), still-picture cameras, audio recorders (reel-to-reel, compact-cassette, R-DAT), flexible disk and rigid disk systems, computer tape drives and streamers and voice-logging recorders.

1.2 Basic principles of magnetic recording

Magnetic recording today is based on the same principles [1.4][1.5] as in Poulsen's original recording machine: an electric current, representing the information to be recorded, flows in a coil (Fig. 1.2). Inside the coil there is a core of magnetic material, usually ring-shaped. When the current in the coil is varied, a varying magnetic field is excited in the core. A discontinuity ('gap') at a defined position in the core makes the magnetic field spread outside the recording head, or 'write head', so that it can magnetize a recording medium that bridges the gap – in this case a tape. Later on, the magnetized tape can be drawn past a playback head, or 'read head', in which a varying magnetic flux is excited, and this in turn induces an electrical signal in the coil. This signal represents the original information. In principle the same head can be used for both recording and playback; for the playback or 'read-out' it is also possible to use the magnetoresistance effect [1.7][1.8][1.9][1.10] instead of the conventional electromagnetic induction method. In that case a playback head of a different design is used.

If the preferred direction of magnetization on the tape is parallel to the direction of tape travel, the recording is said to be longitudinal [1.11]. This has long been the conventional form of recording. Through the years the minimum wavelength λ (fig. 1.2b) has been reduced by a factor of 10^3 to 10^4, so that a very much higher information density can now be achieved (Fig. 1.3). To obtain even higher information densities on the tape it will possibly be advantageous to use 'perpendicular recording', with the preferred direction of magnetization perpendicular to the surface of the medium.

Nowadays the recording medium usually consists of a thin magnetizable layer uniformly deposited on a non-magnetic substrate.

Both the magnetizable part of the medium and the core of the magnetic head consist of magnetic material. Their properties, however, have to be very different. This can be demonstrated by means of the *B-H*

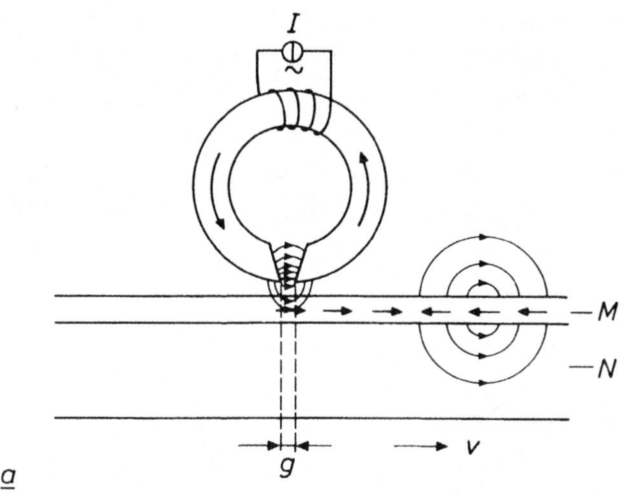

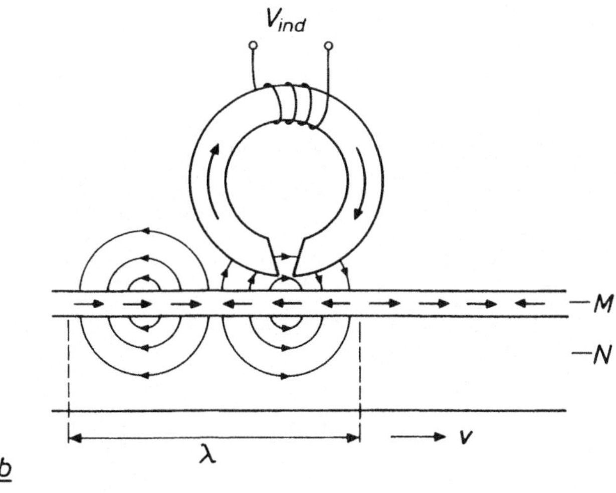

Figure 1.2. Principle of magnetic recording with a ring-shaped write and read head.

a) During recording an alternating current I excites a varying magnetic field in the head. This field emerges from the magnetic circuit at the head gap. A recording medium drawn past this gap is then magnetized in a pattern determined by I.

b) During reproduction the recording medium is again drawn past the head. The magnetization produces a varying magnetic field in the head and an alternating voltage V_{ind} is induced in the coil. The recorded information is recovered from V_{ind}. The magnetic flux and field lines are indicated by arrows. M magnetizable layer, N plastic substrate, v relative velocity, g gap length, λ wavelength of the recorded magnetic pattern.

characteristic, which shows the magnetic flux density B as a function of the magnetic field-strength H (Fig. 1.4). Since the value of B depends not only on the instantaneous value of H, but also on the earlier history of the specimen, the B-H characteristic has the shape of the familiar hysteresis loop. As H increases, so does B, first relatively steeply (until a saturation flux density B_s is reached), and then less steeply. When H

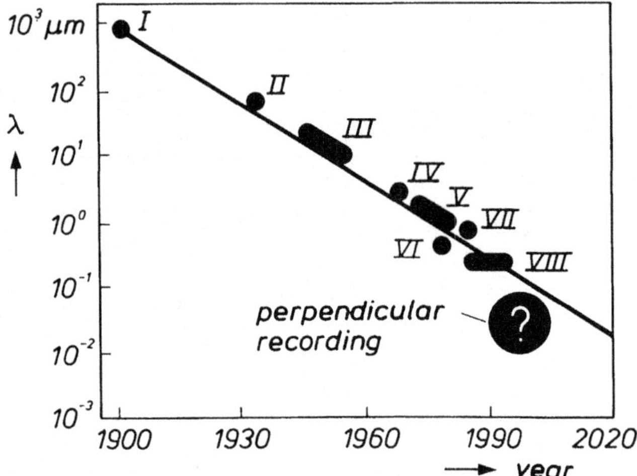

Figure 1.3. As recording media are improved the minimum wavelength of the recorded magnetic pattern decreases and the information density becomes higher. So far longitudinal recording has mainly been used; one of the ways in which the information density can be increased still further is by using perpendicular recording. I Poulsen's steel wire; II magnetic tape; III tape for reel-to-reel recorder; IV compact cassette tape; V video recorder tape (Beta, VHS, V2000); VI video recorder tape (8 mm metal); VII tape for digital video; VIII thin metal tape.

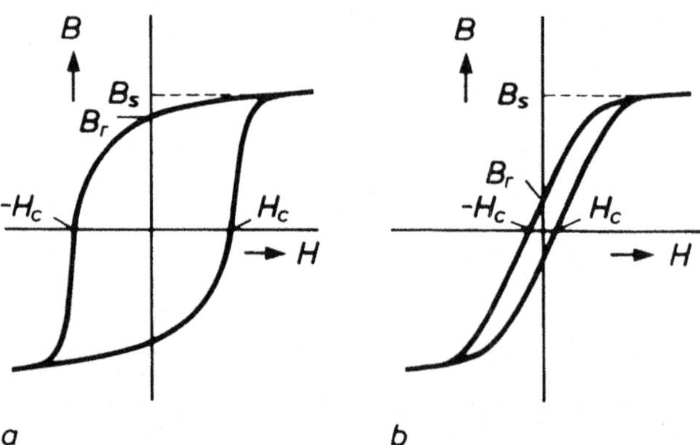

Figure 1.4. Magnetic materials can be characterized by a B-H characteristic or hysteresis loop, where H is the external magnetic field-strength (in A/m) and B is the magnetic flux density (in T). For magnetic recording media B_r and H_c should have *high* values, as in (a), and for the magnetic head material B_s should have a *high* value and B_r and H_c should have *low* values as in (b).

falls to zero, B does not become zero but stays at a remanence value B_r; for B to drop to zero an opposing magnetic field of strength $-H_c$ has to be applied ('c' for coercivity). In the magnetic layer of the tape we want a *high* value of B_r, because this gives a high magnetization after recording. We also want a *high* value of H_c (fig. 1.4a), because this gives a high degree of insensitivity to the demagnetizing effect of adjacent parts of the tape and to any interfering fields that may be present. In the material of the magnetic head we want B_s to have a high value, but B_r and H_c to have *low* values (fig. 1.4b) to ensure that the instantaneous characteristics of the magnetic head will be virtually independent of signals recorded or played back earlier.

The *gap* in the magnetic material of the head is of crucial importance in the magnetic circuit, since this is where the magnetic coupling takes place between the head and the medium (see fig. 1.2). During recording the 'fringing field' at the head gap is responsible for magnetizing the medium; during playback as much of the external magnetic field of the medium as possible should follow the magnetic circuit of the head, and as little as possible should be 'short-circuited' via the gap. This is done by using a material of high relative permeability (μ_r) for the head and making the cross-section of the head at the gap (A_g) smaller than the cross-section of the rest of the magnetic circuit (A_m).

1.2.1 Core permeability

The importance of the relative permeability μ_r of the magnetic head and of the cross-sectional ratio A_m/A_g in the recording process can easily be demonstrated. For simplicity we assume that the gap of a magnetic head contains only a uniform magnetic field H_g, and that the fringing field H_f is negligible.

We first consider a closed ring of magnetic material with a coil of N turns wound round it, with a current I in the coil (fig. 1.5a). The magnetic field strength H_m in the ring is then given by

$$\oint H_m dl = NI. \qquad (1.1)$$

If the length of the centre line through the ring is L, we have:

$$H_m = NI/L. \qquad (1.2)$$

If we now make a gap of length d in this ring (fig. 1.5b) and reduce the cross-section at the gap from A_m to A_g, then, neglecting H_f, we find from (1.1):

$$g\,H_g + (L - g)\,H_m = NI. \qquad (1.3)$$

Also, the total magnetic flux in the gap is the same as in the rest of the ring, so that

$$\Phi_g = \Phi_m, \qquad (1.4)$$

or (using the magnetic flux densities B_g and B_m)

$$B_g\,A_g = B_m\,A_m \qquad (1.5)$$

or

$$\mu_0 H_g A_g = \mu_0 \mu_r H_m A_m. \qquad (1.6)$$

Combining (1.6) and (1.3) we arrive at

$$H_g = \frac{NI}{g + (L - g)\,A_g/\mu_r A_m}. \qquad (1.7)$$

At given values of N, I, L and g the magnetic field strength in the gap is highest when μ_r and the ratio A_m/A_g are as large as possible. In practical situations, where the fringing field H_f neglected above is the crucial quantity, the relation between H_f and the quantities μ_r and A_m/A_g is much the same as for H_g in equation (1.7).

1.2.2 Gap length

The characteristics of a magnetic head depend to a very great extent on the length g of the gap [1.12] (see Fig. 1.5b). The gap length is in fact one of the most important factors determining the maximum signal frequency that can be handled in magnetic recording; in this respect the playback process is more critical than the recording process. To make this clear let us refer again to fig. 1.2, If we are recording a signal frequency of f Hz and the medium (a tape) is travelling at a relative velocity of v m/s with respect to the magnetic head, then a periodic magnetic pattern is recorded on the tape with a wavelength λ of v/f m.

In the recording process the magnetization of the tape after passing the head is mainly determined by the magnetic field at the end of the gap (the 'trailing edge'), rather than the field over the whole length of the gap. During the playback process, on the other hand, the gap plays a significant part over its whole length. If the wavelength of the magnetic pattern on the tape is exactly equal to the gap length, both ends of the gap will always be opposite to places on the tape with the same magnetization. There will then be no varying magnetic flux in the magnetic head and no signal voltage will be induced. The corresponding signal frequency will consequently not be detected by the playback head and the same applies to every integer multiple of this frequency. To achieve the highest possible information density (corresponding to the smallest possible λ) it is therefore necessary to minimize the gap length. It is self-evident that reducing the *width* of the gap can also increase the maximum information density per unit area of the tape: the smaller the gap width (or 'track width') the more tracks can be accommodated side by side on a given area of the tape.

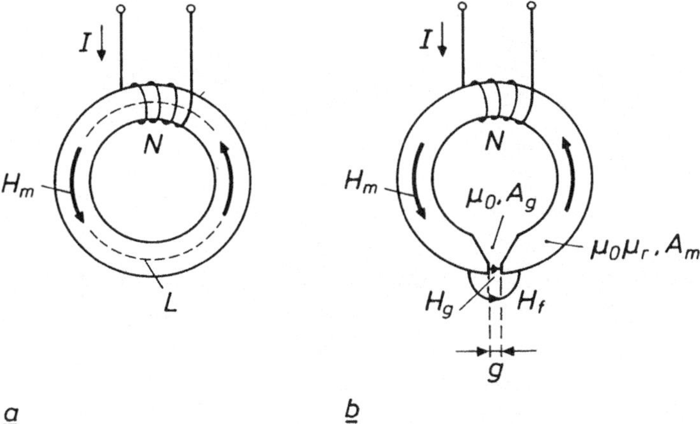

Figure 1.5. a) Closed ring of magnetic material with a coil wound round it. A current I flows in the coil. H_m magnetic field-strength; L length of the centre-line through the ring; N number of turns. b) Simplified model of a ring-shaped magnetic head. g gap length; H_m, H_g, H_f magnetic field-strength; $\mu_0\mu_r$, μ_0 magnetic permeability; A_m, A_g cross-section.

1.2.3 Magnetic head types

The consequences of what has been said in the foregoing about the shape of magnetic heads can be seen from the diagrams in Fig. 1.6. Two relatively modern magnetic heads are illustrated. Their outside dimensions are roughly the same (0.2 mm × 3 mm × 3 mm). The head in fig. 1.6a is made from the magnetic material MnZn ferrite and is used for recording analog video signals in the V2000 system and the VHS system. The head in fig. 1.6b is also of MnZn ferrite and has not only a much smaller gap length but also a much smaller gap width, 10 μm instead of 23 μm, so that a narrower magnetic track can be recorded. This head is designed for recording digital video signals. The use of a narrower track means that both the useful signal and the background-noise signal will be weaker at playback. They are not decreased in proportion, however: every time the gap width is halved there is a 3 dB reduction in the signal-to-noise ratio.

With reduced gap widths it is also possible to produce multiple heads, so that a number of magnetic tracks can be recorded on a medium simultaneously but independently. Techniques of this type are very suitable for recording digital signals, since it is easy to obtain a number of parallel binary signals. Applications have also been found for analog signals, however [1.10]. These multiple heads are often made by photo-

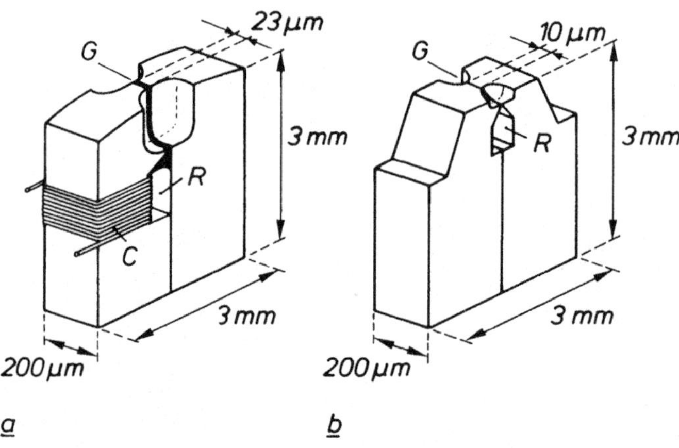

Figure 1.6. Two examples of modern magnetic heads for recording video signals.
  a) Magnetic head designed for analog video signals and used in the V2000 system and the VHS system.
  b) Magnetic head designed for digital video signals. This head has a gap G of even smaller length and width. R coil chamber. (For clarity the coil C has been omitted.)

lithographic processes, as used in the manufacture of integrated circuits. They are then referred to as 'thin-film heads'. An example is shown in Fig. 1.7. The tape is supposed to be travelling in the direction perpendicular to the plane of the photograph here. We shall not consider multiple heads further in this article.

One of the ways of increasing information density in magnetic recording is to increase the coercivity of the magnetic material of the recording medium. To record on a material of higher coercivity it is necessary, however, to use a head material that gives a higher saturation flux density, particularly near the gap [1.13], where the cross-section of the magnetic material is smaller and the magnetic flux density B is higher (see section 1.2.2). Philips therefore uses a material such as NiFe on both sides of the gap and an Fe-Al-Si alloy (known as 'Sendust'), while the rest of the magnetic circuit is made of ferrite. The saturation flux density is then 0.8 T to 1.0 T compared with 0.5 T for MnZn-ferrite. A magnetic head made on this principle is illustrated in Fig. 1.8a. Heads of this type are referred to as 'metal-in-gap heads' [1.14].

Another way of producing a magnetic head for high-coercivity tape is illustrated in fig. 1.8b. This is a 'sandwich head', with the thickness of the magnetic circuit nowhere more than 18 μm. This magnetic layer consists of an amorphous ribbon of an FeCo alloy with a saturation flux density of 0.8 to 1.0 T. It is enclosed between two much thicker parts of non-magnetic material. The manufacture of these types of heads on laboratory scale is described by Verbunt in [1.15].

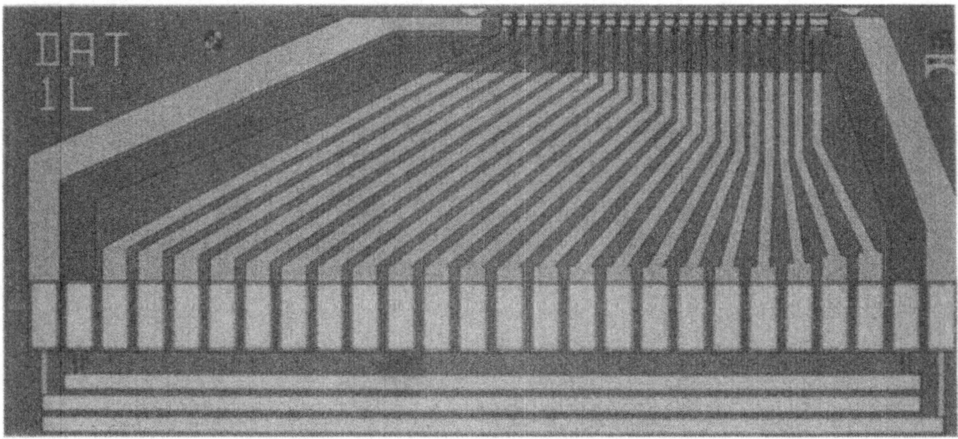

Figure 1.7. Example of a 22-track magnetic head of the thin-film type. The manufacturing technology is similar to that used for producing semiconductor chips. The horizontal dimension of the head shown here is only about 5 mm.

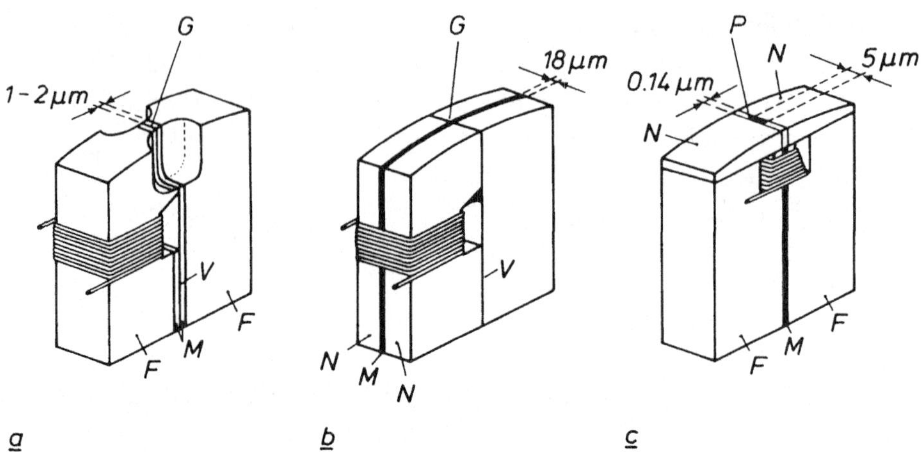

Figure 1.8. Three types of magnetic heads designed to give a higher information density per unit area of the recording medium. They have the same outside dimensions as the heads in fig. 6 (0.2 mm × 3 mm × 3 mm). a) Metal-in-gap head. b) Sandwich head. c) Magnetic head for perpendicular recording. The heads in (a) and (b) are designed for high-coercivity media; these heads therefore have a higher than usual saturation flux density (0.8 to 1.0 T). The head in (c) does not have the well-known structure of an almost completely closed magnetic circuit with a narrow gap; the magnetic circuit is shaped rather like a letter W capped by two non-magnetic pieces. Reading and writing are performed with a very thin soft-magnetic pole *P* between the two non-magnetic pieces. *F* ferrite; *G* gap; *M* magnetic material; *N* non-magnetic material; *V* gap filler material (non-magnetic).

In the future perpendicular recording may play an important role in increasing the information density. In this method, as mentioned, the recording medium is magnetized in the direction perpendicular to its surface. The magnetic head now required is no longer an almost completely closed circuit; in the example shown in fig. 1.8c the magnetic circuit of the head is shaped rather like a 'W' and is covered at the top with a block of non-magnetic material. A magnetic head of this type is known as a 'single-pole head' and requires special types of recording media. A considerable amount of research on perpendicular recording is now under way [1.16].

1.3 Scope of this book

The magnetic recorder has many system and technology aspects: mechanical, chemical, information-theoretical, lubricational and wear, solid-state, many body, electronic-system and last but not least electromagnetic aspects. It is this wide range of disciplines that makes this field so interesting.

In this book an attempt is made to contribute especially to analytical models and theories considering the electromagnetic aspects sometimes in an alternative way. The playback process in particular is suited to analytical description and therefore highly detailed treatment. Most of the calculations are restricted to 'short' wavelengths. This means that side-write and side-read effects have been left out of account.

Today's developments in magnetic recording, using more magnetic layers, demand a description of a multi-layer problem. It depends on the process, playback or recording, whether this can be done analytically or not. Each layer has quite different magnetic characteristics: hard or soft, perpendicularly- or longitudinally-oriented, isotropically or not, having low or high permeability. To find the requirements regarding these characteristics one should be able to calculate the behaviour of a head-tape system as a function of the parameters associated with the above characteristics. The first step will be the derivation of an expression for the flux from a permanently magnetized tape that penetrates an idealized head. This problem is solved for a five-layer system in the second chapter.

However, heads are not ideal. Does this mean that the derived expressions are useless? No, an alternative reciprocity theorem can be derived which expresses the read flux in terms of the flux penetrating an idealized head and the actual magnetic scalar potential of the non-ideal head. After this has been done for a general three-dimensional head in the time domain for reciprocal and dispersion-free (frequency-independent) head materials in chapter 3, the theorem is extended to both non-reciprocal and frequency-dependent head materials in chapter 4.

The main property of a head, its sensitivity, can be formulated elegantly with the aid of the alternative reciprocity theorem. This formulation is independent of the type of head. This sensitivity can usually be divided into two parts, a frequency-dependent part, called the efficiency and a wavelength-dependent part, which we shall call the generalized gap loss function. The formulation of these functions can be and has been chosen independently of the type of head or application, in contrast to the choices made by several other authors. In chapter 5 we attempt to give some insight into the accuracy and feasibility of this division. Also for probe heads on perpendicularly oriented double-layer media the above formalism (in fact the usual choice for ring heads on single-layer media) turned out to be very convenient. In this way a good understanding of the sensitivity and auxiliary-pole effects of this type of head could be obtained and weak points in earlier probe-head designs could easily be eliminated, as will become clear in chapter 9.

Next the formulations are applied to video heads, several kinds of non-ideal gaps and the thin-film head. The efficiency has not been taken into account in any of these applications.

Analytical expressions for the efficiency of 3-dimensional heads are quite difficult to obtain and therefore we chose to derive them for various types of heads separately in later chapters. In these chapters they are also applied to several experimental situations. In chapter 8 the efficiency and electrical impedance of ring heads are considered. The efficiency and generalized gap loss function of various probe heads for perpendicular recording on double-layer media are, as already mentioned, the subjects of chapter 9. In chapter 11 the efficiency of thin-film heads, especially of the yoke type of magneto-resistive heads, is calculated. Many aspects of a special type of ring head, namely the metal-in-gap head, being investigated all over the world nowadays, are discussed in chapter 12. A totally new type of head, with many interesting aspects regarding its efficiency, bandpass characteristic and electrical impedance, the band-pass head, is described extensively in the last chapter.

In chapters 4 and 5 the expression for the flux that would penetrate the ideal head, derived in chapter 2, is not yet applied in the alternative reciprocity theorem. This is carried out in chapter 6. Calculations of the read flux are then possible for different choices of types of permanent magnetization in the tape. Examples are perpendicular, longitudinal or isotropic types of orientation with different magnetization amplitude profiles into the depths of the coating. With the aid of simple magnetization profiles, we investigated the influence of e.g. the backlayer of a double-layer medium and the anisotropy and permeability of the coating on the read flux. The choices, however, for the permeability, anisotropy, maximum permanent magnetization and coercivity are physically not free. Knowledge of the interdependence of these parameters in for example particulate tapes is important in answering questions concerning final choices of tapes for specific heads. Since this is a very complicated many-body problem, an approach only is given in the last part of this chapter. Nevertheless the insight thus obtained stimulates further discussion.

A complete description of the write process is burdened with the same problems. No attempt was made to realise such a complete description. The write model we used in chapter 7 is simple compared to the work of others [1.17, 1.18, 1.19], with the advantage of avoiding (CPU) time-consuming numerical calculations. As a drawback the model does not describe the actual magnetization pattern in the tape, and is restricted to longitudinal recording. But for calculations of the read flux in longitudinal recording for heads with different saturation magnetization and gap length, and tapes with given remanence and coercivity, the calculations have proven to be valuable. Part of this chapter is devoted to the work of others.

Although the major part of this book is theoretical, many practical issues are treated in the course of the descriptions. A new measuring method and a new measuring principle are also described.

The first concerns a quick and very simple measurement of the efficiency of a ring head and is described at the end of chapter 8. The latter gives a graceful principle for separating accurately but easily the write and read performance of an arbitrary head and can be found in chapter 10. In later chapters this principle is frequently used in analyzing recording/playback measurements.

At the end of this introduction I would like to make both a remark and an apology: the extent of many calculations and my personal need for completeness, so that derivations and experiments can be verified by the reader, in combination with the large number of subjects in this book, made it necessary to write in a comprehensive way. It is therefore recommended that considerable attention be paid to the introduction of each chapter, and the thread of arguments presented there be kept in mind.

References

[1.1]   O. Smith, *Some possible forms of phonograph*, The Electrical World, 161 (1888).

[1.2]   V. Poulsen, *Das Telegraphon*, Annales der Physik, Vol. 3, 754-760 (1900).

[1.3]   A. van Maaren, *Bandrecording*, Uitgeverij de Muiderkring N.V., Bussum, the Netherlands (1960).

[1.4]   W.K. Westmijze, *The principle of the magnetic recording and reproduction of sound*, Philips Tech. Rev., Vol. 15, 84-96 (1953/54).

[1.5]   W.K. Westmijze, *Studies on magnetic recording (thesis)*, published in six parts in Philips Res. Rep. Vol. 8, 148-157, 161-183, 245-255, 255-269, 343-354, 354-366 (1953). Reprints of these articles can be found in [1.6].)

[1.6]   M. Camras (ed.), *Magnetic tape recording* (Benchmark papers in acoustics, Vol. 20), Van Nostrand Reinhold Company, New York (1985).

[1.7]   W.J. van Gestel, F.W. Gorter and K.E. Kuijk, *Read-out of a magnetic tape by the magneto-resistance effect*, Philips Tech. Rev., Vol. 37, 42-50 (1977).

[1.8]   W.F. Druyvesteyn, J.A.C. van Ooyen, L. Postma, E.L.M. Raemaekers, J.J.M. Ruigrok, J. de Wilde, *Magnetoresistive heads*, IEEE Trans. Magn., Mag-17, 2884-2889 (1981).

[1.9]   J.J.M. Ruigrok, *Analytic description of magnetoresistive read heads*, J. Appl. Phys, Vol. 53, 2599-2601 (1982).

[1.10] M.G.J. Heijman, J.H.W. Kuntzel and G.H.J. Somers, *Multiple-track magnetic heads in thin-film technology*, Philips Tech. Rev., Vol. 44, no. 6 (1988).

[1.11] It is not easy to make an accurate analysis of the magnetization processes both in and between the head and the recording medium. In the past greatly enlarged scale models have been a useful aid to understanding. See for example: D.L.A. Tjaden and J. Leyten, A 5000:1 scale model of the magnetic recording process, Philips Tech. Rev., Vol. 25, 319-329 (1963/64). (A reprint of this article is also given in [1.6].)

[1.12] The gap length is the dimension of the gap in the direction of relative movement of the magnetic head and the medium during recording and playback; the dimension in the direction perpendicular to the medium is called the gap height, and the dimension in the direction perpendicular to the other two is called the gap width. In the examples given here the gap width is greater than the gap length.

[1.13] F.J. Jeffers, R.J. McClure, W.W. French and N.J. Griffith, *Metal-in-gap record head*, IEEE Trans. Magn., Mag-18, 1146-1148 (1982).

[1.14] C.W.M.P. Sillen, J.J.M. Ruigrok, A. Broese van Groenou and U. Enz, *Permalloy/Sendust metal-in-gap head*, IEEE Trans. Magn. Mag-24, 1802-1804 (1988).

[1.15] J.P.M. Verbunt, *Laboratory-scale manufacture of magnetic heads*, Philips Tech. Rev., Vol. 44, no. 5, (1988).

[1.16] V. Zieren, J.J.M. Ruigrok, M.J. Piena, S.B. Luitjens, C.W.M.P. Sillen and J.P.M. Verbunt, *Efficiency improvement of one-sided probe heads for perpendicular recording on double-layer media*, IEEE Trans. Magn., Mag-23, 2479-2481 (1987).

[1.17] I.B. Ortenburger and R.I. Potter, *A self-consistent calculation of the transition zone in thick particulate media*, J. Appl. Phys., Vol. 50, 2393-2395 (1979).

[1.18] H.N. Bertram, *Geometric effects in the magnetic recording process*, IEEE Trans Magn., Mag-20, 468-478 (1984).

[1.19] A. Eiling, *High-density magnetic recording: theory and practical considerations*, J. Appl. Phys., Vol. 62, 2404-2418 (1987).

Chapter 2

Calculations of the field near magnetized anisotropic layered media including backlayer and gap smear

2.1 Introduction

The uniformity and smoothness of evaporated but also particulate media are such that tapes can be treated to a very good approximation as layered structures. The fields in actual head-tape systems are however affected by the non-layered structure of an actual head. Especially in the region of a non-zero gap the system deviates from a layered one.

The calculations presented in this chapter relate to layered structures only, and thus exclude non-ideal heads, i.e. heads with finite dimensions, non-infinite permeable material, non-zero gap length or non-flat surface facing the tape. Nevertheless, the hypothetical head gives a good description of many qualitative or even quantitative observations on practical, i.e. non-ideal, heads.

Description of the configuration

The infinite structure with dependences only in the y-direction is given in Fig. 2.1. The characteristics of the regions are as follows:
- Region 1 is air with permeability μ_0.
- Region 2 is a backlayer with anisotropic permeability tensor $\hat{\boldsymbol{\mu}}_2 \mu_0$.
- Region 3 is the magnetic coating with anisotropic permeability tensor $\hat{\boldsymbol{\mu}}_3 \mu_0$.
- Region 4 represents the region between the magnetic coating and the magnetic head and may include an air film, a thin film of varnish from the tape [2.1], magnetically dead layers from head or tape surface and a contribution from the surface roughnesses of head and tape.
- Region 5 is a layer of anisotropic magnetic material on top of the

head. This layer may represent gap smear [2.1], i.e. a magnetic smear layer over the gap.

The magnetic materials 2 and 5 are assumed to be linear, time-invariant, reciprocal and locally reacting in their magnetic behaviour. The anisotropy axis is assumed to lie in the longitudinal (x) or perpendicular (y) direction. The magnetization changes are described with the aid of the following (complex) tensor permeability:

$$\boldsymbol{\hat{\mu}}(\omega) = \begin{bmatrix} \hat{\mu}_x(\omega) & 0 \\ 0 & \hat{\mu}_y(\omega) \end{bmatrix} \tag{2.1}$$

where $\hat{\mu}_x \equiv |\hat{\mu}_x| e^{j\varphi_{\mu_x}}$ and $\hat{\mu}_y \equiv |\hat{\mu}_y| e^{j\varphi_{\mu_y}}$, with $|\hat{\mu}|$ being the amplitude factor and $e^{j\varphi}$ the phase factor.

The choice of zero non-diagonal elements ensures that the materials are reciprocal, see Section 4.2.2, and simplifies the calculations.

The above assumptions are also assumed to hold for the permanently magnetizable material of region 3 for the reversible part of the hysteresis curve that is traversed during the read process.

Introduction of complex permeabilities for regions fixed with the region containing the permanent magnetization is not necessary in the present calculations. Phase lags mainly occur in media with high permeability that move relative to the permanently magnetized medium as a consequence of the radial frequency ($\omega = kv$, where v is the velocity)

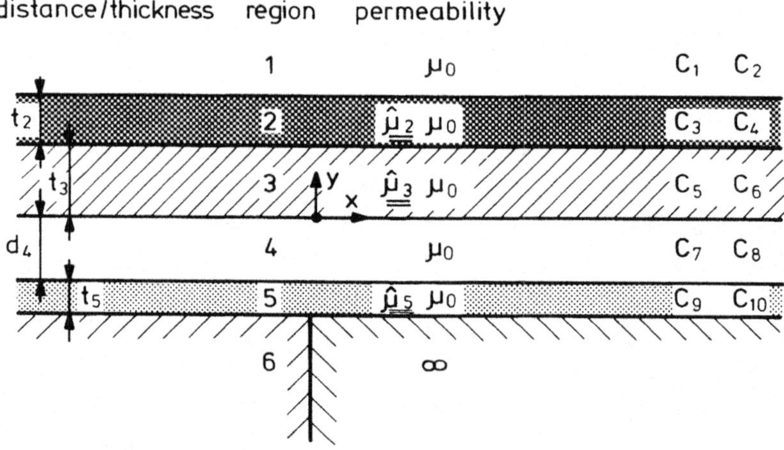

Fig. 2.1. Definition of the two-dimensional configuration. The two constants in every region follow later on from the second order differential equations describing the potential distribution in the configuration.

dependence of the magnetization process due to dispersion of the material. This can be the case for layer 5 on top of the head and for the head material 6 itself if of finite permeability. For an ideally-reflecting head surface, i.e. a smooth semi-infinite zero-gap head with infinite permeability, the tape only feels a zero-frequency field, since the uniform reflection and the permanently magnetized tape travel with the same speed as the permeable coating and backlayer. In the case of non-ideally reflecting head surfaces, as pointed out in chapter 3, the tape also feels reflected fields that contain non-zero frequency components because of non-uniform reflections. An analytical solution of the above problem in the case of a permeable tape, or even a numerical solution, has not yet been given, as far as we know. For a tape without dispersion an analytical formalism will be worked out in chapter 3, which is closely connected with the field calculations in this chapter. The relatively simple formalism makes possible exact calculations of read fluxes of non-ideal heads with dispersion-free tapes by using the results of the ideal-head calculations in this chapter. When the tape is assumed to be free of dispersion, the permeability tensor is frequency independent and real. Although this will be assumed in all following chapters, we will treat the problem in this chapter with complex permeabilities for all magnetic layers and not only for the magnetic-smear layer 5 for reasons of uniformity and for the sake of completeness.

Field calculations on layered structures have been carried out before by Wallace [2.2] for a single-layer medium with permeability μ_0, by Westmijze [2.3] for a single layer of isotropic material in his 'Studies on magnetic recording', and by Tjaden [2.4] for a single-layer anisotropic medium in his study of the 'anhysteretic contact duplication process'. We have adopted Tjaden's 'elegant' method for the more complicated situation of Fig. 2.1, using complex variables. In addition we have tried to write all our results in the same way as Tjaden did. This was a troublesome task for the complicated structure we describe because of the many equivalent forms that are possible to express the results. However, besides the comprehensive form of the results, it provides the advantage of easy comparison of our limiting cases with his results. It also makes it easier to gain insight into the effects of the backlayer 2 and the magnetic-smear layer 5 on top of the head.

2.2 Field calculations

As no electrically conducting materials or dielectric materials are present in the structure, the electric current density $j = 0$ and the time derivative of the electric displacement, $\partial D/\partial t$, can be neglected in $\nabla \times H = j + \partial D/\partial t$ for the 'low' frequencies (MHz), finite permeabilities and small relevant dimensions of the structures (mm) we will consider. (This quasi-static approach means that the wavelengths of propagation of the EM waves at the considered frequency are large compared to the relevant dimensions of the head.) As a consequence

$$\nabla \times H = 0. \tag{2.2}$$

So a scalar potential Ψ can be introduced according to

$$H \equiv -\nabla\Psi \tag{2.3}$$

where $\nabla \equiv \left(\dfrac{\partial}{\partial x}, \dfrac{\partial}{\partial y} \right)$.

Further

$$\nabla \cdot B = \mu_0 \, \nabla \cdot (H + M) = 0. \tag{2.4}$$

Since $M = 0$ in the regions 1 and 4, Ψ in these regions satisfies

$$\nabla^2 \Psi = 0 \tag{2.5}$$

where ∇^2 is the Laplacian, i.e. $(\partial^2/\partial x^2 + \partial^2/\partial y^2)$.

In the regions 2 and 5 where $B = \mu_0\mu H$, in which the relative tensor permeability μ is given in (2.1), i.e. $B_x = \mu_0\mu_x H_x$ and $B_y = \mu_0\mu_y H_y$, we have

$$\mu_x \frac{\partial^2\Psi}{\partial x^2} + \mu_y \frac{\partial^2\Psi}{\partial y^2} = 0. \tag{2.6}$$

Region 3 needs special attention since it is assumed that here a recorded permanent magnetization M_p is present. With $B = \mu_0 (\mu H + M_p)$ and (2.4) we have

$$\mu_x \frac{\partial^2 \Psi}{\partial x^2} + \mu_y \frac{\partial^2 \Psi}{\partial y^2} = \nabla \cdot M_p. \tag{2.7}$$

The differential equations are only valid within the different regions where no jumps in the material parameters have been assumed. So the system of linear differential equations has to be supplemented by boundary conditions. The necessary boundary conditions for the calculations here are:

1) The quantities

$$\Psi, \ H_x = \frac{-\partial \Psi}{\partial x} \ \text{and} \ B_y = \mu_0(H_y + M_y) = \mu_0\left(M_{py} - \mu_y \frac{\partial \Psi}{\partial y}\right)$$
$$\text{(2.8a + b + c)}$$

are continuous across boundary surfaces across which the susceptibility shows a finite jump.

Furthermore

2) $$\Psi \rightarrow 0 \qquad\qquad \text{as } y \rightarrow \infty \tag{2.9}$$

and

3) $$H_x = - \frac{\partial \Psi}{\partial x} \rightarrow 0 \tag{2.10}$$

upon approaching a medium of infinite permeability, which would be valid if e.g. the backlayer (region 2) were idealized in the calculations.

The linear equations can be solved by using the spatial Fourier transformation or by expansion of the quantities, if periodic, into Fourier series with coefficients that are functions of the wave number $k = 2\pi/\lambda$. In the latter case, complex x-independent functions denoted by $\hat{M}_p$, $\hat{\Psi}$, $\hat{H}$ etc. are introduced according to the definition

$$F(x) = \text{Re}\{\hat{F}e^{jkx}\}, \tag{2.11}$$

where $\hat{F} = |\hat{F}| \ e^{j\varphi}$, with $|\hat{F}|$ being the amplitude factor and $e^{j\varphi}$ the phase factor of the considered harmonic component $F = |\hat{F}| \cos(kx + \varphi)$ in the series expansion of the periodic quantity.

In the case of non-periodical quantities, the complex Fourier transform, $\boldsymbol{F}(k, y)$, replaces $\hat{\boldsymbol{F}}(y)$, see also chapter 4. For application to frequency characteristics etc., it is convenient to consider $\hat{\boldsymbol{F}}$ as a function of the angular wave number, i.e. $\hat{\boldsymbol{F}} = \hat{\boldsymbol{F}}(k, y)$.

With the definition of the complex permeability tensor in (2.1) for the equations (2.5), (2.6) and (2.7) describing respectively the regions 1 and 4, 2 and 5 and 3 , this leads to:

regions 1 and 4 with isotropic permeability μ_0

$$-k^2 \hat{\Psi}(k, y) + \frac{\partial^2 \hat{\Psi}(k, y)}{\partial y^2} = 0 \tag{2.12}$$

regions 2 and 5 with anisotropic permeability

$$-\hat{\mu}_x(\omega) k^2 \hat{\Psi}(k, y) + \hat{\mu}_y(\omega) \frac{\partial^2 \hat{\Psi}(k, y)}{\partial y^2} = 0 \tag{2.13}$$

region 3 with permanent magnetization and anisotropic permeability

$$-\hat{\mu}_x(\omega) k^2 \hat{\Psi}(k, y) + \hat{\mu}_y(\omega) \frac{\partial^2 \hat{\Psi}(k, y)}{\partial y^2} = jk\hat{M}_{px}(k, y) + \frac{\partial \hat{M}_{py}(k, y)}{\partial y} .$$

$$\tag{2.14}$$

In the derivations the following definitions for the complex anisotropy and permeability parameter β and μ will be used:

$$\hat{\beta} \equiv (\hat{\mu}_x/\hat{\mu}_y)^{1/2} \tag{2.15}$$

$$\hat{\mu} \equiv (\hat{\mu}_x \hat{\mu}_y)^{1/2}. \tag{2.16}$$

For zero phase shifts these definitions coincide with the real definitions for β and μ in Tjaden's calculations.

In all following calculations in this book we assume $k \neq 0$. We shall consider the solution for $k > 0$ only; however, the calculations remain valid for $k < 0$ as well if one reads $|k|$ instead of k on the right-hand side of all following equations in this book. The general forms of $\hat{\Psi}$ in the regions 1 and 4 obeying Eq. (2.12) then are

$$\hat{\Psi}_1 = \hat{C}_1 e^{ky} + \hat{C}_2 e^{-ky} \tag{2.17}$$

$$\hat{\Psi}_4 = \hat{C}_7 e^{ky} + \hat{C}_8 e^{-ky}. \tag{2.18}$$

The general forms of $\hat{\Psi}$ in the regions 2 and 5 obeying Eq. (2.13) are

$$\hat{\Psi}_2 = \hat{C}_3 e^{\hat{\beta}_2 ky} + \hat{C}_4 e^{-\hat{\beta}_2 ky} \tag{2.19}$$

$$\hat{\Psi}_5 = \hat{C}_9 e^{\hat{\beta}_5 ky} + \hat{C}_{10} e^{-\hat{\beta}_5 ky}. \tag{2.20}$$

It can be derived with the method of 'variation of parameters' as carried out in appendix 2.1 that the solution of $\hat{\Psi}$ obeying (2.14) in region 3, where permanent magnetization is present, is of the form

$$\hat{\Psi}_3 = \hat{C}_5 e^{\hat{\beta}_3 ky} + \hat{C}_6 e^{-\hat{\beta}_3 ky} + \hat{I}_{sc}(y)/\hat{\mu}_3 \tag{2.21}$$

where $\hat{I}_{sc}(y)$ equals the following integral:

$$\hat{I}_{sc}(y) = \int_0^y \left\{ j\hat{M}_{px}(y')\sinh(\hat{\beta}_3 k(y - y')) \right.$$

$$\left. + \hat{\beta}_3 \hat{M}_{py}(y')\cosh(\hat{\beta}_3 k(y - y')) \right\} dy'. \tag{2.22}$$

First the general solution of the potential $\hat{\Psi}$ in region 4 will be derived using the seven boundary conditions, two at each of the three interfaces 1-2, 2-3 and 3-4 and one condition at infinity. Seven of the eight constants C_1-C_8 can thus be eliminated so that one relation between the constants C_7 and C_8 of region 4 remains. The advantage of this procedure is that for different conditions at the head side of the tape the above-mentioned relation between C_7 and C_8 can be used as a starting point for further calculations of complete problems.

Elimination procedure

Because of condition (2.9), $\hat{C}_1 = 0$. Application of the boundary condition (2.8b) combined with Eq. (2.11) for the longitudinal (x) components and application of Eq. (2.8c) for the perpendicular (y) components lead to

boundary 1-2 at $y = t_2 + t_3$

$$\hat{C}_2 e^{-k(t_2 + t_3)} = \hat{C}_3 e^{\hat{\beta}_2 k(t_2 + t_3)} + \hat{C}_4 e^{-\hat{\beta}_2 k(t_2 + t_3)} \tag{2.23}$$

$$-\hat{C}_2 e^{-k(t_2 + t_3)} = \hat{C}_3 \hat{\mu}_2 e^{\hat{\beta}_2 k(t_2 + t_3)} - \hat{C}_4 \hat{\mu}_2 e^{-\hat{\beta}_2 k(t_2 + t_3)} \tag{2.24}$$

boundary 2-3 at $y = t_3$

$$\hat{C}_3 e^{\hat{\beta}_2 k t_3} + \hat{C}_4 e^{-\hat{\beta}_2 k t_3} = \hat{C}_5 e^{\hat{\beta}_3 k t_3} + \hat{C}_6 e^{-\hat{\beta}_3 k t_3} + \frac{\hat{I}_{sc}(t_3)}{\hat{\mu}_3}. \tag{2.25}$$

$$\hat{C}_3 \mu_2 e^{\hat{\beta}_2 k t_3} - \hat{C}_4 \hat{\mu}_2 e^{-\hat{\beta}_2 k t_3} = \hat{C}_5 \hat{\mu}_3 e^{\hat{\beta}_3 k t_3} - \hat{C}_6 \hat{\mu}_3 e^{-\hat{\beta}_3 k t_3} + \hat{I}_{cs}(t_3) \tag{2.26}$$

boundary 3-4 at $y = 0$

$$\hat{C}_5 + \hat{C}_6 = \hat{C}_7 + \hat{C}_8 \tag{2.27}$$

$$\hat{\mu}_3 \hat{C}_5 - \hat{\mu}_3 \hat{C}_6 = \hat{C}_7 - \hat{C}_8 \tag{2.28}$$

where $\hat{I}_{sc}(t_3)$ is defined by Eq. (2.22) for $y = t_3$ and $\hat{I}_{cs}(t_3)$ analogously by

$$\hat{I}_{cs}(t_3) \equiv \frac{1}{\hat{\beta}_3 k} \left(\frac{\partial \hat{I}_{sc}(y)}{\partial y} \right)_{y = t_3} - \frac{\hat{M}_{py}(t_3)}{k}$$

$$= \int_0^{t_3} \left\{ j\hat{M}_{px}(y') \cosh(\beta_3 k(t_3 - y')) \right.$$

$$\left. + \hat{\beta}_3 \hat{M}_{py}(y') \sinh(\hat{\beta}_3 k(t_3 - y')) \right\} \, dy'. \tag{2.29}$$

In the derivation of the equality in (2.29) use is made of the relation

$$\frac{\partial}{\partial y} \int_0^y f(y, y') dy' = f(y, y) + \int_0^y \frac{\partial f(y, y')}{\partial y} \, dy', \tag{2.30}$$

which is valid for functions $f(y, y')$ that are continuously differentiable.

The procedure for eliminating C_1 to C_6 is straightforward. Subsequently writing the results in a comprehensive form was a troublesome task, since the aim was to arrive at expressions with the same structure as in Tjaden's [2.4] paper; numerous alternative expressions are possible which are often hard to convert into each other.

The shortest notation we found that fulfills the requirement is

$$\hat{C}_7 + \hat{R}\hat{C}_8 = -\hat{H}_f/k \tag{2.31}$$

where $\hat{R}$ and $\hat{H}_f$ are the following constants:

$$\hat{R} \equiv \frac{\sinh(\hat{\beta}_3 k t_3 + \hat{\alpha}_{32} - \hat{\alpha}_3)}{\sinh(\hat{\beta}_3 k t_3 + \hat{\alpha}_{32} + \hat{\alpha}_3)} \tag{2.32}$$

$$\hat{H}_f \equiv \frac{k \sinh \hat{\alpha}_3}{\sinh(\hat{\beta}_3 k t_3 + \hat{\alpha}_{32} + \hat{\alpha}_3)} \times$$

$$\int_0^{t_3} \left\{ j\hat{M}_{px}(y) \cosh(\hat{\beta}_3 k(t_3 - y) + \hat{\alpha}_{32}) \right.$$

$$\left. + \hat{\beta}_3 \hat{M}_{py}(y) \sinh(\hat{\beta}_3 k(t_3 - y) + \hat{\alpha}_{32}) \right\} dy \tag{2.33}$$

with

$$\hat{\alpha}_3 \equiv \tfrac{1}{2}\ln \frac{\hat{\mu}_3 + 1}{\hat{\mu}_3 - 1} \tag{2.34}$$

$$\hat{\alpha}_{32} \equiv \tfrac{1}{2}\ln \frac{\hat{\mu}_{32} + 1}{\hat{\mu}_{32} - 1} \tag{2.35}$$

$$\hat{\mu}_{32} \equiv \frac{\hat{\mu}_3}{\hat{\mu}_2} \coth(\hat{\beta}_2 k t_2 + \hat{\alpha}_2) \quad (\leqslant \mu_3 \text{ if real}) \tag{2.36}$$

$$\hat{\alpha}_2 \equiv \tfrac{1}{2}\ln \frac{\hat{\mu}_2 + 1}{\hat{\mu}_2 - 1}. \tag{2.37}$$

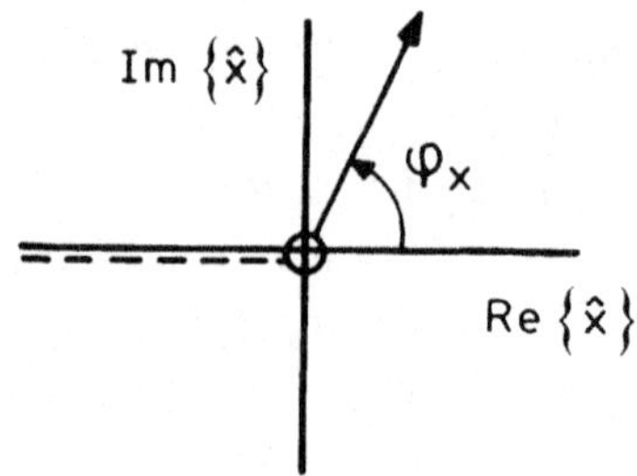

Fig. 2.2. Definition of φ_x in the $\hat{x}$ plane with exclusion of the cut (---) and the origin (o).

The appropriate definition of the ln function is

$$\ln \frac{\hat{x}}{\hat{y}} = \ln \frac{|\hat{x}|}{|\hat{y}|} + j(\varphi_x - \varphi_y) \tag{2.38}$$

with $-\pi < \varphi_{x,y} \leq \pi$, see Fig. 2.2.

2.3 Real coating and backlayer permeability

As remarked in the introduction, it is sufficient to introduce real permeabilities in regions fixed with the permanently magnetized material, i.e. the regions 2 and 3.

It is important to note that not all real definitions of the preceding functions in Eqs. (2.32)-(2.38) can be used with impunity in all cases of real permeabilities since μ_{32} can be smaller than 1 for which the value of the real ln function in (2.35) is undefined. By using the complex definition stated in (2.38) it can easily be shown that for real permeabilities μ_2 and μ_3 and $\mu_{32} < 1$ we may adopt instead of $\hat{a}_{32} \equiv \frac{1}{2}\ln \left| \dfrac{\mu_{32} + 1}{\mu_{32} - 1} \right| + j\pi$ the real definition of ln in

$$\alpha'_{32} \equiv \frac{1}{2}\ln \left| \frac{\mu_{32} + 1}{\mu_{32} - 1} \right| \tag{2.39}$$

while changing all hyperbolic functions containing α_{32} according to

$$\begin{aligned}
\sinh(\alpha_{32} + \varphi) &\rightarrow j \cosh(\alpha'_{32} + \varphi) \\
\cosh(\alpha_{32} + \varphi) &\rightarrow j \sinh(\alpha'_{32} + \varphi).
\end{aligned} \tag{2.40}$$

Since only combinations like $\sinh(\alpha'_{32} + \varphi)/\cosh(\alpha'_{32} + \varphi)$ etc. occur, j can be omitted in the results.

In conclusion, when the complex definitions of the ln and hyperbolic functions are used one needs not to distinguish between $\mu_{32} > 1$ and $\mu_{32} < 1$.

Since μ_2 and μ_3 in practice will in any case be taken real, the real definitions, which are more convenient and easier to compute, may be preferable, but in that case if $\mu_{32} < 1$ part of the result, (2.32), (2.33) and (2.35), will have to be changed according to (2.39) and (2.40). In all following examples and calculations real μ_2 and μ_3 will be assumed.

2.4 Real coating permeability and no backlayer

In the limit of $\mu_2 \downarrow 1$, i.e. no backlayer,

$$\mu_{32} = \mu_3 \Rightarrow \alpha_{32} = \alpha_3 \tag{2.41}$$

and so

$$R = \frac{\sinh(\beta_3 k t_3)}{\sinh(\beta_3 k t_3 + 2\alpha_3)} \tag{2.42}$$

and

$$\hat{H}_\mathrm{f} = \frac{k \sinh \alpha_3}{\sinh(\beta_3 k t_3 + 2\alpha_3)} \times$$

$$\int_0^{t_3} \left\{ \mathrm{j}\hat{M}_{\mathrm{px}}(y) \cosh(\beta_3 k(t_3 - y) + \alpha_3) \right.$$

$$\left. + \beta_3 \hat{M}_{\mathrm{py}}(y) \sinh(\beta_3 k(t_3 - y) + \alpha_3) \right\} \mathrm{d}y. \tag{2.43}$$

This case corresponds to Tjaden's case in [2.4]. The definition of $\hat{H}_\mathrm{f}$ differs slightly from his, due to the introduction of complex variables. Of course results must be equal, including their phases. This is checked with the example in the following section, also carried out by Tjaden for a tape without a backlayer. The example shows the physical meaning of $\hat{H}_f$.

2.5 Free-space surface-field amplitude $|\hat{H}_f|$

In this and following examples complete problems as mentioned below Eq. (2.22) will be solved. For these purposes it is necessary to complete the problem by assumptions concerning the region(s) beyond region 4.

For the present assumption of free space 'beyond' region 4, cf. Fig. 2.3, no reflected wave can be present in region 4 and consequently $\hat{C}_8$ must be zero. Eq. (2.31) thus reduces to $\hat{C}_7 = -\hat{H}_f/k$. The field in region 4 follows from (2.18) and (2.3)

$$\hat{H}(y) = (\hat{H}_x(y), \hat{H}_y(y)) = \hat{H}_f \, e^{-k|y|} \, (j, 1). \qquad (2.44)$$

From this result it is clear that $|\hat{H}_f|$ is the free-space surface $(y = 0^-)$-field amplitude in region 4 of both the x and the y component. Only the (spatial) phase differs ninety degrees, cf. sketch in Fig. 2.3; in free space x and y amplitudes are equal.

It is noted that the preceding remarks hold independently of the presence of a backlayer (etc.).

It is easily checked now that the phases of $\hat{H}_x$ and $\hat{H}_y$ also equal those found with Tjaden's expressions [2.4] when starting with the same M_{px} and M_{py}.

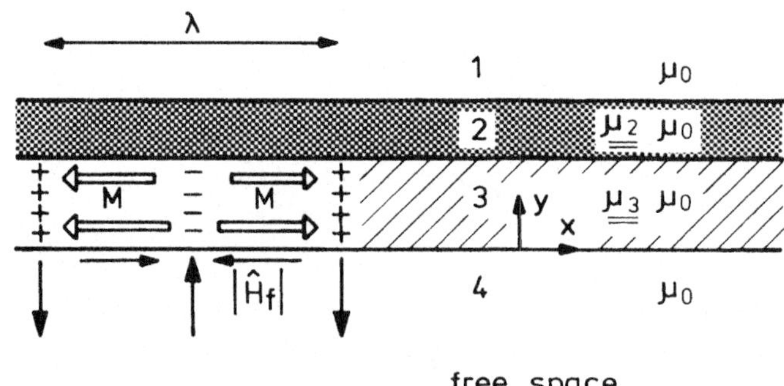

Fig. 2.3. Tape in free space.

2.6 Remagnetization parameter R and factor F_R

The physical meaning of the parameter R in (2.32) becomes clear from the following example where infinitely permeable material is present on the coating side of the tape, cf. Fig. 2.4. The boundary condition at $y = -d_4$ reads $\Psi = $ constant $= 0$. $\hat{C}_7$ and $\hat{C}_8$ then follow from (2.18) and (2.31). $\hat{H}(y)$ in region 4 follows as a result from (2.18), (2.11) and (2.3).

$$\hat{H}(y) = (\hat{H}_x(y), \hat{H}_y(y)) =$$
$$= \frac{2\hat{H}_f}{e^{kd_4} - Re^{-kd_4}} \left(j\sinh(k(y + d_4)), \cosh(k(y + d_4))\right) \quad (2.45)$$

For $y = -d_4$, H_x will be zero of course and

$$\boxed{\hat{H}(-d_4) = F_R 2\hat{H}_f\, e^{-kd_4}\, (0, 1)}\,, \quad (2.46)$$

where $(0, 1)$ is the unit vector in the y direction and

$$F_R \equiv \frac{1}{1 - Re^{-2kd_4}}.$$

Here $\mu_0 H_y(-d_4)$ is the density of the flux that penetrates into the infinitely permeable material. At large distances d_4 where $F_R \to 1$ this flux density is twice as large as in the free-space case, as follows from (2.44) and (2.46). This has its analogy in transmission-line theory, where short-circuiting of a long line gives a current through the short-circuited wire which is twice the current through the perfectly matched termination.

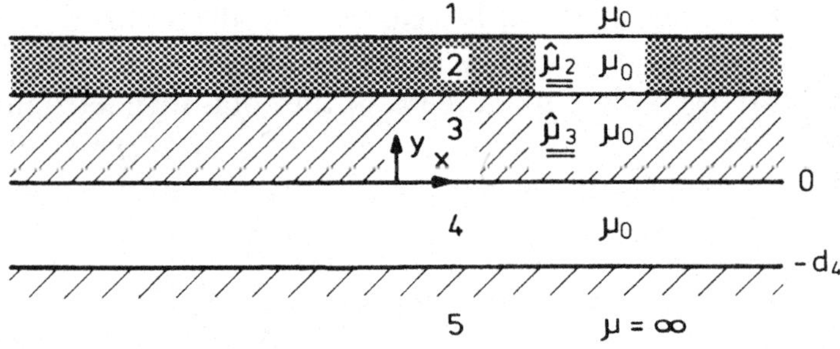

Fig. 2.4. Tape at a distance d_4 from an ideal soft-magnetic material. This material may represent a highly-permeable zero-gap head.

The factor F_R must be due to reversible changes of magnetization in the 'flux source' (tape), i.e. in the coating plus backlayer. Since $F_R > 1$ when $R \neq 0$ (i.e. μ_2 or $\mu_3 \neq 1$) for finite d_4, the total flux leaving the tape is increased by the presence of the head (here region 5). Physically this means that

- the total magnetization in the coating is increased or
- the opposing* longitudinal component of the magnetization in the backlayer is decreased, see Fig. 2.5a, or
- the co-operating* perpendicular component of the magnetization in the backlayer is increased, see Fig. 2.5b.

This becomes more evident from the following cases:

$$\mu_2 = \mu_3 = 1 \qquad \Rightarrow R = 0 \text{ and } F_R = 1 \qquad\qquad (2.47)$$

$$\mu_3 = \infty \qquad \Rightarrow R = 1 \text{ and } F_R = \frac{1}{1 - e^{-2kd_4}} \qquad\qquad (2.48)$$

$$\mu_2 = \infty, \mu_3 = 1 \quad \Rightarrow R = e^{-2kt_3} \text{ and } F_R = \frac{1}{1 - e^{-2k(t_3 + d_4)}}.$$

$$(2.49)$$

For all (real) values of permeabilities, including μ of the mirror

$$0 \leq R \leq 1. \qquad\qquad (2.50)$$

In the case of (2.47), the magnetization in the tape is fixed ($F_R = 1$) since the permeability equals μ_0 everywhere in the tape and no increase of the magnetization can take place.

Case (2.48) shows a strong increase of the magnetization ($F_R \gg 1$) when $kd_4 \ll 1$ due to the high μ_3 of the coating.

Case (2.49) also shows an increase of the total magnetization owing to a *decrease* of the longitudinal component or an *increase* of the perpendicular component of the total magnetization in the backlayer due to the high μ_2. The factor $e^{-2k(t_3 + d_4)}$ in (2.49) accounts for the distance

---

* Note that the backlayer is disadvantageous for longitudinal recording and advantageous for perpendicular recording.

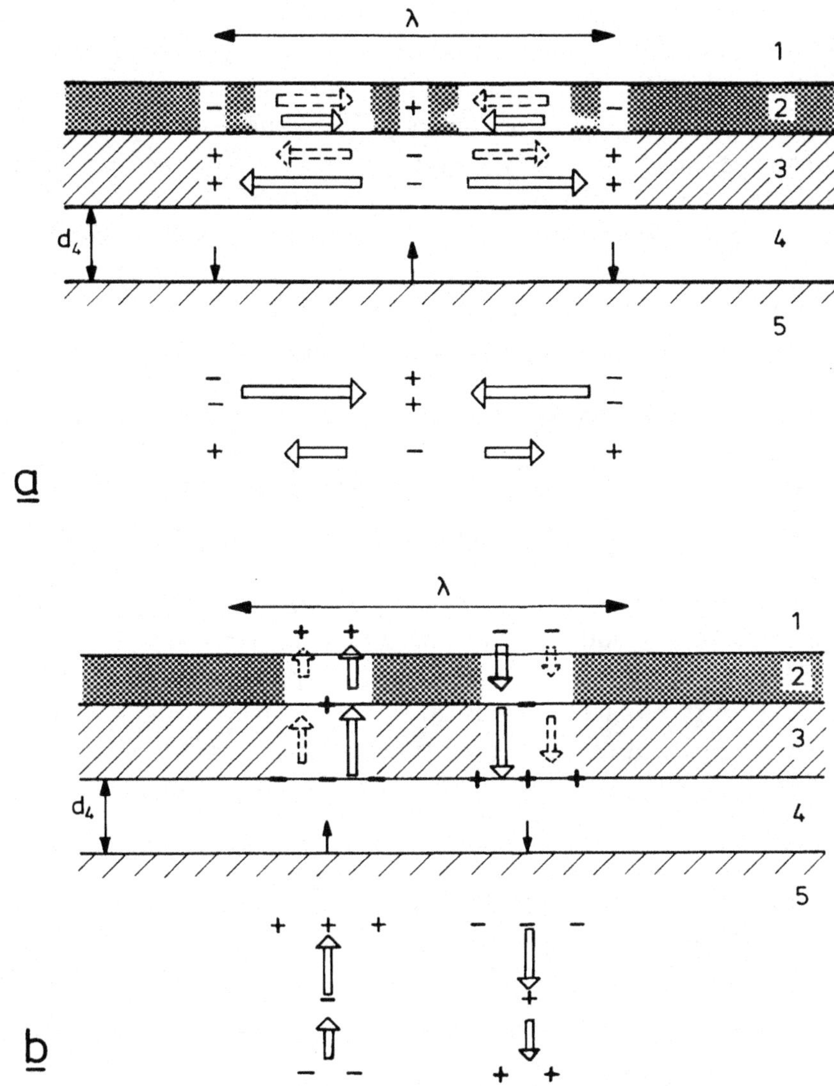

Fig. 2.5. Sketch of the remagnetization in the coating and backlayer. Not the 'real' magnetic charges but the 'fictitious' mirror charges are depicted in the ideally-reflecting head and above the boundary 3-2 with the (partly) reflecting backlayer. However the influence of the mirror image in the head on the magnetization in the backlayer is shown not by drawing the fictitious multiple mirror images from head to backlayer to head etc., but simply by changing the magnetization in the demagnetized state (dashed arrow) in the absence of the head to that in the partly remagnetized state (full arrow) due to the presence of the reflecting head.

$\rightarrow$: fields

$\Rightarrow$: magnetizations

$===\!\!\Rightarrow$: demagnetized situation when $d_4 \rightarrow \infty$ (free space)

$=\!\!\Rightarrow$: partly remagnetized situation in the case of finite d_4

a) Longitudinal magnetization

b) Vertical magnetization

between the mirror surface (infinitely permeable material at the coating side) and the backlayer.

In the cases (2.48) and (2.49) and all other cases different from (2.47) it is clear that remagnetization takes place in the tape.

For the above reasons we will call $F_R = 1/(1 - \mathrm{Re}^{-2kd_4})$ the remagnetization factor and R the remagnetization parameter.

Since the remagnetization decreases with increasing head-tape distance d_4, the distance losses will differ from the usual e^{-kd_4}. The factor DL that describes the distance loss of the ideal head, i.e. the read flux Φ^R penetrating the head at a head-tape distance d_4 divided by the flux penetrating the head when $d_4 = 0$, or the ratio of the corresponding perpendicular fields, follows from Eq. (2.46) and equals

$$DL \equiv \frac{\Phi^R(d_4)}{\Phi^R(d_4 = 0)} = \frac{1 - R}{1 - \mathrm{Re}^{-2kd_4}} \, \mathrm{e}^{-kd_4} \, . \qquad (2.51)$$

An increased distance loss occurs at small distances $d_4 < \lambda/4\pi$; see Fig. 2.6. With the aid of the curves in Fig. 2.6 (I and II) and the comments in the figure captions an attempt is made to give a comprehensive overview of the most important influences that tape parameters

Fig. 2.6.I. The distance loss factor DL in the case of a single-layer permeable medium and no gap smear according to (2.51).

a) DL as a function of head-tape distance d_4.
       $t_3 = 0.5$ µm.
b) DL as a function of head-tape distance d_4.
       $t_3 = 3$ µm.
c) DL as a function of the wavenumber $k \equiv 2\pi/\lambda$.
       $t_3 = 0.5$ µm.
d) DL as a function of the wavenumber $k \equiv 2\pi/\lambda$.
       $t_3 = 3$ µm.

1) $\mu_3 = 1$ $\beta_3 = 1$
2) $\mu_3 = 2$ $\beta_3 = 1/2, 1, 2$
3) $\mu_3 = 4$ $\beta_3 = 1/4, 1/2, 1, 2, 4$
4) $\mu_3 = 8$ $\beta_3 = 1/8, 1/4, 1/2, 1, 2, 4, 8$
5) $\mu_3 = 16$ $\beta_3 = 1/16, 1/8, 1/4, 1/2, 1, 2, 4, 8, 16$

The distance loss is independent of the location of the magnetization in the tape. For $\mu_3 = 1$ the usual e^{-kd_4} results, so that DL has much the same value as a function of d_4 and of k. When $\mu_3 \neq 1$ there is an exchange between head and tape which influences the distance loss, especially at short distances. E.g. for $\mu_3 = 1.75$, $\beta_3 \approx 1$, $t_3 = 3$ µm and $\lambda = 0.7$ µm, the distance loss is 1.67 dB per 13 nm increase of the head-to-tape distance when $d_4 \rightarrow 0$, $DL = 1.1$ dB/13 nm when $d_4 = 0.1$ µm and $DL = 1$ dB/13 nm when $d_4 \rightarrow \infty$. At high μ_3 there is much demagnetization in the tape which is not compensated by remagnetization in the tape when the 'normalized' distance kd_4 is too large ($kd_4 > 1$). A small $\beta_3 kt_3$ ($\lesssim 3$) decreases the distance losses within the tape noticeably. This favours the low-wavenumber regions of the curves with small β_3 in c) and d) and favours the full curves of constant wavenumber and small β_3 in a) but especially in c) since t_3 is small there.

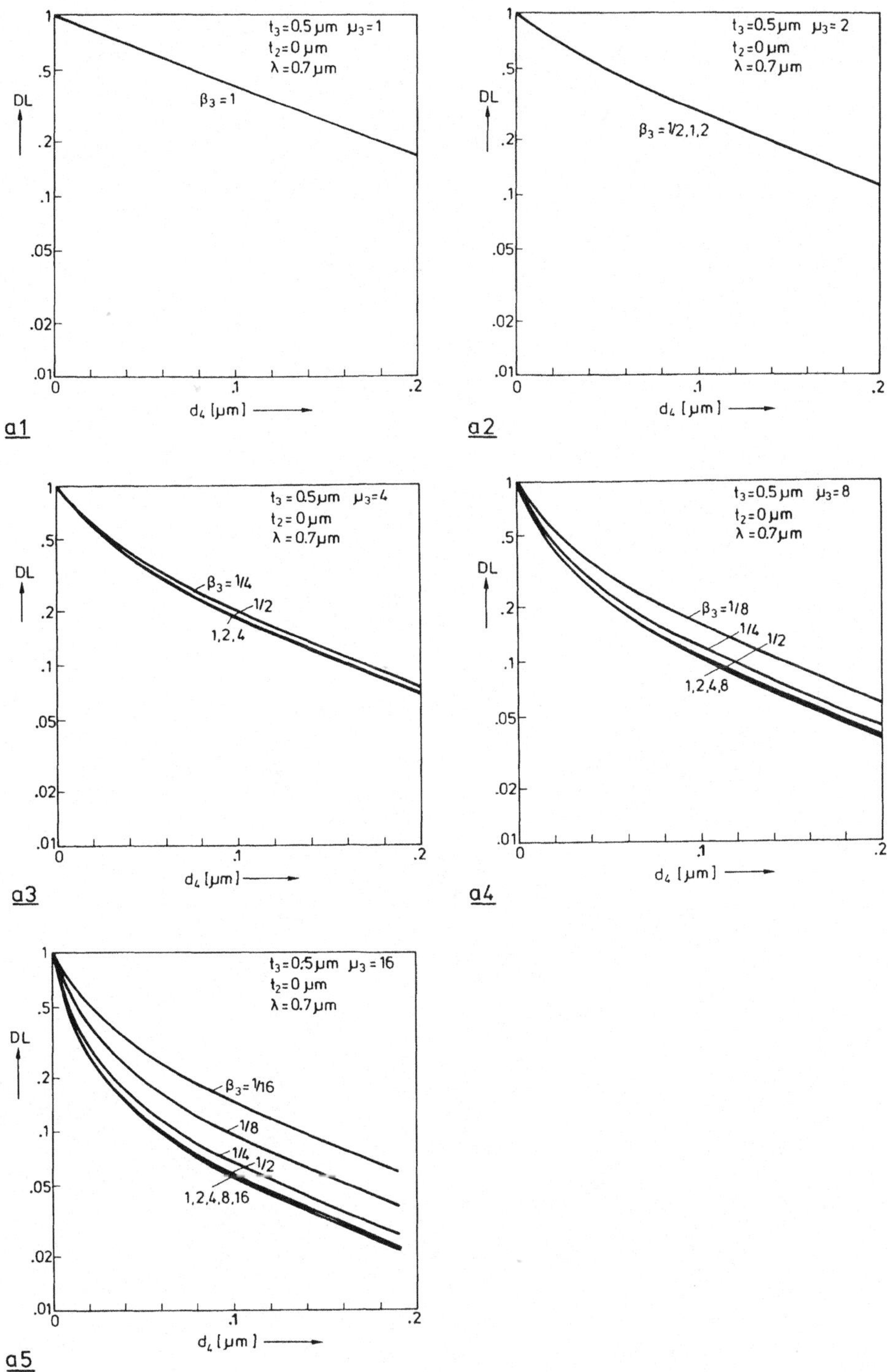

t_3=0.5μm μ_3=1
t_2=0μm
λ=0.7μm
DL
β_3=1
d_4 [μm]
a1
t_3=0.5μm μ_3=2
t_2=0μm
λ=0.7μm
DL
β_3=1/2,1,2
d_4 [μm]
a2
t_3=0.5μm μ_3=4
t_2=0μm
λ=0.7μm
DL
β_3=1/4
1/2
1,2,4
d_4 [μm]
a3
t_3=0.5μm μ_3=8
t_2=0μm
λ=0.7μm
DL
β_3=1/8
1/4 1/2
1,2,4,8
d_4 [μm]
a4
t_3=0.5μm μ_3=16
t_2=0μm
λ=0.7μm
DL
β_3=1/16
1/8
1/4 1/2
1,2,4,8,16
d_4 [μm]
a5

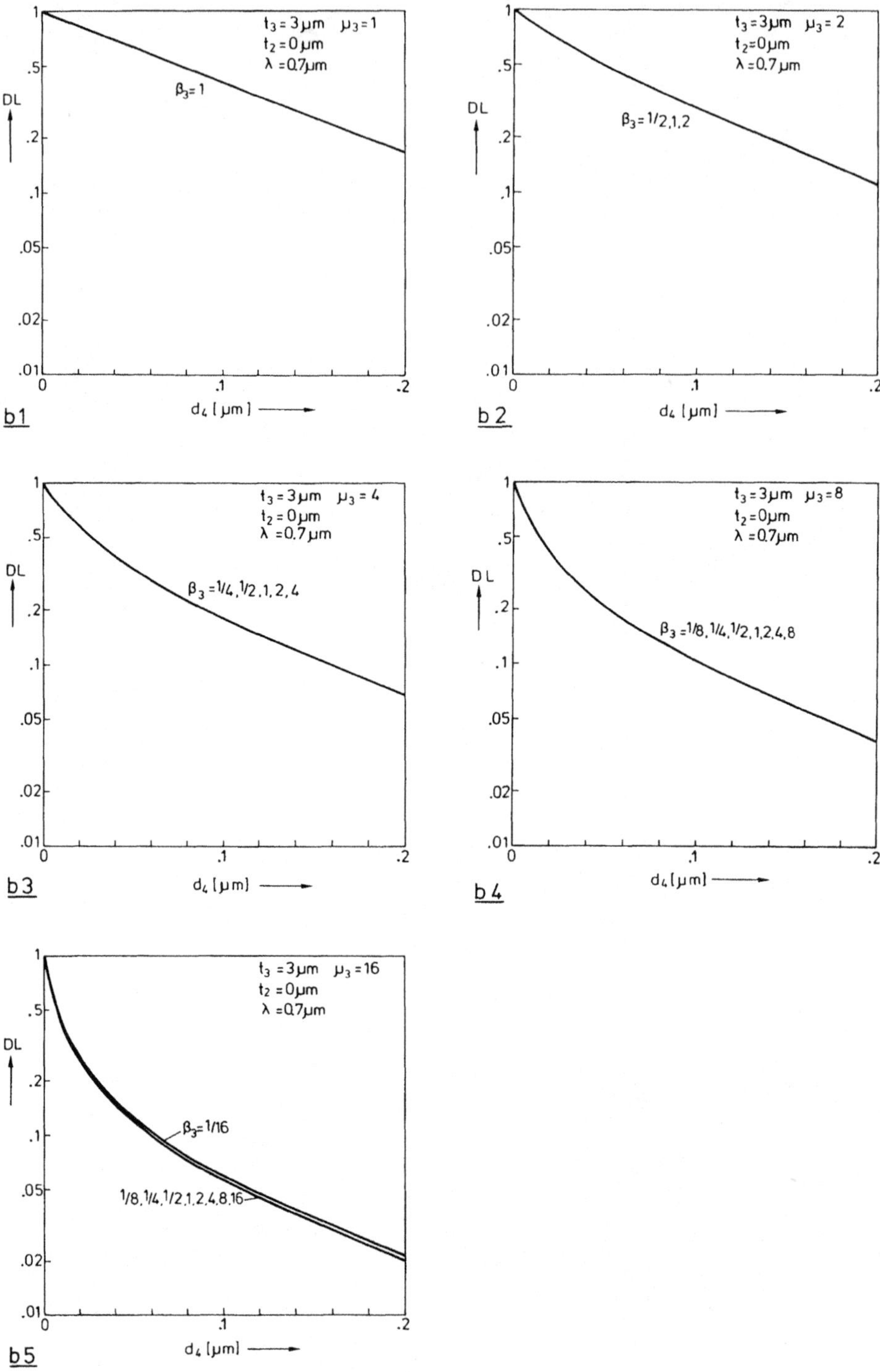
b1
t_3 = 3 μm μ_3 = 1
t_2 = 0 μm
λ = 0.7 μm
β_3 = 1
DL
d_4 [μm]

b2
t_3 = 3 μm μ_3 = 2
t_2 = 0 μm
λ = 0.7 μm
β_3 = 1/2,1,2
DL
d_4 [μm]

b3
t_3 = 3 μm μ_3 = 4
t_2 = 0 μm
λ = 0.7 μm
β_3 = 1/4,1/2,1,2,4
DL
d_4 [μm]

b4
t_3 = 3 μm μ_3 = 8
t_2 = 0 μm
λ = 0.7 μm
β_3 = 1/8,1/4,1/2,1,2,4,8
DL
d_4 [μm]

b5
t_3 = 3 μm μ_3 = 16
t_2 = 0 μm
λ = 0.7 μm
β_3 = 1/16
1/8,1/4,1/2,1,2,4,8,16
DL
d_4 [μm]

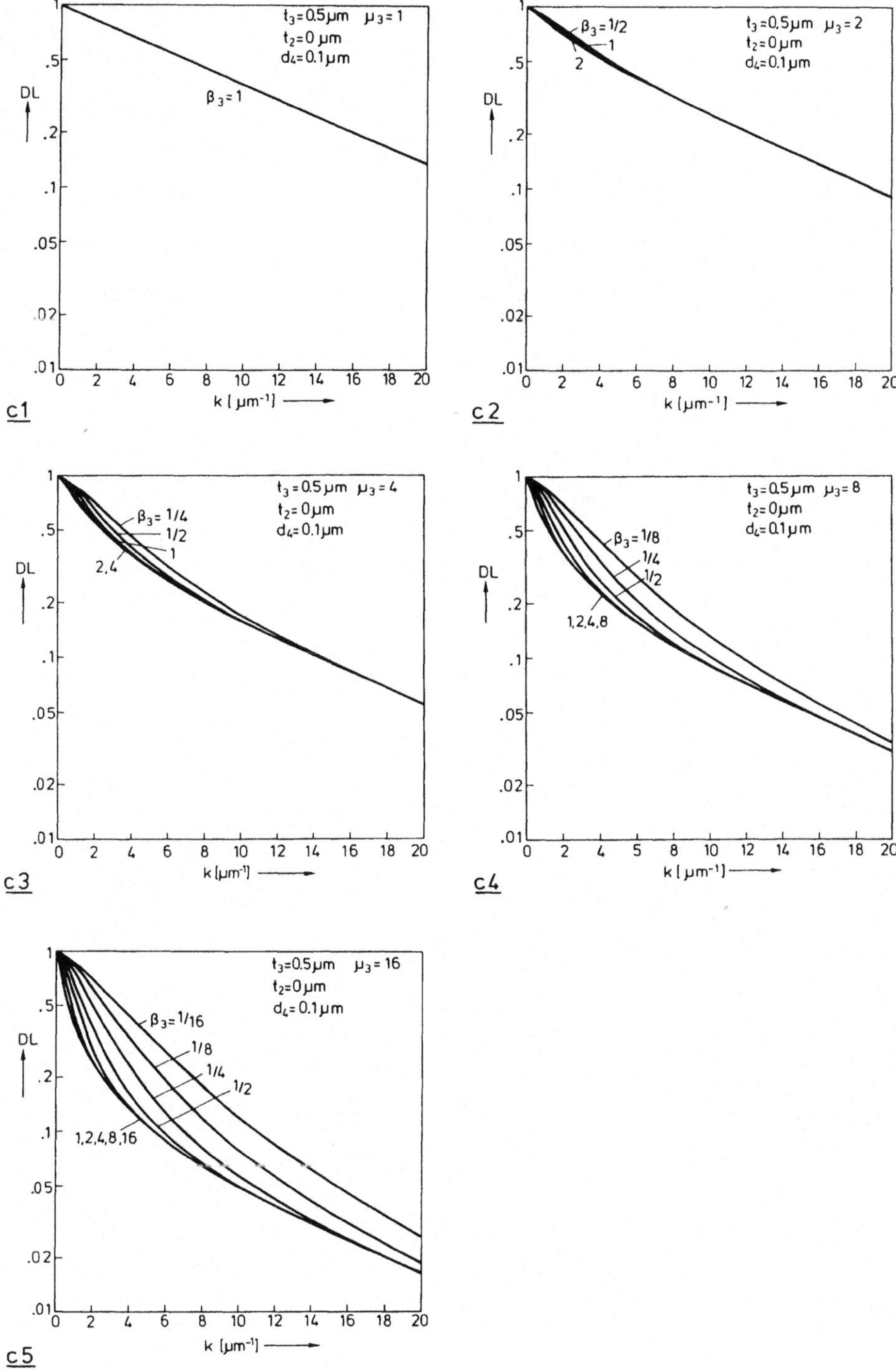

DL
t₃=0.5µm µ₃=1
t₂=0 µm
d₄=0.1µm
β₃=1
k [µm⁻¹]
c1

DL
t₃=0.5µm µ₃=2
t₂=0µm
d₄=0.1µm
β₃=1/2
1
2
k [µm⁻¹]
c2

DL
t₃=0.5µm µ₃=4
t₂=0µm
d₄=0.1µm
β₃=1/4
1/2
1
2,4
k [µm⁻¹]
c3

DL
t₃=0.5µm µ₃=8
t₂=0µm
d₄=0.1µm
β₃=1/8
1/4
1/2
1,2,4,8
k [µm⁻¹]
c4

DL
t₃=0.5µm µ₃=16
t₂=0µm
d₄=0.1µm
β₃=1/16
1/8
1/4
1/2
1,2,4,8,16
k [µm⁻¹]
c5

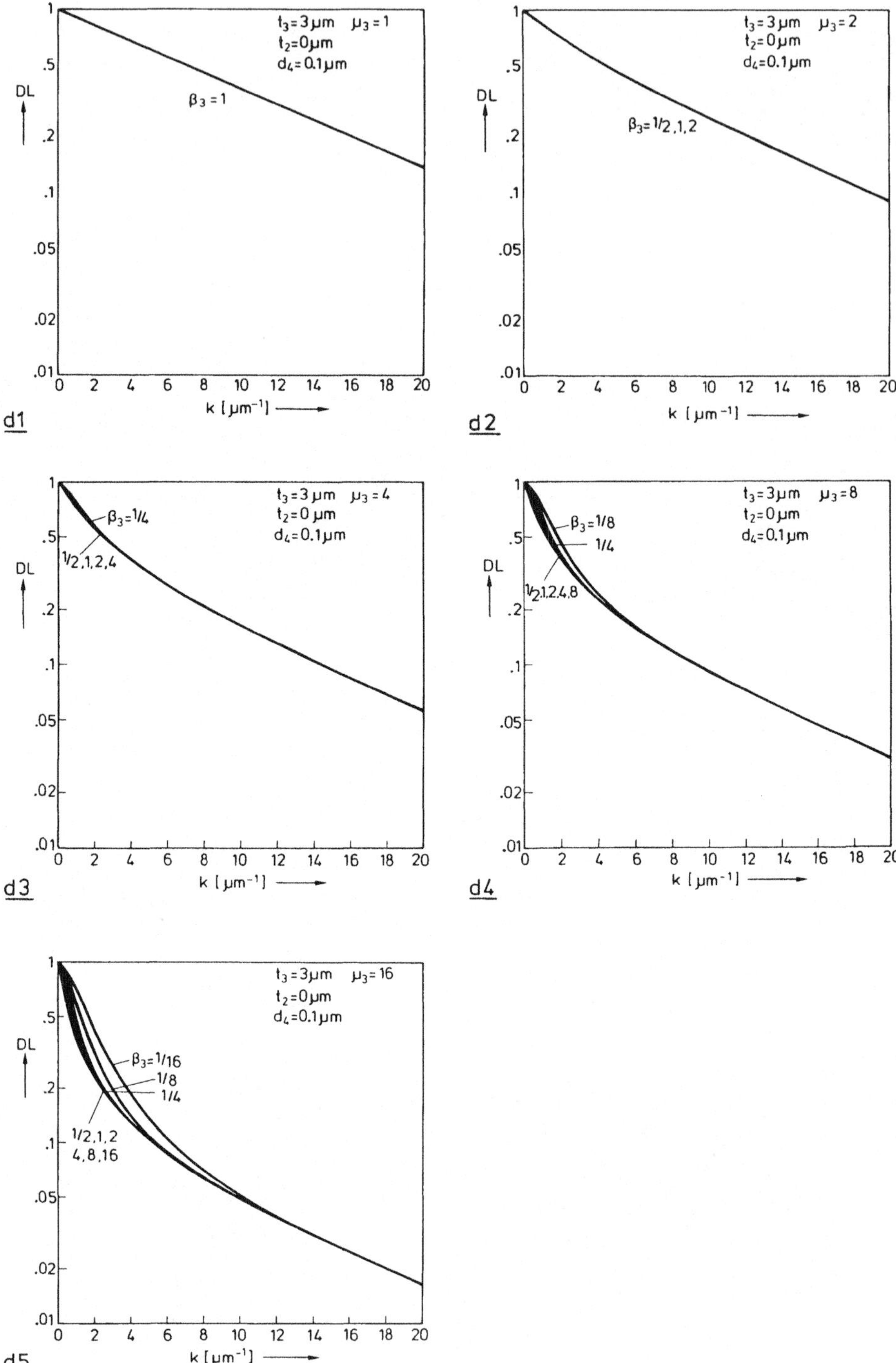

DL
t3=3µm µ3=1
t2=0µm
d4=0.1µm
β3=1
k [µm⁻¹]
d1

DL
t3=3µm µ3=2
t2=0µm
d4=0.1µm
β3=1/2,1,2
k [µm⁻¹]
d2

DL
t3=3µm µ3=4
t2=0µm
d4=0.1µm
β3=1/4
1/2,1,2,4
k [µm⁻¹]
d3

DL
t3=3µm µ3=8
t2=0µm
d4=0.1µm
β3=1/8
1/4
1/2,1,2,4,8
k [µm⁻¹]
d4

DL
t3=3µm µ3=16
t2=0µm
d4=0.1µm
β3=1/16
1/8
1/4
1/2,1,2
4,8,16
k [µm⁻¹]
d5

Fig. 2.6.II. The distance loss factor DL in the case of a double-layer permeable medium and no gap smear according to (2.51). Coating: $t_3 = 0.5$ μm, $\mu_3 = 4$, $\beta_3 = 1$.
a) DL as a function of head-tape distance d_4.
     $\lambda = 0.7$ μm $\mu_2 = 1000$ $\beta_2 = 1$
b) DL as a function of head-tape distance d_4.
     $\lambda = 10$ μm $\mu_2 = 1000$ $\beta_2 = 1$
c) DL as a function of the wavenumber $k \equiv 2\pi/\lambda$.
     $\mu_2 = 10$ $\beta_2 = 1$
d) DL as a function of the wavenumber $k \equiv 2\pi/\lambda$.
     $\mu_2 = 100$ $\beta_2 = 1$
e) DL as a function of the wavenumber $k \equiv 2\pi/\lambda$.
     $\mu_2 = 1000$ $\beta_2 = 1$
f) DL as a function of the wavenumber $k \equiv 2\pi/\lambda$.
     $\mu_2 = 1000$ $\beta_2 = 1/4$
g) DL as a function of the wavenumber $k \equiv 2\pi/\lambda$.
     $\mu_2 = 1000$ $\beta_2 = 4$

The DL is hardly influenced by the backlayer when the 'normalized' distance to the backlayer $k(t_3 + d_4)$ is large, cf. a), but noticeably when $k(t_3 + d_4) < 1$; see b). See also the influence of the backlayer parameters in the low-wavenumber regions of c), d) and e) for increasing values of the backlayer permeability, and f), e) and g) for increasing values of β_2. Although t_3 (and d_4) are rather small, there is no noticeable influence (drawback) of the backlayer on the distance loss in the most interesting, i.e. higher wavenumber, regions.

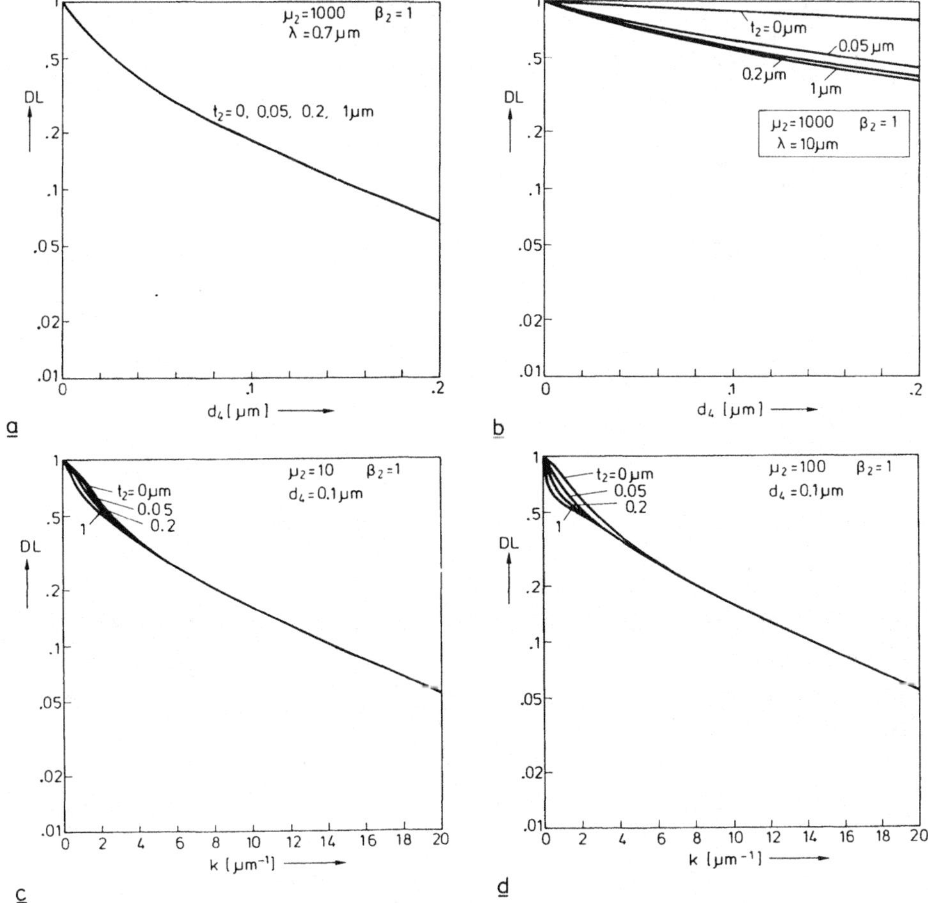

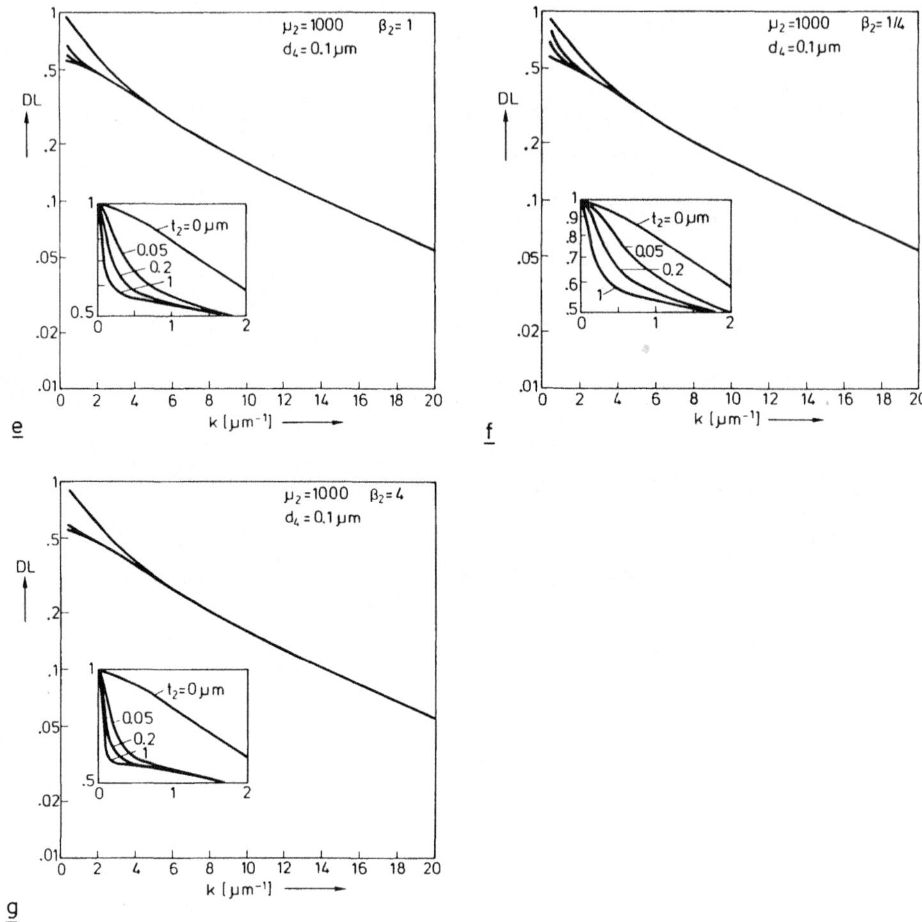

(via the remagnetization parameter R) and head-tape distance (via the interaction distance d_4) have on the distance loss. In this context, reference may also be made to an article by Bertram on reversible permeability effects [2.5]. This article can be seen as an extension of the calculations in the first section of Tjaden's article [2.4]. In Bertram's and Tjaden's calculations no backlayers were assumed. We shall return to Bertram's results in chapter 6 where absolute values of fluxes instead of ratios are calculated.

From (2.46) we can conclude as follows when comparing tapes with different reversible permeabilities:

– All short wavelengths (relative to $2\pi d_4$) are read with the same sensitivity when the free-space field $|\hat{H}_f|$ is equal.

–   Long wavelengths (relative to $2\pi d_4$) are read $1/(1 - R)$ times stronger than in the case of a tape with a relative permeability $\mu_3 = 1$ with the same free-space field $|\hat{H}_f|$.

However, it is not realistic to assume equal free-space fields. The assumption, for example, of equal permanent magnetization changes the above results. This will be worked out in section 6.3.

From the types of symmetries in Fig. 2.5 it is also clear that, compared to the free-space situation without image charges, the factor 2 in Eq. (2.46) arises for the perpendicular component of the field approaching the interface at $y = -d_4$ with the infinitely permeable material, and that the longitudinal component vanishes at the interface.

It is remarkable that the remagnetization factor F_R does not depend on the location or distribution of the permanent magnetization in the coating.

2.7 Remagnetization as a function of k at finite μ_2 and μ_3

Two examples will be discussed. In the first example $\mu_2 = 1$ and μ_3 is a parameter, while k (or λ) is the variable. In the second example $\mu_3 = 1$.

$\mu_2 = 1$, i.e. no backlayer

Then R is given by (2.42) and from this equation follows the range of possible R values:

$$0 \leqslant R \leqslant \frac{\mu_3 - 1}{\mu_3 + 1}. \tag{2.52}$$

The lower limit is reached in the limit of low frequencies and the upper limit at high frequencies $\beta_3 k t_3 \gg 1$, which holds for almost the whole range of wavelengths in video recording. The corresponding range of F_R reads

$$1 \leqslant F_R < \frac{\mu_3 + 1}{2} \tag{2.53}$$

where the lower limit is reached in the limit of low and high frequencies, but the higher value can only (approximately) be reached in the region where both $\beta_3 k t_3 \gg 1$ *and* $\mu_3 k d_4 \ll 1$ if this region exists, i.e. when $\beta_3 t_3 \gg \mu_3 d_4$. In a video recorder with metal powder (MP) tape this region exists since $t_3 \simeq 3$ μm, $\beta_3 = 0.5\text{-}1$, $\mu_3 = 1.5\text{-}2$ and $d_4 \simeq 0.1$ μm. In the 'middle' of this region, where very roughly

$$k \simeq \frac{1}{\sqrt{\beta_3 \mu_3 t_3 d_4}} \quad \text{or} \quad \lambda \simeq 2\pi\sqrt{\beta_3 \mu_3 t_3 d_4}, \tag{2.54}$$

the remagnetization is maximum; see Fig. 2.7. For high values of μ_3 the approximation for the location of the maximum (2.54) is quite accurate, in spite of the fact that the upper limit in (2.53) may not be approached.

For $t_3 = 3$ μm, cf. Fig. 2.7a, the increase is visible in the whole frequency range of interest, since $\beta_3 k t_3 = 1$ at about $\lambda = 10$ μm and $k d_4 = 1$ at about $\lambda = 0.6$ μm. When the coating is very thin, e.g. $t_3 = 0.5$ μm (see Fig. 2.7b) the demagnetizing field is already small in the free-space situation. Consequently the remagnetization is small.

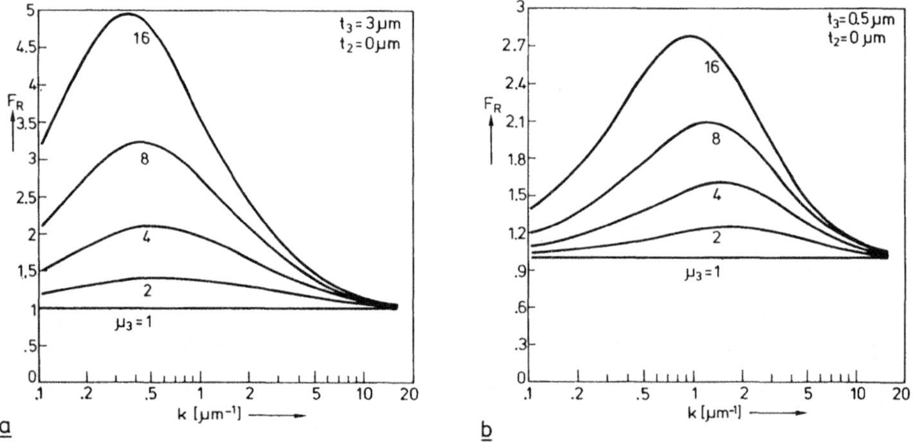

Fig. 2.7. The remagnetization factor F_R in the case of a single-layer permeable medium as a function of the wave number k. The corresponding wavelength range is 63 μm – 0.31 μm. The head-tape distance $d_4 = 0.1$ μm and $\beta_3 = 1$.
a) $t_3 = 3$ μm
b) $t_3 = 0.5$ μm.

$\mu_3 = 1$, i.e. no reversible effects in the coating

Reversible effects can now take place only in the backlayer, i.e. further away from the head. Eq. (2.32) reduces in the limit $\mu_3 \downarrow 1$ with the aid of (2.34)-(2.36) to

$$R = \frac{1 - \mu_{32}}{1 + \mu_{32}} \, e^{-2kt_3} \tag{2.55}$$

with

$$\mu_{32} = \frac{1}{\mu_2} \coth(\beta_2 k t_2 + \alpha_2) \leqslant 1.$$

The 'range' of R reads

$$0 \leqslant R < \frac{\mu_2 - 1}{\mu_2 + 1} \, e^{-2kt_3} \tag{2.56}$$

when $\beta_2 k t_2 \gg 1$.

The corresponding range of the remagnetization factor F_R follows with the aid of equation (2.46) defining F_R and reads

$$1 \leqslant F_R < \frac{\mu_2 + 1}{2} \tag{2.57}$$

where the upper value of F_R can be approximated only when there is a region where both $\beta_2 k t_2 \gg 1$ and $\mu_2 k(t_3 + d_4) \ll 1$. This implies, since $\beta_2 \leqslant \mu_2$, that t_2 must be much larger than $t_3 + d_4$. In the middle of this region, where roughly

$$k \simeq \frac{1}{\sqrt{\beta_2 \mu_2 t_2 (t_3 + d_4)}} \quad \text{or} \quad \lambda \simeq 2\pi\sqrt{\beta_2 \mu_2 t_2 (t_3 + d_4)}, \tag{2.58}$$

the remagnetization (in fact the demagnetization of the magnetization in the backlayer) is maximum; cf. Fig. 2.8. For large μ_2 this is a good approximation for the location of the maximum, in spite of the fact that the upper limit in (2.57) is far from being approached.

In the examples of Fig. 2.7 and Fig. 2.8 an extreme value of remagnetization occurs at the highest frequencies at which distance losses were not yet too large. It does not occur at $k = 0$ since in the limit of low wavenumbers demagnetization fields vanish already in free space and consequently remagnetization cannot take place when the permeabilities are finite. In the cases (2.48) and (2.49) the permeabilities are infinite and the maxima are reached at $k = 0$.

Assuming for a moment that μ_2 and μ_3 are equal, the remagnetization due to the backlayer would be smaller than the remagnetization due to

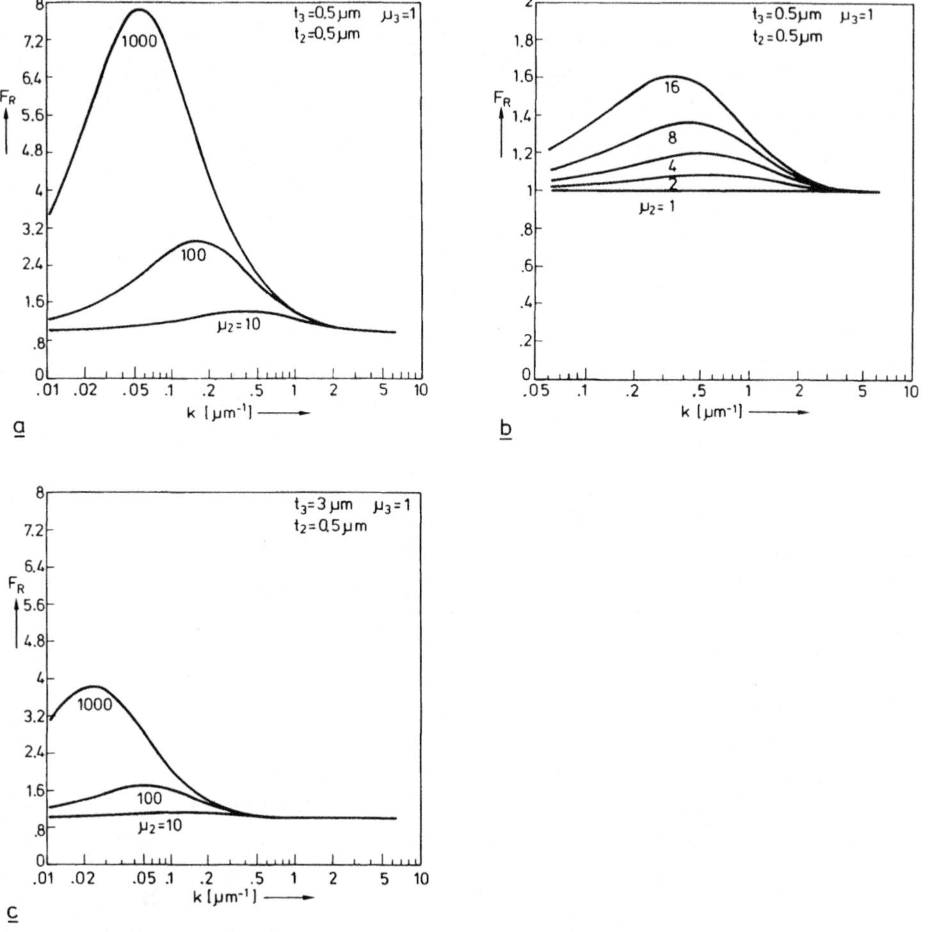

Fig. 2.8. The remagnetization factor F_R in the case of a double-layer medium as a function of the wavenumber k. The head-tape distance $d_4 = 0.1$ μm, $\mu_3 = 1$, $\beta_2 = \beta_3 = 1$ and $t_2 = 0.5$ μm.
a) and b) $t_3 = 0.5$ μm
c) $t_3 = 3$ μm.

the coating, since the backlayer is further away from the reflecting head. This is visible when comparing Figs. 2.7b and 2.8b. The larger the coating thickness, the larger the initial demagnetization and thus the larger the remagnetization in the coating (compare Figs. 2.7a and b) but the smaller the remagnetization due to the backlayer (compare Figs. 2.8c and a). In fact the coating (permeability) and backlayer (permeability) act identically. The difference in qualitative behaviour is due to a different location relative to the head and not to the permanently magnetized layer. Quantitative differences are caused by differences in location as well as in permeability. In this context it is important to realize that, fundamentally, the location of the permanent magnetization pattern does not play any role in the present results! *Thus, although locally different magnetization patterns will create different demagnetizing fields, and when the head is present different remagnetizing fields, the distance-loss function will not change. Only the location and permeability characteristics of the layers determine the distance-loss factor*!

2.8 Extension with magnetic layer on top of idealized zero-gap head

In this section we investigate the influence which a magnetic layer on top of an idealized zero-gap head has upon the playback performance. The magnetic layer may represent gap smear [2.1]. The general result of this calculation includes almost all cases investigated before in the literature. This result will be used as the starting point of many calculations and discussions throughout this book, a great deal of which contains new results. The present configuration was sketched in Fig. 2.1.

Calculations

The relation between $\hat{C}_7$ and $\hat{C}_8$ in (2.31) is now determined by the magnetic layer to tape distance d_4 and the thickness t_5, permeability μ_5 and anisotropy β_5 of the smear layer 5. First of all, the conditions at the boundary 4-5 according to the conditions (2.8b) and (2.8c) read

$$\hat{C}_7 e^{-kd_4} + \hat{C}_8 e^{kd_4} = \hat{C}_9 e^{-\hat{\beta}_5 kd_4} + \hat{C}_{10} e^{\beta_5 kd_4} \qquad (2.59)$$

$$\hat{C}_7 e^{-kd_4} - \hat{C}_8 e^{kd_4} = \hat{C}_9 \hat{\mu}_5 e^{-\hat{\beta}_5 kd_4} - \hat{C}_{10} \hat{\mu}_5 e^{\hat{\beta}_5 kd_4} \qquad (2.60)$$

and according to the boundary condition stated in (2.10), valid at the boundary with the ideal zero-gap head,

$$\hat{C}_9 e^{-\hat{\beta}_5 k(d_4 + t_5)} + \hat{C}_{10} e^{\hat{\beta}_5 k(d_4 + t_5)} = 0. \tag{2.61}$$

The constants $\hat{C}_7$ and $\hat{C}_8$ in (2.31) can be expressed in $\hat{C}_9$ and $\hat{C}_{10}$ by way of (2.59) and (2.60). After this is done, $\hat{C}_9$ or $\hat{C}_{10}$ can be eliminated in the resulting expression by using (2.61), so that $\hat{C}_{10}$ or $\hat{C}_9$ is known. The result for $\hat{C}_{10}$ is

$$\hat{C}_{10} =$$

$$= \frac{\hat{H}_f\, e^{-\hat{\beta}_5 k(d_4 + t_5)}}{k\hat{\mu}_5 \cosh(\hat{\beta}_5 k t_5)\left\{\left[1 + \dfrac{\tanh(\hat{\beta}_5 k t_5)}{\hat{\mu}_5}\right]e^{k d_4} - R\left[1 - \dfrac{\tanh(\hat{\beta}_5 k t_5)}{\hat{\mu}_5}\right]e^{-k d_4}\right\}} \tag{2.62}$$

and $\hat{C}_9$ follows from (2.61).

The complex potential in region 5 which is generally described in (2.20), is now known. The complex normal component of magnetic induction that penetrates the head, $\hat{B}_y(-d_4 - t_5) = -\mu_0\mu_{y5}(\partial\hat{\Psi}_5/\partial y)_{y = -d_4 - t_5}$, can now be evaluated, giving

$$\hat{B}_y(-d_4 - t_5) =$$

$$= \frac{2\mu_0\hat{H}_f}{\cosh(\hat{\beta}_5 k t_5)\left\{\left[1 + \dfrac{\tanh(\hat{\beta}_5 k t_5)}{\hat{\mu}_5}\right]e^{k d_4} - R\left[1 - \dfrac{\tanh(\hat{\beta}_5 k t_5)}{\hat{\mu}_5}\right]e^{-k d_4}\right\}} \;. \tag{2.63}$$

2.9 Distance loss *DL* and gap-smear loss *GSL*

The flux entering an idealized head is proportional to the complex conjugate, see chapter 4, of $\hat{B}_y$ in (2.63). So the factor describing the loss for an idealized head follows for head-tape distance (d_4) variations from

$$
DL \equiv \frac{\Phi(d_4)}{\Phi(d_4 = 0)}
$$

$$
= \frac{\left[1 + \dfrac{\tanh(\hat{\beta}_5^* k t_5)}{\hat{\mu}_5^*}\right] - R\left[1 - \dfrac{\tanh(\hat{\beta}_5^* k t_5)}{\hat{\mu}_5^*}\right]}{\left[1 + \dfrac{\tanh(\hat{\beta}_5^* k t_5)}{\hat{\mu}_5^*}\right]e^{k d_4} - R\left[1 - \dfrac{\tanh(\hat{\beta}_5^* k t_5)}{\hat{\mu}_5^*}\right]e^{-k d_4}} \tag{2.64}
$$

and for variations in magnetic layer thickness t_5 from

$$
GSL \equiv \frac{\Phi(t_5)}{\Phi(t_5 = 0)}
$$

$$
= \frac{e^{k d_4} - R e^{-k d_4}}{\cosh(\hat{\beta}_5^* k t_5)\left\{\left[1 + \dfrac{\tanh(\hat{\beta}_5^* k t_5)}{\hat{\mu}_5^*}\right]e^{k d_4} - R\left[1 - \dfrac{\tanh(\hat{\beta}_5^* k t_5)}{\hat{\mu}_5^*}\right]e^{-k d_4}\right\}}.
$$

$$
\tag{2.65}
$$

Both factors depend on the tape characteristics via the remagnetization parameter R. In Figs. 2.9 and 2.10 results of (2.64) and (2.65) are plotted for a dispersion-free smear layer 5, i.e. real μ_5 and β_5.

When R is zero, due to a tape-permeability equal to 1, i.e. fixed magnetization in the tape, or when the normalized 'interaction' distance $k d_4 \gg 1$, the tape dependence vanishes. The usual factor describing distance loss $e^{-k d_4}$ then remains and (2.65) relaxes to

$$
GSL = \frac{1}{\cosh(\hat{\beta}_5^* k t_5)\left[1 + \dfrac{1}{\hat{\mu}_5^*}\tanh(\hat{\beta}_5^* k t_5)\right]} \tag{2.66}
$$

which can be rewritten to read

$$GSL = 2 \; \frac{\hat{\mu}_5^*}{\hat{\mu}_5^* + 1} \cdot \frac{e^{-\hat{\beta}_5^* k t_5}}{1 + [(\hat{\mu}_5^* - 1)/(\hat{\mu}_5^* + 1)]e^{-2\hat{\beta}_5^* k t_5}} . \qquad (2.67)$$

From this expression $e^{-\hat{\beta}_5^* k t_5}$ described the (idealized) transport of the plane wave $\sim \sin(kx)$ in the magnetic layer 5 when reflections at the boundary with the soft-magnetic head are left out of account. The factor 2 results when, analogously to the factor 2 in (2.46), the reflection at the boundary between region 4 and the smear layer is maximum i.e. $|\mu_5| \gg 1$, but also $\text{Re}\{2\beta_5 k t_5\} \gg 1$ is now required so that the complete reflection at the boundary with the soft-magnetic head will not interfere (in a destructive way) with this reflection; see also the comment to (2.46). The approximation for large $|\hat{\mu}_5|$, which holds for all values of R and d_4 including $R = 1$ and $d_4 = 0$,

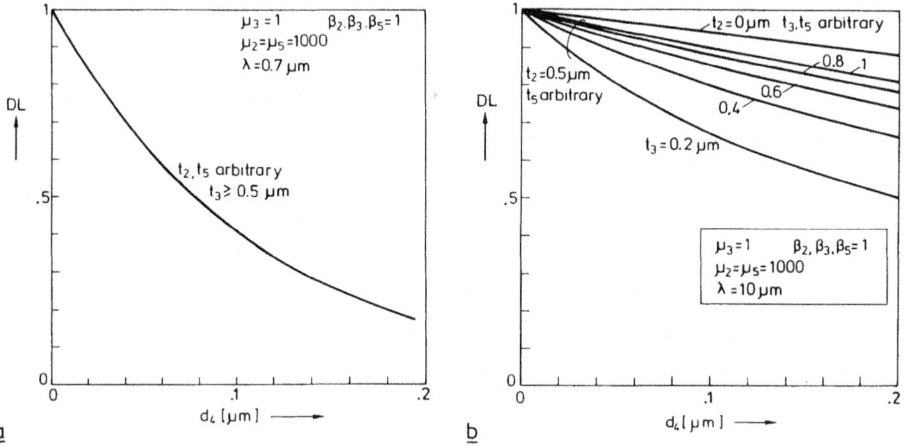

Fig. 2.9. The distance-loss factor DL as a function of head-tape distance d_4 at different backlayer and gap-smear layer thicknesses.
a) $\lambda = 0.7$ μm.
No influence of the back and gap-smear layer is visible. The backlayer does not influence the distance loss since its normalized distance to the head $kd_4 + \beta_3 k t_3$ is too large (> 1) and thus no interaction between head + gap-smear layer and permeable backlayer occurs. On the other hand the permeability of the coating was assumed to be zero, so that the coating does not interact with the head and gap-smear layer either. As a result the usual e^{-kd_4} is approached.
b) $\lambda = 10$ μm.
Now $kd_4 + \beta_3 k t_3 < 1$ and a clear influence of the coating thickness t_3 and backlayer on the distance loss is visible. A reduction of t_3 or increase of t_2 does increase the remagnetization, especially at small (normalized) distances. This has the consequence that the dependence on head-tape distance variations increases at the longer wavelengths. The thickness of the assumed highly-permeable gap-smear layer does not influence the distance loss since it acts as a part of the ideally soft-magnetic zero-gap head and d_4 was taken constant. The interaction has thus not been altered by the presence of the highly-permeable gap-smear layer.

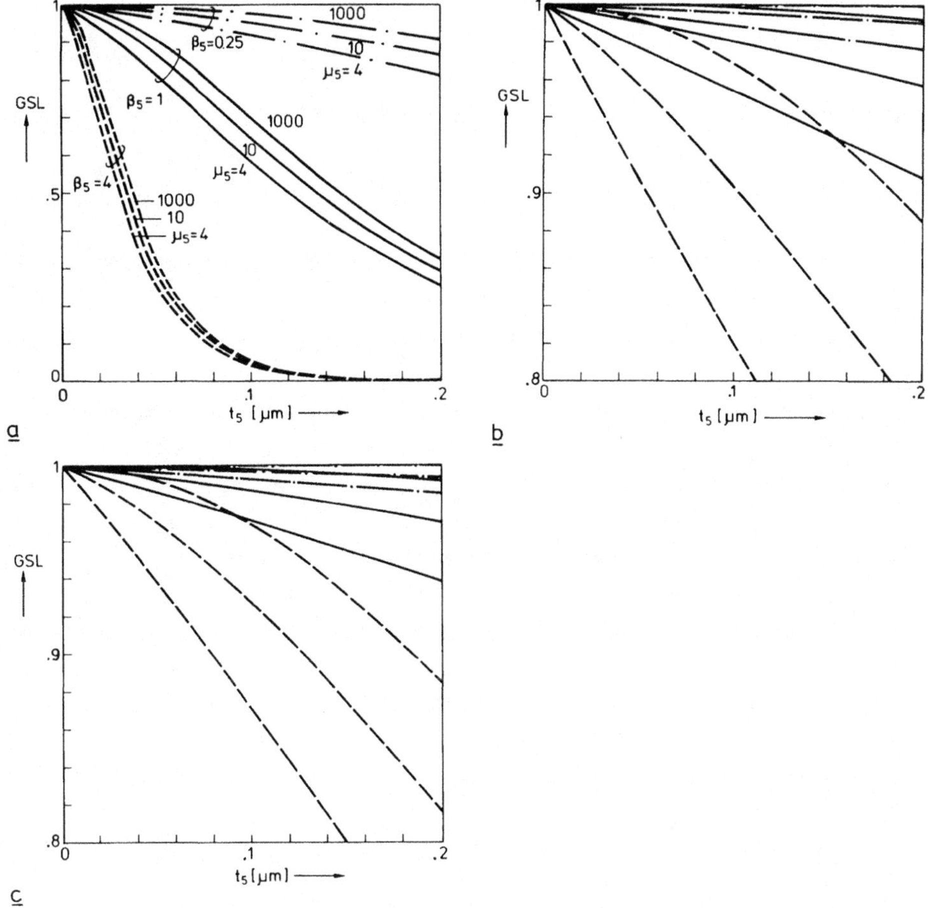

Fig. 2.10. The gap-smear-loss factor *GSL* as a function of the gap-smear layer thickness t_5.
$t_2 = 0.5$ μm, $\mu_2 = 1000$, $\beta_2 = 1$
$\mu_3 = 2$, $\beta_3 = 2$
$d_4 = 0.1$ μm
The μ_5 and β_5 values are indicated in *a* and also apply in the same order to *b* and *c*.
a) $\lambda = 0.7$ μm, $t_3 \geq 0.5$ μm.
The normalized interaction distance to the low-permeable coating $kd_4 = 0.9$ and to the highly permeable backlayer $kd_4 + \beta_3 kt_3 \geq 10$ are large enough for interaction effects to be neglected. Consequently a change in one of the tape parameters, e.g. in the thickness t_3, does not influence the *GSL*. A reduction of the anisotropy constant of the gap-smear layer β_5 reduces the transmission losses in that layer.
b) $\lambda = 10$ μm, $t_3 = 0.5$ μm.
The *GSL* decreases strongly compared to a) because of the reduction of the wavenumber $k = 2\pi/\lambda$, analogous to the distance loss. Note the different vertical scale compared to a.
c) $\lambda = 10$ μm, $t_3 = 3$ μm.
Because of the large λ there is strong interaction of the head with both coating and backlayer. The transmission loss (for small μ_5) in this layer is less pronounced when this interaction decreases, for instance as a result of increasing the distance to the backlayer by way of t_3, analogously to the distance loss ($\mu_4 = 1$) for increasing values of t_3 in Fig. 2.9b. In the present case $\mu_3 \neq 1$ ($\mu_3 = 2$) there is also an interaction with the coating. This interaction in contrast increases with t_3. Obviously the reduced interaction with the highly permeable backlayer is more important. For large μ_5 the tape cannot influence the transmission loss in the gap-smear layer.

$$GSL = \frac{1}{\cosh(\hat{\beta}_5^* kt_5)}, \qquad (2.68)$$

then relaxes to

$$GSL = 2e^{-\hat{\beta}_5^* kt_5}. \qquad (2.69)$$

Fewer losses during the transport occur if $\mathrm{Re}\{\hat{\beta}_5\}$ decreases, since a decrease of $\hat{\beta}_5 \equiv (\hat{\mu}_x/\hat{\mu}_y)^{1/2}$, i.e. reduction of $\hat{\mu}_x$ relative to $\hat{\mu}_y$, will benefit the transport in the direction of the head, and will reduce flux closure in the x-direction.

Fig. 2.11 illustrates the effects of $\beta_5 < 1$ and $\mu_5 > 1$ in the case of a real permeability tensor, i.e. dispersion-free layer 5. In the sketch the (wavelength-dependent) characteristic distance $\lambda_c \equiv \lambda/(2\pi\beta)$ is introduced; when reflections do not play a role, signals drop to 1/e part over every distance equal to λ_c, as indicated in the figure.

When $\mu_5 = \beta_5 = 1$ then the GSL describes nothing more than the usual increase of the distance loss due to an increase of the distance d_4 to $d_4 + t_5$. From Fig. 2.10 it follows that GSL increases when μ_5 increases or β_5 decreases. This has a remarkable consequence since $\mu_5 \gg 1$ and $\beta_5 \simeq 1$ usually; the loss due to gap smear is considerably lower than the increase of the distance loss would be if d_4 were increased by the thickness of the smear layer.

Concluding remarks concerning DL and GSL

From the previous calculations the following can be concluded:
1) A magnetic layer on top of an idealized head causes extra losses. This loss decreases when the anisotropy constant β_5 decreases and permeability μ_5 increases.
2) The same magnetic layer on top of the idealized head, but now *instead of a part t_5 of the distance d_4* between head and tape, *may reduce the losses*!

This last conclusion seems to contradict the idea that a magnetic layer on top of the head must shield the head from information on the tape. However, for idealized zero-gap heads the calculations are valid, i.e. there is no shielding. The question arises as to whether the shielding concept is perhaps valid for non-ideal, i.e. practical heads.

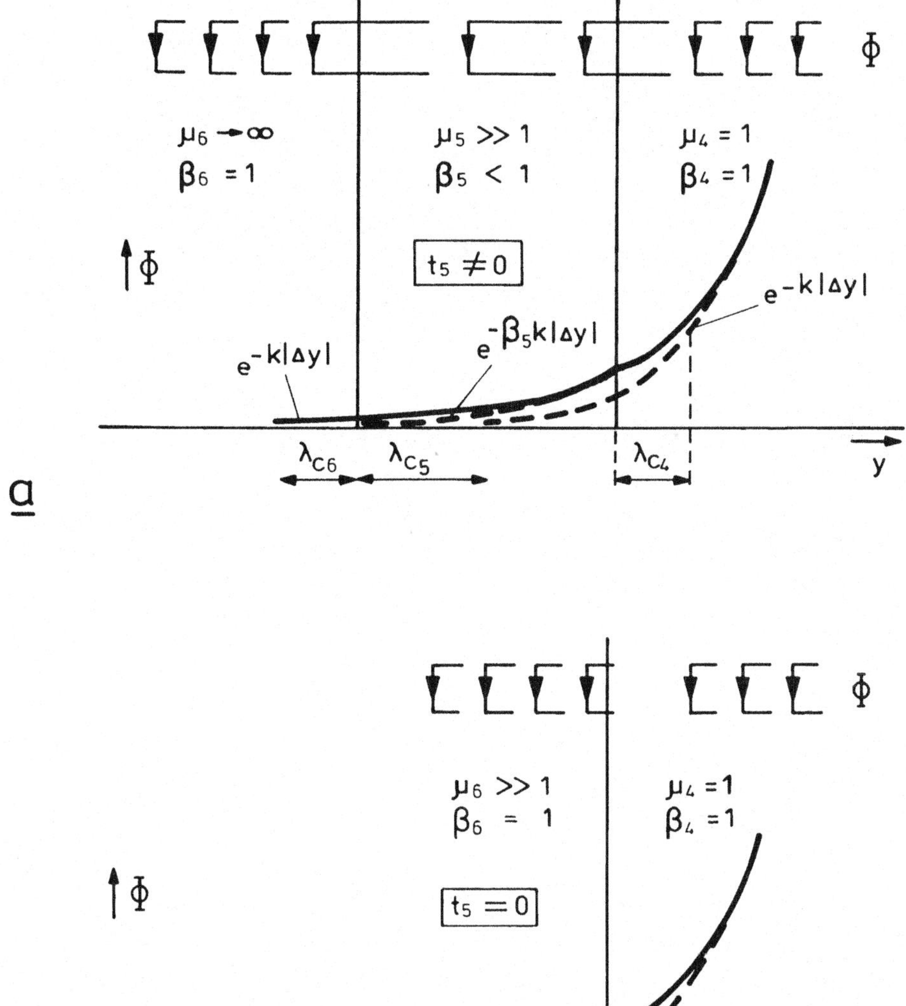

Fig. 2.11. Sketch of the flux decrease in the surrounding of the magnetic layer t_5.
Most of the flux that arrives within a characteristic distance $\lambda_c \equiv \lambda/(2\pi\beta)$ from an interface with a material having a much higher permeability will cross the boundary instead of only a part $1/e$. This happens twice in succession in b, first at the interface 4-5 where $\mu_5 \gg \mu_4$ and then at the interface 5-6 where $\mu_6 \gg \mu_5$.
In regions with $\beta < 1$, i.e. $\mu_x < \mu_y$, the transport of flux in the $-y$ direction increases since flux closure in the x direction decreases.

Before answering this question in the affirmative in a later chapter on a 'magnetic-smear layer on top of an actual head', a new reciprocity theorem is derived that relates the head's sensitivity to the magnetic potential in a fictitious write situation on an arbitrary (closed) surface enclosing at least the winding of the head.

This alternative reciprocity theorem will also be used in calculations of the absolute value of the flux Φ. It is noted that no absolute values of fluxes entering the head or entering the read coil of the head have been calculated so far. Only ratios of fluxes in the case of idealized heads could be (and have been) calculated in this chapter.

Appendix 2.1　General solution of Eq. (2.14)

Eq. (2.14) will be solved by the method of 'variation of parameters'. First the second order differential equation (2.14) is rewritten with the aid of the definitions of β and μ in (2.15) and (2.16) to read

$$\frac{\partial^2 \hat{\Psi}}{\partial y^2} - \hat{\beta}_3^2 k^2 \hat{\Psi} = \frac{j\hat{\beta}_3 k \hat{M}_x(y)}{\hat{\mu}_3} + \frac{\hat{\beta}_3}{\hat{\mu}_3} \frac{\partial \hat{M}_y(y)}{\partial y}. \tag{a2.1}$$

Introduction of $\hat{k}_1 \equiv \hat{\beta}_3 k$, $\hat{k}_2 \equiv j\hat{\beta}_3 k/\hat{\mu}_3$ and $\hat{k}_3 \equiv \hat{\beta}_3/\hat{\mu}_3$ yields

$$\frac{\partial^2 \hat{\Psi}}{\partial y^2} - \hat{k}_1^2 \hat{\Psi} = \hat{k}_2 \hat{M}_x + \hat{k}_3 \frac{\partial \hat{M}_y}{\partial y}. \tag{a2.2}$$

The reduced differential equation is

$$\frac{\partial^2 \hat{\Psi}}{\partial y^2} - \hat{k}_1^2 \hat{\Psi} = 0. \tag{a2.3}$$

This homogeneous equation resembles Eq. (2.12) and has the well-known solution, see Eqs. (2.17) and (2.18):

$$\hat{\Psi} = \hat{C}_5' e^{\hat{k}_1 y} + \hat{C}_6' e^{-\hat{k}_1 y}. \tag{a2.4}$$

In order to find one particular solution of Eq. (a2.1), and thus arrive at the general solution, the constant(s) will be thought to be functions of y. Although the variation of only one constant is enough, it is easier to define instead a relation between the constants at a later stage.

Differentiation of $\hat{\Psi}$ now gives

$$\frac{\partial \hat{\Psi}}{\partial y} = \frac{\partial \hat{C}_5'}{\partial y} \, e^{\hat{k}_1 y} + \frac{\partial \hat{C}_6'}{\partial y} \, e^{-\hat{k}_1 y} + \hat{k}_1 \hat{C}_6' e^{-\hat{k}_1 y}. \tag{a2.5}$$

The simplest way to find the solution of the differential equation (a2.2) is to introduce here the following relation between $\hat{C}_5'$ and $\hat{C}_6'$:

$$\frac{\partial \hat{C}_5'}{\partial y} \, e^{\hat{k}_1 y} + \frac{\partial \hat{C}_6'}{\partial y} \, e^{-\hat{k}_1 y} \equiv 0, \tag{a2.6}$$

since this choice will reduce Eq. (a2.2) to a first order differential equation. This follows after another time differentiation of Eq. (a2.5) with the condition given in Eq. (a2.6) and substitution of the result in Eq. (a2.2). The result is

$$\hat{k}_1 \frac{\partial \hat{C}_5'}{\partial y} \, e^{\hat{k}_1 y} - \hat{k}_1 \frac{\partial \hat{C}_6'}{\partial y} \, e^{-\hat{k}_1 y} = \hat{k}_2 \hat{M}_x + \hat{k}_3 \frac{\partial \hat{M}_y}{\partial y}. \tag{a2.7}$$

Elimination of $\hat{C}_6'$ with the aid of Eq. (a2.6) yields

$$\frac{\partial \hat{C}_5'}{\partial y} = \frac{\hat{k}_2}{2\hat{k}_1} \, \hat{M}_x e^{-\hat{k}_1 y} + \frac{\hat{k}_3}{2\hat{k}_1} \, \frac{\partial \hat{M}_y}{\partial y} \, e^{-\hat{k}_1 y} \tag{a2.8}$$

and so $\hat{C}_6'$ follows with the aid of Eq. (a2.6) from

$$\frac{\partial \hat{C}_6'}{\partial y} = - \frac{\hat{k}_2}{2\hat{k}_1} \, \hat{M}_x e^{\hat{k}_1 y} - \frac{\hat{k}_3}{2\hat{k}_1} \, \frac{\partial \hat{M}_y}{\partial y} \, e^{\hat{k}_1 y}. \tag{a2.9}$$

After integration of Eq. (a2.8) and Eq. (a2.9) and partial integration of the integral part corresponding to the last term in (a2.8) and in (a2.9), the following expressions are obtained:

$$\hat{C}_5'(y) = \hat{C}_5 + \frac{\hat{k}_2}{2\hat{k}_1} \int_0^y \hat{M}_x(y')e^{-\hat{k}_1 y'}dy'$$

$$+ \frac{\hat{k}_3\hat{M}_y(y')e^{-\hat{k}_1 y}}{2\hat{k}_1} + \frac{\hat{k}_3}{2} \int_0^y \hat{M}_y(y')e^{-\hat{k}_1 y'}dy' \qquad \text{(a2.10)}$$

$$\hat{C}_6'(y) = \hat{C}_6 - \frac{\hat{k}_2}{2\hat{k}_1} \int_0^y \hat{M}_x(y')e^{\hat{k}_1 y'}dy'$$

$$- \frac{\hat{k}_3\hat{M}_y(y')e^{\hat{k}_1 y}}{2\hat{k}_1} + \frac{\hat{k}_3}{2} \int_0^y \hat{M}_y(y')e^{\hat{k}_1 y'}dy', \qquad \text{(a2.11)}$$

where $\hat{C}_5$ and $\hat{C}_6$ are integration constants.

Substitution in Eq. (a2.4) and applying the definitions of the sinh and cosh functions to the result yields

$$\hat{\Psi}_3 = \hat{C}_5 e^{\hat{\beta}_3 ky} + \hat{C}_6 e^{-\hat{\beta}_3 ky} + \frac{1}{\hat{\mu}_3} \int_0^y \left\{ j\hat{M}_x(y')\sinh(\hat{\beta}_3 k(y-y')) \right.$$

$$\left. + \hat{\beta}_3\hat{M}_y(y')\cosh(\hat{\beta}_3 k(y-y')) \right\}dy' \qquad \text{(a2.12)}$$

where the original constants of Eq. (a2.1) are again used.

References

[2.1] Jorgensen, Finn, *The complete handbook of magnetic recording*, Blue Ridge Summit: Tab Books Inc., 1980, p. 203.

[2.2] Wallace, R.L., *The reproduction of magnetically recorded signals*, The Bell System Technical Journal, 1145-1173 (1951).

[2.3] Westmijze, W.K., *Calculation of the fields in and around the tape*, Philips Res. Rep. 8, 255-269 (1953).

[2.4] Tjaden, D.L.A. and Rijckaert, Albert M.A., *Theory of anhysteretic contact duplication*, IEEE Trans. Magn., vol. MAG-7, 532-536 (1971).

[2.5] Bertram, H. Neal, *Anisotropic reversible permeability effects in the magnetic reproduce process*, IEEE Trans. Magn., vol. MAG-14, 111-118 (1978).

Chapter 3

An alternative expression for the read flux in magnetic recording theory

With the aid of an appropriate reciprocity relation for quasi-static magnetic fields an alternative expression for the read flux of a magnetic recording head is derived. It expresses the read flux in terms of the distribution of the magnetic scalar potential of the head's write field in the plane that coincides with the front plane of the head and the normal component of a certain magnetic flux density, that is related to the recorded magnetization pattern.

This chapter recently appeared as a paper [3.1]. The derivation in the present chapter is more straightforward; it makes no use of previously derived expressions for the read flux, in contrast to the proof given in the paper. The present proof of the alternative expression is elegant, because it is very short and comprehensive and it leads directly to the 'common' as well as to the 'alternative' expression for the read flux in magnetic recording theory.

3.1. Introduction

In this chapter we derive a novel expression for the magnetic flux linked to the read coil of a magnetic recording head when a magnetization pattern recorded in a multilayered medium is present in front of the head. The expression follows from a reciprocity relation for quasi-static magnetic fields that differs from the usual one [3.2], [3.3]. It expresses the read flux in terms of the distribution of the magnetic scalar potential of the head's write field in the plane that coincides with the front plane of the head and the normal component of a certain magnetic flux density, that is related to the recorded magnetization pattern. The scalar potential that occurs in the expression is the one that is associated with the magnetic field of the head in the presence of a multilayered structure when a current of unit strength flows through the coil (cf. the head field definition proposed by Mallinson and Bertram [3.4]). The relevant normal component of the magnetic flux density is the one that would be caused by the recorded permanent magnetization pattern in the multilayered recording medium, when infinite permeability is assumed in the halfspace behind the head's front plane. The expression for the read flux is applicable to both longitudinal and perpendicular recording. The expression holds for three-dimensional configurations; its two-dimensional approximation is presented as well.

3.2. Description of the configuration

The configuration consists of a magnetic reproduce head with reading coil, in the neighbourhood of which electric and/or magnetic shields may be present. These parts of the configuration occupy a bounded domain in space (Fig. 3.1).

The nomenclature pertaining to the different subdomains is shown in Table 3.1.

If parts of the reproduce head are of infinite permeability they are included in V_1 and if shields of finite permeability are present they are included in V_μ. If parts of the head are perfectly conducting they are included in V_2. The domains V_μ, V_1 and V_2 are non-overlapping. The domains V_p and V_i may be overlapping.

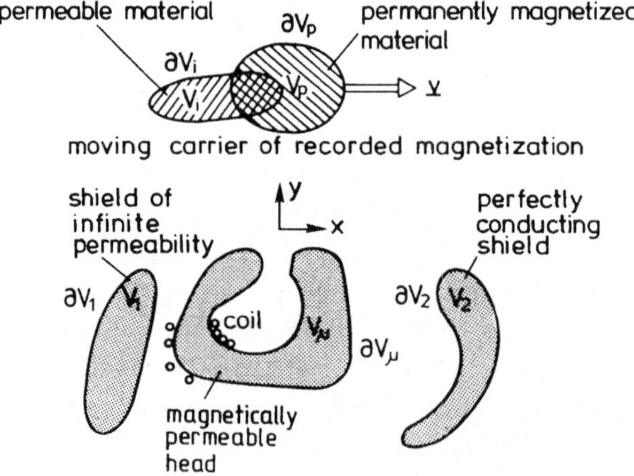

Fig. 3.1. Configuration consisting of recording head, coil, shields and moving carrier of magnetization.

Table 3.1. Nomenclature pertaining to the different subdomains in the magnetic-recording configuration.

domain	boundary surface	physical property
V_μ	∂V_μ	recording head of permeable material
V_J	∂V_J	coil, capable of carrying external current
V_1	∂V_1	magnetic shield of infinite permeability
V_2	∂V_2	perfectly conducting shield
V_p	∂V_p	moving piece of matter permanently magnetized
V_i	∂V_i	moving piece of matter with induced, reversible magnetization

To locate position in space we employ the coordinates $\{x, y, z\}$ with respect to an orthogonal Cartesian reference frame with origin 0 and three mutually perpendicular base vectors $\{e_x, e_y, e_z\}$ of unit length each. In the indicated order, the three base vectors form a right-handed system. The reference frame is such that the head structure is located in the halfspace $y < 0$, while the motion of the permanent magnetization pattern take place in the direction of increasing x (Fig. 3.1). The position vector r is given by

$$r = xe_x + ye_y + ze_z. \tag{3.1}$$

The Cartesian reference frame is at rest with respect to the head structure.

3.3. The reciprocity relation

We employ, as customary in magnetic recording theory, the quasi-static approximation of the field equations as in chapter 2. Frequency-independent materials will be assumed. To derive our reciprocity relation we first consider a general bounded domain $\mathcal{V}$ with sufficiently smooth boundary $\partial\mathcal{V}$, where no electric current is present. The magnetic field equations in this domain $\mathcal{V}$ are:

$$\nabla \times \boldsymbol{H} = \boldsymbol{0}, \tag{3.2}$$

$$\nabla \cdot \boldsymbol{B} = \boldsymbol{0} \tag{3.3}$$

while

$$\boldsymbol{B} = \mu_0(\boldsymbol{H} + \boldsymbol{M}), \tag{3.4}$$

where μ_0 is the permeability in vacuum, $\boldsymbol{H}$ the magnetic field strength, $\boldsymbol{B}$ the magnetic flux density and $\boldsymbol{M}$ the magnetization. In view of (3.2) a scalar potential $\Psi(\boldsymbol{r})$ exists, such that:

$$\boldsymbol{H} = -\nabla\Psi. \tag{3.5}$$

Here, Ψ is a single-valued function of position apart from an additive constant. This constant is usually chosen such that Ψ vanishes at infinity when the domain $\mathcal{V}$ extends to infinity. In order to specify the constitutive relations we separate the magnetization into a field-dependent, induced part $\boldsymbol{M}_i$ and a field-independent, permanent part $\boldsymbol{M}_p$:

$$\boldsymbol{M} = \boldsymbol{M}_i + \boldsymbol{M}_p. \tag{3.6}$$

Now, for the reciprocity relation to hold, the media in $\mathcal{V}$ have to be linear, time-invariant, reciprocal and instantaneously and locally reacting in their magnetic behaviour.

For this kind of media, we have:

$$\boldsymbol{M}(\boldsymbol{r}) = \boldsymbol{K}(\boldsymbol{r}) \cdot \boldsymbol{H}(\boldsymbol{r}) \tag{3.7}$$

where $\boldsymbol{K}$ denotes the magnetic susceptibility, a symmetrical tensor (due to the material being reciprocal, see also section 4.2.1) of rank two. Equation (3.7) applies to anisotropic media; for isotropic media it reduces to:

$$\boldsymbol{M}(\boldsymbol{r}) = K(\boldsymbol{r})\boldsymbol{H}(\boldsymbol{r}) \tag{3.8}$$

where $K(\boldsymbol{r})$ is the scalar susceptibility.

At surfaces across which the properties of the media show abrupt changes, the field equations have to be supplemented by boundary conditions. These conditions are:

$\{v \times H, v \cdot B \text{ and } \Psi\}$ continuous across surfaces where the susceptibility
shows a finite jump   (3.9)

$v \times H \to 0$ (and hence $\Psi \to$ a constant) upon approaching the boundary
of a medium of infinite permeability   (3.10)

$v \cdot B \to 0$ upon approaching the boundary of a medium of infinite
conductivity   (3.11)

where v denotes the unit vector normal to the relevant surface.

A reciprocity relation interrelates two admissible 'states' that could
be present in one and the same domain $\mathcal{V}$ (Fig. 3.2). Let $\mathcal{V}$ be the
bounded domain interior to the closed surface $\partial\mathcal{V}$ and let v be the unit
vector along the normal to $\partial\mathcal{V}$, pointing away from $\mathcal{V}$. The field quantities in the two states 'a' and 'b' are denoted by the superscripts a and
b, respectively.

We now consider the expression for $\nabla \cdot (\Psi^a B^b - \Psi^b B^a)$. Using some
formulas of vector analysis this expression is rewritten as:

$$\nabla \cdot (\Psi^a B^b - \Psi^b B^a)$$

$$= (\nabla\Psi^a) \cdot B^b + \Psi^a(\nabla \cdot B^b) - (\nabla\Psi^b) \cdot B^a - \Psi^b(\nabla \cdot B^a). \quad (3.12)$$

With (3.2), (3.3), (3.5), (3.6) and (3.7) we arrive at:

$$\nabla \cdot (\Psi^a B^b - \Psi^b B^a) = -\mu_0(H^a \cdot M_p^b - H^b \cdot M_p^a). \quad (3.13)$$

Integration of (3.13) over the domain $\mathcal{V}$ and application of Gauss'
divergence theorem to the resulting left-hand side leads to the appropriate reciprocity relation:

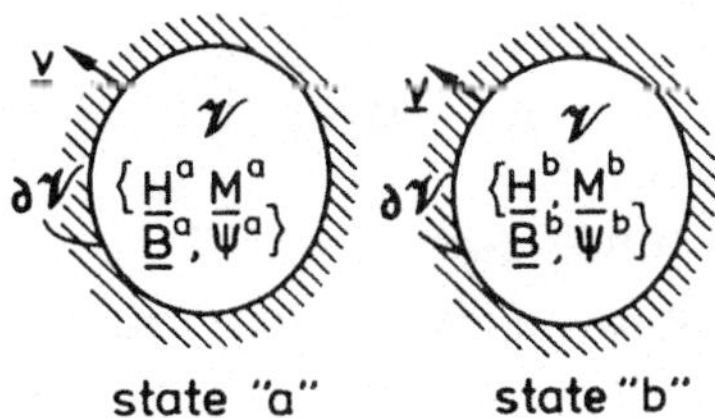

Fig. 3.2. Illustration of the reciprocity relation for quasi-static magnetic fields.

$$-\int_{\partial \mathcal{V}} \mathbf{v} \cdot (\Psi^a \mathbf{B}^b - \Psi^b \mathbf{B}^a)\mathrm{d}A = \mu_0 \int_{\mathcal{V}} (\mathbf{H}^a \cdot \mathbf{M}_p^b - \mathbf{H}^b \cdot \mathbf{M}_p^a)\mathrm{d}V \quad . \quad (3.14)$$

The following surfaces do not contribute to the left-hand side of (3.14):

(a) interfaces within $\mathcal{V}$ where $\mathbf{K}$ jumps by a finite amount;

(b) closed boundary surfaces of domains with infinite permeability (on these surfaces Ψ is a constant and $\oint \mathbf{v} \cdot \mathbf{B}\mathrm{d}A = 0$ according to Gauss' theorem, since $\nabla \cdot \mathbf{B} = 0$);

(c) boundary surfaces of domains with infinite conductivity ($\mathbf{v} \cdot \mathbf{B} = 0$ on such surfaces);

(d) surfaces 'at infinity', since for structures occupying a bounded (finite) domain in space we have $\Psi = \mathbb{O}(|r|^{-2})$ as $|r| \to \infty$ and $\mathbf{B} = \mathbb{O}(|r|^{-3})$ as $|r| \to \infty$ and hence $\int \mathbf{v} \cdot \Psi \mathbf{B}\mathrm{d}A = \mathbb{O}(|r|^{-3})$ as $|r| \to \infty$;

(d') in the 2-dimensional approximation, boundary curves 'at infinity' do not contribute, since for 2-dimensional structures occupying a bounded domain in the 2-dimensional space ($\mathbb{R}2$), $\Psi = \mathbb{O}(|r|^{-1})$ and $\mathbf{B} = \mathbb{O}(|r|^{-2})$ and hence $\int \mathbf{v} \cdot \Psi \mathbf{B}\mathrm{d}r = \mathbb{O}(|r|^{-2})$ as $|r| \to \infty$;

(e) closed boundary surfaces completely surrounding the structure, because of (d) and since volumes where permanent magnetization is absent do not contribute to the right-hand side of (3.14);

(e') in the 2-dimensional approximation, closed boundary curves completely surrounding the structure.

3.4. Derivation of the alternative expression for the read flux

In this section we shall investigate the implication of the reciprocity relation derived in Section 3.3 for the magnetic recording situation described in Section 3.2.

First we shall give a new very short, but complete, derivation of the common reciprocity theorem.

To get rid of the second term in the right-hand side of (3.14), one usually identifies the state 'a' with a 'writing' state, choosing $I^a \neq 0$ while $\mathbf{M}_p^a = \mathbf{0}$. In the following this will be assumed. The associated fields and potentials are denoted $\mathbf{H}^a$ and Ψ^a.

The state 'b' will be referred to as the actual reading state, i.e. $M_p^b = M_p^R$, while $I^b = 0$.

First it must be realised that the quantities in the *left-hand side* of (3.14), Ψ^b and B^b, although being associated with their source M_p^R, still have a lot of freedoms. *Everywhere outside V_p, boundary conditions can be freely chosen* as long as they are physically admissable.

In the second place it must be realised that *the right-hand term is constant*, independent of the chosen domain V, *as long as V_p is an interior domain of V*, and H^a and M_p^b are not changed.

With these freedoms in mind, it is possible to find two situations, the first of which equalizes the left-hand side of (3.14) to the actual read flux Φ^R, without changing the value of the right-hand side (so proving the common reciprocity theorem in an alternative way), the second situation delivers on the left-hand side the desired alternative for the common read flux expression on the right-hand side.

Situation 1: Read flux Φ^R

Choose V to be the domain exterior to a curved slice with thickness t that encloses the surface A_R through which the read flux needs to be known, see Fig. 3.3. Let, in state 'a', a current I^a flow through the

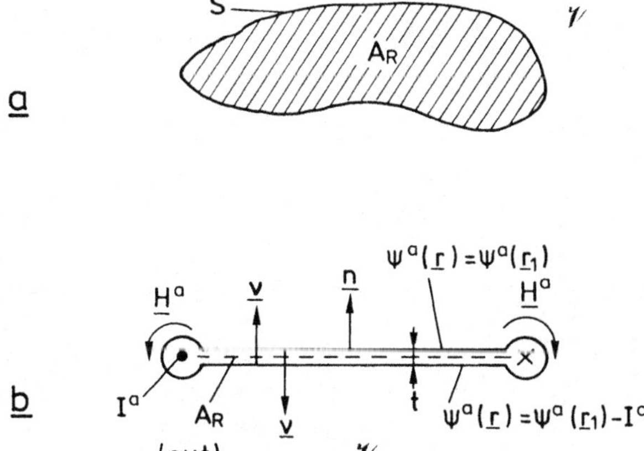

Fig. 3.3. Surface spanning the 'coil' ($\bullet$ **x**) and the cut (- - - - -).
a) Top view of unbounded surface A_R through which the read flux needs to be known.
b) Side view of closed surface surrounding A_R. The domain that is in this way excluded from V forms a thin slice with thickness t.

closed boundary curve S of A_R. The unit vector $\boldsymbol{n}$ along the normal to the surface A_R forms a right-handed system with the chosen direction of circulation along S, see Fig. 3.3b.

As $t \to 0$, the second term in the left-hand expression of (3.14) vanishes, since Ψ^b is continuous over the slice since $I^b = 0$ and $\boldsymbol{v} \cdot \boldsymbol{B}^a$ only changes its sign for $t \to 0$.

In the first term, $\Psi^a(\boldsymbol{r})$ jumps by an amount of I^a amperes, while $\boldsymbol{v} \cdot \boldsymbol{B}^b$ changes its sign for $t \to 0$.

Hence the left-hand integral reduces to $I^a \int_{A_R} \boldsymbol{n} \cdot \boldsymbol{B}^b \mathrm{d}A$ and (3.14) becomes:

$$\Phi^R \equiv \int_{A_R} \boldsymbol{n} \cdot \boldsymbol{B}^b \mathrm{d}A = \mu_0 \int_{V_p} \boldsymbol{h}^a \cdot \boldsymbol{M}_p^R \, \mathrm{d}V, \tag{3.15}$$

where $\boldsymbol{h}^a \equiv \boldsymbol{H}^a / I^a$ is introduced.

Situation 2: Alternative reciprocity theorem(s)

Next, choose Ψ^b constant on a closed surface, $\delta \mathcal{V}_1$, surrounding S and denote the corresponding $\boldsymbol{B}^b$ by $\boldsymbol{\mathcal{B}}^b$. Let $\mathcal{V}$ now be the domain exterior to the domain encompassed by this closed surface S. The 'outer' boundary surface of $\mathcal{V}$, $\delta \mathcal{V}_2$, is chosen so far away that it completely surrounds the domain where permanent magnetization is present. In view of (e) below (3.14), this 'outer' closed boundary surface does not contribute to the integral at the left-hand side of (3.14). The inner closed boundary surface, $\delta \mathcal{V}_1$, does not contribute to the second part of the integral at the left-hand side in view of (b) below (3.14). Hence (3.14) reduces to

$$-\int_{\partial \mathcal{V}} \boldsymbol{v} \cdot \psi^a \boldsymbol{\mathcal{B}}^b \mathrm{d}A = \mu_0 \int_{V_p} \boldsymbol{h}^a \cdot \boldsymbol{M}_p^R \mathrm{d}V \stackrel{(3.15)}{=} \Phi^R , \tag{3.16}$$

where $\delta \mathcal{V}_1$ is replaced by $\delta \mathcal{V}$ and $\psi^a \equiv \Psi^a / I^a$.

In a recording situation we usually deal with a layered medium in which the source of $\boldsymbol{v} \cdot \boldsymbol{\mathcal{B}}^b$, i.e. the permanent magnetization $\boldsymbol{M}_p^R$, is located. The head is usually located at one side of the medium. When we choose the surface where $\Psi^b \equiv constant$ parallel to this layered structure and in between the medium and the head, it is possible, for several 'layered' permanent magnetization structures, to find analytical expressions for $\mathbf{v} \cdot \mathfrak{B}^{\mathrm{b}}$, since the problem is only one-dimensional then, see chapter 2.

Choosing $\mathcal{V}$ according to this idea, and realising that at infinity (see (d) below Eq. (3.14)) there are no contributions, yields:

$$\boxed{ -\int_{\text{plane}} \mathbf{v} \cdot \psi^{\mathrm{a}} \mathfrak{B}^{\mathrm{b}} \mathrm{d}A = \Phi^{\mathrm{R}} } \qquad (3.17)$$

Equation (3.17) holds in case the coil of the read head is located in the halfspace in which no permanent magnetization is present. The boundary surface between the two halfspaces is denoted by 'plane' in (3.17). This situation is depicted in Fig. 3.4, where this plane is denoted as the plane $y = 0$. In state 'b' the lower halfspace where the coil was present can be envisaged as being filled with infinitely permeable material. The field that emanates from the permanently magnetized material and penetrates this lower halfspace perpendicularly, has a magnetic flux density $\mathfrak{B}_y^{\mathrm{b}} = -\mathbf{v} \cdot \mathfrak{B}^{\mathrm{b}}$ at the plane $y = 0$.

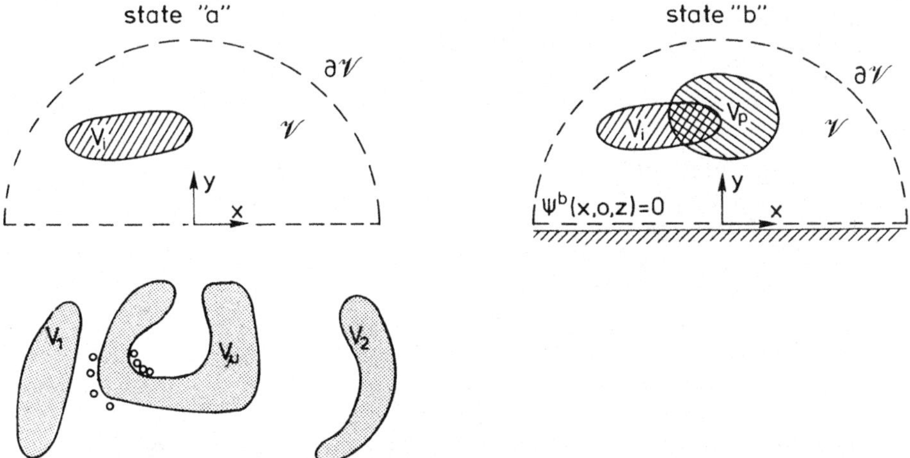

Fig. 3.4. Illustration of the two states in the alternative expression for the reciprocity integral (3.17).

In state 'a' the scalar potential $\Psi^{\mathrm{a}}(x, o, z)$ due to a current I in the coil is calculated in the presence of the part of the medium with induced reversible magnetization, V_{i}.

In state 'b' the magnetic flux density component $\mathfrak{B}_y(x, o, z)$ that would be caused by the permanently magnetized part of the medium, V_{p}, in the presence of V_{i} and the infinitely permeable halfspace $y < 0$, is calculated.

In practice V_{i} and V_{p} form a layered structure that is bounded in space, e.g. the 5-fold layered structure in Chapter 2, and the xz plane will then be chosen parallel to these layers, see Chapter 4 for applications.

All formulations so far have been in three-dimensional space. The corresponding two-dimensional approximation of (3.17) is:

$$\Phi^R = W \int_{-\infty}^{\infty} \psi^a \mathcal{B}_y^b \, dx \quad , \tag{3.18}$$

when a track width W is assumed.

3.5. Conclusion

In equation (3.17) we have derived an expression for the read flux that consists of an integral of the product of a magnetic scalar potential and the normal component of the magnetic flux density of a field related to the actual field in the read situation. The scalar potential characterizes the head and the material with reversible magnetization in front of the head. This potential is independent of the magnetization pattern containing the information to be retrieved by the head and is in accordance with the proposed definition for the head field [3.4]. The normal component of the magnetic flux density that further occurs, depends only on the tape configuration and is the same for every head configuration. Observe that in the expression for the read flux the scalar potential only acts as a 'weighting function' for the magnetic flux density, the latter being determined by the tape only. In view of this, equation (3.17) or (3.18) looks like a promising tool for comparing the efficiencies of different heads in the read situation including, if necessary, the permeability effects of the tape.

References

[3.1] Ruigrok, Jaap J.M. and Quak, Dirk, '*An alternative expression for the read flux in magnetic recording theory*', IEEE Trans. Magn., Mag-23, 1764-1766, (1987).

[3.2] De Hoop, A.T., *Motional influences on magnetic reproduction: An analysis based on the reciprocity theorem*, IEEE Trans. Magn., Mag-18, 758-762, March (1982).

[3.3] Geurst, J.A., *The reciprocity principle in the theory of magnetic recording*, Proc. IEEE, Vol. 51, 1573-1577, (1963).

[3.4] Mallinson, J.C. and H. Neal Bertram, *On the characteristics of pole-keeper head fields*, IEEE Trans. Magn., Mag-20, 721-723 (1984).

Chapter 4

Further considerations concerning the alternative reciprocity theorem

4.1 Introduction

In chapter 2 only ratios of fluxes could be calculated and no absolute values of read fluxes, since it was unclear which 'weighting functions' would translate the calculated normal component of magnetic induction B_y on the surface of an idealized head to the read flux Φ of the actual head. One of the first candidates, if existing, for this weighting function was the scalar potential ψ^a, i.e. the potential of the actual head on the same surface where B_y is calculated. In chapter 3 this was proved to be right.

This so-called alternative reciprocity theorem will be extended in different directions in this chapter. These extensions include:
-   non-single-sided head structures
-   reciprocal head materials
-   frequency-dependent head materials.

The inclusion of the frequency-dependent effects demands a formulation of the alternative reciprocity theorem in the k space and the ω space and an adapted definition of the dimensionless scalar potential, as can be carried out by Fourier transforms. Another method, applicable to periodic and harmonic signals, makes use of complex time- and x-independent quantities as defined in (2.11) for a harmonic signal or one harmonic component (of a Fourier *series*) of a periodic signal. This method has the advantage that the δ functions, that result for periodical signals from the Fourier *transform*, vanish. Both methods will be worked out in this chapter. Fourier transformed quantities will be denoted by $f(k)$, $f(\omega)$ etc. The complex x- or time-independent quantities for the harmonic component with wavenumber k_1, or with radial frequency ω_1 will be denoted by $\hat{f}(k_1)$ and $\hat{f}(\omega_1)$.

4.2 The extended form of the alternative reciprocity theorem in the x, y, z space

4.2.1 Introductory remarks

In chapter 3 the alternative reciprocity theorem has been derived. This expression for the read flux, stated in (3.17) for single-sided head structures (with at least the read coil on one side) as for example head a in Fig. 4.1, is

$$\Phi^{R} = \int_{y=0} \psi^{a}\mathcal{B}_{y}^{b}\mathrm{d}A \qquad (4.1)$$

and holds if the materials of head and tape are instantaneously reacting, linear and reciprocal. These requirements were necessary because in the derivation of (4.1) use has been made of the relation

$$\boldsymbol{H}^{a} \cdot \boldsymbol{M}_{i}^{b} = \boldsymbol{H}^{b} \cdot \boldsymbol{M}_{i}^{a}, \qquad (4.2)$$

($\boldsymbol{M}_{i}$ is the reversible part of the magnetization) by means of which (3.12) could be simplified to (3.13). Eq. (4.2) is a result of (3.7), i.e.

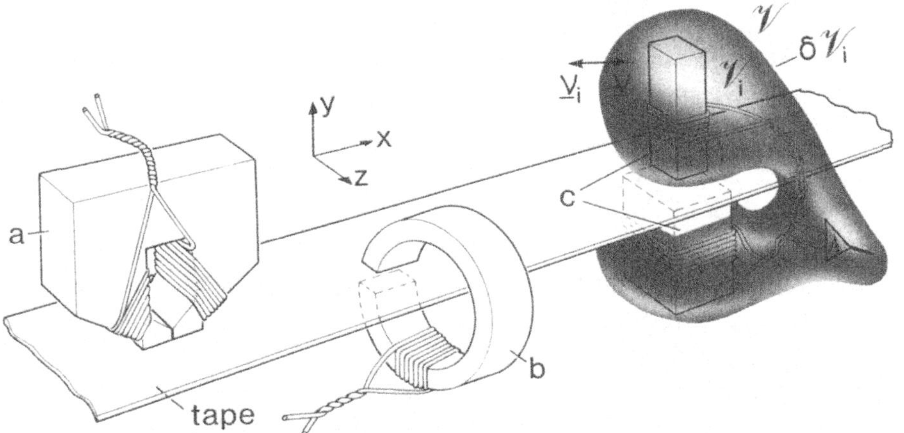

Fig. 4.1. Three, partly-hypothetical, three-dimensional head structures whose responses to a magnetized tape can be calculated by way of the general form of the alternative reciprocity theorem.
a. Conventional one-sided head structure.
b. Hypothetical two-sided head structure with coil on one side of the tape.
c. Hypothetical two-sided head structure with coils on both sides of the tape with convenient choice of the regions V and V_{i}.

$$M_i = KH \qquad (4.3)$$

which is valid for linear and instantaneously reacting materials when K is a symmetrical tensor, i.e. the material is reciprocal.

The definition of $\mathcal{B}_y$, see chapter 3, is such that it coincides with the B_y component calculated analytically in chapter 2, e.g. B_y from (2.63). The reason is that in chapter 2 all non-zero k components $\hat{\psi}(k, y)$ of the potential Ψ on the flat head surface have been taken zero, because of the assumption there of the ideal zero-gap head. This is consistent with taking $\Psi^b(r) = 0$ on the flat surface $y = 0$ in the fictitious state 'b' in chapter 3 in the definition of $\mathcal{B}_y^b$ in (4.1).

The dimensionless potential ψ^a in (4.1) serves as a weighting function for the 'ideal' magnetic induction component $\mathcal{B}_y^b$, as e.g. calculated in chapter 2. For a given head-tape system the potential ψ^a can be calculated numerically or approximated analytically, except for a constant. (It is noted that adding a constant C to ψ^a does not contribute to ϕ^R in (4.1) since

$$\oiint C\boldsymbol{v} \cdot \mathcal{B}\mathrm{d}A = C \oiint \boldsymbol{v} \cdot \mathcal{B}\mathrm{d}A = C \iiint \boldsymbol{\nabla} \cdot \mathcal{B}\mathrm{d}V = 0$$

as a consequence of $\boldsymbol{\nabla} \cdot \mathcal{B} = 0$.)

Only in some idealized cases can exact explicit expressions or series be derived analytically; see e.g. references [5.5] and [5.6]. In most cases approximations from for instance numerical results (see e.g. [5.20] and [5.21]), are sufficient, at least for qualitative results. This is further discussed in the next chapter.

4.2.2 Non-reciprocal materials

For non-reciprocal materials Eq. (4.2) is not valid by definition and neither is (4.1). However it is possible to define the fictitious (write) situation, state 'a', such that a relation like (4.2) remains.

The appropriate choice is to let the magnetic materials in state 'a' have a susceptibility tensor K^T that is the transposed of the actual susceptibility tensor K in the read situation. So it is only necessary to interchange the non-diagonal elements of the tensor K, which is supposed to be known, and add a superscript T to all quantities in state 'a'. As a result

$$H^{\mathrm{Ta}} \cdot M_{\mathrm{i}}^{\mathrm{b}} = H^{b} \cdot M_{\mathrm{i}}^{\mathrm{Ta}}. \qquad (4.4)$$

The above fictitious state 'Ta' also obeys Maxwell's equations and those of vector analysis. Thus the final result simply reads

$$\boxed{\Phi^{\mathrm{R}} = \int_{y=0} \psi^{\mathrm{Ta}} \mathcal{B}_{\mathrm{y}}^{\mathrm{b}} \, \mathrm{d}A} \,. \qquad (4.5)$$

For reciprocal materials Eq. (4.5) is equivalen to Eq. (4.1) since then $K^{\mathrm{T}} = K$ and consequently $\psi^{\mathrm{Ta}} = \psi^{\mathrm{a}}$. In other words, state 'Ta' is then state 'a'.

4.2.3 Non-single-sided head structures

For the non-single-sided head structure b in Fig. 4.1 Eq. (4.1) is still valid. Parts of the head b then have to be taken into account in the calculation of $\mathcal{B}_{\mathrm{y}}^{\mathrm{b}}$. This disturbs the layered structure in state 'b' so that it is a troublesome task to calculate $\mathcal{B}_{\mathrm{y}}^{\mathrm{b}}$. It would then be necessary to calculate $\mathcal{B}_{y}^{b}$ numerically for all frequencies of interest. All the benefits of the alternative reciprocity theorem would vanish. To avoid this problem and to make the calculations possible on heads with coils on both sides of the tape like head C in Fig. 4.1, a more general form of the alternative reciprocity theorem will first be stated. This expression follows from (3.14) when instead of $\Psi^{b}(r) \equiv 0$ on the $y = 0$ plane, $\Psi^{b}(r) \equiv C$ is chosen on an arbitrary closed surface ∂V_{i} enclosing at least the read coil(s) of the head, and is written (see also (3.16))

$$\Phi^{\mathrm{R}} = \oiint_{\partial V_{\mathrm{i}}} v_{\mathrm{i}} \cdot \psi^{\mathrm{a}} \mathcal{B}^{\mathrm{b}} \, \mathrm{d}A \qquad (4.6)$$

where
- ∂V_{i} describes the closed surface of the domain V_{i} that encompasses at least the read coil(s) and is *exterior* to the domain V that encompasses at least the permanently-magnetized parts of the tape. V_{i} is not necessarily the complement of V!
- v_{i} is the unit vector normal to V_{i} and directed outward (i.e. $v_{\mathrm{i}} = -v$).
- $\mathcal{B}^{\mathrm{b}}$ is the (fictitious) magnetic induction of state 'b' and $v_{\mathrm{i}} \cdot \mathcal{B}^{\mathrm{b}}$ is consequently the normal (outward) component of this magnetic induction on ∂V_{i}, due to the magnetization in the tape when

$M_p^b = M_p^R$ if the potential on $\partial \mathcal{V}_i$ is zero.

– ψ^a is the potential of the head, in a fictitious (write) situation where the tape is present and $M_p^b = \mathbf{0}$, divided by the total current NI if the actual head is excited by a current of I ampère through the N turns of the read coil. Hence ψ^a is a dimensionless quantity or weighting factor.

Cuts of $\partial \mathcal{V}_i$ with the head structure are thus allowed, but usually complicate the calculations.

It is most appropriate to choose $\mathcal{V}_i$ such that a one-dimensional approach of $v_i \cdot \mathcal{B}^b(r)$ is possible. For the head-tape systems b and c depicted in Fig. 4.1 it is evident that $\mathcal{V}_i$ is chosen such that $\partial \mathcal{V}_i$ contains at least two flat surfaces parallel to the tape, one just in between the tape and lower-pole surface and one between tape and upper-pole surface. The resulting two-dimensional configuration of state 'b' to calculate the normal component $v_i \cdot \mathcal{B}^b$ is sketched in Fig. 4.2. The appropriate alternative reciprocity theorem thus reads

$$\Phi^R = \iint_{y = y_{p_1}} \psi^a \mathcal{B}_y^b \, dx dz - \iint_{y = y_{p_2}} \psi^a \mathcal{B}_y^b \, dx dz \ . \tag{4.7}$$

In practical calculations it is not necessary to use a closed surface or one infinite plane (or two in the case of reading coils or only head

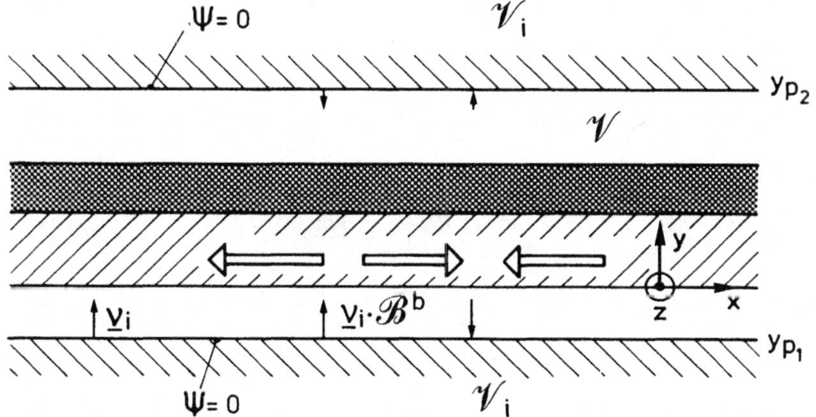

Fig. 4.2. The appropriate choice of domains for two-sided head structures. For a layered tape, vanishing z dependences and Fourier-transformed x dependences a one-dimensional approach is then obtained.

structures on both sides of the tape). For instance in the configurations
b and c of Fig. 4.1 the chosen planes just above and below the tape do
not necessarily need to exceed the written track width or exceed an
area a few wavelengths larger than the surfaces of the head configurations facing the tape. Outside the track width $\mathscr{B}_y^b$ approaches zero,
while in the area further than a few wavelengths from gap surfaces in
reasonable contact with the tape, the potential ψ^a in the (fictitious)
write situation (the weighting function) is approximately constant over
a wavelength, i.e. the contribution of $\mathscr{B}_y^b$ to the integral averages out
in those regions. For further discussion of this latter condition with
regard to bulk video heads and thin-film heads see the sections 5.4.1
and 5.10 repectively.

4.3 The alternative reciprocity theorem in k space

In recording situations where the tape moves with constant velocity
in the $+x$-direction (W is the track width), we can write $\mathscr{B}_y^b(x, y, t) =
\mathscr{B}_{0y}^b(x - vt, y)$, hence

$$\Phi^R(t) = W \int_{-\infty}^{\infty} \psi^a(x, y_p)\,\mathscr{B}_{0y}^b(x - vt, y_p)\mathrm{d}x. \tag{4.8}$$

The integral expresses the cross-correlation between the two real-
valued functions ψ^a and $\mathscr{B}_{0y}^b$. The use of the Fourier transforms in
evaluating such integrals is advantageous.

Transformation of both sides of the expression from the t to the
ω-domain through

$$f(\omega) \equiv \int_{-\infty}^{\infty} f(t)\mathrm{e}^{-\mathrm{j}\omega t}\mathrm{d}t \tag{4.9}$$

$$\left(\text{and hence inversely } f(t) \equiv \frac{1}{2\pi} \int_{-\infty}^{\infty} f(\omega)\mathrm{e}^{\mathrm{j}\omega t}\mathrm{d}\omega\right) \tag{4.10}$$

while using the definition for transformation from x to the k domain

$$f(k) \equiv \int_{-\infty}^{\infty} f(x) \mathrm{e}^{-jkx} \mathrm{d}x \qquad (4.11)$$

$$\left(\text{and hence inversely } f(x) \equiv \frac{1}{2\pi} \int_{-\infty}^{\infty} f(k) \mathrm{e}^{jkx} \mathrm{d}k \right) \qquad (4.12)$$

yields

$$\boxed{\Phi^{\mathrm{R}}(\omega) = \frac{W}{v}\, \psi^{\mathrm{a}}(k, y_{\mathrm{p}})\, \mathscr{B}_{\mathrm{y}}^{*\mathrm{b}}(k, y_{\mathrm{p}}, t = 0)}\;, \qquad (4.13)$$

where $\omega \equiv kv$.

The addition '$t =$' is necessary in our notation, in order to avoid confusion with $\omega = 0$. In the shorter, but equivalent, notation $\mathscr{B}_{0\mathrm{y}}^{\mathrm{b}}(k, y_{\mathrm{p}})$ no confusion is possible.

In the actual evaluation of Eq. (4.13) the cross-correlation theorem [4.1] for real-valued functions $\psi^{\mathrm{a}}(x)$ and $\mathscr{B}_{\mathrm{y}}^{\mathrm{b}}(x)$ has been used.

ψ^{a} is, physically, dependent on k (geometric effects) and on $\omega \equiv kv$ (dispersion effects). This is worked out in the next chapter.

In (4.13) the choice of the 'plane' $y = y_p$ is free as long as it is lying in between the tape and the coil. This leads to the following 'reciprocity relation':

$$f(k, y_{\mathrm{p_a}}, y_{\mathrm{p_b}}) \equiv \frac{\psi^{\mathrm{a}}(k, y_{\mathrm{p_a}})}{\psi^{\mathrm{a}}(k, y_{\mathrm{p_b}})} = \frac{\mathscr{B}_{\mathrm{y}}^{*\mathrm{b}}(k, y_{\mathrm{p_b}}, t = 0)}{\mathscr{B}_{\mathrm{y}}^{*\mathrm{b}}(k, y_{\mathrm{p_a}}, t = 0)}. \qquad (4.14)$$

This means that the decay of the actual potential (or field) in the $+y$-direction equals the decay of the fictitious magnetic flux density at the boundary surface of the infinite permeability region when this boundary surface is moved in the $-y$-direction. This statement can be used in the harmonic analysis on layered, i.e. one-dimensional, structures. Expressions for the numerator and denominator in the right-hand side of (4.14) are given in chapter 2. (See for example (2.46), or, in the case of gap smear, (2.65).)

4.4 Release from the requirement 'dispersion free'

4.4.1 Introductory remarks

In order to let $H^a(r, t) \cdot M^b(r, t)$ equal $H^b(r, t) \cdot M^a(r, t)$ in the derivation of the alternative reciprocity theorem (see Section 4.2.1) we assumed instantaneously reacting magnetic materials. This means that the head and the tape materials have to be *free from dispersion* (i.e. the permeability is frequency independent). Being dispersion-free in the video frequency range is a too heavy requirement for almost all head materials. It is worthwhile to investigate whether this requirement is necessary when frequency components are treated 'separately' and how $\Phi^R(\omega)$ and $\Phi^R(t)$ can then be expressed.

4.4.2 Heads showing dispersion

The symmetrical tensor K that describes the relation between the induced magnetization M_i and the field H in both state 'a' and state 'b' is now assumed to be frequency-dependent because of dispersion, i.e.

$$M_i(r, \omega) = K(r, \omega) H(r, \omega) \tag{4.15}$$

where M and H are temporal Fourier transforms and K must be complex (in view of Kramers-Kronig causality relations, see chapter 8). This relation replaces (3.7) which was necessary to arrive at (3.13) (see also Section 4.2.1) but (3.13) required materials without dispersion.

In Appendix 4.1 it is derived that for a dispersion-free tape and a head showing dispersion the following expressions hold in the quasi-static approximation where wave travel times are neglected:

$$\Phi^R(\omega) = \frac{W}{v} \, \psi^a(k, y_p, \omega) \, \mathcal{B}_y^{*b}(k, y_p, t = 0) \tag{4.16}$$

$$\Phi^R(t) = \frac{W}{2\pi v} \int_{-\infty}^{\infty} \psi^a(k, y_p, \omega) \, \mathcal{B}_y^{*b}(k, y_p, t = 0) \, e^{j\omega t} d\omega \tag{4.17}$$

where

- k changes with ω according to $k = \omega/v$,
- $\psi^{\mathrm{a}}(k, y_{\mathrm{p}}, \omega) \equiv \Psi^{\mathrm{a}}(k, y_{\mathrm{p}}, \omega)/I^{\mathrm{a}}(\omega)$,
- $I^{\mathrm{a}}(\omega)$ is the Fourier transform of an excitation current $I(t)$ that contains all frequencies, e.g. the impulse function $\delta(t)$, (otherwise $\psi^{\mathrm{a}}(k, y_{\mathrm{p}}, \omega)$ is undefined for parts of the frequency spectrum);
- $\psi^{\mathrm{a}}(k, y_{\mathrm{p}}, \omega)$ is the potential in the plane $y = y_{\mathrm{p}}$ parallel to the tape and lying between tape and coil caused by the (fictitious) excitation current, Fourier transformed both from x to k space and from t to ω space.

The weighting function ψ^{a} in (4.16) will in general have a complex frequency dependence, in contrast to ψ^{a} in (4.13).

The choice of the plane $y = y_{\mathrm{p}}$ does not influence the result, but is conveniently taken to be the plane that coincides with the head surface facing the tape.

Methods to calculate and measure Ψ^{a} will be described in later chapters.

4.4.3 Tapes showing dispersion

The reversible material of the tape, depicted by the domain V_{i} in Chapter 3, travels with the same speed as the permanently magnetized domain V_{p}. This means that the field due to the permanent magnetization M_{p}, as felt by V_{i}, is a dc field. However, when reflections on a non-ideally-reflecting head surface play a role, non-zero frequency components appear. Non-ideally reflection comprises both non-parallellism (with the tape surface) and finite reflectivity (including reflections that change with place, as is the case at a non-zero gap). The non-zero frequency components mainly occur where the potential ψ^{a} changes abruptly, e.g. at the gap area, since part of the reflection is missing there. Since the largest contributions to the flux Φ^{R} come precisely from places where ψ^{a} changes most drastically, these non-zero frequency components have to be treated appropriately, i.e. taken into account as felt by the tape. (On the contrary, the head only feels fields with frequency $\omega = kv$, as in the case of a dispersion-free tape.) So simply using $\mathbf{K}(\mathbf{r}, \omega = 0)$ for the tape material in the calculations of ψ^{a} and $\mathcal{B}^{\mathrm{b}}$ is probably a too rough approximation of the reality when the tape shows considerable dispersion in the considered frequency range. It is doubtful whether any simple formalism exists similar to (4.16) to describe this case. We shall therefore consistently assume that the tape does not show dispersion.

4.5 The use of complex time- and x-independent quantities in the alternative reciprocity theorem

For periodic signals, especially sinusoidal, it is more convenient to use complex x- and t-independent quantities as defined in (2.11) instead of Fourier transforms. This practice will frequently be adopted in the present book.

The relation between the Fourier transform $f_{k_1}(k)$ of a real sinusoidal signal $f(x) = A(k_1) \cos\left(k_1 x + \varphi(k_1)\right)$ and the corresponding complex x-independent quantity $\hat{f}(k_1) = A(k_1)\, e^{i\varphi(k_1)}$ is

$$f_{k_1}(k) = \pi\delta(k - k_1)\hat{f}(k_1) + \pi\delta(k + k_1)\hat{f}^*(k_1) \qquad (4.18)$$

where $\delta(k - k_1)$ is the Dirac delta function (impulse) defined in

$$\int_{-\infty}^{\infty} \delta(k - k_1)\, g(k)\, \mathrm{d}k = g(k_1). \qquad (4.19)$$

Eq. (4.18) is easily verified by calculating the inverse Fourier transform of $f_{k_1}(k)$ while using (4.19).

The alternative reciprocity theorem in the k space for instantaneously reacting materials was given in (4.13) as a relation between the Fourier transformed $\Phi^{\mathrm{R}}(\omega)$ and $\mathcal{B}_y^{*\mathrm{b}}(k, y_{\mathrm{p}}, t = 0)$ and $\psi^{\mathrm{a}}(k, y_{\mathrm{p}})$. With the aid of (4.18), the analogous relation between $f_{\omega_1}(\omega)$ and $\hat{f}(\omega_1)$, the relation $\delta(\omega) \equiv \delta(kv) = \delta(k)/|v|$, the fact that $\psi^{\mathrm{a}}(x)$ is a real function and the assumption of positive v (4.13) can be rewritten for one sinusoidal signal as

$$\hat{\Phi}^{\mathrm{R}}(\omega_1) = W\psi^{\mathrm{a}}(k_1, y_{\mathrm{p}})\,\hat{\mathcal{B}}_y^{*\mathrm{b}}(k_1, y_{\mathrm{p}}, t = 0). \qquad (4.20)$$

It is noted that ψ^{a} cannot be written as a complex x-independent quantity $\hat{\psi}^{\mathrm{a}}$ in the sense of (2.11) since $\psi^{\mathrm{a}}(x)$ is not a periodic function of x.

Analogously for a head that shows dispersion (4.16) can be rewritten for one sinusoidal component as

$$\boxed{\hat{\Phi}^{R}(\omega_1) = W\psi^{a}(k_1, y_p, \omega_1)\,\hat{\mathcal{B}}_y^{*b}(k_1, y_p, t=0)} \quad . \qquad (4.21)$$

And, as must be expected from (4.21), the inverse Fourier transform (4.17) simplifies for one sinusoidal component to

$$\boxed{\Phi^{R}(t) = W\,\mathrm{Re}\{\psi^{a}(k_1, y_p, \omega_1)\,\hat{\mathcal{B}}_y^{*b}(k_1, y_p, t=0)\}} \quad . \qquad (4.22)$$

Note that when $\omega_1 \neq 0$, $\psi^{a}(k_1, y_p, \omega_1) \equiv \Psi^{a}/I^{a} = \hat{\Psi}^{a}/\hat{I}^{a} \equiv \hat{\psi}^{a}(k_1, y_p, \omega_1)$ as e.g. follows from the equivalent of (4.18) for ω instead of k applied to both $\Psi^{a}(k_1, y_p, \omega_1)$ and $I^{a}(\omega_1)$. The symbol $\hat{}$ above ψ^{a} always denotes a time- and not an x-independent quantity.

Equation (4.21), or for dispersion-free heads (4.20), forms a simple and very useful starting point for calculations of responses to sinusoidal and periodic magnetization patterns. On the other hand, equation (4.16), or (4.13), is a good starting point for non-periodic magnetization patterns, for instance in the calculation of impulse responses.

Appendix 4.1 The alternative reciprocity theorem for a dispersion-free tape and a head showing dispersion

In Chapter 3 all quantities were functions of x, y, z and t, but were abbreviated to $\psi(r)$ etc. Instead of starting with the expression in Chapter 3, $\nabla \cdot (\Psi^{a}(r, t)B^{b}(r, t) - \Psi^{b}(r, t)B^{a}(r, t))$, we now start with

$$\nabla \cdot (\Psi^{a}(r, \omega)B^{b}(r, \omega) - \Psi^{b}(r, \omega)B^{a}(r, \omega)) \qquad (a4.1)$$

where Ψ^{a} and B^{b} are the Fourier transforms from the t to the ω domain. The introduction of ω instead of t does not have any consequences in the derivations in Chapter 3 as long as

$$M_i(r, \omega) = K(r, \omega)H(r, \omega). \qquad (a4.2)$$

This equation replaces equation (3.7) in Chapter 3, so that (3.13) and all following expressions hold, but with Fourier-transformed quantities instead of functions that are time-dependent or constant in time. Note that the approximation remains quasi static, i.e. retardation effects are neglected, see sections 3.3 and 2.2.

Equation (3.16) now reads

$$\Phi^{R}(\omega) = \int \psi^{a}(\boldsymbol{r}, \omega)\, \boldsymbol{v} \cdot \boldsymbol{\mathscr{B}}^{b}(\boldsymbol{r}, \omega)\mathrm{d}A \qquad (a4.3)$$

where
- $\psi^{a}(\boldsymbol{r}, \omega) \equiv \Psi^{a}(\boldsymbol{r}, \omega)/I^{a}(\omega)$
- $I^{a}(\omega)$ is the Fourier transform of a fictitious excitation current $I(t)$ containing all frequencies, e.g. $\delta(t)$, (otherwise $\psi^{a}(\boldsymbol{r}, \omega)$ is undefined for parts of the spectrum);
- $\Psi^{a}(\boldsymbol{r}, \omega)$ is the Fourier transform of the potential due to the fictitious excitation current.

The use of the common reciprocity theorem in the proof of the alternative expression for the read flux as we did in [3.1], instead of the direct proof in chapter 3, would require the common reciprocity theorem both to be rewritten for harmonic signals and proven to be valid for a head showing dispersion. In fact, an expression has been derived by Geurst [4.2] for the case of a retardation-free head showing dispersion, but with a permanently magnetized tape without induced magnetization, i.e. $\mu_{\mathrm{rtape}} = 1$; see his Eq. 34.

Now assume that the tape has a velocity v in the $+x$ direction. The ω as felt by the fixed head equals kv, where k is the wavenumber $2\pi/\lambda$ of the considered magnetization component with wavelength λ in the tape. The k and ω have real values. For all magnetic material, including the dispersion-free tape material ($\boldsymbol{K}(\boldsymbol{r}, \omega_{1}) = \boldsymbol{K}(\boldsymbol{r}, \omega_{2})$), equation (a4.2) is valid and so too is (a4.3). Since we can write in this particular case $M_{\mathrm{p}}^{b}(\boldsymbol{r}, t) = M_{0}^{b}(x-vt, y, z)$, $\boldsymbol{\mathscr{B}}^{b}$ can be written as

$$\boldsymbol{\mathscr{B}}^{b}(x, y, z, t) = \boldsymbol{\mathscr{B}}_{0}^{b}(x-vt, y, z). \qquad (a4.4)$$

So the Fourier transforms to the k and the ω domains equal

$$\boldsymbol{\mathscr{B}}^{b}(k, y, z, t = 0) \equiv \int_{-\infty}^{\infty} \boldsymbol{\mathscr{B}}^{b}(x, y, z, t = 0)\,\mathrm{e}^{-jkx}\mathrm{d}x$$

$$= \int_{-\infty}^{\infty} \boldsymbol{\mathscr{B}}_{0}^{b}(x, y, z)\,\mathrm{e}^{-jkx}\mathrm{d}x, \qquad (a4.5)$$

and
$$\mathcal{B}^{b}(x=0, y, z, \omega) \equiv \int_{-\infty}^{\infty} \mathcal{B}^{b}(x=0, y, z, t)\,\mathrm{e}^{-\mathrm{j}\omega t}\mathrm{d}t$$

$$= \int_{-\infty}^{\infty} \mathcal{B}_{0}^{b}(-vt, y, z)\,\mathrm{e}^{-\mathrm{j}\omega t}\mathrm{d}t$$

$$= \frac{1}{v} \int_{-\infty}^{\infty} \mathcal{B}_{0}^{b}(x, y, z)\,\mathrm{e}^{\mathrm{j}kx}\mathrm{d}x$$

$$\equiv \mathcal{B}^{*b}(k, y, z, t=0)/v \tag{a4.6}$$

where $-x/v$ has been substituted for t and $k = \omega/v$ is used.

Equation (a4.6) gives the relation, for the particular case of a constant velocity v in the $+x$-direction, between the Fourier transform from the x to the k domain and the Fourier transform from the t to the ω domain of the function $\mathcal{B}^{b}(x, y, z, t)$. (In the results we write $x = 0$ (or $t = 0$) instead of 0 only, in order to avoid confusion with the condition $k = 0$ (or $\omega = 0$)).

In the same way, but now substituting vt' for $vt-x$ and consequently $\omega t' + kx$ for ωt, we find

$$\mathcal{B}^{b}(x, y, z, \omega) \equiv \int_{-\infty}^{\infty} \mathcal{B}_{0}^{b}(x - vt, y, z)\,\mathrm{e}^{-\mathrm{j}\omega t}\mathrm{d}t$$

$$= \mathrm{e}^{-\mathrm{j}kx} \int_{-\infty}^{\infty} \mathcal{B}_{0}^{b}(-vt', y, z)\,\mathrm{e}^{-\mathrm{j}\omega t'}\mathrm{d}t'$$

$$\equiv \mathrm{e}^{-\mathrm{j}kx}\mathcal{B}^{b}(x=0, y, z, \omega) \tag{a4.7}$$

Hence Eq. (a4.3) becomes

$$\Phi^{R}(\omega) = \int \psi^{a}(x, y, z, \omega)\,\boldsymbol{v} \cdot \mathcal{B}^{b}(x=0, y, z, \omega)\,\mathrm{e}^{-\mathrm{j}kx}\mathrm{d}A. \tag{a4.8}$$

In the plane $y = y_{p}$ and in the 2-dimensional approximation, where the entire configuration is uniform in z, (a4.8) can be written as

$$\Phi^{\mathrm{R}}(\omega) = W\mathscr{B}^{\mathrm{b}}_{\mathrm{y}}(x=0, y_{\mathrm{p}}, \omega) \int_{-\infty}^{\infty} \psi^{\mathrm{a}}(x, y_{\mathrm{p}}, \omega)\,\mathrm{e}^{-\mathrm{j}kx}\mathrm{d}x, \qquad (a4.9)$$

in which we have omitted the argument z and assumed a trackwidth W.
This can be rewritten, using (a4.6), in the following way:

$$\Phi^{\mathrm{R}}(\omega) = \frac{W}{v}\,\psi^{\mathrm{a}}(k, y_{\mathrm{p}}, \omega)\mathscr{B}^{*\mathrm{b}}_{\mathrm{y}}(k, y_{\mathrm{p}}, t=0). \qquad (a4.10)$$

This equation is a generalization of Eq. (4.13). Here the weighting function $\psi^{\mathrm{a}}(k, y_{\mathrm{p}}, \omega)$ is also a function of the angular frequency $\omega = kv$, where k is the angular wavenumber of the considered spatial harmonic in the tape, v the velocity of the tape in the $+x$-direction and ω the angular frequency of the flux and (by definition) of the excitation current I^{a} which defines $\psi^{\mathrm{a}}(k, y_{\mathrm{p}}, \omega)$. When dispersion is not neglected $\psi^{\mathrm{a}}(k, y_{\mathrm{p}}, \omega)$ will be complex, since $\mathbf{K}(\mathbf{r}, \omega)$ must then be complex. When $\psi^{\mathrm{a}}(k, y_{\mathrm{p}}, \omega)$ is known as a function of both ω and k, $\Phi^{\mathrm{R}}(t)$ can be calculated for any permanent magnetization distribution in the tape from the inverse transformation:

$$\Phi^{\mathrm{R}}(t) = \frac{W}{2\pi v} \int_{-\infty}^{\infty} \psi^{\mathrm{a}}(k, y_{\mathrm{p}}, \omega)\mathscr{B}^{*\mathrm{b}}_{\mathrm{y}}(k, y_{\mathrm{p}}, t=0)\,\mathrm{e}^{\mathrm{j}\omega t}\mathrm{d}\omega, \qquad (a4.11)$$

where k changes with ω according to $k = \omega/v$.

(Instead of $\mathscr{B}^{\mathrm{b}}_{\mathrm{y}}(k, y_{\mathrm{p}}, t=0)$ one might also use the shorter notation $\mathscr{B}^{\mathrm{b}}_{0\mathrm{y}}(k, y_{\mathrm{p}})$.)

In conclusion: in the case of a dispersion-free tape, in the quasi-static approximation, Eqs. (a4.10) and (a4.11) hold.

References

[4.1] See most mathematical handbooks on integral transforms, e.g.: Blok, H, *Integraal Transformaties in de Electrotechniek*, Delft University of Technology, Sept. 1973.
[4.2] Geurst, J.A., *The reciprocity principle in the theory of magnetic recording*. Proceedings of the IEEE, Vol. 51, no. 11, 1573-1577 (1963).

Chapter 5

Sensitivity functions, gap loss functions and the efficiency

5.1 Introduction

In the previous chapter various expressions for the flux encompassed by the filamentary coil of a reading head have been derived on the basis of the alternative reciprocity theorem. Besides a constant factor, these expressions contain the product of a magnetic scalar potential ψ^a on a surface and the perpendicular component of a magnetic induction $\mathcal{B}_y^b$. ψ^a characterizes the head and the soft-magnetic properties of the tape (state 'a'). $\mathcal{B}_y^b$ characterizes the 'hard' (permanent magnetization) and the 'soft' (non-zero susceptibility) properties of the tape. For heads showing dispersion, the frequency dependences of the head were incorporated in the adapted definition of ψ^a, see (4.21).

In this chapter the expressions of the time derivative of the flux, the induction voltage, are first derived. Roughly, the potential ψ^a will then turn out to be replaced by its spatial derivative, the longitudinal component of the field h_x^a, as expected. This Fourier transformed longitudinal field can therefore be interpreted as the (voltage) sensitivity function of the head *(V)SF*. For analogous reasons the potential ψ^a could have been called the flux sensitivity function *FSF*. They will be defined more precisely in this chapter.

The sensitivity function can often be divided into a frequency-dependent factor due to dispersion in the head (usually called the efficiency η) and a wavelength-dependent factor (usually called the gap loss factor or function, *GLF*). Our treatment of this division in this chapter is general. As we do not need to distinguish in definitions of sensitivity functions for ring heads or probe heads, applied for longitudinal or perpendicular recording and vice versa, we similarly do not need to, and will not, distinguish between efficiency definitions and *GLF* definitions for any of these combinations. Moreover our, for longitudinal recording ordinary, efficiency definition (and corresponding *GLF* definition) turned out to be very relevant in the description and design of probe type heads on double layer media, see chapter 9.

An accurate split of *SF* into a wavelength-independent $\eta(\omega)$ and a frequency-independent $GLF(k)$ is, however, not always possible.

The above topics together with calculation and discussion of *GLF*s for different types of ring heads and many types of distorted gaps form the subject of this chapter. These applications include asymmetrical TFHs, different types of gradual gaps and (statistically described) irregular gaps.

Some comparisons of the analytical results with the results of a finite-element method (FEM) are carried out. The influence of the permeability of the tape on the *GLF* is also briefly discussed, and illustrated with a finite-element solution.

A general expression for $\mathcal{B}_y^b$ was derived in chapter 2. For idealized (unit-efficiency) heads the description is thus complete, and for various situations the results can be discussed. This will be carried out in the following chapter.

Analytical expressions for efficiencies are usually not easily obtained. For various types of heads efficiency expressions will therefore be derived in later chapters.

5.2 Sensitivities

5.2.1 Voltage sensitivity

The relation between $\psi^a(x, y_p)$ and $h_x^a(x, y_p)$ according to (2.3) for the one-dimensional case is

$$\psi^a(x, y_p) = - \int_{-\infty}^{x} h_x^a(x', y_p)\,dx' \tag{5.1}$$

since $\psi^a(x = -\infty, y_p) \equiv 0$.

Fourier transformation of this integral yields [5.1]

$$\psi^a(k, y_p) = - \frac{h_x^a(k, y_p)}{jk} - \pi h_x^a(k = 0, y_p)\,\delta(k). \tag{5.2}$$

The last term is zero, since also $\psi^{\mathrm{a}}(x = +\infty, y_{\mathrm{p}}) = 0$ and consequently, according to (4.10), the dc component is given by

$$h_{\mathrm{x}}^{\mathrm{a}}(k = 0, y_{\mathrm{p}}) = \int_{-\infty}^{\infty} h_{\mathrm{x}}^{\mathrm{a}}(x, y_{\mathrm{p}})\,\mathrm{d}x = \psi^{\mathrm{a}}(x = -\infty, y_{\mathrm{p}}) - \psi^{\mathrm{a}}(x = \infty, y_{\mathrm{p}}) = 0.$$

$$(5.3)$$

and (5.2) reduces to

$$h_{\mathrm{x}}^{\mathrm{a}}(k, y_{\mathrm{p}}) = -\mathrm{j}k\psi^{\mathrm{a}}(k, y_{\mathrm{p}}). \tag{5.4}$$

Hence, (4.20), or analogously (4.21), can be expressed as

$$\hat{\Phi}^{\mathrm{R}}(\omega) = \mathrm{j}Wh_{\mathrm{x}}^{\mathrm{a}}(k, y_{\mathrm{p}})\,\hat{\mathscr{B}}_{\mathrm{y}}^{*\mathrm{b}}(k, y_{\mathrm{p}}, t = 0)/k. \tag{5.5}$$

For different wavenumbers k, i.e. different λ, but equal velocity v, $\omega = kv$ and the complex output voltage $\hat{V}^{\mathrm{R}} = -\mathrm{j}\omega\hat{\Phi}^{\mathrm{R}}$ reads[□]

$$\hat{V}^{\mathrm{R}}(\omega) = vWh_{\mathrm{x}}^{\mathrm{a}}(k, y_{\mathrm{p}})\,\hat{\mathscr{B}}_{\mathrm{y}}^{*\mathrm{b}}, \tag{5.6}$$

where we omit the arguments of $\hat{\mathscr{B}}_{\mathrm{y}}^{*\mathrm{b}}$, given in (5.5), for simplicity.

5.2.2 Definition of sensitivity functions

The physical meaning of the dimensionless $h_{\mathrm{x}}^{\mathrm{a}}(k, y_{\mathrm{p}})$ is now quite clear. It represents the 'voltage' sensitivity as a function of the wavelength (and the frequency in the case of heads showing dispersion) of the head (including the tape if permeable). More precisely, the (voltage) sensitivity function, abbreviated to $(V)SF$, will be defined as

$$(V)SF \equiv \frac{h_{\mathrm{x}}^{\mathrm{a}}(k, y_{\mathrm{p}})}{N} = \frac{H_{\mathrm{x}}^{\mathrm{a}}(k, y_{\mathrm{p}})}{NI}. \tag{5.7}$$

[□] Note that $V^{\mathrm{R}} = -\mathrm{d}\Phi^{\mathrm{R}}/\mathrm{d}t$ and not $-N\mathrm{d}\,\Phi^{\mathrm{R}}/\mathrm{d}t$, because Φ^{R} was defined as the total flux enclosed by the filamentary (N-turn) coil, which equals N times the average flux per turn. Analogously $h_{\mathrm{x}}^{\mathrm{a}}(x, y_{\mathrm{p}})$ equals N times the average field per turn at a current of 1 A (and has the dimension m^{-1} in contrast to its Fourier transform $h_{\mathrm{x}}^{\mathrm{a}}(k, y_{\mathrm{p}})$ which is dimensionless).

Hence (5.6) becomes

$$\boxed{\hat{V}^{R}(\omega) = vNW\, SF(k, y_{\mathrm{p}})\hat{\mathcal{B}}_{y}^{*\mathrm{b}}}\ .\tag{5.8}$$

Because of (4.20), $\psi^{\mathrm{a}}(k, y_{\mathrm{p}})/N$ can analogously be called the 'flux sensitivity function' *FSF*.

Equation (4.20) or (5.6) or (5.8) is an extremely simple expression for the read flux or voltage, valid for all one-dimensional problems.

5.2.3 Comparison with the common reciprocity theorem

The usual reciprocity theorem

$$\Phi^{R}(t) = \mu_0 \int \boldsymbol{h}^{\mathrm{a}}(\boldsymbol{r})\boldsymbol{M}_{\mathrm{p}}(\boldsymbol{r}, t)\mathrm{d}\boldsymbol{r}\tag{5.9}$$

can be worked out along the same lines, leading to

$$\hat{\Phi}^{R}(\omega) = \mu_0 W \int \left\{ h_{x}^{\mathrm{a}}(k, y)\hat{M}_{\mathrm{p}x}^{*\mathrm{b}}(k, y, t = 0) + h_{y}^{\mathrm{a}}(k, y)\hat{M}_{\mathrm{p}y}^{*\mathrm{b}}(k, y, t = 0)\right\}\mathrm{d}y.$$

$$\tag{5.10}$$

Expression (5.5) gives supplementary insight and has two advantages over Eq. (5.10):
- there is no integration;
- there is only one term instead of two.

The price to be paid for these advantages is a necessary calculation of $\hat{\mathcal{B}}_{y}^{\mathrm{b}}(k)$, which includes an integration of the magnetization over y (see chapter 2). Whether (5.10) is preferable depends on the application.

5.2.4 Gap-null wavelength and sensitivity coefficients for longitudinal and perpendicular magnetization components

From Eq. (5.5) or (5.6) it can directly be concluded that the sensitivity function does not distinguish between parts of $\hat{\mathcal{B}}_{y}^{\mathrm{b}}$ originating from longitudinal and vertical (permanent) magnetization components. This means that the *gap-null wavelength will be equal for perpendicular and* *longitudinal components* of magnetization, since physically $\hat{\mathscr{B}}_y(k)/\hat{M}_{py}(k)$ and $\hat{\mathscr{B}}_y(k)/\hat{M}_{px}(k)$ are always finite. This does not necessarily mean that the head is as sensitive to the $\hat{M}_{py}$ component as it is to the $\hat{M}_{px}$ component, since the relation between $\hat{\mathscr{B}}_y$ and the one or the other component may perhaps differ. That this may indeed happen follows from a comparison between (5.5) and (5.10), leading to the following expression for $\hat{\mathscr{B}}_y^{b*}$:

$$\hat{\mathscr{B}}_y^{b*}(y_p, k, t = 0) = -j\mu_0 k \int_{(y_1)}^{\infty} \left\{ \frac{h_x^a(k, y)}{h_x^a(k, y_p)} \, \hat{M}_{px}^{b*}(k, y, t = 0) \right.$$

$$\left. + \frac{h_y^a(k, y)}{h_x^a(k, y_p)} \, \hat{M}_{py}^{b*}(k, y, t = 0) \right\} dy. \tag{5.11}$$

In general the coefficients of the x and y component of M differ.

Since the field $h^a(r)$ is irrotational in the considered region and independent of z,

$$\frac{\partial h_x^a(x, y)}{\partial y} = \frac{\partial h_y^a(x, y)}{\partial x}. \tag{5.12}$$

With the aid of (4.12) and the one-to-one characteristic of the Fourier transform, (5.12) can be rewritten as

$$h_y^a(k, y) = \frac{1}{jk} \frac{\partial}{\partial y} h_x^a(k, y). \tag{5.13}$$

So, equal sensitivity is obtained for both permanent magnetization components if, and only if, in the region of permanent magnetization

$$\left| h_x^a(k, y) \right| = \left| \frac{1}{k} \frac{\partial}{\partial y} h_x^a(k, y) \right|. \tag{5.14}$$

analogously to the expressions (2.19) and (2.20), h_x^a has the form

$$h_x^a = C_1 e^{\beta k y} + C_2 e^{-\beta k y} \tag{5.15}$$

hence

$$h_y^a = \frac{1}{k} \frac{\partial}{\partial y} h_x^a = \beta C_1 e^{\beta ky} - \beta C_2 e^{-\beta ky} \qquad (5.16)$$

The first term on the right side represents the 'reflections' at the backside of the coating.

Assume absence of reflections, i.e. the coating is infinitely thick or at least $\beta_3 t_3 \gg \lambda/2\pi$, then

- both 'sensitivity' coefficients in (5.11) are equal when $\beta = 1$ ($\mu_x = \mu_y$), i.e. no preferential orientation
- the perpendicular sensitivity coefficient is largest when $\beta > 1$ ($\mu_x > \mu_y$), i.e. preferential orientation (easy axis) in y direction
- the longitudinal sensitivity coefficient is largest when $\beta < 1$ ($\mu_y > \mu_x$), i.e. preferential orientation (easy axis) in x direction.

Thus *the sensitivity for the longitudinal components is larger than for the vertical components when the preferential orientation (easy axis) is in the x direction and vice versa for a preferential orientation in the y direction.*

5.2.5 General remarks on the sensitivity function

It must be noted that the *SF* is not a characteristic of the head only. The coating permeability, the possible presence of a backlayer etc., also influence the *SF*, and so does the choice of y_p, especially at short wavelengths. The plane y_p can best be defined as the plane that just coincides with the mechanical surface of the head facing the tape. This means that a dead layer on top of the head, or hollow around the gap (see later chapters), are included in the *SF*. However, depending on the aim, other definitions can be chosen for the plane y_p, e.g. the interface with the permanently magnetized material such that any air-film and dead layer that may be present in the coating (including roughness) are also included, leading to an *SF* of the recording/playback system.

5.3 The sensitivity function *SF* as a product of the generalized gap loss function *GLF* and the efficiency η

Often the *SF* of a head (-tape system) can be written as the product of a *wavelength-independent* factor η and a *wavelength-dependent* factor due to (local) 'geometric' effects of the head, when only the wavelengths of interest are considered. For instance, the strong geometric effects of the gap region often determine the sensitivity ratio for all different wavelengths of interest, and the function for this part of the wavelength spectrum is therefore called the gap loss function *GLF*. Sometimes, as described for instance in chapters 9, 12 and 13, the k- or λ-dependent factor is not only caused by geometric effects of the gap, so that the name geometric loss or still better generalized (gap) loss factor is closer to what is meant with *GLF*.

The efficiency η and generalized loss factor *GLF* are not yet completely defined. It is even unsure whether it is always possible to split or divide *SF* into functions η and *GLF* with the above mentioned properties, especially when this product is required to describe all relevant wavelengths in a particular application. The split, if it is possible, has the advantage that the sensitivity for wavelengths longer than those of practical interest does not need to be considered.

Firstly we will introduce general definitions for η and *GLF*. Then we will discuss under which conditions the division of *SF* into these functions, $SF \simeq \eta \cdot GLF$, is sufficiently accurate.

Since in the alternative reciprocity theorem ψ^{a} acts like a weighting factor, it is obvious that ψ^{a} will play a central role in the definition of η and *GLF*. With the aid of the following definitions

$$\eta \equiv \left| \psi^{a}(x_{r}) - \psi^{a}(x_{\ell}) \right| / N \leqslant 1 \tag{5.17}$$

and

$$GLF(k) \equiv \frac{1}{\eta^{a}N} \int_{x_{\ell}}^{x_{r}} - \frac{\partial \psi^{a}}{\partial x} \, e^{-jkx} \mathrm{d}x, \tag{5.18}$$

a split

$$\boxed{SF \simeq \eta \cdot GLF(k)} \tag{5.19}$$

is possible. In the present quasi-static description η is independent of ω. For a head showing dispersion η becomes frequency dependent as will be discussed in detail in chapter 8, while *GLF* usually remains unaffected in a very good approximation.

The split is only useful when such a choice can be made for the boundaries x_ℓ and x_r that the accuracy of the split, $a(k) \equiv |SF - \eta \cdot GLF|/SF$, i.e.

$$a(k) = \frac{\left| \int_{-\infty}^{x_\ell} \frac{\partial \psi^a}{\partial x} e^{-jkx} dx + \int_{x_r}^{\infty} \frac{\partial \psi^a}{\partial x} e^{-jkx} dx \right|}{\left| \int_{-\infty}^{\infty} \frac{\partial \psi^a}{\partial x} e^{-jkx} dx \right|}, \qquad (5.20)$$

is sufficient for the wavelengths of interest. For instance for video applications, $\lambda = 0.5 \sim 10$ μm, one may require that this accuracy is always better than 0.1. An accurate result can only be expected when x_r and x_ℓ are chosen in the neighbourhood of the right and left sides of the region near the tape with the most important potential drop (a combination of steepness and size), if such a region exists.

For quick estimates equation (5.20) is unpractical. To get rid of this problem, consider the potential around the perimeter of e.g. the video-like head, as sketched in Fig. 5.1a. For simplicity no phase differences are assumed around the perimeter (no dispersion), so η is real. This potential can be transformed to pole-surface ($y = 0$) level, parallel to the tape, as sketched in Fig. 5.1b. Fig. 5.1a shows that almost all loss of potential outside the gap region, which is responsible for the η, takes place in the region where the cross-sectional area is small. In chapter 6 this is worked out in detail. The potential at pole-surface level in Fig. 5.1b also shows the potential decrease just outside the outer corners of the head, causing the well-known secondary-gap effects at (very) long wavelengths. This and any potential function can be approximated by straight lines as depicted in Fig. 5.1c. For the present purpose of deriving a reliable rule of thumb for the accuracy of the split, it is safest to choose the straight lines tangential to the steepest parts of the potential curve, so that especially the high-frequency contributions of those regions cannot be underestimated.

After some mathematics the straight-line approximation results in the following approximation for the *SF*:

$$SF \simeq -\sum_i \Delta\psi_i^a \text{sinc}\,(g_i/\lambda) e^{-jkx_i}/N. \qquad (5.21)$$

Every region contributes to the sensitivity function a sinc function (the Fourier transform of the rectangular field distribution within each region see Fig. 5.1d) times a phase factor, caused by the distance between regions.

With the aid of the approximation (5.21) a simple and safe (worst case) expression can easily be derived for the accuracy of the split or for the range of λ's for which the accuracy is sufficient. Select the region

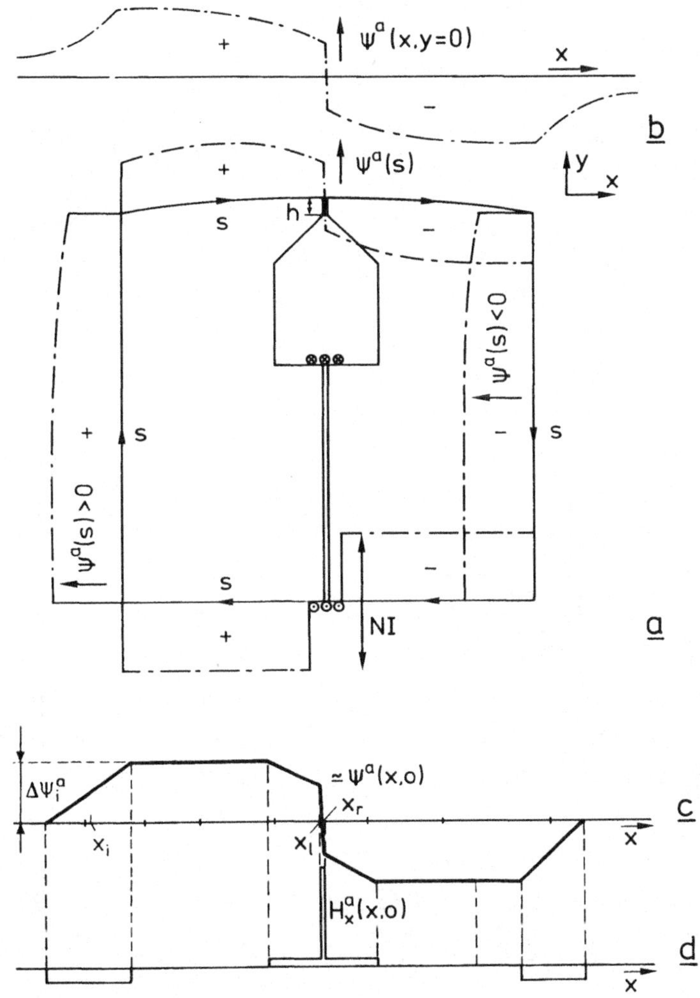

Fig. 5.1. Sketches of the potential distribution of a video head:
a) around the perimeter
b) at pole-surface ($y = 0$) level
c) straight-line approximation
d) corresponding right-angles approximation of H_x field.

with the second-steepest change in potential and assume that all loss of the dimensionless potential ψ^a/N, i.e. $1-\eta$, vanishes with that steepness. This results in a region with length $g' = (1-\eta)/$steepness over which this drop takes place. Since the potential increase at the outer corners of the head might be the second steepest (e.g. for a thin film head) and the increase of ψ^a at the corners may exceed $1-\eta$, it is safer to use $g' = 1/$steepness. An accuracy better than a for the split in the case of a head with only one main gap then requires that

$$\frac{|\sin (kg'/2)|}{kg'/2} \lesssim a\eta \; \frac{|\sin (kg/2)|}{kg/2}. \qquad (5.22)$$

In Fig. 5.2 these functions are shown for a certain ratio g'/g. The worst-case situations are the extrema of the function sinc (g'/λ), i.e. $\sin (kg'/2) \simeq 1$. Substitution of this value in (5.22) yields as a rule of thumb

$$a(k) \lesssim \frac{1}{\eta |\sin (kg/2)|} \cdot \frac{g}{g'} \qquad (0 < kg/2 < \pi) \qquad (5.23)$$

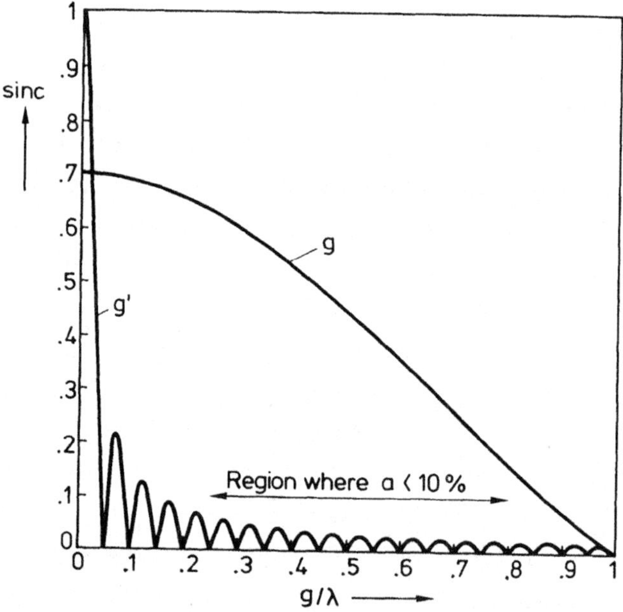

Fig. 5.2. The region $1/b < g/\lambda < b/(b + 1)$ where the accuracy of the split, $a = bg/\eta\pi g'$, is at least 0.1. The 'gap length' g' represents the (worst case) potential drop outside the main gap g. $g' = 20\,g$ and $\eta = 0.7$ such that $b = 4$ and $1/4 < g/\lambda < 4/5$.

where wavelengths of interest are assumed to be longer than g, i.e. $\sin(kg/2)$ is always positive. In this range (5.23) can just be fulfilled for two values of λ, while for all values in between a higher worst-case accuracy is obtained. In a good approximation the range of λ's for which the accuracy $a(k)$ never exceeds a follows from (5.23), hence

$$\left(1 + \frac{1}{b}\right) g \lesssim \lambda \lesssim bg \qquad (5.24)$$

when $b \equiv \eta \, a \, \pi \, g'/g \gg 1$.

This simple rule of thumb gives information about the usefulness ($b \gg 1$?) as well as the frequency range (does the wavelength range encompass the range of wavelengths of practical interest?) of the split when a certain accuracy a is required. When the required a is too small or g/g' too large, (5.23) results in a very small wavelength range (a split might be meaningless) or no wavelength range at all (a split might be impossible). (We say *might*, because we simplified the accurate expression for $a(k)$ in (5.20) drastically; (5.23) and (5.24) describe 'worst case' situations.)

Practical values for e.g. a video head are $\eta = 0.5$, $g = 0.30$ μm and $g'/g \simeq 300$. When $a = 0.1$ ($\simeq 1$ dB) is required, then $b = 15 \, \pi \gg 1$ and so the interesting λ range may not exceed the range $(1 + \frac{1}{45})g \leqslant \lambda \leqslant 45g$ or 0.3 μm $\leqslant \lambda \leqslant 14$ μm which encompasses all wavelengths of interest in video recording.

Criteria for heads with more than one main gap, see e.g. chapters 9, 12 and 13, can be derived analogously from (5.21). Because the phase factors in (5.21) have to be taken into account when more than two regions are used in the calculations, this can easily result in unpractical criteria.

The *GLF* as well as the η are allowed to be functions of frequency (when the wavelength is kept constant, i.e. only the velocity v is changed) in the case of a head that shows dispersion. However, since the *GLF* is usually a result of a very local geometry of the head, permeability changes if present in that region do not influence the local potential distribution noticeably, and dispersion effects can therefore be disregarded. The efficiency η on the other hand is often dependent on frequency, especially at video frequencies (0.5 − 5 MHz), since η is strongly determined by the magnetic material of the head, which usually shows dispersion in the video-frequency range.

5.4 Video head

5.4.1 Introductory remarks

The flux as a function of frequency through the read coil of a video head with infinite permeability but non-zero gap length will be calculated. As a first approximation for the potential ψ^a a linearly increasing potential over the gap is assumed, as depicted in Fig. 5.3. This is equivalent to the Karlqvist approximation. For the deep-gap field and for the field at the surface of a head where the hypothetical surrounding would have a permeability $\mu = 0$, this assumption would be exact. At the dashed $y = 0$ line outside the gap region in Fig. 5.3 it is also assumed that the potential changes gradually and very slowly in comparison to wavelengths of practical interest. This is a very good approximation for all wavelengths of interest in video recording read by any modern video head, since the longest wavelengths of interest are far below 100 μm and the distance d_e from the edge of a video head to the tape, represented by the dashed line, is more than 100 μm. In the case where the written track on the tape is much wider than the track width of the read head W, it will be assumed that $W \gg h$ (where h is the gap height defined in Fig. 5.1), so that the z-dependence of ψ^a can be neglected in (4.1). When this is not true, it is assumed that the written track on the tape equals (or is smaller than) and coincides with the track width of the read head, so that the z-dependence in ψ^a can again be neglected.

An adequate quantitative description of the long wavelength behaviour, including these side-fringing effects, can only be given by using results of numerical calculations; see [5.2]. We will neglect side-fringing effects, i.e. restrict the calculations to shorter wavelengths.

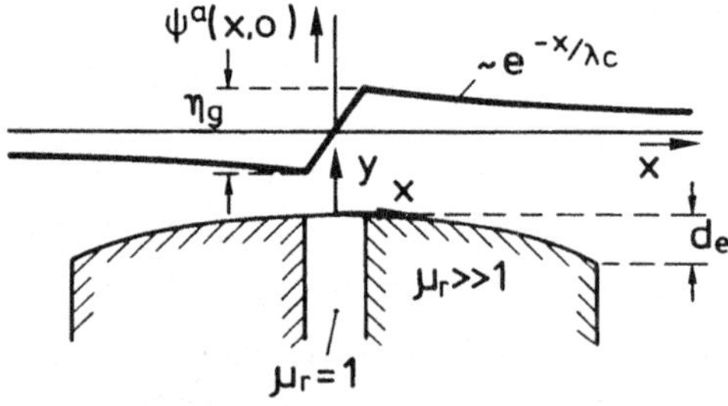

Fig. 5.3. Sketch of the slowly-changing potential of a video-like head. The choice of an exponential decay is arbitrary but the use of one or the other slow function is necessary in the direct application of the alternative reciprocity theorem in the x space. $N = 1$ for simplicity.

The 'drop' of the dimensionless potential ψ^a over the gap is by definition η_g, as also depicted in Fig. 5.3. The dimensionless potential ψ^a at $y = 0$ is approximated by

$$\psi^a/N = \begin{cases} -(\eta_g/2)e^{(x + g/2)/\lambda_c} & \text{for } x \leqslant -g/2 \\[2mm] (x/g)\eta_g & \text{for } |x| \leqslant g/2 \\[2mm] (\eta_g/2)e^{-(x - g/2)/\lambda_c} & \text{for } x \geqslant g/2 \end{cases} \qquad (5.25)$$

where a characteristic length $\lambda_c \gg \lambda/2\pi$ is introduced (over every λ_c, the potential drops by a factor $e \simeq 2.7$).

The slowly changing functions are arbitrarily chosen, but this choice will not influence the results notably because of $\lambda/2\pi \ll \lambda_c$. The magnetization component on the tape with wavenumber $k \equiv 2\pi/\lambda$ causes a normal component $\mathscr{B}_y^b$ in (4.1) that can be expressed over the effective track width W (coinciding with the track width of the read head) by

$$\mathscr{B}_y^b = |\hat{\mathscr{B}}_y^b|\cos(kx + \varphi_{B_y}). \qquad (5.26)$$

With the above expression it is implicitly assumed that the tape parameters as well as the permanent magnetization in the tape are uniform in the z-direction over at least the effective track width $W \geqslant t_2 + t_3 + d_4 + t_5$, see definitions in Fig. 2.1. The last assumption is also necessary to assure the absence of a z-dependence of $\mathscr{B}_y^b$ on the head surface and is always fulfilled in practice.

With the previous assumption concerning the z-dependence of ψ^a equation (4.1) can be simplified to

$$\Phi = W \int_{-\infty}^{\infty} \psi^a \mathscr{B}_y^b dx. \qquad (5.27)$$

This integral can be evaluated directly, but time-consumingly, by substituting equations (5.25) and (5.26). Although we are mainly interested in the result for $\lambda_c \rightarrow \infty$, this does not simplify the direct method, since the integral in (5.27) is not defined for $\lambda_c = \infty$, but only for $\lambda_c \rightarrow \infty$.

After this direct method we will also use the Fourier transformation to arrive more quickly at the result for $\lambda_c = \infty$ without the problem of undefined integrals.

5.4.2 Solution for Φ via x space

Substitution of Eqs. (5.25) and (5.26) into (5.27) yields

$$\Phi = -C_1 \int_{-\infty}^{-g/2} e^{(x + g/2)/\lambda_c} \cos(kx + \varphi_{B_y}) \, dx$$

$$+ \frac{2C_1}{g} \int_{-g/2}^{g/2} x \cos(kx + \varphi_{B_y}) \, dx$$

$$+ C_1 \int_{g/2}^{\infty} e^{-(x - g/2)/\lambda_c} \cos(kx + \varphi_{B_y}) \, dx \qquad (5.28)$$

where $C_1 \equiv N |\hat{\mathcal{B}}_y^b| (\eta_g/2) W$.

By splitting $\cos(kx + \varphi_{B_y})$ into $\cos(kx) \cos \varphi_{B_y} - \sin(kx) \sin \varphi_{B_y}$ and by suitable substitutions, these integrals can all be transformed into standard integrals that can be found in e.g. [5.3]. After (time-consuming) calculations the result reads

$$\Phi = \frac{-2C_1}{k} \left[\left(1 + \frac{\lambda_n}{1 + \lambda_n^2} \frac{kg}{2} \right) \frac{\sin(kg/2)}{kg/2} - \frac{\cos(kg/2)}{1 + \lambda_n^2} \right] \sin \varphi_{B_y} \qquad (5.29)$$

where $\lambda_n \equiv \lambda/(2\pi\lambda_c)$ is the normalized wavelength. In the limit of large λ_c ($\lambda_c \gg \lambda$ already suffices in practice!),

$$\Phi = \frac{-2C_1}{k} \frac{\sin(kg/2)}{kg/2} \sin \varphi_{B_y}. \qquad (5.30)$$

The same result would have been obtained if, instead of the e-power, any other slowly decreasing function had been chosen. Also potential functions that first stay constant or increase slowly (with respect to the steepness of the potential in the gap) to an amount not much larger than the potential over the gap, and that further behave comparably to the previous example, deliver the same approximation. See e.g. the potential of the video head sketched in Fig. 5.1, where with the aid of a 'rule of thumb' it was concluded that regions outside the gap region did not contribute to the head sensitivity in the wavelength range of interest. (The meaning of g' in section 5.3 is comparable with the meaning of λ_c in this section.)

If the tape moves with constant velocity v in the $+x$ direction, we can write $\varphi_{B_y} = \varphi_0 - 2\pi(vt/\lambda) = \varphi_0 - \omega t$, where φ_0 is the phase at the instant $t = 0$ and ω the radial frequency. The flux then equals the well-known expression

$$\Phi(t) = N\eta_g |\hat{\mathcal{B}}_y^b| (W/k) \, \frac{\sin(kg/2)}{kg/2} \sin(\omega t - \varphi_0) \qquad (5.31)$$

for wavelengths smaller than the characteristic length λ_c. According to the arguments in the introduction of this section this is true for all the interesting wavelengths in all today's video heads.

So we can conclude that edge effects play no role in video recording today. Only side-fringing effects play a role at the longest wavelengths of interest in video recording but can only be analysed numerically.

5.4.3 Solution for Φ via k space

The *SF* of a head with the potential of the head in Fig. 5.3, but with zero gap and λ_c (or g') $\rightarrow \infty$, follows easily with the aid of (2.3) and (4.19) and the transformation (4.11), giving

$$h^a(x)/N = -\eta_g \delta(x) \Leftrightarrow h^a(k)/N = -\eta_g \qquad (5.32)$$

and analogously for the non-zero gap head of Fig. 5.3 with $\lambda_c \rightarrow \infty$ it follows that

$$h^a(x)/N = -(\eta_g/g) \cdot \Pi(x/g) \Longleftrightarrow h^a(k)/N = -\eta_g \, \frac{\sin(kg/2)}{kg/2}, \qquad (5.33)$$

where $\Pi(x/g)$ is the right-angle function with height 1 and width g around $x = 0$, which transforms to the gap loss function, and η_g is the efficiency.

Substitution of (5.33) in (5.5) and expressing $\hat{\mathcal{B}}_y^{b*}$ as a product of an amplitude and phase factor yields

$$\hat{\Phi}(\omega) = -jWN\eta_g \frac{\sin(kg/2)}{kg/2} |\hat{\mathcal{B}}_y^b| \, e^{-j\varphi_0/k} \tag{5.34}$$

and thus

$$\Phi(t) = N\eta_g |\hat{\mathcal{B}}_y^b| (W/k) \frac{\sin(kg/2)}{kg/2} \sin(\omega t - \varphi_0) \, . \tag{5.35}$$

The same result as in (5.31) is obtained, but without the time-consuming calculations necessary to avoid the non-defined integrals for $\lambda_c \to \infty$ in the preceding section.

The Fourier method works much quicker since $\lambda_c \to \infty$ can be introduced at an early stage before integrals have to be calculated. However as a consequence the result given in for instance (5.35) is only valid for finite wavelengths, i.e. (much) smaller than λ_c. It is physically impossible to make λ_c infinite, since ψ^a must equal zero at infinity. This means that the (non-zero) dc response of the Karlqvist head (i.e. a head with linearly increasing potential at the gap and $\psi^a/N = +1/2$ of $-1/2$ for $x \geqslant g/2$ and $x \leqslant -g/2$ respectively) is unphysical. This problem is discussed by Mallinson [5.4] too, who also suggests some simple head fields that show a zero dc response.

5.5 Ideal non-zero gap head and tape with unit permeability

5.5.1 Exact expressions for *GLF*s of idealized gaps

The first to calculate the exact *GLF* of a head, without the presence of a permeable tape, was Westmijze [5.5]. He considered three cases:
a) The infinitely thin head, cf. Fig. 5.4a:

$$GLF = \frac{1}{\pi} \int_{-\pi/2}^{\pi/2} \cos\left(\frac{kg}{2} \sin \tau\right) d\tau \equiv J_0\left(\frac{kg}{2}\right) \tag{5.36}$$

b) The semi-infinite head, cf. Fig. 5.4b:

$$GLF = \frac{1}{\pi} \int_0^{\pi/2} \tan v \, \sin\left\{\frac{2}{\pi}\,\frac{kg}{2}\left(\frac{\pi}{2} - v + \tan v\right)\right\} dv$$

$$+ \frac{1}{\pi} \sin \frac{kg}{2} \int_0^1 \frac{u}{1-u^2}\, e^{\frac{2}{\pi}\,\frac{kg}{2}\left(u + \frac{1}{2}\ln\frac{1-u}{1+u}\right)} du \tag{5.37}$$

$$\equiv S\left(\frac{kg}{2}\right)$$

c) The infinite head, cf. Fig. 5.4c:

$$GLF = \frac{\sin(kg/2)}{kg/2} \equiv \text{sinc } g/\lambda. \tag{5.38}$$

Westmijze used conformal mapping and the Schwarz-Christoffel transformation to arrive at his results (5.36) and (5.37) respectively. The solution to the half-infinite head is not a simple integral expression

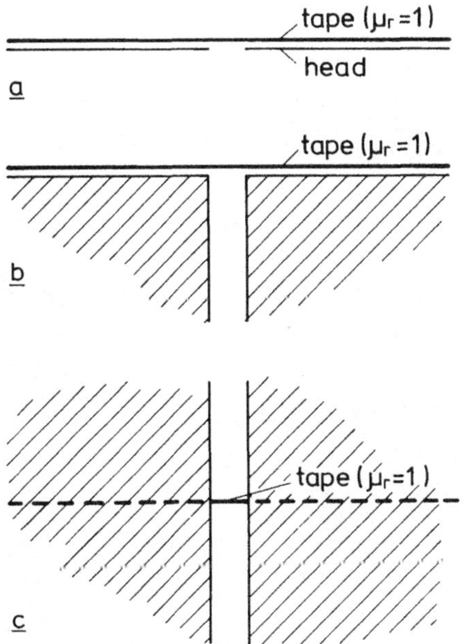

Fig. 5.4. The three head types considered by Westmijze:
a) infinitely-thin head
b) semi-infinite head
c) infinite head.

and is hard to compute numerically because of the singularities in both integrals. Only in the second integral can the singularity easily be avoided.

Fan [5.6] used a superposition of Fourier terms to describe the potential in the gap. His zero wavenumber term coincides with the linearly increasing Karlqvist potential. The result is a *GLF* that reads sin $(kg'/2)/$ $(kg/2)$, where g' however is a function of the wavelength and can be written as the sum of the gap length g (due to the Karlqvist term) and an infinite series of correction terms:

$$g' = g\left(1 + \sum_{n=1}^{\infty} \frac{(-1)^n a_n n\pi}{1 - (n\lambda/g)^2}\right).$$

(5.39)

Depending on head-tape distance a number of correction terms have to be included [5.7]. The first six coefficients, after Baird [5.8], are given in Table 5.1.

Table 5.1 Coefficients in the GLF in the case of unit relative permeability tape after Baird.

a_1	-0.08054
a_2	$+0.02327$
a_3	-0.01108
a_4	$+0.00651$
a_5	-0.00429
a_6	$+0.00305$

5.5.2 Approximations for *GLF* of idealized semi-infinite gap

Szczech [5.9] at al. used rational functions with three coefficients as rather accurate approximations of the field component(s) in the case of unit tape permeability:

$$H_x^a(x, 0) = H_x^a(0, 0)\left\{0.835 + \frac{0.0433}{(0.512)^2 - (x/g)^2}\right\} \qquad |x| \leq g/2$$

(5.40a)

$$H_x^a(x, 0) = 0 \qquad\qquad |x| > g/2$$

(5.40b)

The three coefficients were obtained from a least-squares fit with a numerically obtained finite-difference solution.

Integration of Szczech's expression gives the (dimensionless) potential:

$$\psi^{\mathrm{a}}(x, 0) = \mp N\eta_{\mathrm{g}}\left(0.69\,\frac{|x|}{g} - 0.035 \ln \frac{1 - 1.953|x|/g}{1 + 1.953|x|/g}\right) \qquad |x| \leqslant g/2$$

$$(5.41\mathrm{a})$$

$$\psi^{\mathrm{a}}(x, 0) = \mp 0.5\, N\eta_{\mathrm{g}} \qquad\qquad\qquad\qquad\qquad |x| \geqslant g/2$$

$$(5.41\mathrm{b})$$

The upper sign is valid for positive and the lower sign for negative values of x. The normalized field $h_{\mathrm{x}}^{\mathrm{a}}(0, 0) \equiv H_{\mathrm{x}}^{\mathrm{a}}(0, 0)/I^{\mathrm{a}}$ in (5.40) is chosen to be $0.827\, N\eta_{\mathrm{g}}/g$. As a result the dimensionless potential difference over the gap equals $N\eta_{\mathrm{g}}$ in (5.41). This choice is necessary because of the introduction of η_{g} in (5.41) and the definition of η_{g} in (5.17).

The Fourier transformation of the field according to (5.18) delivers the sensitivity function:

$$SF = \eta_{\mathrm{g}}\, GLF = \eta_{\mathrm{g}}\left(0.69\,\frac{\sin(kg/2)}{kg/2} + 0.07\{\cos(0.512kg)[\mathrm{Ci}(1.012kg)\right.$$

$$\left.-\mathrm{Ci}(0.012kg)] + \sin(0.512kg)[\mathrm{Si}(1.012kg) - \mathrm{Si}(0.012kg)]\}\right). \qquad (5.42)$$

In the evaluation of this integral use is made of transformation 31 on p. 43 in reference [5.10]. The sine and cosine integrals Si and Ci can be expanded, according to ref. [5.11], in the following series:

$$\mathrm{Si}(x) \equiv \int_0^x \frac{\sin t}{t}\, dt = \sum_{n=0}^{\infty} \frac{(-1)^n x^{2n+1}}{(2n+1)!(2n+1)} \qquad (5.43)$$

$$\mathrm{Ci}(x) \equiv -\int_x^{\infty} \frac{\cos t}{t}\, dt = \gamma + \ln x + \sum_{n=1}^{\infty} \frac{(-1)^n x^{2n}}{(2n)!2n}\,, \qquad (5.44)$$

where $\gamma \equiv$ Euler's constant $= 0.57721$.

With the aid of these expansions it is easily verified that $GLF(k = 0) = (0.69 + \ln 1.012 - \ln 0.012) = 1$.

A more practical choice than using Fan's results, Fourier-transforming Szczech's results, or using the difficult expression of Westmijze's case 2), is the following approximate expression:

$$GLF \simeq \tfrac{1}{2}\{J_0(\pi g/\lambda) + \mathrm{sinc}(g/\lambda)\} \qquad (5.45)$$

for tapes with permeability μ_0 or tapes far enough away from the head's surface. The structure of the approximation of the GLF in (5.45) is similar to that in (5.42), but simpler.

In Fig. 5.5, a comparison is made between the exact function and this arithmetic mean of the results of the infinitely thin and infinite head. The similarity is striking.

Another well known approximation, using the gap-null wavelength $\lambda_0 \approx 1.13\,g$ calculated from (5.37) in the Karlqvist result is

$$GLF = \mathrm{sinc}(1.13g/\lambda). \qquad (5.46)$$

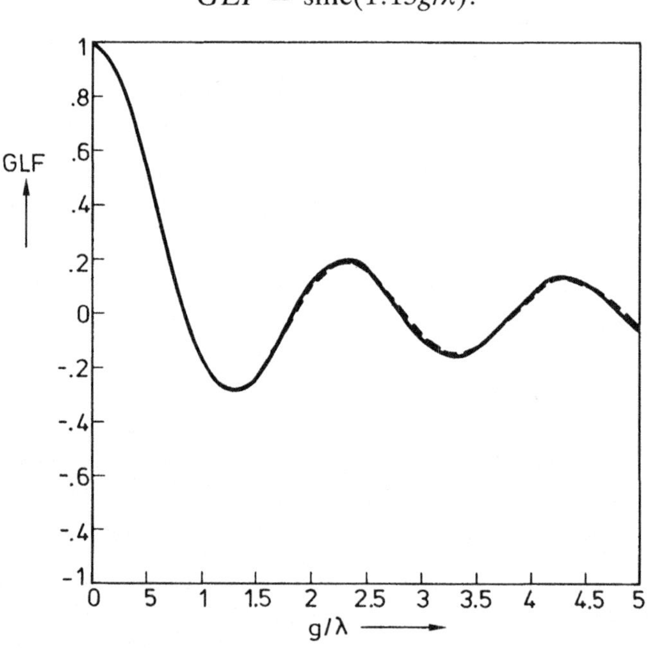

Fig. 5.5. The striking similarity between the proposed, easy-to-compute gap loss function (5.45) and the exact, but hardly computable GLF (5.37) of the semi-infinite head with tape of unit permeability. —— approximation, ——— exact.

Our approximation (5.45) has the advantage over the sinc approximation and over using the first few terms of Fan's expression, that even beyond the first gap-null frequency the approximation is (very) good.

5.5.3 Potential at larger distances

In evaluating the potential $\psi(x, y)$ at distances y from the head surface on which the potential $\psi(x, 0)$ corresponding to one of the above GLFs is defined, use can be made of (see the end of App. 5.1)

$$\psi(x, y) = \frac{y}{\pi} \int_{-\infty}^{\infty} \frac{\psi(x', 0)}{y^2 + (x' - x)^2} \, \mathrm{d}x'. \tag{5.47}$$

The field in the tape (per ampere) follows by differentiation from this expression, and must be known when using the usual reciprocity theorem. The above integral can then only be avoided by using Fourier transforms. See in this context also (7.5) in chapter 7 and section 9.2.4 'Transformation to main-pole surface level'. The use of the alternative reciprocity theorem only requires the potential at pole-surface level and avoids the use of the above, often complicated, integral even when the direct method (without Fourier transform) is used.

5.6 Influence of tape permeability on GLF

In the limit of very high tape permeability the tape surface has a constant magnetic potential ψ^a; cf. Fig. 5.6a. The x-component of the field $H_x(x, 0)$ clearly is largest at the corners of the magnetic material, especially when the head-tape distance to gap length ratio d_4/g is very low. The outer corners of the head, far away from the gap, are assumed to be separated far from the tape, so that they do not contribute to the head's high frequency response. In the limit of zero head-tape distance, at the gap region, the field H_x can be determined directly with the aid of only boundary conditions (cf. Fig. 5.6b). Evidently the result is

$$H_x^a(x, 0) = \frac{\eta^a N I^a}{2} \left\{ \delta(x + g/2) + \delta(x - g/2) \right\} \tag{5.48}$$

where $\eta^a N$ is defined according to (5.17) as the absolute value of the potential difference per Ampère, $|\Delta\psi^a|$, across the gap.

The Fourier transform of this field delivers the *GLF*; see (5.18):

$$GLF = \frac{1}{\eta^a} \int_{-\infty}^{\infty} \frac{H_x^a(x,0)}{NI^a} \, e^{-jkx} dx = \cos(kg/2). \qquad (5.49)$$

This *GLF* has gap nulls at $kg/2 = (n - \frac{1}{2})\pi$, i.e. the first null at exactly $\lambda_1 = 2g$ (i.e. $f_1 = v/2g$) instead of about g in the case of a unit relative tape permeability (cf. Fig. 5.7).

The second gap null occurs at $3f_1$ instead of $2f_1$. The second and following maxima are as strong as the *GLF* at dc. Although assuming material with infinite permeability is unphysical and so to a lesser extent is the assumption $d_4/g \ll 1$, the previous example indicates that:

−   The shift of the gap null is only a function of d_4/g and μ_r, i.e. $\lambda_1 = f(d_4/g, \mu_r)$, if the magnetic-coating thickness $\gg g$. When d_4/g decrea-

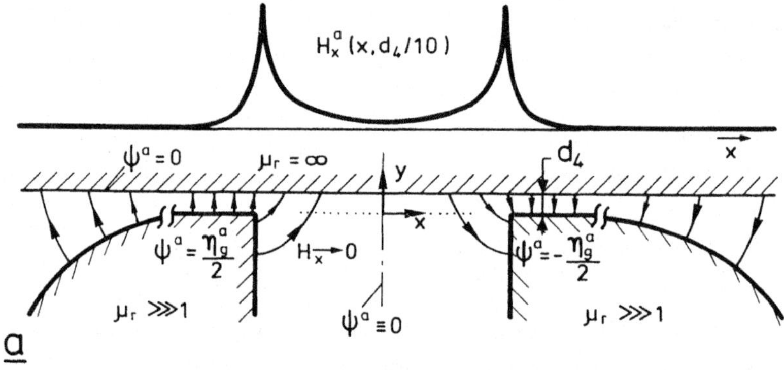

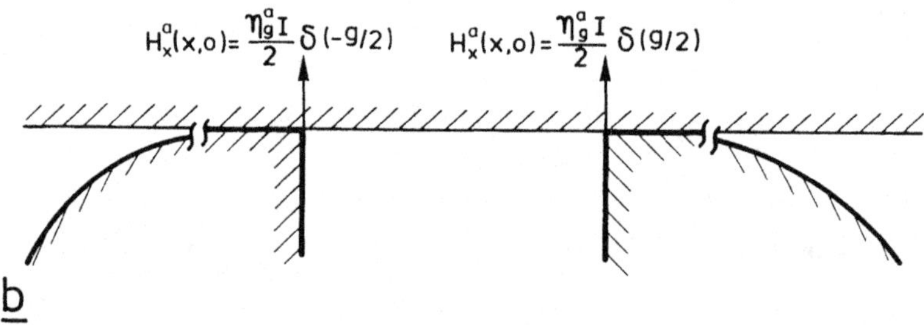

Fig. 5.6. Sketch of the concentration of the sense field H_x^a at the gap corners when the tape permeability is high and the head-tape distance is small. $N = 1$ for simplicity.
a) Finite head-tape distance $d_4 = g/12$. The sense field H^a at distance $y = d_4/10$ has been numerically calculated. The flux lines are sketched (⇀).
b) Zero head-tape distance. The shape of the sense field follows directly from the boundary conditions.

ses and μ_r increases, the δ functions are approached more and more, i.e. at lower frequencies the gap null appears and stronger second and following maxima will be observed.

Fig. 5.8 shows the shift of the gap-null to a lower frequency when a double layer (DL) medium is used. The thickness of the soft-magnetic NiFe backlayer t_2 and the thickness of the hard-magnetic coating t_3 are both about 0.4 μm. Since $\mu_3 \simeq 1$, $\mu_2 \simeq 140 \gg 1$ and $\mu_2 t_2 \gg g = 3.5$ μm, the tape can be approximated by a tape with an infinite permeability

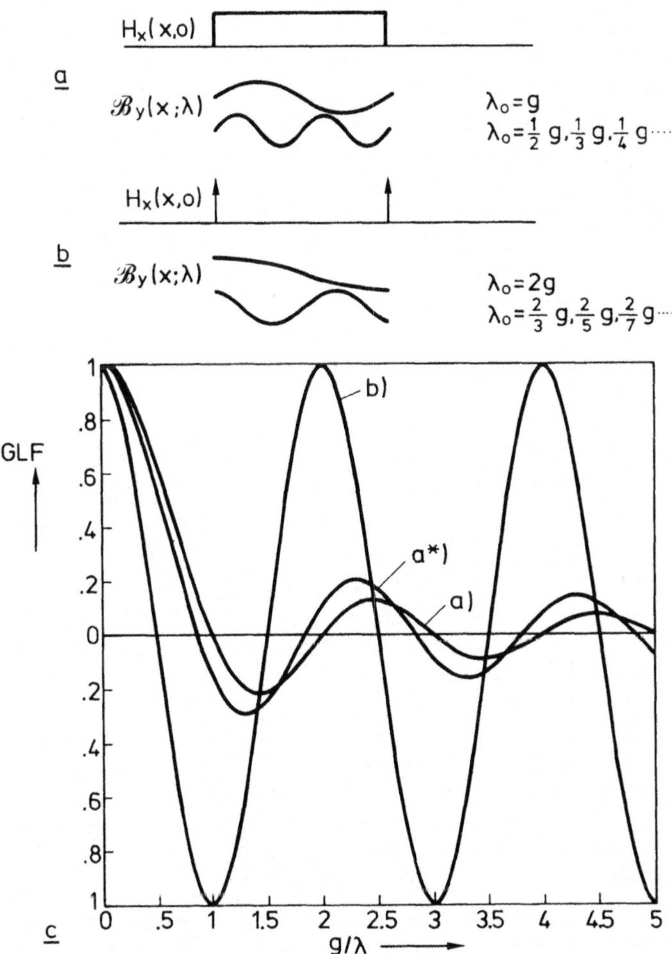

Fig. 5.7. Increase of the gap-null wavelength.
a) Unit tape permeability.
b) Infinite tape permeability and zero head-tape distance.
c) Corresponding sensitivity functions. The accurate approximation according to (5.45) is indicated by the asterix, the other is the well-known Karlqvist approximation.

at a distance of 0.5 µm (coating thickness plus a head-coating distance of approx. 0.1 µm), i.e. distance/g = 0.143, for which a shift of the gap-null wavelength λ_0 from 1.13 g ≃ 4.0 µm (CrO_2, ME and MP with μ_3 ≃ 1) to 1.45 g ≃ 5.0 µm (DL) must be expected [5.12]. This is in reasonable agreement with the experimental values:
- λ_0 = 4.04 µm for the low-permeability and thin ME coating,
- λ_0 = 4.11 µm for the low-permeability but think MP coating,
- λ_0 = 4.76 µm for the DL medium.

Measurements on the same tapes using a head with a small gap of 0.37 µm did, as expected, not show visible shifts of the gap-null.

In reference [5.12] many theoretical details can be found concerning the dependence of reproducing gap null on the medium permeability and the head-tape distance. The gap null is determined there numerically and it is shown analytically that only the geometric mean permeability, as defined in (2.16), influences the gap nulls. Fan [5.6], with the aid of a Fourier method, also calculated the responses of heads at

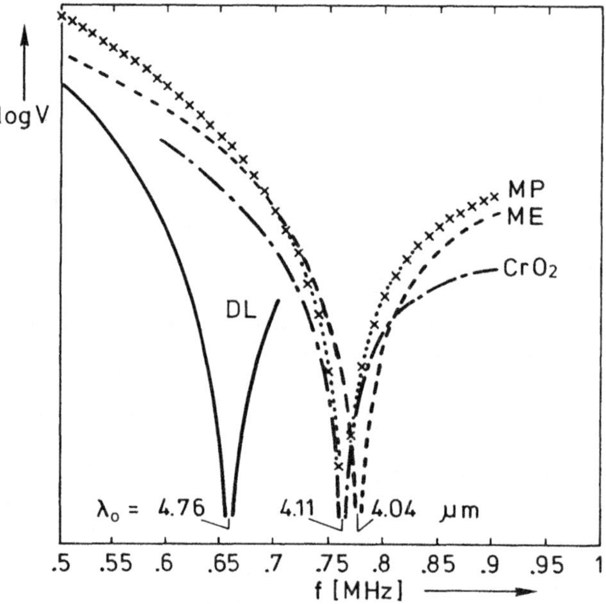

Fig. 5.8. Gap-null measurements on various tapes with a 3.5 µm-gap head. Head-tape distance d_4 ≃ 0.1 µm and v = 3.14 m/s.
CrO_2 = Chromium-dioxide tape with t_3 = 4.6 µm
ME   = Metal-evaporated tape with t_3 = 0.18 µm
MP   = Metal powder tape with t_3 = 3.6 µm
DL   = Double-layer tape with t_2 ≃ t_3 ≃ 0.4 µm and μ_2 ≃ 140.
The MH curves of the tapes, except for the DL tape, are shown in chapter 5. (The Log(V) scales differ for the different tapes.)

different tape permeabilities, but his limiting case, i.e. $\mu_r \to \infty$, contains an error; cf. also [5.12]. In addition, Fan concludes that spacing and coating thickness will not influence the shift of the gap null. This is incorrect, as may be evident from the previous example; cf. Fig. 5.6, and the numerical results of Bertram [5.12].

5.7 Gradual gap

5.7.1 Introductory remarks

Although the transition between the gap and core material can be very sharp, it is not guaranteed that the magnetic transition will be as sharp. Possible reasons for this are:
–   local stresses due to mechanical surface treatments (grinding, polishing),
–   chemical interactions before the gap is formed,
–   diffusion of gap material into the core [5.13].

It seems almost impossible to describe these effects theoretically. Rajeva et al. [5.13] attempted to do so for a glass-gap ferrite head by assuming that the widening of the non-magnetic gap due to only the diffusion equals the integral of the glass concentration in the ferrite, i.e. the total glass content. However, the non-magnetic gap length 'must' be defined by

$$\boxed{\; g \equiv \int_{\text{gap area}} \frac{1}{\mu_r(x)}\, dx \;}, \qquad (5.50)$$

as will be shown later on (5.57). Only for very high glass concentrations, say > 0.5, there is reason to believe that $1/\mu_r(x)$ approximates the glass content (or porosity). Rajeva et al. [5.13] measured by means of a scanning-electron microscope (SEM) a clear increase of the diffusion length and the glass concentration in the diffusion layer with the time of residence on the glass-bonding temperature (580°C), 140°C above the softening temperature of the glass. They do not, however, make any comparison of their assumed gap widening, proportional to the glass content in the ferrite, with the actual gap widening, e.g. by measuring the shift of the gap-null wavelength too.

The influence of mechanical surface treatments on the (local) stresses is described by Knowles [5.14]. For stresses in a glass-bonded head see Tang [5.15].

It is not our aim to describe one of the above-mentioned phenomena, but rather to correlate the actual degradation in the output of a head with assumed gradual-permeability profiles $\mu_r(x)$ caused by one or more of those phenomena or the like.

5.7.2 Strategy

The relation between the deep-gap field and the (x component of the) permeability simply reads

$$H_{dg}(x) = \frac{B_{dg}}{\mu_0 \mu_r(x)}. \tag{5.51}$$

Since $\mathbf{\nabla} \cdot \mathbf{B} = 0$, B_{dg} cannot change over distances that are small compared to the track width and gap height. As a consequence $H_{dg}(x)$ has the same profile as $1/\mu_r(x)$.

The x-component of the field at the head surface $y = -d_4$ will differ from the deep-gap field since outside the head the permeability is μ_0 instead of zero. In an exact analytical calculation it is not possible to take this into account. Only in ideal circumstances ($\mu_{core} = \infty$, $\mu_{gap} = \mu_{upper\ half\text{-}space} = $ constant) can use be made of conformal transformations, in particular the Schwarz-Christoffel transform. Therefore one of the following analytical approaches will be used.

5.7.3 Karlqvist-like approach

Assume that the deep-gap field H_{dg} is present at the head surface. Then for different assumptions concerning $1/\mu_r(x)$ or $h_{dg}(x)/\eta N$ ($= H_{dg}(x)/(\eta NI)$), the *GLF* will be determined with the aid of (5.18), written as

$$GLF(k) = \int_{x_l}^{x_r} \frac{h_{dg}(x)}{\eta N} e^{-jkx} dx. \tag{5.52}$$

Because of the definition of the efficiency, see (5.17), the functions must satisfy the condition

$$GLF(k = 0) = \int_{x_\ell}^{x_r} \frac{h_{\mathrm{dg}}(x)}{\eta N} \, \mathrm{d}x = 1. \qquad (5.53)$$

These functions only need to describe satisfactorily the field in the neighbourhood of the gap, not infinitely far away. Thus, in contrast to the field of an actual head, for which $\int_{-\infty}^{+\infty} h_x \mathrm{d}_x = 0$, h_{dg} can be chosen such that

$$\eta \equiv \int_{x_\ell}^{x_r} \frac{h_{\mathrm{dg}}(x)}{N} \, \mathrm{d}x = \int_{-\infty}^{\infty} \frac{h_{\mathrm{dg}}(x)}{N} \, \mathrm{d}x, \qquad (5.54)$$

as will be done. Consequently the integration interval in (5.52 + 53) can be extended to $\pm \infty$. The actual deep-gap field has in common with the actual sense field that (5.20) holds for video heads and frequencies, so that η and $GLF(k)$ are meaningful quantities (see Sect. 5.3).

The above strategy is in fact a generalization of the Karlqvist approximation, in which the head's sense field is approximated by the deep-gap field of a semi-infinite head. Therefore the subscript dg will be changed to the K of Karlqvist.

5.7.4 Realistic approach

For the correction in the realistic GLF, corresponding to the actual sense field of the head (i.e. the field at the surface of the head, which deviates a little from the deep-gap or Karlqvist field) the same correction can be chosen in the k-domain as follows for the deep-gap or Karlqvist field.

It is convenient to choose the zero points in the gap loss function of the assumed ideal Karlqvist head GLF_{iK}, to coincide with those in the gap loss function of the gradual Karlqvist head (GLF_{gK}).

When the fields of the gradual Karlqvist heads are restricted to those that can be written as a convolution of the ideal Karlqvist field $h_{\mathrm{iK}}(x)$ and some other finite correction function $f_{\mathrm{c}}(x)$, this will always be fulfilled, since then Fourier transformation yields

$$GLF_{\mathrm{gK}}(k) = GLF_{\mathrm{iK}}(k) \cdot f_{\mathrm{c}}(k). \qquad (5.55)$$

Hence the correction function we require in the k-domain equals

$f_c(k)$. Since $h_{gK}(x)$ is (usually) symmetric, $f_c(x)$ must be chosen symmetric too. When $f_c(x) \to \delta(x)$, $GLF_{gK}(k) \to GLF_{iK}(k)$.

$GLF(k = 0) = 1$ for any GLF, see (5.53). This substituted in (5.55) gives

$$f_c(k = 0) = \int_{x_\ell}^{x_r} f_c(x)\,\mathrm{d}x = 1, \qquad (5.56)$$

Hence for the correction function any suitable 'GLF' can be chosen.

5.7.5 The η and SF of a gradual gap compared to an idealized gap

Substitution of (5.51) into (5.53) yields

$$\eta = \frac{B_{dg}g}{\mu_0 NI} \qquad (5.57)$$

with

$$g \equiv \int_{x_\ell}^{x_r} \frac{1}{\mu_r(x)}\,\mathrm{d}x,$$

the generalized definition of the gap length. So defined, η retains its usual form as known for a sharp-gap head. In addition, heads with equal g but different $1/\mu_r(x)$ profiles keep the same gap reluctance, R_g, since

$$R_g = \frac{1}{A_g} \int_{x_\ell}^{x_r} \frac{1}{\mu_0 \mu_r(x)}\,\mathrm{d}x, \qquad (5.58)$$

where A_g is the gap surface.

Thus when, except for the $1/\mu_r(x)$ profile, heads with equal g are similar (i.e. equal permeabilities and geometry elsewhere), then the flux efficiencies η_Φ are equal and the (field) efficiencies η are equal (this is carried out in more detail in Sect. 8.1.5.2). This is precisely the case when heads with different gradual gaps are constructed from the same sharp-gap head by way of the correction functions. So η is not changed and consequently the sensitivity function $SF_{gK}(= \eta \cdot GLF_{gK})$ of an actual gradual-gap head follows by multiplying the sensitivity function of the same head having the assumed ideal gap, SF_{iK}, by the proper $f_c(k)$.

Next the correction function can be applied to others than the *GLF* or *SF* of the ideal Karlqvist head. For example, it can be applied to the actual *GLF* of an ideal semi-infinite head as given by Westmijze; see (5.37), or its approximation (5.45).

Summarizing, we have

$$\boxed{GLF_g(k) \simeq GLF_i(k) \cdot f_c(k)} \qquad (5.59a)$$

$$\boxed{SF_g(k) \simeq SF_i(k) \cdot f_c(k)} \qquad (5.59b)$$

$$\boxed{\eta_g = \eta_i.} \qquad (5.59c)$$

The above approach means that it is assumed that the actual field of a head having a disturbed (gradual) deep-gap field, described by the convolution of the ideal *deep-gap* field with a (symmetric) correction function $f_c(x)$, is approximated well by the convolution of the sense field at the *surface* of an ideal-gap head with the correction function $f_c(x)$ of the deep-gap field. It is equivalent to saying that the changes in the frequency spectrum of the head's sense field, due to the gradual $\mu_r(x)$ profile, approach those of the deep-gap field.

5.7.6 Correction functions

Suitable symmetric functions that fulfil (5.56) are listed below.

a) *The rectangle function*

$$f_c(x) = \frac{1}{\Delta g} \, \Pi(x/\Delta g) \equiv \begin{cases} 1/\Delta g & |x| \leq \Delta g/2 \\ \\ 0 & |x| > \Delta g/2 \end{cases} \qquad (5.60)$$

with standard deviation $\sigma = \Delta g/2$ defined by the variance $\sigma^2 = \int_{-\infty}^{\infty} f_c(x)x^2 dx$, and mean width Δg, defined as $\left(\int_{-\infty}^{\infty} f_c(x)dx\right)/f_c(0)$. Fourier transformation yields

$$f_c(k) = \text{sinc}(\Delta g/\lambda). \tag{5.61}$$

Fig. 5.9 shows the convolution with the ideal Karlqvist field and curve a in Fig. 5.10 gives the Fourier transform of the correction function. As shown by curve a in Fig. 5.11, $\mu_r(x)$ increases hyperbolically.

b) *The triangle function*

$$f_c(x) = \frac{1}{\Delta g} \Lambda(x/\Delta g) \equiv \begin{cases} (1 - |x|/\Delta g)/\Delta g & |x| \le \Delta g \\ 0 & |x| > \Delta g \end{cases} \tag{5.62}$$

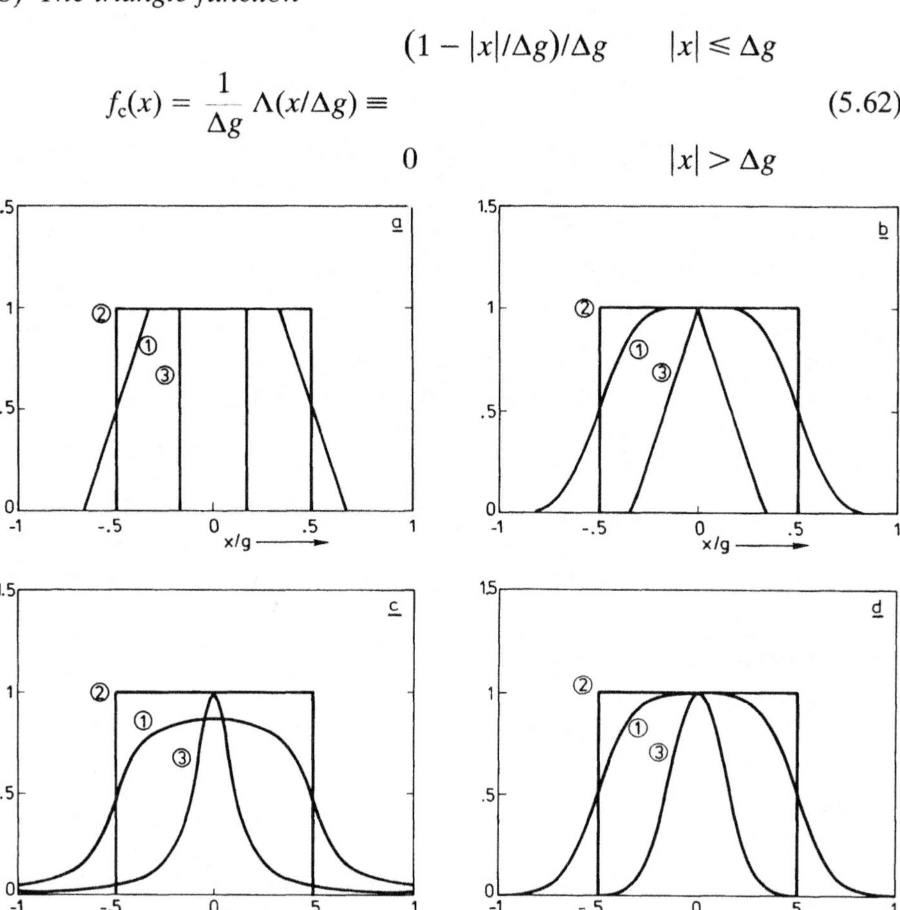

Fig. 5.9. Four models for a gradual Karlqvist 'field' h_{gK} (curves 1) composed by way of a convolution of the ideal Karlqvist 'field' h_{iK} (curves 2) with a correction function f_c (curves 3). All mean widths are chosen $\Delta g = g/3$.
a. Rectangle correction function
b. Triangle correction function
c. Arctan gradual gap
d. Gaussian correction function
curves 1: plot of the function $g \cdot h_{gK}(x)/\eta N = 1/\mu_r(x)$
curves 2: plot of the function $g \cdot h_{iK}(x)/\eta N$
curves 3: plot of the function $\Delta g \cdot f_c(x)$

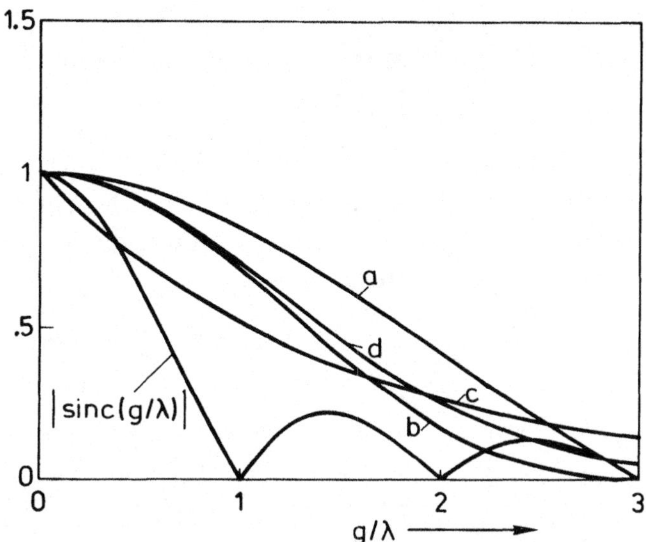

Fig. 5.10. The gradual-gap loss factor $f_c(k)$ for the different models. Curves a, b, c and d correspond to those in Fig. 5.9. Multiplication of curve a, b, c or d with $\mathrm{sinc}(g/\lambda)$ gives the desired gradual gap loss function GLF_g

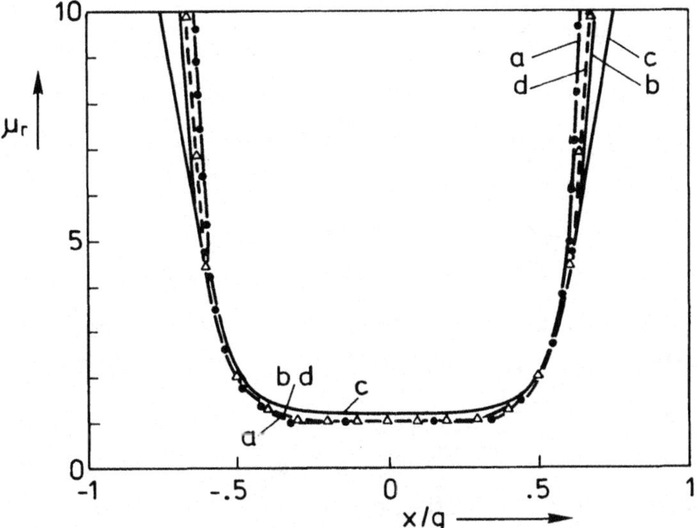

Fig. 5.11. The permeability in the gap region (x-component) corresponding to the assumptions concerning the gradual deep-gap (Karlqvist) field. The permeabilities in a, b, c and d correspond to the gradual deep-gap fields in Fig. 5.9 a, b, c and d.

with standard deviation $\sigma = \Delta g/\sqrt{6}$ and mean width Δg. This function just equals the convolution of the above rectangle with itself, hence

$$f_c(k) = \mathrm{sinc}^2(\Delta g/\lambda). \tag{5.63}$$

See Fig. 5.9b and Figs. 5.10 and 5.11 curve b. The permeability changes more gradually than in the case of the rectangle function. This causes less high-frequency output.

c) *The arctan model*

$$f_c(x) = \frac{1}{\pi} \cdot \frac{\Delta g/\pi}{x^2 + (\Delta g/\pi)^2} \tag{5.64}$$

with standard deviation $\sigma = \infty$ and mean width Δg. The Fourier transformation follows from a contour integration in the lower half z-plane $(z = x + jy)$:

$$f_c(k) = e^{-k\Delta g/\pi}. \tag{5.65}$$

See Fig. 5.9c and Figs. 5.10 and 5.11 curve c. The reason for the name 'arctan model' becomes clear in the application of this model, see Sect. 5.7.9.

d) *The Gaussian function*

$$f_c(x) = \frac{1}{\Delta g}\, e^{-\frac{1}{2}\{x/(\Delta g/\sqrt{2\pi})\}^2} \tag{5.66}$$

with standard deviation $\sigma = \Delta g/\sqrt{2\pi}$ and mean width Δg. Fourier transformation yields

$$f_c(k) = e^{-(k\Delta g)^2/4\pi}. \tag{5.67}$$

See Fig. 5.9d and Fig. 5.10 curve d.

More functions follow e.g. by mutual and self-convolutions of these functions.

The first two correction functions will introduce extra zero points in the GLF_{gK}, in contrast to the last two. Since $\Delta g \ll g$ in most circumstances, these zero points appear at frequencies far above the first gap-null frequency.

5.7.7 Comparison with finite-element solution

In order to check the reliability of the proposed analytical approach, we also carried out a numerical calculation for one rather extreme case. For this purpose the numerical program package Maggy 2 was used [5.16] which employs the finite-element method to solve two-dimensional magnetostatic problems. The gradual gap is sketched in Fig. 5.12a by way of the $1/\mu_r(x)$ curve. Over the total gap 52 meshes were used. The mesh density was chosen largest near to the 'geometrical' gap (0.3 μm) edges. Near to these edges the residues of the singularities of a sharp gap are visible. In the y direction just above and below the head surface about the same mesh density was chosen. The field was determined at 1/2 mesh distance, i.e. 0.00225 μm, above the head surface. With the aid of a fast-Fourier transform (FFT) the corresponding GLF was calculated. Since the FFT we used required field values at 2^n equidistant points, where n is an arbitrary integer, first a splinefunction routine was used to generate these points. (By the spline function all first and second derivatives are continuous.) This numerically obtained

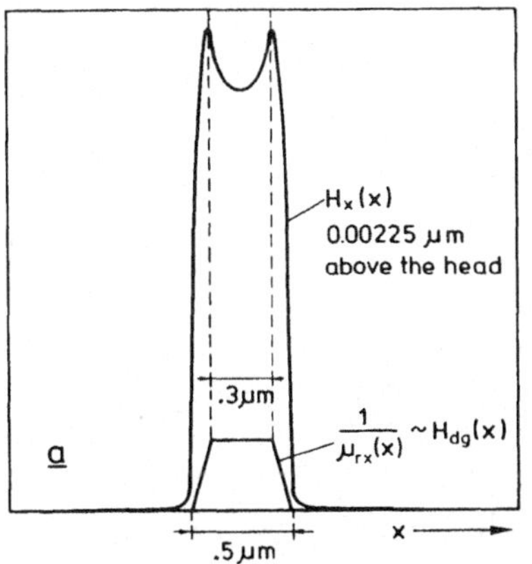

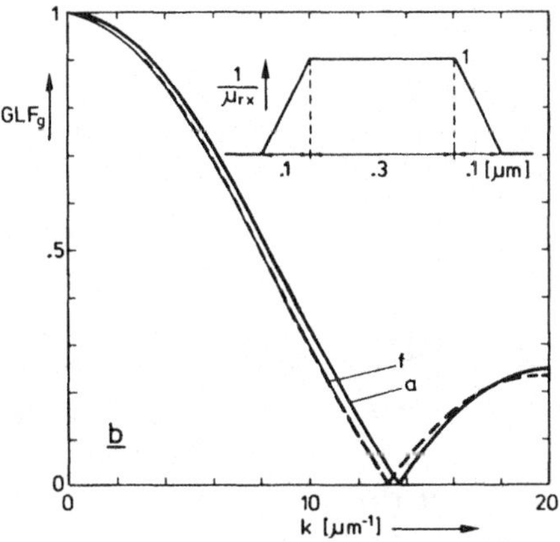

Fig. 5.12. Verification of the proposed analytical approach.
a) Field just above the head surface obtained by a finite-element method. The assumed deep-gap field $H_{dg}(x)$ is drawn on a different scale.
b) curve f: GLF corresponding to finite-element solution.
curve a: Analytical approximation of GLF.

GLF is plotted in Fig. 5.12b curve f after a correction for the head-tape distance of $0.00225\ \mu m$ is made. Curve a gives the result of the analytical approach according to (5.59a), with substitution of the correction function (5.61) and the accurate approximation (5.45) for the *GLF*. The similarity is very satisfactory.

5.7.8 Application to scratched gap surfaces

Model 'a' is suited in calculations of the effects of compressive stress due to microscopic scratches on the gap surfaces when the magnetostriction constant $\lambda < 0$.

According to measurements of Knowles [5.14] this compressive stress σ_c decreases approximately linearly with the depth, d, into the material, at least for the Mn-Zn-Ti ferrite he considered. The local effects are only relevant when the induced anisotropy energy $\lambda\sigma_c$ strongly exceeds the crystal anisotropy energy $K \equiv \frac{1}{2}\mu_0 H_{an} M_s$. When $\lambda < 0$, the gap plane then becomes an easy plane and flux transport to the gap takes place by rotation magnetization with a relative permeability

$$\mu_{rx} = M_s/H_{an}^i \tag{5.68}$$

where the induced anisotropy field H_{an}^i is related to the stress-induced anisotropy energy $\lambda\sigma_c$ by

$$\lambda\sigma_c = \tfrac{1}{2}\mu_0 H_{an}^i M_s \tag{5.69}$$

Thus

$$\mu_{rx} = \frac{(\mu_0 M_s)^2}{2\lambda\sigma_c\mu_0}. \tag{5.70}$$

Knowles found an approximately linear decreasing compressive stress σ_c with the depth d after diamond polishing $\sigma_c = 600 - 120\ d/10^{-6}$ [MPa], until it reaches zero at a depth d of $5\ \mu m$ in this approximation. (In fact the stress perpendicular to the grinding direction reported by Knowles was 20% higher, and parallel to the grinding direction 20% lower than given by this expression). For $\lambda \approx -0.8 \times 10^{-6}$ (for simplicity assumed constant in the gap plane) and $\mu_0 M_s = 0.5\,T$, (5.70) leads to $\mu_{rx} = 2$ at the surface and $\mu_{rx} \rightarrow \infty$ at $5\ \mu m$ depth.

Gap-surface machining of present-day video heads (Syton polishing) is much more friendly than Knowles' diamond polishing. However, an approximately linear dependence of compressive stress with depth may still be expected and thus $h_{dg}(x) \sim 1/\mu_{rx}(x)$ decreases linearly with the depth into the core.

This means that the effects of stresses on the *GLF* caused by scratches can be described by the superposition of two *GLF*s:

- a gradual GLF_g according to model 'a' with

  $g = g_{opt} + \Delta g$

  $\Delta g = $ stress depth

  $g_{opt} = $ 'optical' or mechanical length of the gap;

- an ideal GLF_i with $g = g_{opt}$ and $\Delta g = 0$, see Fig. 5.13.

The ratio between the two *GLF*s corresponds to the ratio between the surface of the two sensitivity functions, while the total surface must be normalized to one, according to (5.53). Hence

$$GLF = \frac{(1 + \Delta g/g_{opt})GLF_g + (\mu_{rmin} - 1)GLF_i}{\Delta g/g_{opt} + \mu_{rmin}} . \tag{5.71}$$

For different $\Delta g/g_{opt}$ and μ_{rmin} values the GLF is plotted in Fig. 5.14. The curve for $\mu_{rmin} = 3$ and $\Delta g/g = 1.5$ corresponds to the situation depicted in Fig. 5.13.

5.7.9 Discussion and application of the arctan model

Among the different models for the gradual gap, the model with the infinite standard deriviation, i.e. the 'arctan' model c, shows the most dramatic effects in the wavelength range of interest $0 < g/\lambda < 0.5$.

The convolution of $f_c(x)$ with the ideal Karlqvist field (normalized to about 1 at $x = 0$) $gh_{iK}/\eta N = 1/\mu_r(x)$ yields

$$\frac{g \cdot h_{gK}(x)}{\eta N} = \frac{1}{\pi} \int_{-\infty}^{\infty} \frac{(g \cdot h_{iK}(x')/\eta N) \cdot \Delta g/\pi}{(x' - x)^2 + (\Delta g/\pi)^2} \, dx' \tag{5.72}$$

$$= \frac{1}{\pi} \left(\arctan \frac{x + g/2}{\Delta g/\pi} - \arctan \frac{x - g/2}{\Delta g/\pi} \right).$$

The relation between h_{gK} and h_{iK} as expressed by this equation equals the relation between the same field components on $\Delta g/\pi$-differing distances in the free half-space above a head; see (a5.10) in App. 5.1. This is consistent with the result (5.65). Moreover, because of the fixed (Hilbert-transform) relation between y and x components in the free half-space (see (a5.15)), the y-component of the field also equals the y-component of the ideal-gap head at a $y = \Delta g/\pi$ larger distance.

The complete equivalence with a head at a $\Delta g/\pi$ larger distance makes this choice very suitable for a description of the effects of a gradual gap on both the recording and playback. This description will be given

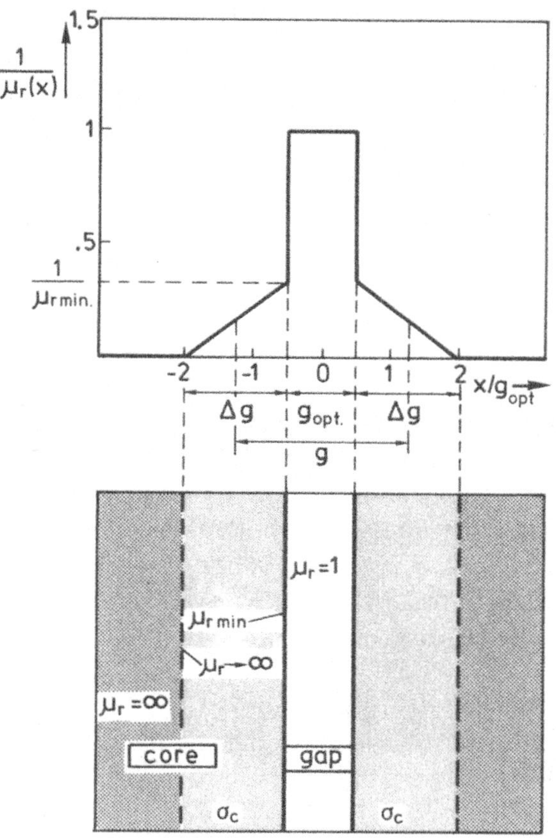

Fig. 5.13. The $1/\mu_r(x)$ profile due to compressive stress after (diamond) polishing of the surface.

in chapter 7. The choice of this arctan model for the gradual gap was inspired by the success of Miyata and Hartel [5.17] with this approximation for a transition in pulse-recording calculations.

5.7.10 Exponential gradual gap

The deep-gap field of an exponential gradual gap is proportional to the $1/\mu_{rx}(x)$ curve defined in Fig. 5.15a. Outside the optical gap, μ_{rx} increases according to $\mu_{rx} = \exp\big((|x|-g_{opt}/2)/(\Delta g/2)\big)$. The gap length g is defined in (5.50). Here Δg is defined as the gap widening relative

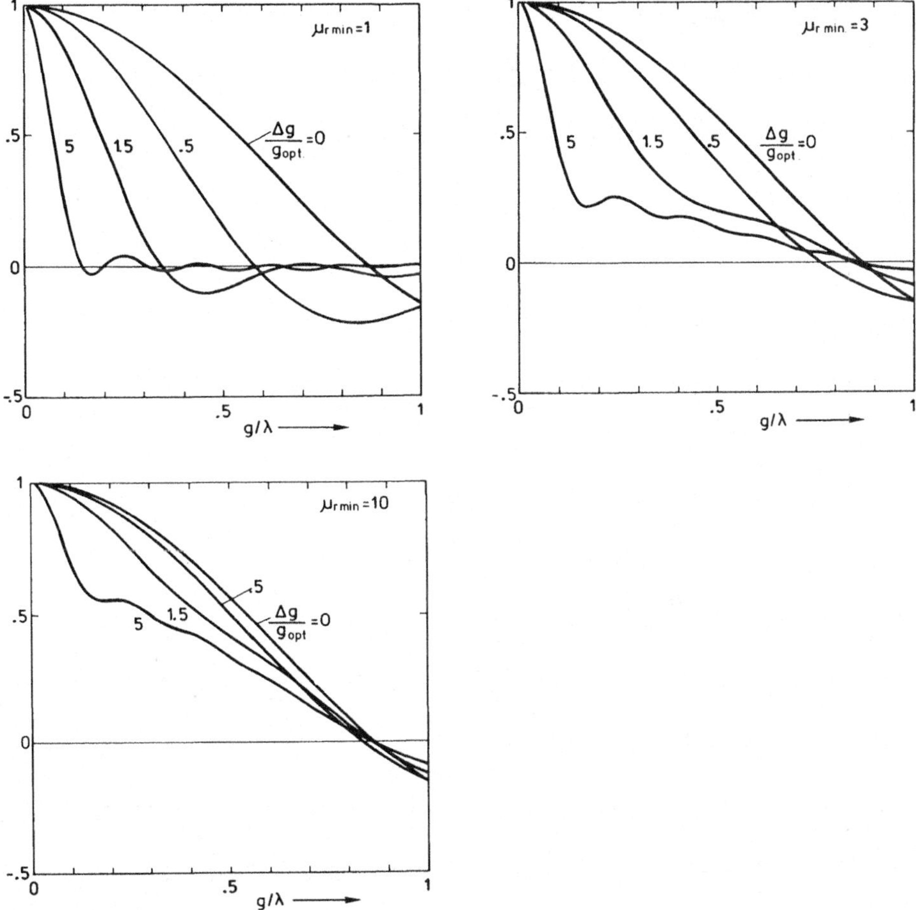

Fig. 5.14. *GLF*s according to (5.71), when the accurate approximation (5.45) for a sharp gap without interaction with a non-unit relative permeability tape is used for *GLF*$_i$ (and for the *GLF*$_i$ included in *GLF*$_g$ too). See also Fig. 5.13, which corresponds to the curve for $\mu_{rmin} = 3$ and $\Delta g/g_{opt} = 1.5$.

to g_{opt}, i.e. $\Delta g = g - g_{opt}$. In the other models a-d, where it was possible to describe the gap with the aid of correction functions in contrast to the present case, Δg was defined as the mean width of that correction function, which does not necessarily equal the gap widening. However, observation of Fig. 5.9, while realizing that $\Delta g = g/3$ was chosen there, shows that choosing $g_{opt} \simeq 2g/3$ is reasonable, especially for the triangle function, where other assumptions are unrealistic.

Hence Δg can, for most practical purposes, be interpreted as the above defined gap widening.

Fourier transformation of the properly normalized $1/\mu_{rx}(x)$ curve, defined by the $\mu_{rx}(x)$ profile above, leads to the following ('Karlqvist') *GLF* for this gradual gap, see also the derivation of (5.100) later on:

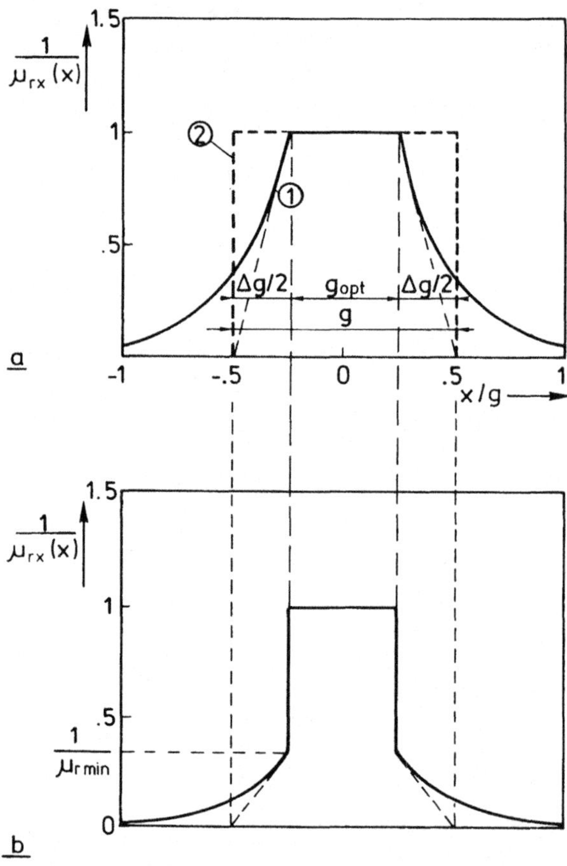

Fig. 5.15. The $1/\mu_{rx}(x)$ profile of the exponential gradual gap; $\mu_{rx}(x) = \mu_{rmin}e^{(|x| - g/2)/(\Delta g/2)}$.
a) $\mu_{rmin} = 1$
b) $\mu_{rmin} > 1$.

$$GLF_g = \frac{1}{g}\left[g_{opt}\,\text{sinc}\!\left(\frac{g_{opt}}{\lambda}\right) + \frac{\Delta g}{1 + (k\Delta g/2)^2}\cos\!\left(\frac{kg_{opt}}{2}\right)\right.$$

$$\left. - \frac{0.5\,k(\Delta g)^2}{1 + (k\Delta g/2)^2}\sin\!\left(\frac{kg_{opt}}{2}\right)\right]. \tag{5.73}$$

When $\Delta g \ll g_{opt}$ it must be expected that the accuracy of this expression enhances when instead of the first ('Karlqvist') term $\text{sinc}(g_{opt}/\lambda)$ the Westmijze expression (5.37) or its accurate approximation (5.45) is used.

The first term corresponds to the rectangular part of the (Karlqvist) field, see Fig. 5.15a. The *sum* of the two other terms corresponds to the remaining exponentially decaying parts and follows after straightforward calculations.

For the following applications the $1/\mu_r(x)$ function is extended to the one depicted in Fig. 5.15b, analogously as in the 'application of model a'. Everything that is said there, in particular the expression for the resulting GLF (5.71), consequently holds. This extended model can be used when the gap widening is caused by diffusion of gap material into the surfaces of the core that are in contact with the gap as described in Sect. 5.7.1. Δg is related to the diffusion length.

Also in case of stresses relaxating from the gap, caused by a difference in expansion coefficients of (hot-pressed) gap and magnetic core or by different lattice constants in case of sputtered gaps or gap-claddings, the assumption of an exponentially-increasing permeability approaches reality. Δg is then related to the relaxation length which is in turn related to the track width W. Since $W \gg \lambda_{max}$ with λ_{max} the longest wavelength of interest in video recording, this works out in an efficiency loss, see the description of the metal-in-gap head (MIG) in chapter 12 and the calculations concerning efficiency losses due to stresses near the gap in section 8.1.6.3.

5.7.11 Interaction with the tape

When there is interaction with the tape (i.e. μ_2 and/or $\mu_3 \neq 1$), the corresponding GLF^a of the sharp-gap head can be corrected as proposed in Sect. 5.7.4.

However, for a particular choice of the gradual gap, namely the arctan approximation, see section 5.7.6, a probably better approach

will be used. The sense field of the gradual Karlqvist head is then
completely equivalent to that of the ideal Karlqvist head at a larger
distance, as explained in Sect. 5.7.9. (This equivalent distance is inversely proportional to the steepness of the gap edges.) In this case it is
reasonable to assume that this equivalence also holds quite well when
there is interaction with the tape (may be this can be shown to be
exact). This means that:
–   everywhere in the reproduction-loss expression d_4 must be increa-
    sed by the equivalent distance;
– GLF^a is the gap-loss function of the ideal head at a distance 'd_4 +
    equivalent distance' from the non-unit relative permeability tape.

This choice leads to a slightly smaller gap-null wavelength than the
first proposal (Sect. 5.7.4), because the distance to the tape is larger
(see in this context also section 5.6).

In the case of no head-tape interaction it is completely equivalent to
the first proposal.

In the phenomenological description of the write process in chapter
7 the introduction of a gradual gap only requires a simple adaptation
of the recording model if the arctan approach is chosen.

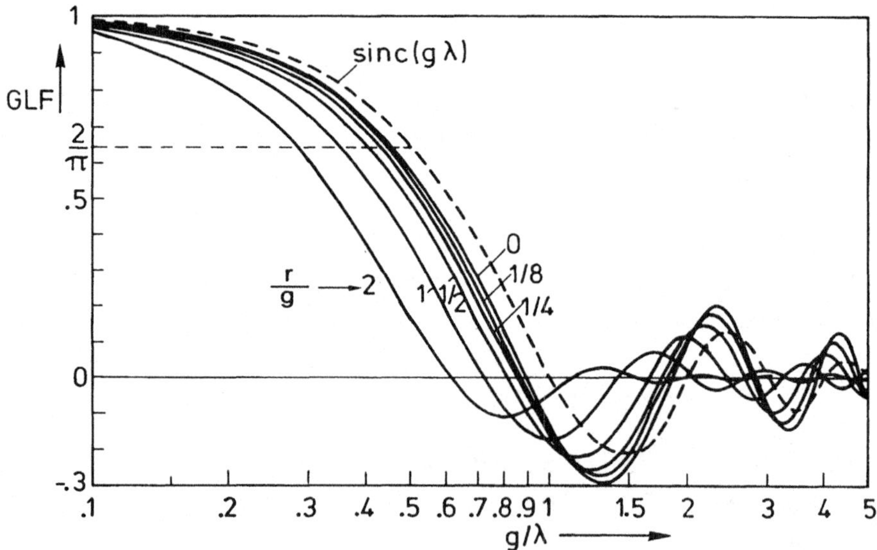

Fig. 5.16. The *GLF*, after Duinker, of a head with rounded gap edges for different ratios of radius
of curvature *r* to gap length *g*.

5.8 Rounded gap-edges

For another kind of non-ideal gap, namely one with rounded gap edges, but further $\mu_{\text{core}} = \infty$, it is possible to use the Schwarz-Christoffel transform. This has been carried out by Duinker [5.18]. The resulting complicated *GLF* contains integrals that cannot be written in terms of elementary functions. The numerical results, after Duinker, are shown in Fig. 5.16.

5.9 Irregular gaps

5.9.1 Introductory remarks

In Fig. 5.17 different types of gap irregularities are shown. In 'a' the irregularities on both sides of the gap are in phase. So the gap length is constant and only the gap-centre position changes erratically. In 'b' the irregularities on both sides of the gap, described by the coordinates x_1 and x_2 in Fig. 5.18, are assumed to be independent (statistically). The gap length then changes simultaneously with the gap-centre position. In the following calculations of the gap-irregularity loss (*GIL*) it is assumed that the changes in the gap length and gap-centre positions are so gradual with respect to the gap length that the fringing field and so the problem can be treated two-dimensionally.

5.9.2 Case 'a': in phase irregularities

When $\Delta x(z)$ describes the variations in the gap centre position and $p(\Delta x)$ is the probability-density function, i.e. $p(\Delta x)\mathrm{d}\Delta x$ is the probability of finding the gap centre position between Δx and $\Delta x + \mathrm{d}\Delta x$ and $\int_{-\infty}^{\infty} p(\Delta x)\mathrm{d}\Delta x = 1$, then (4.8) results in

$$\Phi(t) = W \int_{-\infty}^{\infty} \mathscr{B}_{0y}^{\text{b}}(x-vt)\left[\int_{-\infty}^{\infty} p(\Delta x)\psi^{\text{a}}(x-\Delta x)\mathrm{d}\Delta x\right]\mathrm{d}x, \tag{5.74}$$

where the argument y_{p} is omitted for simplicity.

First applying the cross-correlation theorem, by analogy with the

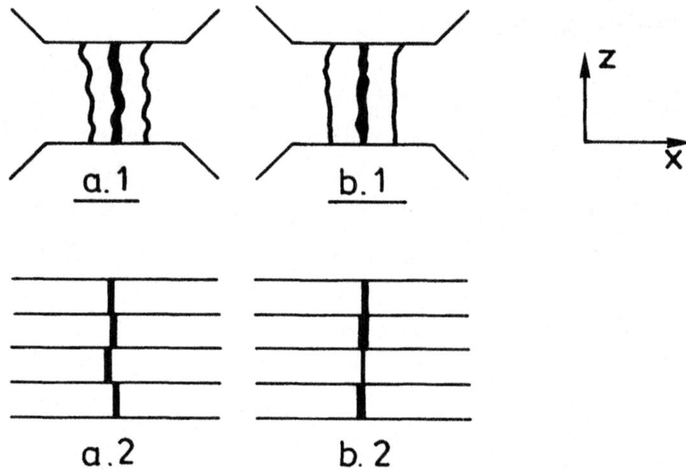

Fig. 5.17. Different kinds of gap irregularities.
a. Deviations from straightness of the gap centre, described in case 'a'.
b. Deviations from straightness of the gap-magnetic material interface, described in case 'b'.
1. Ferrite or metal-in-gap (video) head.
2. Laminated amorphous-ribbon (video) head.

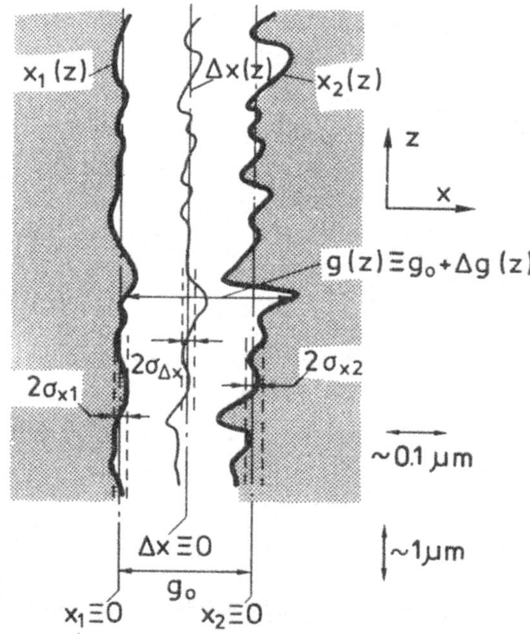

Fig. 5.18. Irregular gap due to statistically-independent irregularities on both sides of the gap (x scale is compressed) as used in the calculations.
x_1, x_2 and Δx describe the interface and gap-centre positions, all relative to their average positions. The σ's are the standard deviations of the above quantities.
g is the gap length and $\Delta g(z)$ the deviation at position z from the average gap length g_0.

derivation of (4.13), and afterwards the convolution theorem to the function of x between the brackets, yields

$$\Phi(\omega) = \frac{W}{v}\, p(k)\, \mathscr{B}_y^{b*}(k, t = 0)\, \psi^a(k) \,, \qquad (5.75)$$

where $k \equiv \omega/v$.

For heads showing dispersion, the argument t in $\psi^a(\boldsymbol{r}, t)$, see appendix 4.1, can not be omitted in (5.74) and leads to $\psi^a(k, \omega)$ in (5.75), see also the difference between (4.16) and (4.13).

So probability-density functions used for the gap position have the same characteristic ($\int_{-\infty}^{\infty} p(\Delta x)\mathrm{d}\Delta x = 1$) and work out the same (output multiplied by $p(k)$) as the correction functions in the description of the gradual gaps!

Application with Gaussian probability densities

For a Gaussian probability function, the following gap-irregularity loss factor *GIL* equal to (5.67) with standard deviation σ substituted for $\Delta g/\sqrt{2\pi}$ results:

$$GIL = \mathrm{e}^{-(k\sigma)^2/2} \,. \qquad (5.76)$$

This is a well-known result, derived by Mallinson [5.19] in a less general way. For other choices of probability functions the results of the models a, b and c in section 5.7.6 can directly be used.

When a separate write head with an irregular gap is used extra phase errors are introduced by the write head. For independent irregularities in write and read head the total gap irregularity losses are found by adding, quadratically, the standard deviations, i.e. multiplying both loss factors.

5.9.3 Case 'b': independent irregularities

For video heads that have small track widths, the above described gap irregularity can be neglected. Only for a laminated head, see Fig. 5.17a.2, is case 'a' more or less possible. For ferrite and metal-in-gap

(MIG) video heads there are more likely to be irregularities of the type 'b.1', since both gap surfaces are machined separately before the gap is formed. In Fig. 5.18 this situation is sketched in more detail, and quantities used in the further calculations are defined. In this case of no dependence between x_1 and x_2 (defined as in (5.80) between Δx and Δg instead of x_1 and x_2) it can easily be calculated that the standard deviation $\sigma_{\Delta g}$ of the gap-length variations Δg is $\sqrt{2}$ times the standard deviation of the gap-surface irregularities σ (when equal irregularities on both sides are assumed, $\sigma \equiv \sigma_{x1} = \sigma_{x2}$). This factor is $1/\sqrt{2}$ for the gap-centre variations Δx.

Because Δg and Δx are of the same origin, namely the roughnesses x_1 and x_2 as also shown by the following expressions,

$$\Delta g = x_2 - x_1 \tag{5.77a}$$

$$\Delta x = \frac{x_2 + x_1}{2}, \tag{5.77b}$$

a close relationship can be expected between Δx and Δg. How close this is will turn out to be dependent on the particular choice for the probability density describing the irregular gap surfaces. The flux expression reads in this general case

$$\Phi(t) = W \iint_{-\infty}^{\infty} p(\Delta x, \Delta g) \left\{ \int_{-\infty}^{\infty} \mathscr{B}_{0y}^{b}(x - vt) \right.$$

$$\left. \times \psi^{a}(g_0 + \Delta g, x - \Delta x) dx \right\} d\Delta x d\Delta g, \tag{5.78}$$

where the dependence of ψ^{a} on the gap length is now expressed by the first argument of ψ^{a}. For the definition of $p(\Delta x, \Delta g)$ see note [5.26].

Also generally, when $p(x_1) = p(x_2) \equiv p(x)$,

$$p(\Delta g) = \int_{-\infty}^{\infty} p(x)\,p(x + \Delta g)dx \tag{5.79a}$$

$$p(\Delta x) = \int_{-\infty}^{\infty} p(x)p(2\Delta x - x)dx. \tag{5.79b}$$

Only when it turns out, for a particular choice of $p(x)$, that Δg and Δx are independent of each other, then

$$p(\Delta x, \Delta g) = p(\Delta x)p(\Delta g) \tag{5.80}$$

by definition (or $p(\Delta x/\Delta g) = p(\Delta x)$).

This expression simplifies the evaluation of the integral (5.78), but even without this the integral can be simplified drastically by Fourier transforming (5.78).

First rearranging terms gives

$$\Phi(t) = \int_{-\infty}^{\infty} \mathcal{B}_{0y}^{b}(x-vt)\, f(x)\mathrm{d}x \tag{5.81a}$$

with

$$f(x) = \iint_{-\infty}^{\infty} p(\Delta x, \Delta g)\psi^{a}(g_0 + \Delta g, x-\Delta x)\mathrm{d}\Delta x\, \mathrm{d}\Delta g. \tag{5.81b}$$

Fourier transformation, using the cross-correlation theorem, leads to

$$\boxed{\Phi(\omega) = \frac{1}{v}\, \mathcal{B}_{y}^{b*}(k, t = 0)f(k)} \tag{5.82a}$$

while, according to the convolution theorem,

$$f(k) = \int_{-\infty}^{\infty} p(k, \Delta g)\psi^{a}(g_0 + \Delta g, k)\mathrm{d}\Delta g \tag{5.82b}$$

with $\psi^{a}(g_0 + \Delta g, k) \equiv j\eta^{a}N \cdot GLF^{a}(g_0 + \Delta g, k)/k$ according to (5.2) and (5.19). The efficiency of a head with $\sigma_{\Delta g} = 0$ equals the efficiency of the present head with $\sigma_{\Delta g} \neq 0$ because of equal gap reluctance since $\Delta g(z) = 0$ on average for the head with $\sigma_{\Delta g} \neq 0$. Therefore, the total loss factor due to irregularities reads

$$GIL(k) = \frac{\displaystyle\int_{-\infty}^{\infty} p(k, \Delta g)\, GLF^{\mathrm{a}}(g_0 + \Delta g, k)\, \mathrm{d}\Delta g}{GLF^{\mathrm{a}}(g_0, k)}. \qquad (5.83)$$

This factor has singularities at wavenumbers k_i where the 'ideal' $GLF^{\mathrm{a}}(g_0, k)$ is zero.

Application with Gaussian probability densities

Fortunately, and rather unexpectedly, for the most practical choice of $p(x_1)$ and $p(x_2)$, namely Gaussian functions, it will turn out that Δx is (statistically) independent of Δg! Another advantage of using the Gaussian probability-density function is its steep decrease which means that negative values of the gap length are very improbable and so need not be excluded in the calculations. For other choices, such as the functions of the models a, b and c, it follows that Δx and Δg are interdependent. These results may change in the case of relations other than those given in (5.77a+b) between Δx or Δg and x_1 and x_2. Only the consequences of the Gaussian function will be analysed here.

First $p(\Delta x, \Delta g)$ must be calculated. This can be done by substituting the inverse relations (compared to (5.77) or $(\Delta x, \Delta g) \equiv \mathbf{A}(x_1, x_2)$)

$$x_1 = \Delta x - \Delta g/2 \quad \text{and} \quad x_2 = \Delta x + \Delta g/2$$

or
$$(x_1, x_2) = \mathbf{A}^{-1}(\Delta x, \Delta g) \qquad (5.84)$$

into $p(x_1, x_2)$. In order to let the new function again be a probability density, i.e. $\iint_{-\infty}^{\infty} p(\Delta x, \Delta g)\mathrm{d}\Delta x\mathrm{d}\Delta g = 1$, it must be multiplied by the determinant of $\mathbf{A}^{-1}$, so

$$p(\Delta x, \Delta g) = |\mathbf{A}^{-1}|\, p\big(\mathbf{A}^{-1}(\Delta x, \Delta g)\big) \qquad (5.85)$$

with $\mathbf{A}^{-1} = \begin{pmatrix} 1 & -\tfrac{1}{2} \\ 1 & +\tfrac{1}{2} \end{pmatrix}$ and $|\mathbf{A}^{-1}| = 1$.

As a result

$$p(\Delta x, \Delta g) = p(\Delta x - \Delta g/2, \Delta x + \Delta g/2). \tag{5.86}$$

When $\sigma_{x1} = \sigma_{x2}$, $p(x_1) = p(x_2) \equiv p(x)$ equals

$$p(x) = \frac{1}{\sigma\sqrt{2\pi}} e^{-\frac{1}{2}\left(\frac{x}{\sigma}\right)^2} \tag{5.87}$$

and x_1 and x_2 are independent, then

$$p(x_1, x_2) = p(x_1)p(x_2) = p^2(x). \tag{5.88}$$

Applied in (5.86) this easily leads to

$$p(\Delta x, \Delta g) = \left(\frac{1}{\sigma\sqrt{2\pi}}\right)^2 e^{-\frac{1}{2\sigma^2}\left[(\Delta x - \Delta g/2)^2 + (\Delta x + \Delta g/2)^2\right]}$$

$$= \frac{e^{-\frac{1}{2}\left(\frac{\Delta x}{\sigma_{\Delta x}}\right)^2}}{\sigma_{\Delta x}\sqrt{2\pi}} \cdot \frac{e^{-\frac{1}{2}\left(\frac{\Delta g}{\sigma_{\Delta g}}\right)^2}}{\sigma_{\Delta g}\sqrt{2\pi}} \tag{5.89a+b}$$

with $\sigma_{\Delta x} = \sigma/\sqrt{2}$ and $\sigma_{\Delta g} = \sigma\sqrt{2}$, which is consistent with the general statements mentioned in the text above (5.77).

Now $p(\Delta x)$ and $p(\Delta g)$ follow with the aid of the general relations

$$p(\Delta x) = \int_{-\infty}^{\infty} p(\Delta x, \Delta g)\,\mathrm{d}\Delta g \tag{5.90a}$$

$$p(\Delta g) = \int_{\infty}^{\infty} p(\Delta x, \Delta g)\,\mathrm{d}\Delta x \tag{5.90b}$$

and just equal the two terms in (5.89b). Since obviously $p(\Delta x, \Delta g) = p(\Delta x)p(\Delta g)$, Δx is (statistically) independent of Δg. Use of the relations (5.79a + b) leads to the same results.

It is clear from (5.89a) that only special relations between (x_1, x_2)

and $(\Delta x, \Delta g)$ as e.g. in (5.84, 86 or 77) lead to this result. When for instance $\sigma_{x1} \neq \sigma_{x2}$, Δx is dependent on Δg. This can also be seen from Fig. 5.18, where $\sigma_{x1} \neq \sigma_{x2}$ was chosen. When another probability function is chosen than the Gaussian, e.g. the function (5.64) of the arctan model 'c', an interdependence between Δx and Δg is found already for $\sigma_{x1} = \sigma_{x2}$.

Since $p(\Delta x, \Delta g) = p(\Delta x)p(\Delta g)$, expression (5.83) simplifies to

$$GIL(k) = \frac{p(k) \int_{-\infty}^{\infty} p(\Delta g)\, GLF^{\mathrm{a}}(g_0 + \Delta g, k)\, \mathrm{d}\Delta g}{GLF^{\mathrm{a}}(g_0, k)}. \tag{5.91}$$

The Fourier transformation $p(k)$ of the Gaussian $p(\Delta x)$ reads, analogously to (5.67),

$$p(k) = e^{-(k\sigma_{\Delta x})^2/2} = e^{-(k\sigma)^2/4}. \tag{5.92}$$

The losses related only to this factor $p(k)$ are smaller than those described in (5.76), where the irregularities on both sides of the gap were assumed to be in phase and thus phase errors were maximum.

Substitution of the expression for $p(\Delta x)$ in (5.91) leads to the total GIL including gap loss effects

$$GIL(k) = \frac{e^{-(k\sigma)^2/4}}{2\sigma\sqrt{\pi}\, GLF^{\mathrm{a}}(g_0, k)} \int_{-\infty}^{\infty} e^{-(\Delta g/\sigma)^2/4} GLF^{\mathrm{a}}(g_0 + \Delta g, k)\, \mathrm{d}\Delta g. \tag{5.93}$$

Even for the simplest GLFs this integral must be calculated numerically, possibly after writing it first into a series expansion.

In the range of interest, $\lambda/g_0 > 1.6$ and $\sigma/g_0 > 0.25$, the enhanced gap losses (less than 1 percent) can be disregarded. Hence

$$\boxed{GIL(k) \simeq e^{-(k\sigma)^2/4}} \tag{5.94}$$

which gives considerably smaller losses than in case 'a' (in dB's half as large), where the irregularities on both sides of the gap were in phase. A complete numerical evaluation of (5.93) shows a considerable shift of the gap null to higher wavenumbers when σ/g_0 increases and a disappearance of the gap null for $\sigma/g_0 \gtrsim 0.3$.

5.9.4 Summary and practical considerations

In decibels ($-20\,dB = 0.1$) the gap-irregularity loss factor for Gaussian probability densities with standard deviation σ reads

$$GIL = -343\ C(\sigma/\lambda)^2 \qquad (5.95)$$

where $C \simeq \frac{1}{4}$ for independent gap-edge irregularities
          $C = \frac{1}{2}$ for in phase gap-edge irregularities
          C is twice as large when the gap of the writing head has the
          same irregularities.

The result is plotted in Fig. 5.19. The loss is negligible for small values of σ/λ, becomes noticeable ($>1\,dB$) for $\sigma/\lambda > 1/(18.5\sqrt{C})$, and is extremely large ($>10\,dB$) for $\sigma/\lambda > 1/(5.9\sqrt{C})$.

Deviations from the straightness of the gap centre, i.e. in phase gap-edge irregularities, can usually be neglected over small track widths. For video heads, with the exception of laminated heads, this effect is therefore always negligible. The independent gap edge irregularities

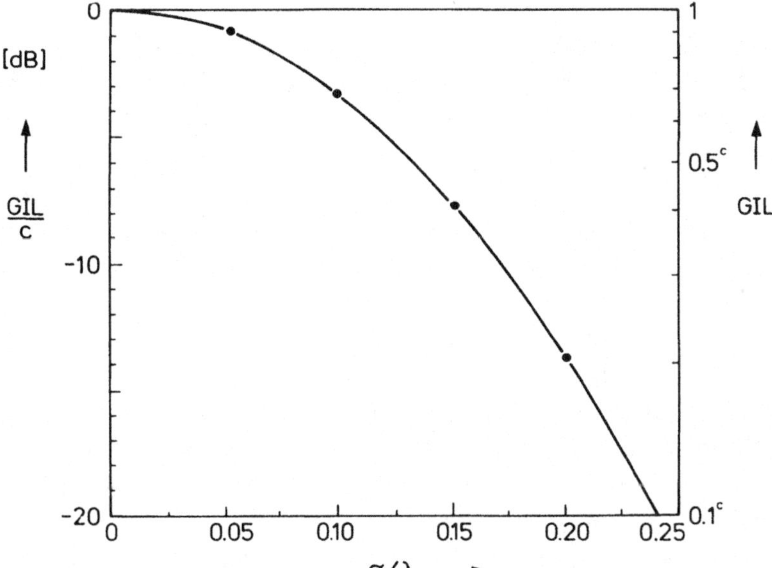

Fig. 5.19. Gap-irregularity loss according to (5.95). The corresponding logarithmic scale is given on the right. The irregularities are assumed to have periodicities that are large compared to the gap length. σ is the standard deviation of the Gaussian distribution describing the irregularity.

however do not depend on the track width. Particularly in video head recording experiments they may play a role because of the very small gap length and hence relatively large roughness of the gap edges compared to the smallest wavelengths applied in experiments. For instance, when the write and read head contribute to the final phase errors, such that $C = 2 \times \frac{1}{4} = \frac{1}{2}$, then 1 dB loss is already obtained for $\sigma/\lambda = \frac{1}{13}$ and 10 dB loss at a wavelength three times smaller.

5.10 Thin-film head

In contrast with conventional video heads as described in Section 5.4, the effect(s) of the outer edge(s) of a thin-film head (TFH) cannot be neglected. There are two main reasons for this:
–   The pole tip length is of the same order as wavelengths of interest, i.e. not very much larger than the gap length.
–   The contact between head and tape at the outer edges of the head is comparable with the contact at the gap region.

5.10.1 General considerations concerning its potential

In Fig. 5.20a a single-turn inductive TFH is schematically drawn. The thin-film poles have 'length' p_1 and p_2 and in between there is a gap with length g. Within the thickness of the poles the magnetic potential changes are negligible. Deep in the gap a cross section of the winding, carrying a current I (with arbitrary direction) perpendicular to the paper, is drawn. The return current flows in the opposite direction at a distance h_w. The positive current direction is chosen in the $-z$ direction.

A cut $(-\cdot-\cdot-)$ is necessary to make the magnetic scalar potential single valued, as in chapter 3. This cut is chosen perpendicular to the flux lines through the winding, such that the lines just to the left and right of the cut are equipotential lines. At the left side the dimensionless potential ψ_3 is chosen to be $-\frac{1}{2}$; this choice and the previous choices of the positive current direction and the cut mean that the dimensionless potential is defined everywhere and independent of the value and sign of I. For instance at the right side of the cut $\psi^a = -\frac{1}{2} - \oint(H/I)\mathrm{d}r = +\frac{1}{2}$. At distances far away compared to the diameter of the winding and the height of the head, $|r| \gg h_w, h_h$, the potential reaches zero.

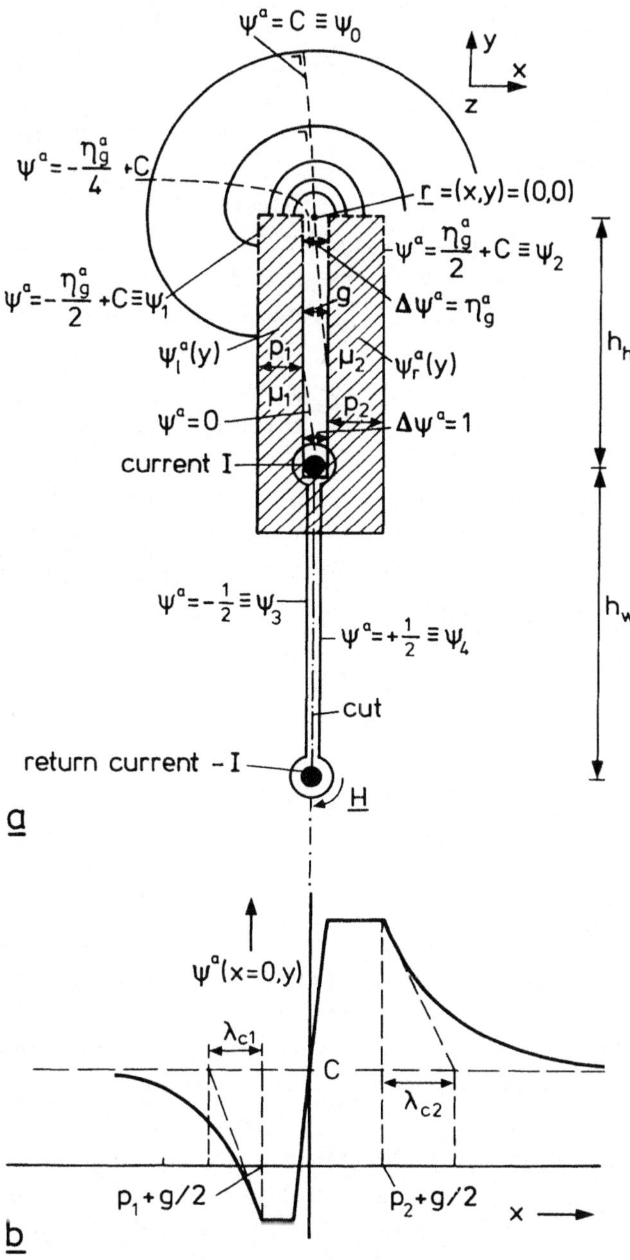

Fig. 5.20. TFH and its potential distribution ψ^a. $N = 1$ for simplicity.
a. Schematic drawing of a TFH ($h \ggg p_1 + p_2 + g$). The flux lines ($-$) near the pole tips approach circular arcs. Some equipotential lines ($----$) are sketched. Because of the limited efficiency and the asymmetry of the head, the equipotential line $\psi^a = C$ through $\underline{r} = \boldsymbol{0}$ deviates slightly from the $x = 0$ line.
b. Sketch of the potential of the TFH at pole-tip level.

The difference between the dimensionless magnetic scalar potential in the right film and in the left film, $\Delta\psi^a = \psi_r^a - \psi_\ell^a$, deep in the gap equals 1. Since the thin-film head (p_1 or p_2 and $g \gg h_h$) acts like a transmission line (see chapter 11 for a detailed description), $\Delta\psi^a$ drops in the positive y direction, from 1 at $y = -h_h$ to η_g^a (by definition) at $y = 0$. The parameter that characterizes the steepness of this decay in transmission lines is the 'characteristic distance', λ_c. For the head this value is called λ_{ch}. For every distance equal to λ_{ch} in the positive y direction $\Delta\psi^a$ drops by a factor e, reflections at $y = 0$ being neglected. This potential drop is a result of the decrease of $|\psi_r^a|$ as well as $|\psi_\ell^a|$ with y. The decrease of $|\psi_\ell^a|$ will however be larger than that of $|\psi^a|$ when $p_1\mu_1 < p_2\mu_2$. This means that the potential in the middle of the gap $\psi^a(x = 0, y)$ increases with y. Hence the equipotential line $\psi^a = C$ through $r = \mathbf{0}$ deviates slightly from the $x = 0$ line and consequently the flux lines do not cross the $x = 0$ plane exactly perpendicularly. Near the winding $\psi^a(x = 0, y = -h_h) = 0$ and at the pole tip $\psi^a(x = 0, y = 0) = C > 0$. For symmetrical heads where $p_1\mu_1 = p_2\mu_2$ and for heads with $\eta = 1$ or $\mu = \infty$, $C = 0$.

The potential in the middle of the gap and the potential at the surfaces of (and within) the left and right pole can be treated as being the *constants*, $\psi_0 = C$, $\psi_1 = -\eta_g^a/2 + C$ and $\psi_2 = \eta_g^a/2 + C$ respectively, in calculations of the field and potential at distances from the origin which are small compared to λ_{ch}.

The flux lines approach circular arcs at distances $|r|$ in the region $p + g/2 \ll |r| \ll \lambda_{ch}, h_h$, where $p \equiv \frac{1}{2}(p_1 + p_2)$. Hence the potential at pole-tip level, $\psi^a(x, y = 0)$, approaches in the above defined region at the left side of the head $(\psi_0 + \psi_1)/2 = C - \eta_g^a/4$ and at the right side of the head $(\psi_0 + \psi_2)/2 = C + \eta_g^a/4$.

The lengths $p_1 + g/2$ and $p_2 + g/2$ are evidently measures of the decay of the potential $\psi^a(x, y = 0)$ near the outer edges of the pole tips, i.e. $x \leqslant -(p_1 + g/2)$ and $x \geqslant p_2 + g/2$. The exact decrease of this potential (for $\mu = \infty$) is found by conformal mapping for $p_1 = p_2 \equiv p$ by Potter [5.20] and for the general case including $p_1 \neq p_2$ by Ichiyama [5.21].

When we fit our exponential decays for $p_1 = p_2 \equiv p$ with Potter's hyperbolic decay by choosing the initial steepnesses equal, then we find exponential decays with characteristic distances $\lambda_{c1} = \lambda_{c2} = p + g/2$. Ichiyama found in the case $p_1 \neq p_2$ decays proportional to $p_1 + g/2$ and $p_2 + g/2$ on the left and right side of the pole tips respectively. Hence we have $\lambda_{c1} = p_1 + g/2$ and $\lambda_{c2} = p_2 + g/2$. If λ_{ch} is not very much greater than $p_1 + g/2$ and $p_2 + g/2$, the potential decay over the height of the head has to be taken into account and the potential can no longer be calculated by conformal mapping. When $\lambda_{ch} \lesssim h_h$, then the efficiency of the head $\eta \ll 1$ and the response (GLF) to harmonics with wavelengths comparable to λ_{ch} are influenced by the decay of ψ_ℓ^a and ψ_r^a.

The potential $\psi^a(x, y = 0)$, at distances $|r|$ from the origin which are large compared to h_h but small compared to h_w (if this region exists), approaches $(\psi_3 + 0)/2 = -\frac{1}{4}$ at the left side and $(\psi_4 + 0)/2 = +\frac{1}{4}$ at the right side of the head.

Summarizing the approximations and asymptotic values for the potential at pole tip level $(y = 0)$ gives: $\psi^a(x, 0) =$

$$
\psi^a(x, 0) =
\begin{cases}
0 & \text{if } -x \gg h_w \\[4pt]
-\frac{1}{4} & \text{if } h_h \ll -x \ll h_w \\[4pt]
-(N\eta_g^a/4)[e^{+(x + p_1 + g/2)/\lambda_{c1}} + 1] + C & \text{if } \min(\lambda_{ch}, h_h) \gg -x \geqslant p_1 + g/2 \\[4pt]
-N\eta_g^a/2 + C & \text{if } g/2 \leqslant -x \leqslant p_1 + g/2 \\[4pt]
(x/g)N\eta_g^a + C & \text{if } |x| \leqslant g/2 \\[4pt]
+N\eta_g^a/2 + C & \text{if } g/2 \leqslant +x \leqslant p_2 + g/2 \\[4pt]
+(N\eta_g^a/4)[e^{-(x - p_2 - g/2)/\lambda_{c2}} + 1] + C & \text{if } \min(\lambda_{ch}, h_h) \gg +x \geqslant p_2 + g/2 \\[4pt]
+\frac{1}{4} & \text{if } h \ll +x \ll h_w \\[4pt]
0 & \text{if } +x \gg h_w,
\end{cases}
\tag{5.96}
$$

where $\min(\lambda_c, h_h)$ means the smallest of the two. In (5.96) $\lambda_{c1} \simeq p_1 + g/2$ and $\lambda_{c2} \simeq p_2 + g/2$ if the height of the head h_h and the characteristic distance of the head $\lambda_{ch} = \sqrt{\mu_{r1}p_1 g \times \mu_{r2}p_2 g/(\mu_{r1}p_1 g + \mu_{r2}p_2 g)}$ are both much larger than $p_1 + g/2$ and $p_2 + g/2$. In the expression for λ_{ch}, derived in chapter 11, μ_r is the *relative* permeability.

5.10.2 Approximation of potential for video applications

The maximum wavelengths of interest in video recording, $\lambda_{max} \approx 5$ µm, are usually smaller than h_h and λ_{ch}. For the response on harmonics of the signal with these wavelengths the incompletely defined potentials in the regions where $|x| \gtrsim \min(\lambda_{ch}, h_h)$ are thus of no importance. Since the constant C occurs in *all* the remaining relevant regions, C can be omitted. Another important point is that the track width is usually larger than the maximum wavelength of interest, such that our two-dimensional approach suffices. Hence for video purposes the potential can be approximated by

$$
\psi^{a}(x, 0) =
\begin{cases}
-(N\eta_{g}^{a}/4)[e^{+(x + p_1 + g/2)/\lambda_{c1}} + 1] & \text{if } -x \geq p_1 + g/2 \\[4pt]
-N\eta_{g}^{a}/2 & \text{if } g/2 \leq -x \leq p_1 + g/2 \\[4pt]
(x/g)N\eta_{g}^{a} & \text{if } |x| \leq g/2 \\[4pt]
+N\eta_{g}^{a}/2 & \text{if } g/2 \leq +x \leq p_2 + g/2 \\[4pt]
+(N\eta_{g}^{a}/4)[e^{-(x - p_2 - g/2)/\lambda_{c2}} + 1] & \text{if } +x \geq p_2 + g/2. \quad (5.97)
\end{cases}
$$

When the thin-film head is processed on a ferrite substrate [5.22], then $p_1 \rightarrow \infty$ and (5.97) simplifies to

$$
\psi^{a}(x, 0) =
\begin{cases}
-N\eta_{g}^{a}/2 & \text{if } -x \geq g/2 \\[4pt]
(x/g)N\eta_{g}^{a} & \text{if } |x| \leq g/2 \\[4pt]
+N\eta_{g}^{a}/2 & \text{if } g/2 \leq +x \leq p + g/2 \\[4pt]
+(N\eta_{g}^{a}/4)[e^{-(x - p - g/2)/\lambda_{c}} + 1] & \text{if } +x \geq p + g/2,
\end{cases}
$$
$$(5.98)$$

where $p_2 \equiv p$, and $\lambda_{c} \simeq p + g/2$ if h_{h} and $\lambda_{ch} = \sqrt{\mu_{r}pg} \gg p + g/2$.

This requirement is about equivalent to $\mu_{r} \gg p/g$ since $g < p$ in all practical heads.

For all known TFH configurations ($p/g = 2 - 25$, $\mu_{r} > 500$ and $h \geq 20\,\mu\text{m}$) the different requirements previously mentioned are fulfilled. Thus for video and other high density purposes the potentials stated in (5.97) for a TFH processed on a non-magnetic substrate and stated in (5.98) for a TFH processed on a magnetic substrate suffice.

5.10.3 Approximation of *SF* for video applications

The impulse response and sensitivity function can now be calculated by using the (alternative) reciprocity theorem. On the basis of the hyperbolic approximation of Potter, and using the usual reciprocity theorem, the impulse response has been calculated analytically by Labunov et al. [5.23] (published in Russian). The resulting expression however is very lengthy. They report a good agreement between the theoretical impulse response (isolated magnetic transition) and experimental data. The potential stated in (5.97) or (5.98) based on the exponential approximation delivers simple expressions for the sensitivity functions. With the aid of (2.3), (4.11), (5.7), (5.97) and a few mathematics, we find

$$SF = -\eta_g^a \left[\frac{\sin(kg/2)}{kg/2} - \tfrac{1}{4}\left(\frac{e^{jk(p_1 + g/2)}}{1 - jk\lambda_{c1}} + \frac{e^{-jk(p_2 + g/2)}}{1 + jk\lambda_{c2}} \right) \right] . \qquad (5.99)$$

This complex expression relaxes for symmetrical heads with $p_1 = p_2 \equiv p$ and $\lambda_{c1} = \lambda_{c2} \equiv \lambda_c \simeq p + g/2$ to the real valued SF:

$$SF = -\eta_g^a \left(\frac{\sin(kg/2)}{kg/2} - \tfrac{1}{2} \frac{\cos(k(p + g/2)) - k(p + g/2)\sin(k(p + g/2))}{1 + k^2(p + g/2)^2} \right)$$

$$(5.100)$$

and for a TFH processed on a magnetic substrate, i.e. $p_1 \rightarrow \infty$, $p \equiv p_2$ and $\lambda_c \equiv \lambda_{c2}$:

$$SF = -\eta_g^a \left(\frac{\sin(kg/2)}{kg/2} - \tfrac{1}{4} \frac{e^{jk(p + g/2)}}{1 + jk\lambda_c} \right) . \qquad (5.101)$$

Equation (5.100) is consistent (for $p = 0$) with (5.73) (for $\Delta g = g_{opt}$), besides a factor $-\tfrac{1}{2}$ before the 2nd and 3rd term due to half as large and reversed exponential fields in the present case compared to those in Fig. 5.15a.

The sensitivity functions (SFs) of the asymmetrical heads given in (5.99) and (5.101) contain an imaginary part. This implies that, in contrast with symmetrical heads, these SFs generally do not contain zeros since the real and imaginary parts may both *only by accident* be zero at the same wavenumber k. Because the first term in (5.99) and (5.101) due to the gap usually predominates in the high frequency region where the gap nulls occur, *the effect of the asymmetry is a 'gap minimum' instead of a gap null.*

5.10.4 Work of others and final remarks

Equations (5.99) and (5.100) can also be distilled from a paper by Ichiyama [5.21]. However Ichiyama used in his calculations a 33% slower exponential decay for his potential near the head, and he also included a long-range exponential decay to the zero potential at infinity, which must be due to the presence of a return current. This extra term is only of importance at very long wavelengths. In addition this extra term closely depends on the location of the return current, as may be clear from Fig. 5.20; e.g. when $h_w \gg h_h$ is not valid no region exists where $\psi^a \to \frac{1}{4}$ as in region IV in Fig. 5.20. Ichiyama gives no information about the presence or location of a return current, and therefore his very long wavelength results cannot be discussed quantitatively. In any case, his very long wavelength response can only be valid for one particular location of the return current and, one value of η_g^a (1 in his case). The 33% smaller steepness and 23% higher asymptotic value of his approximated potential near the pole tips compared to Potter's (and thus ours) cannot be explained by another choice for the return current or another configuration far away from the tape surface. The difference is probably due to the inaccuracies of the different approximations.

Asymmetric, infinitely permeable recording heads have also been studied by Baker and Koziol [5.24]. They emphasize the reduction of the number of grids in numerical calculations of the Fourier transform of the potential $\Psi(k)$ from e.g. the potential distribution alongside the periphery of a two-dimensional head. First they use a boundary integral method to find the potential in the region outside the head. In a recent article Baker [5.25] used the method developed in [5.24] in an investigation of the effects of a sublayer on the frequency and pulse response of an asymmetric, infinitely permeable TFH. He shows good correspondence with the experimental frequency response and reasonable correspondence of the theoretical pulse response with the vertical field component. Uniform magnetization distributions through the thickness of the coating were assumed. The fact that the above simplifications (uniform magnetization and infinite head permeability or $\eta = 1$) already resulted in good correspondences, for long wavelength effects too, must be due to the long gap length of 0.87 μm of his experimental head, such that $\eta \to 1$. For video or other very high density purposes (short gap length) the long wavelength effects would not have been calculated accurately, but are usually of no interest. The correspondence of the high frequency response with theoretical results is so very good since it is not the Karlqvist approximation that is used for the potential in the gap region but the exact potential at pole-surface level.

In conclusion: in spite of the reported successes for infinitely permeable heads, it is necessary for a general description of short- and long-wavelength effects of TFHs with finite permeability to include the efficiency of the head and the location of the return current or location of the magnetoresistive element in the case of a magneto-resistive head (see chapter 11). In the case of high tape permeability it can directly be concluded from Fig. 5.20 that the efficiency η_g^a decreases and the potential decay at the corners of the pole tips is steeper. The reason is that a tape with a high-permeability coating or backlayer will tend to form a constant potential surface (zero when $p_1 = p_2$ because of symmetry). The potential at pole surface level just beside the poles is then forced to change quickly to the potential in the tape (see also section 5.6) where in the limit of $\mu_r \to \infty$ and $d_4 \to 0$ delta (δ) functions occurred.

Appendix 5.1 Field relationships and Hilbert transforms

The imaginary part $V(x, y)$ of the complex potential

$$W(z) = U(z) + jV(z) \text{ with } z \equiv x + jy \tag{a5.1}$$

is chosen to represent the two-dimensional potential in the upper-half plane (vacuum, $y > 0$) above the two-dimensional head, where W is stated to be analytic, i.e. $dW/dz = \partial W/\partial x = \partial W/\partial(jy)$, or

$$\frac{dW}{dz} = \frac{\partial U}{\partial x} + j\frac{\partial V}{\partial x} = \frac{1}{j}\left(\frac{\partial U}{\partial y} + j\frac{\partial V}{\partial y}\right). \tag{a5.2}$$

Comparing real and imaginary parts and introducing $\mathbf{H} = -\nabla V$, we have

$$H_x \equiv -\frac{\partial V}{\partial x} = \frac{\partial U}{\partial y} \tag{a5.3}$$

$$H_y \equiv -\frac{\partial V}{\partial y} = -\frac{\partial U}{\partial x}. \tag{a5.4}$$

Cauchy-Riemann's differential equations are thus obtained.

Another differentiation yields Laplace's differential equation in vacuo:

$$\frac{\partial^2 V}{\partial x^2} + \frac{\partial^2 V}{\partial y^2} = 0, \tag{a5.5}$$

so that the use of V as static potential in the 'empty' space above the head is justified.

The 'trick' now is to consider the following two integrals (one with the $+$ and one with the $-$ sign)

$$\oint \frac{dW(z')}{dz'} \left(\frac{1}{z' - x - jy} \pm \frac{1}{z' - x + jy} \right) dz' \tag{a5.6a}$$

along a closed contour (in contour clockwise direction), consisting of the real axis and supplemented by a semicircle in the upper halve plane, which for $y > 0$ delivers according to Cauchy

$$2\pi j \left(\frac{dW(z')}{dz'} \right)_{z' \equiv x' + jy' = x + jy}. \tag{a5.6b}$$

Since $dW/dz = -H_y - jH_x$ (see (a5.2), (a5.3) and (a5.4)), and $H_{x,y} = \mathcal{O}(R^{-2})$ as $R \to \infty$ in the present 2-dimensional approximation (see (d') below (3.14)), the integral along the semicircle equals $\mathcal{O}(R^{-2})$ as $R \to \infty$. Hence (a5.6a) and (a5.6b) lead to

$$\int_{-\infty \atop (y'=0)}^{\infty} \frac{dW(z')}{dz'} \left(\frac{1}{x' - x - jy} \pm \frac{1}{x' - x + jy} \right) dx' = 2\pi j \left(\frac{dW(z')}{dz'} \right)_{x' = x \atop y' = y}.$$

$$\tag{a5.7}$$

Substitution of $dW/dz = -H_y - jH_x$ according to (a5.2)-(a5.4) gives in the case of the $+$ and $-$ signs in (a5.7) respectively:

$$\int_{-\infty}^{\infty} \{-H_y(x', 0) - jH_x(x', 0)\} \frac{2(x' - x)}{(x' - x)^2 + y^2} dx'$$

$$= 2\pi \{-jH_y(x, y) + H_x(x, y)\} \tag{a5.8}$$

and

$$\int_{-\infty}^{\infty} \{-H_y(x',0) - jH_x(x',0)\} \frac{2jy}{(x'-x)^2 + y^2} \, dx'$$

$$= 2\pi\{-jH_y(x,y) + H_x(x,y)\}. \tag{a5.9}$$

By comparing real and imaginary parts we obtain four relations:
a) two relations between the same field components in different lines of constant y:

$$H_x(x,y) = \frac{y}{\pi} \int_{-\infty}^{\infty} \frac{H_x(x',0)}{(x'-x)^2 + y^2} \, dx' \tag{a5.10}$$

$$H_y(x,y) = \frac{y}{\pi} \int_{-\infty}^{\infty} \frac{H_y(x',0)}{(x'-x)^2 + y^2} \, dx' \tag{a5.11}$$

b) two relations between the x and y components in different or equal lines of constant y:

$$H_x(x,y) = -\frac{1}{\pi} \int_{-\infty}^{\infty} \frac{H_y(x',0)(x'-x)}{(x'-x)^2 + y^2} \, dx' \tag{a5.12}$$

$$H_y(x,y) = \frac{1}{\pi} \int_{-\infty}^{\infty} \frac{H_x(x',0)(x'-x)}{(x'-x)^2 + y^2} \, dx'. \tag{a5.13}$$

In the same line ($y = 0$) this results in the Hilbert transform relation between x and y components:

$$H_x(x,0) = -\frac{1}{\pi} \fint_{-\infty}^{\infty} \frac{H_y(x',0)}{x'-x} \, dx' \equiv \frac{H_y(x,0) * 1/x}{\pi} \tag{a5.14}$$

$$H_y(x,0) = \frac{1}{\pi} \fint_{-\infty}^{\infty} \frac{H_x(x',0)}{x'-x} \, dx' \equiv \frac{-H_x(x,0) * 1/x}{\pi}, \tag{a5.15}$$

where $*$ means convolution and the integrals are Cauchy principal value integrals.

According to the convolution theorem, or analogously to the derivation of (4.12), we obtain after Fourier transformation (operator $\mathcal{F}$):

$$h_x(k, 0) = \frac{1}{\pi}\, h_y(k, 0)\, \mathcal{F}\left\{\frac{1}{x}\right\} = -jh_y(k, 0)\,\mathrm{sign}(k). \qquad (a5.16)$$

This is consistent with (5.13) in the case of vacuum above the head.

When $dW(z')/dz'$ in (a5.6a) is replaced by $W(z')$ one finds instead of (a5.10)

$$V(x, y) = \frac{y}{\pi} \int_{-\infty}^{\infty} \frac{V(x', 0)}{(x' - x)^2 + y^2}\, dx'. \qquad (a5.17)$$

The other relations are not interesting.

Note that the $y = 0$ line can be arbitrarily chosen (in the upper-half plane) above the head!

References

[5.1]   See most mathematical handbooks on integral transforms, e.g.: Blok, H, *Integraal transformaties in de electrotechniek*, Delft University of Technology, Sept. 1973.

[5.2]   Herk, A. van, *Three-dimensional analysis of magnetic fields in recording head-configurations*, Thesis, Delft University of Technology, Department of Electrical Engineering, Delft, the Netherlands.

[5.3]   See e.g.: Carmichael, Robert D. and Smith, Edwin R., *Mathematical tables and formulas*, Dover Publications, Inc., New York.

[5.4]   Mallinson, J.C., *On recording head field theory*, IEEE Trans. Magn., Vol. Mag-10, 773-775 (1974).

[5.5]   Westmijze, W.K., *Field configuration around the gap and the gap-length formula*, Philips Res. Rep. 8, 161-183 (1953).

[5.6]   Fan, George J., *A study of the playback process of a magnetic ring head*, IBM Journal, Oct. (1961).
It should be noted that although Fan's method is sound, he made a mistake in evaluating the gap-null wavelength in the case of infinite tape permeability (and zero head-tape distance). As was noted by Bertram [5.12], his approximation formula underestimates the gap-null wavelength. In addition Fan's footnote 12: '*It should be noted that the shift of the gap null is not dependent on tape thickness and head/tape distance separation*', is quite incorrect (see section 5.6).

[5.7]   Middleton, B.K. and Davies, A.V., *Gap effects in head field distributions and the replay process in longitudinal recording*, Journal of the Institution of Electronic and Radio Engineers, Vol. 55, 111-118 (1985).

[5.8]   Baird, A.W., *An evaluation and approximation of the Fan equations describing magnetic fields near recording heads*, IEEE Trans. Magn., Vol. Mag-16, 1350-1352 (1980).

[5.9]   Szczech, Theodore J. et al., *Improved field calculations for ring heads*, IEEE Trans. Magn., Vol. Mag-19, 1740-1744 (1983).

[5.10] Oberhettinger, Fritz, *Fourier Transforms of Distributions and their Inverses*, Academic Press, New York and London, 1973, Table I number 18.

[5.11] Mangulis, *Handbook of Series for Scientists and Engineers*, Academic Press, New York and London, 1965, page 43.

[5.12] Bertram, H. Neal, *Dependence of reproducing gap null on medium permeability and spacing*, IEEE Trans. Magn., Vol. Mag-18, 893-897 (1982).

[5.13] Rajeva, T., Kodjabaschev, P., Marinov, I., *Beurteilung des Einflusses von Glasdiffusion auf die effektive Spaltbreite*, J. Signal AM 10 no. 3, 159-164 (1982), (in Russian). Translated into Dutch by J. Damen, Philips Research Laboratories, Eindhoven, The Netherlands.

[5.14] Knowles, J.E., *The origin of the increase in magnetic loss induced by machining ferrites*, IEEE Trans. Magn., Vol. Mag-11, 44-50 (1975).

[5.15] Tang, T., *Stress analysis of glass-bonded ferrite recording heads*, IBM J. Res. Dev., Vol. 18, 274-278 (1974).

[5.16] Polak, S.J., Beer, A. de, Wachters, A. and Welij, J.S. van., *Maggy 2, a program package for two-dimensional magnetostatic problems*, Int. J. Num. Meth. Engng., Vol. 15, 113-127 (1980).

[5.17] Miyata, J.J. and Hartel, R.R., *The recording and reproduction of signals on magnetic medium using saturation-type recording*, IRE Trans. Electr. Comp., 159-169 (1959).

[5.18] Duinker, S., *Short-wavelength response of magnetic reproducing heads with rounded gap edges*, Philips Res. Repts. 16, 307-322 (1961).

[5.19] Mallinson, J.C., *Gap irregularity effects in tape recording*, IEEE Trans. Magn., Vol. Mag-4, 71 (1968).

[5.20] Potter, Robert I., *Analytic expression for the fringe field of finite pole-tip length recording heads*, IEEE Trans. Magn., Vol. Mag-11, 80-81 (1975).

[5.21] Ichiyama, Y., *Reproducing characteristics of thin film heads*, IEEE Trans. Magn., Vol. Mag-11, 1203-1205 (1975).

[5.22] Druyvesteyn, W.F., van Ooyen, J.A.C., Postma, L., Raemaekers, E.L.M., Ruigrok, J.J.M., and de Wilde, J., *Magnetoresistive heads*, IEEE Trans. Magn., Vol. Mag-17, 2884-2889 (1981).

[5.23] Labunov, V.A., Giro, A.M., Mosolov, V.A., Shukh, A.M., *Analytical calculation of thin-film magnetic head output*, Doklady Akademii Nayk BSSR, 25 (10), 880-883 (1981).

[5.24] Baker, Bill R. and Koziol., Mark R., *Response of asymmetrical thin film heads to vertical and horizontal magnetization*, IEEE Trans. Magn., Vol. Mag-17, 3123-3125 (1981).

[5.25] Baker, Bill R., *Effects of a sublayer on readback from a vertical film*, J. Appl. Phys. 55(6), 2217-2219 (1984).

[5.26] All different probability-density functions are called p for simplicity's sake (as usual in information theory), although this is mathematically incorrect. The arguments of p make clear which probability density is meant. So $p(\Delta x, \Delta g)\mathrm{d}\Delta x\mathrm{d}\Delta g$ is the probability of finding the gap centre position between Δx and $\Delta x + \mathrm{d}\Delta x$ *and* the gap length between $g_0 + \Delta g$ and $g_0 + \Delta g + \mathrm{d}\Delta g$, while $p(\Delta x/\Delta g)\mathrm{d}\Delta x$ is the probability of finding the gap centre position between Δx and $\Delta x + \mathrm{d}\Delta x$ *when* $g = g_0 + \Delta g$.

Chapter 6

Special and limiting cases of the general flux expression

6.1 General flux expression

In this chapter we will start at the most universal result we have derived up to now for the flux. This flux expression follows after substituting the expression for $\hat{H}_\mathrm{f}$ (2.33) and $\hat{B}_\mathrm{y}$ (2.63) in the alternative reciprocity theorem in complex notation (5.5), while using definition (2.39), i.e. substitution of (2.40) into (2.33) and (2.32), and the definitions (2.34), (2.36) and (2.37) for real permeabilities. The result, when the 'reference' plane $y = y_\mathrm{p}$ is chosen at $y_\mathrm{p} = -d_4 - t_5$, is:

$$
\hat{\Phi}^\mathrm{R}(\omega) = \mathrm{j}\, \frac{h_\mathrm{x}^\mathrm{a}(k, -d_4 - t_5, \omega)}{k \cosh(\hat{\beta}_5^* k t_5)}
$$

$$
\times \frac{2\mu_0 W}{\left(1 + \dfrac{\tanh(\beta_5^* k t_5)}{\hat{\mu}_5^*}\right) \mathrm{e}^{k d_4} - R\left(1 - \dfrac{\tanh(\hat{\beta}_5^* k t_5)}{\hat{\mu}_5^*}\right) \mathrm{e}^{-k d_4}}
$$

$$
\times \left[\frac{k \sinh \alpha_3 \displaystyle\int_0^{t_3} \{-\mathrm{j}\hat{f}_{\mathrm{px}}(k, y) + \beta_3 \hat{f}_{\mathrm{py}}(k, y)\}\,\mathrm{d}y}{\genfrac{}{}{0pt}{}{\sinh}{\cosh}(\beta_3 k t_3 + \alpha'_{32} + \alpha_3)} \right] \tag{6.1}
$$

with $\hat{f}_{\mathrm{px}}(k, y) = \hat{M}_{\mathrm{px}}^*(k, y, t = 0)\,\genfrac{}{}{0pt}{}{\cosh}{\sinh}(\beta_3 k(t_3 - y) + \alpha'_{32})$

$$
\hat{f}_{\mathrm{py}}(k, y) = \hat{M}_{\mathrm{py}}^*(k, y, t = 0)\,\genfrac{}{}{0pt}{}{\sinh}{\cosh}(\beta_3 k(t_3 - y) + \alpha'_{32})
$$

$$\text{with} \quad R \equiv \frac{\genfrac{}{}{0pt}{}{\sinh}{\cosh}(\beta_3 kt_3 + \alpha'_{32} - \alpha_3)}{\genfrac{}{}{0pt}{}{\sinh}{\cosh}(\beta_3 kt_3 + \alpha'_{32} + \alpha_3)}$$

$$\alpha_3 \equiv \tfrac{1}{2} \ln \frac{\mu_3 + 1}{\mu_3 - 1}$$

$$\alpha'_{32} \equiv \tfrac{1}{2} \ln \left| \frac{\mu_{32} + 1}{\mu_{32} - 1} \right| = \genfrac{}{}{0pt}{}{\cosh^{-1}}{\tanh^{-1}}(\mu_{32})$$

$$\mu_{32} \equiv \frac{\mu_3}{\mu_2} \cosh(\beta_2 kt_2 + \alpha_2)$$

$$\alpha_2 \equiv \tfrac{1}{2} \ln \frac{\mu_2 + 1}{\mu_2 - 1},$$

where the upper functions are valid when $\mu_{32} > 1$ and lower functions are valid when $\mu_{32} < 1$. The ratio ω/k equals the head-tape velocity, v. The expression between square brackets represents the complex conjugate of the free space surface field $\hat{H}_f$. The results are quite general in the sense that the flux can be calculated for all kinds of permanent-magnetization distributions, choice of a backlayer ($t_2 \neq 0$) or not, and anisotropic permeability (restricted to anisotropy axes in the x and y directions). The limitations of the expression are that the magnetization in the coating must be composed of a permanent part and a reversible and linear part, the latter being responsible for all changes of magnetization during the read process. Retardation must be negligible and the magnetic materials of the tape must in addition be dispersion-free.

6.2 Coating variations

6.2.1 'Thick' media; $\beta_3 kt_3 \gg 1$

The perpendicular component of magnetic induction $\mathcal{B}_y(y = -d_4)$ at the head surface can be written in the limit of large $\beta_3 kt_3$ ($\gg 1$) with the aid of Eqs. (2.46), (2.52) and (2.33) as

$$\hat{\mathscr{B}}_{y}(-d_4) = \frac{2}{\mu_3 + 1} \; \frac{\mu_0 k\, e^{-kd_4}}{1 - \dfrac{\mu_3 - 1}{\mu_3 + 1}\, e^{-2kd_4}} \int_0^{t_3} (j\hat{M}_{\mathrm{px}} + \beta_3 \hat{M}_{\mathrm{py}})\, e^{-\beta_3 ky}\, dy \qquad (6.2)$$

(except for the hypothetical magnetization distribution that is sharply concentrated within a characteristic length $\lambda/(2\pi\beta_3)$ from the back of the coating). For t_3 one may read infinite as well. The influence of the backlayer disappeared.

From this magnetic induction and the alternative reciprocity theorem, or directly from the corresponding limiting case of the general flux expression (6.1), the read flux of the ideal zero-gap head can be calculated. The amplitude of this flux reads

$$\hat{\Phi}^{\mathrm{R}} = \frac{jWN}{k}\, \hat{\mathscr{B}}_y^*(-d_4). \qquad (6.3)$$

The result after substitution of (6.2) in (6.3) equals equation 16 in a paper by Bertram [6.1] (except for the factor N, due to our definition of $\hat{\Phi}^{\mathrm{R}}$, and except for the phase). Bertram indicates an increase in the reproduce flux due to the effect of the anisotropy β if $\beta < 1$, which causes reduced 'distance losses' in the coating. These effects were also included in our calculations in Section 2.6. The loss (in the present limit $\beta_3 kt_3 \to \infty$) in between a considered sinusoidally magnetized layer at y in the coating and a fixed head at distance d_4 is described by the usual $e^{-\beta_3 ky}$ in the tape and e^{-kd_4} outside the tape; see (6.2).

The factor $2/(\mu_3 + 1)$ in (6.2) describes the loss due to demagnetization when the ideally soft magnetic head is not active in the remagnetization process, i.e. $d_4 \to \infty$. The remaining factor $1/[1-(\mu_3 - 1)e^{-2kd_4}/(\mu_3 + 1)]$ is the remagnetization factor F_{R} (see (2.46)) in the present limit $\beta_3 kt_3 \to \infty$. When $d_4 \to 0$, F_{R} equals $(\mu_3 + 1)/2$, so that the demagnetization loss is exactly compensated by remagnetization. This does not mean that the magnetization in the tape is exactly remagnetized, such that $M = M_{\mathrm{p}}$ everywhere. Only for $\mathscr{B}_y$ and thus for the output of the head is the result equivalent to a completely remagnetized state, i.e. $M = M_{\mathrm{p}}$ as if μ_3 were 1, but with an anisotropy constant equal to the actual values β_3 (see (6.2)).

From the fact that the factor before the integral in (6.2) is independent of the mean permeability μ_3 if $d_4 \to 0$ it can be concluded that the output of a head *would not be changed by any other choice of the permeability* assuming

1) $d_4 = 0$;

2) equal permanent magnetization patterns;

3) a normalized coating thickness $\beta_3 k t_3$ much larger than 1 (starting point at the beginning of this section);

4) that the head's sensitivity at the considered wavelengths does not change, and

5) the anisotropy constant β_3 remains the same.

The highest permeability μ_3 offers the possibility of the most extreme values of β_3, since $1/\mu_3 \leqslant \beta_3 \leqslant \mu_3$ (because of $\mu_{3y} \geqslant 1$ and $\mu_{3x} \geqslant 1$ in principle).

From the integral in (6.2) it follows *for fixed μ_3* that

- *for longitudinal magnetization a small β_3 is favourable*;
- *for perpendicular magnetization a large β_3 is favourable* since M_{py} is a decreasing function of y.

The first conclusion is proven here for the case that $\beta_3 k t_3$ remains $\gg 1$. The last conclusion follows from the fact that the integral of the 'distance loss function' or 'weighting factor' in the integrand, $\beta_3 e^{-\beta_3 k}$ equals a constant, $1/k$. Since this function decreases with y, this weighting factor should be as high as possible at $y = 0$, where M_{py} is usually maximal. (This results in a reduced contribution to the read flux of M_{py} deeper in the tape at larger values of y.)

The above conclusions are consistent with the findings in Section 5.2.4.

However, since in practice $d_4 > 0$ μm, an extreme ($\ll 1$ or $\gg 1$) value of β_3 and consequently high value of μ_3 has the drawback of a large distance loss. This distance loss is described by (part of) the factor before the integral. This distance loss is so large for high values of μ_3 because the remagnetization decreases strongly with d_4 when the permeability is high.

Thus a larger $k d_4$ reduces the advantage of an orientation of the medium ($\beta < 1$ for longitudinal and $\beta > 1$ for perpendicular recordings) with respect to a unit relative permeability tape. The advantage may even become a disadvantage, see also section 6.2.5.

At a given μ_3 the most extreme β_3 must be chosen, if only one magnetization component has to be reproduced, and at a given β_3 the lowest possible value of μ_3.

6.2.2 Perpendicularly oriented media; $\beta_3 > 1$

As will be explained in section 6.4, for a perpendicularly oriented medium $\mu_{3x} > \mu_{3y}$ is always valid, i.e. $\beta_3 > 1$. In the extreme case $\beta_3 \gg 1$ and so $\mu_3 \gg 1$, the decrease of M_{py} with y over the extremely short characteristic distance $\lambda/(2\pi\beta_3) \ll t_3$, see (6.2), can be neglected. Then the contribution of the permanent longitudinal component vanishes and the contribution of the permanent perpendicular component approaches a constant (independent of further increases in β_3 since for $\beta_3 kt_3 \gg 1$, $\int_0^{t_3}\beta_3 e^{-\beta_3 ky}dy \to 1/k$) and delivers a read flux

$$\hat{\varPhi}^R = \frac{j\mu_0 WN\hat{M}^*_{py}(y=0^+)}{k}\, e^{-kd_4}\, \frac{2}{\mu_3 + 1}\, \frac{1}{1 - \dfrac{\mu_3 - 1}{\mu_3 + 1}\, e^{-2kd_4}} \tag{6.4}$$

irrespective of whether a backlayer is present or not, since $\beta_3 kt_3 \gg 1$. The first factor is N times the total internal permanent perpendicular flux in the head-facing side of the coating. The second term is the distance loss for a medium with unit relative permeability. The third and fourth factors describe the 'distance' loss, demagnetization and remagnetization. The product of the last three factors forms the present distance loss factor for the medium considered here with non-unit relative permeability,

$$DL = \frac{2e^{-kd_4}}{(1 + e^{-2kd_4}) + \mu_3(1 - e^{-2kd_4})}. \tag{6.5}$$

For $d_4 = 0$ all internal perpendicular flux $(2/\pi) \cdot \mu_0 M_{py} \cdot (\lambda/4) \cdot W = \mu_0 M_{py}W/k$ is reproduced (see also Fig. 6.1). For $kd_4 \ll 1$ the distance loss equals

$$DL \simeq \frac{1}{1 + \mu_3 kd_4}. \tag{6.6}$$

Even for small d_4 this loss can be large when β_3 is chosen (too) high and consequently μ_3 according to the expression $\mu_3 \gtrsim \beta_3$. This is due to

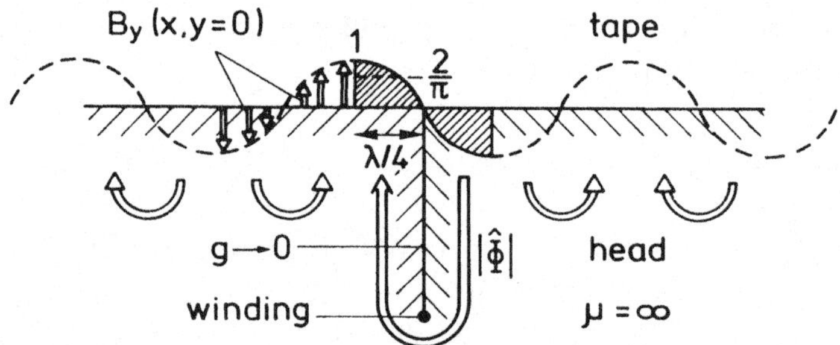

Fig. 6.1. Clarification of the expression for the internal perpendicular flux.
Since there are no demagnetizing fields ($d_4 = 0$ and $\beta_3 \to \infty$) $B_y = M_y = M_{py}$ at the coating-head interface. The amplitude of the flux that is reproduced follows because of symmetry considerations from $\lambda/4$ (and not $\lambda/2$) times the average flux density $(2/\pi)B_y$ times the track width W.

the extreme sensitivity of the medium to demagnetizing fields because of its high permeability μ_3. Only when $d_4 \to 0$ and surface charges at the head side of the coating are completely cancelled by mirror charges in the soft magnetic head *and* β_3 is large enough for volume charges to be located (virtually) at large enough distances, will there be no field in the coating at the head side. Then $\hat{M}_y = \hat{M}_{py}$ and because of continuity of B_y and no field being present in the coating, $\hat{B}_y = \hat{M}_{py}$ and all the internal flux is reproduced.

There is one 'other' trivial case where this happens: $\mu_3 = 1$ instead of $\mu_3 \geq \beta_3 \gg 1$, $\hat{M}_{py}$ constant over at least some characteristic distances $\lambda/2\pi$ instead of the smaller $\lambda/(2\pi\beta_3)$, and $kd_4 \ll 1$ instead of the more difficult $\mu_3 kd_4 \ll 1$.

Which case is most favourable is hard to say in general. It depends on technology (d_4) and effective writing depth; if d_4 can be made very small a perpendicularly oriented medium may be advantageous for perpendicular recording, but if d_4 cannot be made very small $\mu_3 = 1$ may be a better choice when the effective writing depth is large enough.

The present discussion can be continued when the writability of coatings with different permeabilities is considered.

6.2.3 Longitudinally oriented media; $\beta_3 < 1$

Expression (6.2) proves that *at small d_4 the reproduction of the longitudinal magnetization component should have the lowest possible β_3 value*. The validity of (6.2) however was restricted to $\beta_3 kt_3 \gg 1$, i.e. to situations where the reflections at the back of the coating could be neglected. This may be difficult to fulfil especially at small β_3 values.

Therefore we also investigated the general expression (6.1) for $\beta_3 k t_3 \ll 1$, $\beta_3 \ll 1$ and consequently $\mu_3 \geqslant 1/\beta_3 \gg 1$ since $\mu_{3x} \geqslant 1$, assuming no backlayer and no gap smear. After some mathematics* one obtains

$$\hat{\Phi}^{R} = \mu_0 WN \; \frac{e^{-kd_4}}{1 + \dfrac{\mu_{3x}kt_3}{2}\,(1 - e^{-2kd_4})} \int_0^{t_3} \hat{M}^*_{px}\mathrm{d}y. \qquad (6.7)$$

Although (6.7) also follows exactly from equation 14 in Bertram's paper [6.1], Bertram found another expression, his equation 15 being similar to (6.7) but without the denominator. This means that he was wrong in concluding that in this extreme case the total flux is reproduced, except for the effect of head-to-medium spacing e^{-kd_4}.

Instead, the total flux is reproduced, except for the effect of the following larger distance loss

$$DL = \frac{e^{-kd_4}}{1 + \dfrac{\mu_{3x}kt_3}{2}\,(1 - e^{-2kd_4})}. \qquad (6.8)$$

Expressions (6.7) and (6.8) show that the total internal flux is reproduced when $d_4 = 0$.

6.2.4 Interim discussion

The influence of the thickness for $\beta_3 k t_3 \to 0$ in the longitudinal case (see (6.8)) is completely included in the distance loss! In the case of the ideally perpendicularly oriented tape, (see (6.5)), thickness did not play any role, not even as a separate thickness-loss factor as in the case of

---

* Some effort is saved when direct use is made of the expression for a tape without backlayer and a head without gap-smear, expression (2.43), completed with the expressions (5.5), (2.46), (2.42) and $h_x^a(k) = N$ (idealized zero-gap head with N turns).

 It should be realized that $\mu_3 \to \infty$ means $\alpha_3 = 1/\mu_3 \to 0$. With the condition $\beta_3 k t_3 \to 0$ this gives the following useful limits in the derivation leading to (6.7):

$\sinh(2\alpha_3 + \beta_3 k t_3) = 2\alpha_3 + \beta_3 k t_3 = (1 + \mu_{3x}kt_3)/\mu_3$

$\sinh(\beta_3 k t_3) = \beta_3 k t_3$

$\sinh \alpha_3 = \alpha_3 = 1/\mu_3$

$\cosh(\beta_3 k(t_3 - y) + \alpha_3) = 1$

$\beta_3 \sinh(\beta_3 k(t_3 - y) + \alpha_3) = 0.$

a $\mu_3 = 1$ tape without a backlayer. Equations (6.5) and (6.8) look rather similar. The most relevant difference is that μ_3 in the expression for the extreme perpendicularly oriented tape is replaced by $\mu_{3x}kt_3$ for the extreme longitudinally oriented tape. The question arises why μ_3 in the extreme perpendicularly oriented case plays a role similar to that of $\mu_{3x}kt_3$ in the extreme longitudinally oriented case. The answer can best be built up by starting at the situation of zero head-tape distance and then going to finite head-tape distance.

– When $d_4 = 0$, then t_3 does not play any role in the extreme perpendicularly oriented case, since only the magnetization component at the interface with the head is significant. $M_{py} = B_y$ continues as the flux density over the interface with the ideally reflecting soft-magnetic zero-gap head, since $d_4 = 0$ and the characteristic distance $\lambda/2\pi\beta_3$ is very short, so that no demagnetizing fields are generated in the tape. In the extreme longitudinally oriented case, on the other hand, the magnetization over the whole thickness t_3 represents the internal flux which is also reproduced by the ideal head. (The characteristic distance $\lambda/2\pi\beta_3$ is now so large that the magnetization pattern can be thought to be virtually compressed into a thin sheet. Consequently the free-space demagnetizing field is then maximum, but for $d_4 = 0$ this field is completely cancelled by the reflection, i.e. by virtual mirror charges at zero distance.)

– When $d_4 > 0$, a larger μ_3 makes the magnetization very sensitive to the demagnetizing fields in both x and y directions that occur at the interface with the extreme perpendicularly oriented coating when $d_4 > 0$ and increase with d_4. The thickness t_3 has no effect at all on the value of these demagnetizing fields, since $\beta_3kt_3 \to \infty$. In contrast in a longitudinally oriented medium, a large kt_3 increases the fully-longitudinal demagnetizing fields that occur when $d_4 > 0$ and increase with d_4. These longitudinal demagnetizing fields have more effect on M_x when μ_{3x} increases.

6.2.5 Orientation versus unit μ_3

For $d_4 > 0$ the factor that accounts for both thickness t_3 and distance d_4 arises in (6.8) and (6.7). For a uniform magnetization distribution over the coating thickness and $\mu_{3x} = \mu_{3y} = 1$, (and $t_2 = t_5 = 0$) equation (6.1) delivers the well-known separated distance loss e^{-kd_4} and thickness loss $(1-e^{-kt_3})/kt_3$ factors for both horizontal as well as vertical magnetization components. The ratio of the above two read fluxes then gives

$$\frac{\hat{\Phi}^{R}(\mu_{3x}, \mu_{3y} \to \infty)}{\hat{\Phi}_{1}^{R}(\mu_{3x} = \mu_{3y} = 1)} = \frac{kt_{3}}{(1 - e^{-kt_{3}})\left[1 + \frac{\mu_{3x}kt_{3}}{2}(1 - e^{-2kd_{4}})\right]} \cdot \frac{\hat{M}_{px}^{*}}{\hat{M}_{px}^{*} + j\hat{M}_{py}^{*}}$$

(6.9)

where it is assumed that $\beta_{3} \ll 1$ and $\beta_{3}kt_{3} \ll 1$. This means that $\mu_{3y} \ggg$ 1 suffices for the validity of (6.9) when $\mu_{3x} = 1$ is chosen, as long as kt_{3} < 1. For $2kd_{4} \gg 1$, $\hat{M}_{px}^{*} = j\hat{M}_{py}^{*}$ (circular magnetization) and the practical circumstance that $t_{3} \gg d_{4}$ (and $\mu_{3x} \gtrsim 1$) means that this expression approaches $1/\mu_{3x}$! Consequently, at least for the region of short wavelengths where most information can be stored, no significant increase of output can be expected in practical circumstances ($t_{3} \gg d_{4}$) if μ_{3y} can be increased drastically, while $\mu_{3x} = 1$ (most favourable case). In Fig. 6.2.I relation (6.9) when $\mu_{3x} = 1$ and $\hat{M}_{px}^{*} = j\hat{M}_{py}^{*}$ is illustrated with μ_{3x}, t_{3} and d_{4} as parameters, while the ratio is plotted for 'more realistic' values of μ_{3y} in Fig. 6.2.II when $\mu_{3x} = 1$.

For extreme perpendicularly oriented media the ratio becomes, using (6.4) etc.,

$$\frac{\hat{\Phi}^{R}(\mu_{3x} \to \infty, \mu_{3y})}{\hat{\Phi}_{1}^{R}(\mu_{3x} = \mu_{3y} = 1)} = \frac{1}{(1 - e^{-kt_{3}})\left[\frac{1}{2}(1 + e^{-2kd_{4}}) + \frac{\mu_{3}}{2}(1 - e^{-2kd_{4}})\right]}$$

$$\times \frac{j\hat{M}_{py}^{*}}{\hat{M}_{px}^{*} + j\hat{M}_{py}^{*}}$$

(6.10)

where it is assumed that $\beta_{3} \gg 1$ and $\beta_{3}kt_{3} \gg 1$. This means that $\mu_{3x} \ggg 1$ or $\mu_{3} \gg 1$ suffices for the validity of (6.10) when $\mu_{3y} = 1$ is chosen, as long as $kt_{3} \gtrsim 1$. This ratio approaches $1/\mu_{3}$ at (very) short wavelengths where $2kd_{4} \gg 1$ and $2kt_{3} \gg 1$, when again circular magnetization, i.e. $\hat{M}_{px}^{*} = j\hat{M}_{py}^{*}$, is assumed. In the hypothetical situation of no head-tape distance the factor is considerably higher than one for the moderate and longer wavelengths when kt_{3} is small enough. For relatively short head-to-tape distances, however, this factor already approaches zero since μ_{3x} is by definition very large for an extreme per-

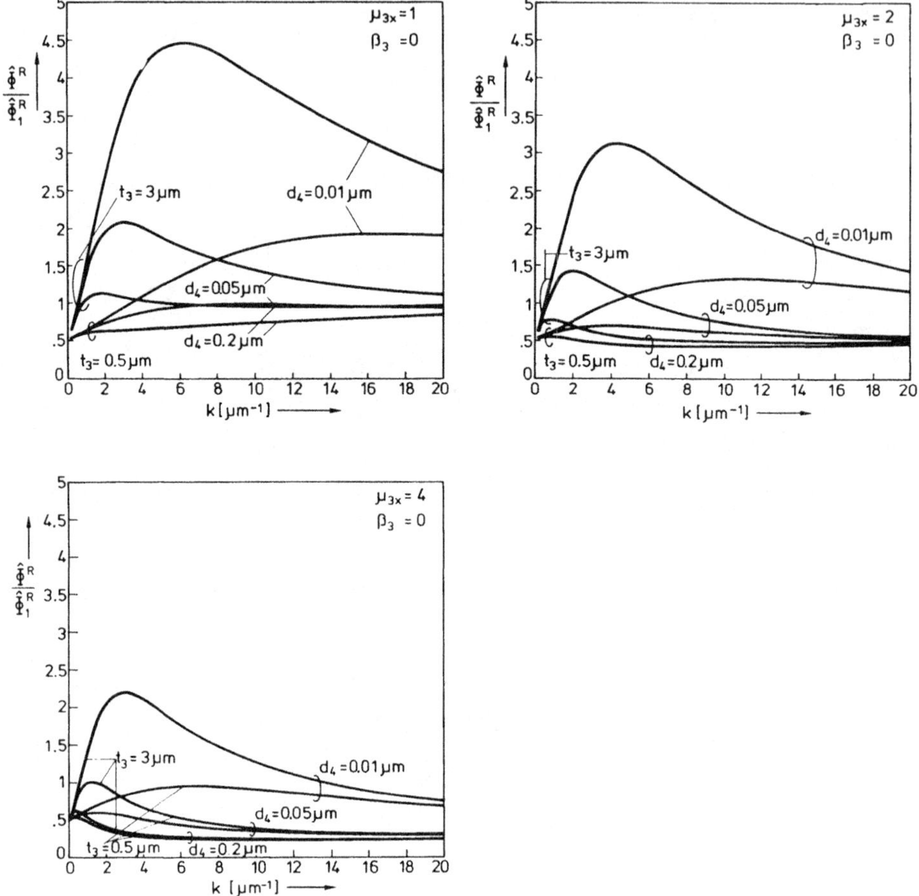

Fig. 6.2.I. The influence of the coating thickness t_3 and head-to-tape distance d_4 on the read flux of an extremely longitudinally oriented tape Φ^R for three different values of μ_{3x}, relative to the read flux of a unit-permeability tape Φ_1^R according to (6.9). Circular permanent magnetization, $\hat{M}_{px}^* = j\hat{M}_{py}^*$ is assumed, such that for very long wavelengths the longitudinally oriented medium gives half the output of the isotropic medium, see (6.9). For (extremely) short wavelengths this ratio is only $1/\mu_{3x}$; see also (6.9). Only for moderate wavelengths (or short wavelengths when t_3 is small and demagnetizing fields of the oriented tape and thickness loss of the unit-permeability tape become relevant at shorter wavelengths) is the extreme longitudinal orientation advantageous when d_4 can be made (very) small. The possible advantage of the longitudinal orientation vanishes when μ_{3x} becomes too high, because the sensitivity to demagnetization fields, which increase with kd_4, is then high.

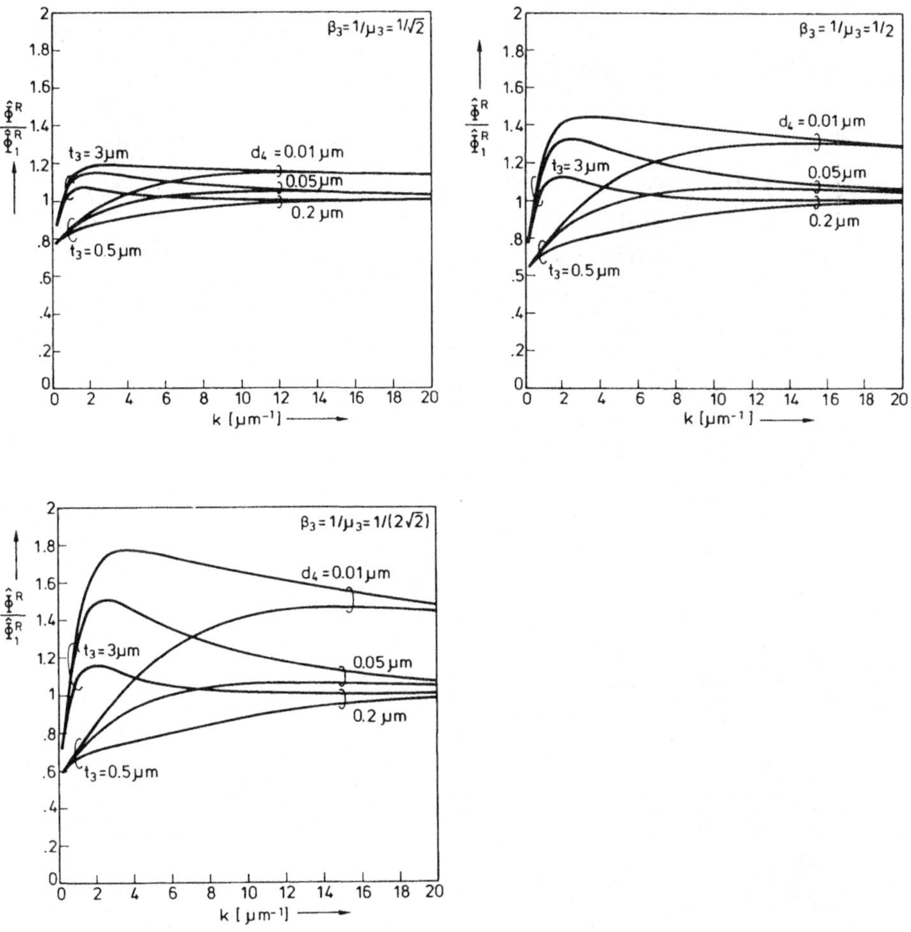

Fig. 6.2.II. As 6.2.I, but for realistic values of the coating permeability μ_3 and anisotropy β_3. For these values an exact representation of the flux, e.g. (6.1) must be used. Uniform magnetization is assumed and $t_2 = t_5 = 0$ µm, so that the much simpler (6.12) or (6.13) can be used. See below.

Perpendicular demagnetizing fields are maximum for the very long wavelengths, $\beta_3 k t_3 \ll 1$. This causes the contribution of the perpendicular permanent magnetization component to be proportional to $1/\mu_{3y} = \beta_3/\mu_3$, while all longitudinal flux is reproduced (see also (6.12) or (6.13)). As a result, the plotted ratio approaches $(1 + 1/\mu_{3y})/2 < 1$ at very long wavelengths. The playback of the moderate- and short-wavelength information is about equally favoured by the longitudinal orientation, especially when the head-tape distance is very small. The moderate wavelengths are further favoured when the thickness of the coating (which is assumed to be uniformly-magnetized!) is large. For very high frequencies the ratio approaches 1 since $\mu_{3x} = 1$; see e.g. (6.9). *The conclusion is that a longitudinally oriented medium is advantageous above an isotropic (unit-relative-permeability) medium when no backlayer ($t_2 = 0$ µm) is present.*

pendicularly oriented medium and hence $\mu_3 \geqslant \sqrt{\mu_{3x}}$ is large. The above ratio is plotted in Fig. 6.3.I with t_3 and d_4 as parameters for a hypothetical permeability, $\mu_3 = 100$, and in Fig. 6.3.II for more realistic values of μ_{3x}, while $\mu_{3y} = 1$.

In the above hypothetical examples unrealistic permanent magnetization distributions are assumed which are, however, physically possible as long as the (demagnetizing) fields do not exceed the coercive field of the tapes. The possibility of writing the assumed homogeneous permanent magnetization patterns in these hypothetical tapes is probably negligible. The examples only serve to show what can be expected from changes in the permeability component of the tape on the playback process, and moreover what cannot be expected from changes in the permeability components.

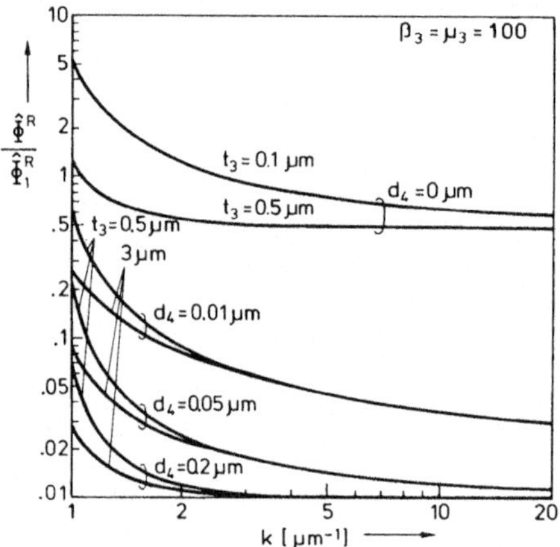

Fig. 6.3.I. The read flux in the case of an extremely perpendicularly oriented tape $\hat{\Phi}^R$, relative to the read flux in the case of a unit-permeability tape $\hat{\Phi}^R$, according to (6.10), for different coating thicknesses t_3 and head-to-tape distance d_4. $\beta_3 = \mu_3 = 100$ is assumed, such that the requirement for using (6.10), $\beta_3 k t_3 \gg 1$, holds for the considered range of moderate to high wavenumbers. The extremely perpendicularly oriented medium is favoured only for extremely short head-to-tape distances and moderate and smaller wavenumbers. For further comment see Fig. 6.3.II.

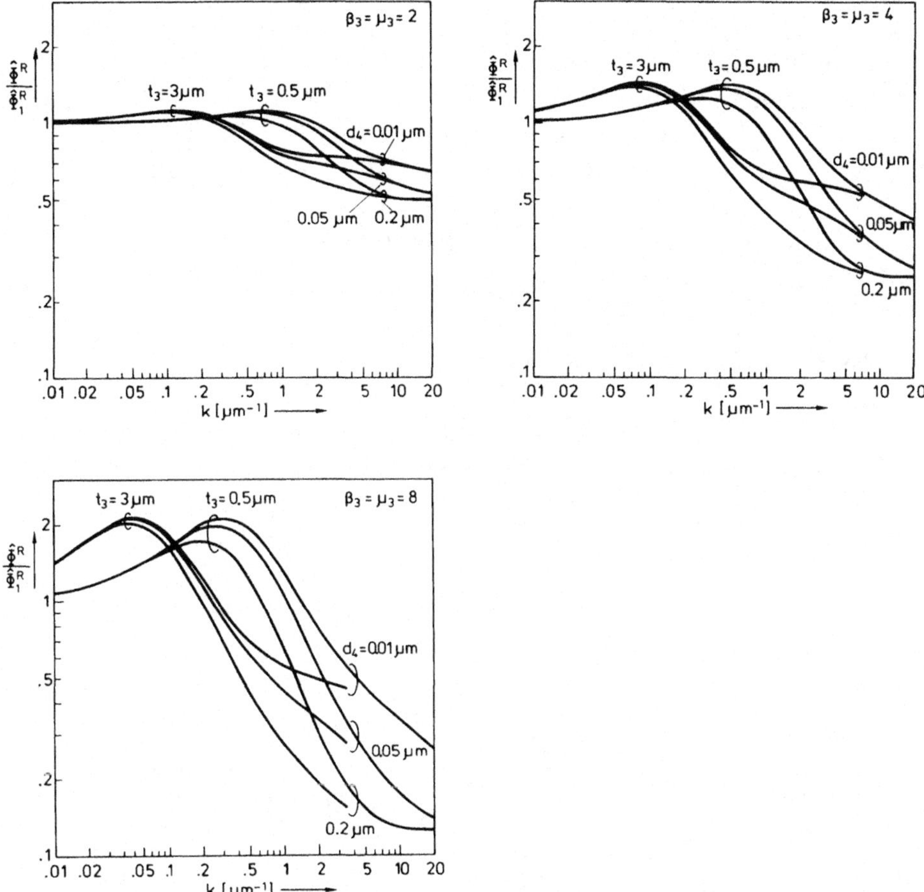

Fig. 6.3.II. As Fig. 6.3.I, but for more realistic values of the coating permeability μ_3 and anisotropy β_3. For these values an exact representation of the flux, e.g. (6.1), must be used. Uniform magnetization components are assumed and $t_2 = t_5 = 0$ µm, so that the much simpler (6.12) or (6.13) can be used. Since $\beta_3 = \mu_3$, i.e. $\mu_y = 1$ is assumed, the contribution of both components is equal for extremely long wavelengths and consequently all curves approach 1 for $\beta_3 k t_3 \ll 1$. This is in contrast with the longitudinally oriented medium where $\beta_3 = 1/\mu_3$ was assumed and, see caption of Fig. 6.2.II, $(1 + 1/\mu_{3y})/2 < 1$ was approached at long wavelengths. Only the playback of rather long wavelength information is favoured by the perpendicular orientation, especially when the head-to-tape distance is small. For the most interesting range of moderate and short wavelengths, however, the playback from perpendicularly oriented media is disadvantageous. In the limit of very short wavelengths, $2kd_4 \gg 1$, the plotted ratios approach $1/\mu_3$. This is caused by demagnetizing effects on the head-facing side of the coating. *The conclusion must be that a perpendicularly oriented medium is disadvantageous to an isotropic (unit-relative-permeability) medium when no backlayer ($t_2 = 0$ µm) is present.*

6.3 Layer-stack variations

Playback results of perpendicular (and to a lesser extent, longitudinal) recording on double layer (DL) and single-layer (SL) media will be derived and briefly discussed in the following sections. For simplicity it will first be assumed that the coating is uniformly magnetized over a certain depth. For this case expressions for the output of various layer stacks, single or double-layer, will be derived and the results compared. Then it will be proven that the results qualitatively in terms of 'what is a better combination' and 'what is worse' are also valid for more-complex layer stacks. Finally it is shown that the latter results are valid for any magnetization distribution in the tape that satisfies the condition $\hat{M}_{\mathrm{px}}(y) = \hat{C}\hat{M}_{\mathrm{py}}(y)$ with $\hat{C}$ being a constant.

6.3.1 Uniform magnetization

It will be assumed that the coating is uniformly permanently magnetized over a depth $t_{\mathrm{x}} \leqslant t_3$ for the horizontal component and over a depth $t_{\mathrm{y}} \leqslant t_3$ for the vertical component. In this case the integral in (6.1) can be solved. The result is

$$\hat{\Phi}^{\mathrm{R}}(\omega) = \frac{h_{\mathrm{x}}^{\mathrm{a}}(y_{\mathrm{p}} = -d_4 - t_5; k, \omega)}{k}$$

$$\times \; \frac{2\mu_0 W}{\cosh(\hat{\beta}_5^* k t_5)\left[\left(1 + \dfrac{\tanh(\hat{\beta}_5^* k t_5)}{\hat{\mu}_5^*}\right)\mathrm{e}^{kd_4} - R\left(1 - \dfrac{\tanh(\hat{\beta}_5^* k t_5)}{\hat{\mu}_5^*}\right)\mathrm{e}^{-kd_4}\right]}$$

$$\times \left[\frac{\frac{\sinh}{\cosh}\alpha_3}{\frac{\sinh}{\cosh}(\beta_3 k t_3 + \alpha'_{32} + \alpha_3)}\left\{\frac{\hat{M}_{\mathrm{px}}^*}{\beta_3}\left(\frac{\sinh}{\cosh}(\beta_3 k t_3 + \alpha'_{32})\right.\right.\right.$$

$$\left.- \frac{\sinh}{\cosh}(\beta_3 k(t_3 - t_{\mathrm{x}}) + \alpha'_{32})\right)$$

$$\left.\left.+ \mathrm{j}\hat{M}_{\mathrm{py}}^*\left(\frac{\cosh}{\sinh}(\beta_3 k t_3 + \alpha'_{32}) - \frac{\cosh}{\sinh}(\beta_3 k(t_3 - t_{\mathrm{y}}) + \alpha'_{32})\right)\right\}\right]. \qquad (6.11)$$

The expression between square brackets represents ($-j$ times the complex conjugate of) the free-space surface field $\hat{H}_f$. The contribution of the longitudinal magnetization component is in phase with $\hat{M}_{px}^*$ ($t = 0$; k) and the contribution of the perpendicular magnetization component leads by 90° relative to $\hat{M}_{py}^*$, as must be expected considering the definition of the flux (especially its direction) in chapter 2.

6.3.2 General single-layer (SL) media

Without a backlayer and without gap smear Eq. (6.11) relaxes for $t_x = t_3$ to

$$\hat{\Phi}_{SL}^R(\omega) = Wt_3\mu_0\hat{M}_{px}^* \left[\frac{2\sinh\alpha_3}{\beta_3kt_3\sinh(\beta_3kt_3 + 2\alpha_3)} \cdot \frac{\sinh(\beta_3kt_3 + \alpha_3) - \sinh\alpha_3}{e^{kd_4} - \mathrm{Re}^{-kd_4}} \right.$$

$$\left. \times\, h_x^a(y_p; k, \omega) \right] + jWt_3\mu_0\hat{M}_{py}^* \left[\frac{2\sinh\alpha_3}{kt_3\sinh(\beta_3kt_3 + 2\alpha_3)} \right.$$

$$\left. \times\, \frac{\cosh(\beta_3kt_3 + \alpha_3) - \cosh\alpha_3}{e^{kd_4} - \mathrm{Re}^{-kd_4}}\, h_x^a(y_p; k, \omega) \right]. \tag{6.12}$$

After straightforward, but time-consuming, calculations this equation can be written in a simpler form without hyperbolic functions of compound arguments:

$$\hat{\Phi}_{SL}^R(\omega) = N\hat{\Phi}_{1x}^* \left[\frac{\tanh(\beta_3kt_3)}{\beta_3kt_3} \right.$$

$$\left. \times\, \frac{\left(1 + \dfrac{1}{\mu_3}\tanh\dfrac{\beta_3kt_3}{2}\right) SF}{e^{kd_4} + \tanh(\beta_3kt_3)\left(\dfrac{1}{\mu_3} + \mu_3\tanh(kd_4)\right)\cosh(kd_4)} \right]$$

$$+ jN\hat{\Phi}_{1y}^* \left[\tanh(\beta_3kt_3) \right.$$

$$\left. \times\, \frac{\left(\dfrac{1}{\mu_3} + \tanh\dfrac{\beta_3kt_3}{2}\right) SF}{e^{kd_4} + \tanh(\beta_3kt_3)\left(\dfrac{1}{\mu_3} + \mu_3\tanh(kd_4)\right)\cosh(kd_4)} \right], \tag{6.13}$$

where the sensitivity function $SF = h_x^a/N$ is introduced and where $\hat{\Phi}_{ix}$ and $\hat{\Phi}_{iy}$ are the longitudinal and perpendicular permanent fluxes in the coating. We will call these the internal fluxes (subscript i). They equal

$$\hat{\Phi}_{ix} = Wt_3\mu_0\hat{M}_{px} \qquad (6.14)$$

$$\hat{\Phi}_{iy} = W\mu_0\hat{M}_{py}/k, \qquad (6.15)$$

(see also Fig. 6.1).

For an isotropic tape ($\beta_3 = 1$) and infinite permeability zero-gap head ($h_x^a = 1$), (6.13) is equivalent to the equations (8) and (11) of Westmijze [6.8].

The term between the brackets can be denoted as the reproduction loss RLx or RLy.
–   RLx includes thickness, distance and gap loss.
–   RLy includes 'thinness', distance and gap loss.

6.3.3 Thickness and thinness loss

We will explain here that the choice of the term thinness loss for perpendicular recording and thickness loss for longitudinal recording is very suitable in combination with the internal fluxes 'defined' in (6.15) and (6.14). The expressions (6.14) and (6.15) are more than definitions. They express the maximum amount of flux that can be extracted from the uniformly permanently magnetized tape.

All *longitudinal* internal flux will leave the tape and reach an idealized zero-gap head when $t_3 \to 0$ (no *thickness* loss) and $d_4 = 0$. In contrast, all the perpendicular internal flux, $(2/\pi) \times (\mu_0 \cdot M_{py}) \times (\lambda/4) \times W$ see Fig. 6.1, will be reproduced when $t_3 \to \infty$ (no *thinness* loss) and $d_4 = 0$.

In the above cases no demagnetizing field occurs at the tape-head interface. At larger head-to-tape distance or greater thickness, reproduction losses occur in the longitudinal case. Reproduction losses occur for the perpendicular magnetization when $d_4 \neq 0$ or t_3 is not infinite. When $d_4 \to \infty$ and $t_3 = 0$ or $t_3 \to \infty$, half of the internal longitudinal or perpendicular flux leaves the tape on the head side, or less when $\mu \neq 1$ and demagnetization is not compensated by remagnetization. (When $d_4 \to \infty$, $H_f \leqslant \frac{1}{2}M_{py}$ in the perpendicular case and $H_f \leqslant \frac{1}{2}kt_3M_{px}$ in the longitudinal case.)

6.3.4 SL media with $\mu_3 = 1$

For $\mu_3 = \beta_3 = 1$ the well-known result

$$\hat{\Phi}^{R}_{SL}(\omega) = N\hat{\Phi}^{*}_{ix}\left[\frac{1 - e^{-kt_3}}{kt_3} e^{-kd_4}SF\right]$$

$$+ jN\hat{\Phi}^{*}_{iy}\left[(1 - e^{-kt_3})e^{-kd_4}SF\right] \qquad (6.16)$$

is obtained.

This demonstrates the thickness- and thinness-loss factors $(1-e^{-kt_3})/kt_3$ and $1-e^{-kt_3}$. The thinness-loss factor is 1 when $kt_3 >> 1$, and approaches 0 when $kt_3 << 1$, in contrast to the thickness-loss factor.

The output voltage in the present limiting case, $\hat{V}^{R}_{SL} = -j\omega\Phi^{R}_{SL}$, equals

$$\hat{V}^{R}_{SL}(\omega) = v2N\mu_0W \cdot e^{-kd_4} \cdot SF$$

$$\times \left[\left\{-j\hat{M}^{*}_{px}(t=0; k) + \hat{M}^{*}_{py}(t=0; k)\right\} \frac{1 - e^{-kt_3}}{2}\right], \qquad (6.17)$$

where the plane $y = y_p$ is located at a distance d_4 from the tape. The expression between square brackets in (6.17) represents the complex conjugate of the free space field $\hat{H}_f$. The contributions from both magnetization components to the amplitude of $\hat{H}_f$ are now equal.

The distance loss equals the well-known

$$DL = e^{-kd_4}, \qquad (6.18)$$

assuming that the sensitivity function does not alter with d_4.

If $\hat{\Phi}_{iy}$ were defined like $\hat{\Phi}_{ix}$, i.e. $\hat{\Phi}'_{iy} = Wt_3\mu_0\hat{M}_{py}$, as in fact Westmijze did, then kt_3 would arise in the denominator between the second brackets in (6.16). Then the expressions between both pairs of brackets would be equal (for $\mu_3 = \beta_3 = 1$). But since $\hat{\Phi}'_{iy}$ does not represent the internal perpendicular magnetization, $(1-e^{-kt_3})/kt_3$ must then not be interpreted as the thickness loss.

A large thickness reduces the thinness loss, so too does a large β_3, since it increases, virtually, the thickness of the coating. A backlayer also reduces the thinness loss, because of a reduction of demagnetizing fields. For zero head-tape distance, idealized soft-magnetic head and infinite permeability backlayer the demagnetizing fields in a uniformly perpendicularly-magnetized coating are completely cancelled and as a consequence all internal flux is reproduced. This will be discussed in more detail in separate sections about double-layer (DL) media. The thinness- and distance-loss factors then equal 1. This would also be true in the hypothetical case of using an infinitely thick coating without a backlayer. Instead of an infinitely thick coating, β_3 can be chosen very large, which has the advantage that the permanent perpendicular magnetization on the head-facing side of the coating is mainly reproduced then, so that a uniform magnetization is not important. The loss that occurs when the head-tape distance differs from zero is however quite different for the above three cases.

6.3.5 Factorization of the reproduction loss

In equation 6.16, where $\mu_3 = \beta_3 = 1$ is assumed, the total reproduction loss is factorized in respectively a thickness- and a thinness-loss factor, a distance-loss factor and a gap-loss factor. This is not exactly possible when part of the tape, for instance the coating, has a non-unit relative permeability. The physical reason is that such a tape interacts with the head and as usual for a spatial interaction it depends on the distance.

In spite of this, Siakkou [6.2] tried to approximate the reproduction-loss factor RL_x for $\mu_3 \neq 1$ by factorizing it into a thickness-, distance- and gap-loss factor (see his equations 32 and 36) in the following way:

$$RL_x \simeq \frac{1 - e^{-\mu_3 kt_3}}{\mu_3 kt_3} \; e^{-kd_4} \; \frac{\sin\left(\dfrac{kg}{2}\sqrt{\dfrac{1.5\mu_3 + 1}{\mu_3 + 1}}\right)}{\dfrac{kg}{2}\sqrt{\dfrac{1.5\mu_3 + 1}{\mu_3 + 1}}}. \tag{6.19}$$

We checked this result without the last gap-loss factor and found a large discrepancy between this (6.19) and the exact result, (6.13) or (6.12); see Fig. 6.4 curves s and e. Siakkou claims an accuracy within 3dB for the interesting range of parameters of metal (film) and oxide (particulate) tapes. The inaccuracy however is much larger than the

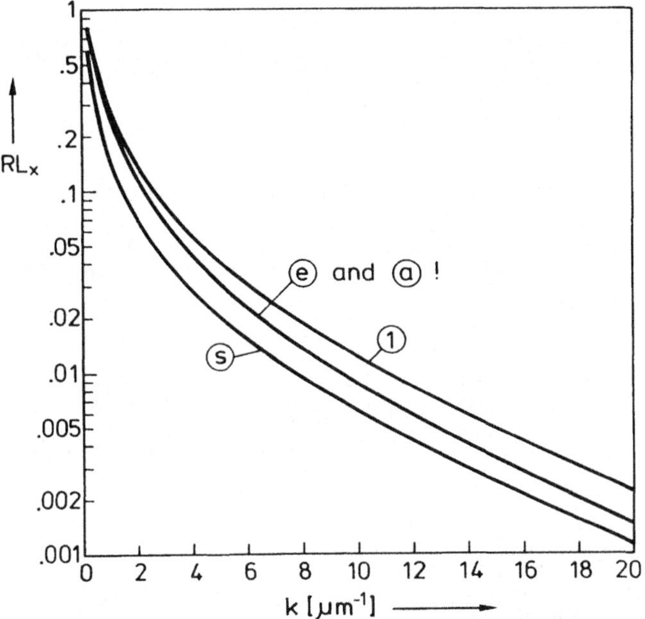

Fig. 6.4. Different approximations for the reproduction-loss factor for longitudinal recording RL_x. $\mu_3 = 2$, $\beta_3 = 1$, $t_3 = 3\mu$m (uniformly magnetized throughout) and $d_4 = 0.1$ μm.
curve ⓔ: exact solution according to (6.12) or (6.13),
curve Ⓢ: Siakkou's factorized approximation,
curve ①: $\mu_3 = 1$ case,
curve ⓐa: our factorized approximation (6.15).
The discrepancy between Siakkou's approximation and the exact solution is of the order of the effect of the permeability itself; see curves Ⓢ and ① relative to ⓔ. Our approximation coincides completely with the exact solution. Even for small k the difference is negligible.

effects of μ_3 itself; see Fig. 6.4 curve *1*. With the aid of the approximation (6.2), valid for $\beta_3 k t_3$ larger than say 1, we arrive at a much better and more general 'factorization'[□] for the reproduction losses:

$$RL_x \simeq \frac{1 - e^{-\beta_3 k t_x}}{\beta_3 k t_x}\left(\frac{2}{\mu_3 + 1}\cdot\frac{1}{1 - \dfrac{\mu_3 - 1}{\mu_3 + 1}e^{-2kd_4}}e^{-kd_4}\right)SF \qquad (6.20)$$

$$RL_y \simeq (1 - e^{-\beta_3 k t_y})\left(\frac{2}{\mu_3 + 1}\cdot\frac{1}{1 - \dfrac{\mu_3 - 1}{\mu_3 + 1}e^{-2kd_4}}e^{-kd_4}\right)SF \qquad (6.21)$$

[□] The factorization is complete when the influence of d_4, t_3 and μ_3 on SF can be neglected. This is allowed for ring heads on particulate single-layer media.

in
$$\hat{\Phi}_{SL}^{R} \equiv N\hat{\Phi}_{ix}^{*}RL_x + jN\hat{\Phi}_{iy}^{*}RL_y \qquad (6.22)$$

with
$$\hat{\Phi}_{ix} = Wt_x\mu_0\hat{M}_{px} \quad \text{and} \quad \hat{\Phi}_{iy} = W\mu_0\hat{M}_{py}/k \qquad (6.23)$$

where

$t_x \le t_3$ is the thickness of the uniformly permanently longitudinally magnetized layer and

$t_y \le t_3$ is the thickness of the uniformly permanently perpendicularly magnetized layer.

Only in the region $\beta_3 kt_3 \lesssim 1$ is the approximation less accurate. The main difference with Siakkou's approximation is that the anisotropy constant β_3 appears in the exponent, instead of the permeability μ_3. This is also better to understand since β_3 is the parameter that influences the 'propagation' of fields.

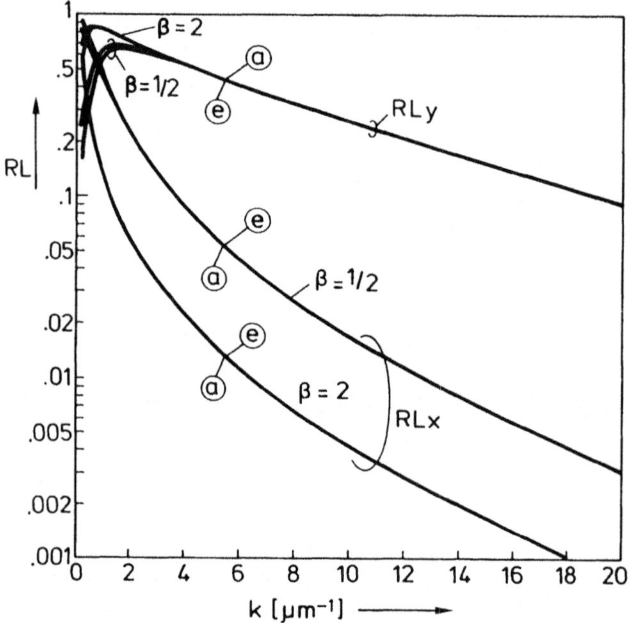

Fig. 6.5. Our factorized approximation for the reproduction-loss factors RL_x and RL_y when $\beta_3 \neq 1$ compared with the exact solution.
$\mu_3 = 2$, $t_3 = 3$ μm (uniformly magnetized throughout) and $d_4 = 0.1$ μm.

ⓔ = exact solution

ⓐ = approximation

The approximation is almost exact, except for a slight difference at very long wavelengths. The increasing thickness loss and decreasing thinness loss with frequency, due to the assumption of uniform magnetization over the coating thickness, causes $RL_x \ll RL_y$ at high frequencies.

The three factors in (6.20) and (6.21) represent respectively the generalized thickness- and thinness-loss, the generalized distance loss and the generalized gap-loss factors. Fig. 6.4 curve a shows the approximation according to (6.20) and (6.21). For $\beta_3 = 1$ both (6.20) and (6.21) are equally accurate. In Fig. 6.5 curve a the approximations (6.20) and (6.21) are shown for $\beta_3 \neq 1$. These curves almost coincide with the exact curves e.

The correction of the gap length in Siakkou's approximation to a maximum of $\sqrt{1.5}$ or 22% for $\mu_3 \to \infty$ is also quite incorrect. Because of the non-zero interaction this correction factor too must be dependent on head-tape distance, tape thickness,etc. With $kd_4 \to 0$ and $\mu_3 \to \infty$ for instance, the correction factor must be 2 (see section 5.6). Probably Siakkou used the wrong results of Fan [6.3].

In spite of the above arguments against Siakkou's approximation (6.19) it is sometimes cited, e.g. by Jorgensen [6.4].

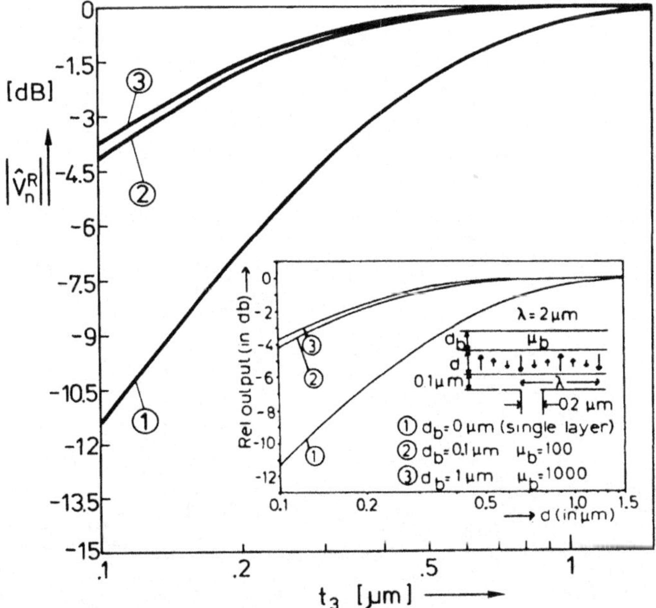

Fig. 6.6. Comparison of our normalized read voltage V_n^R (or flux $\Phi_n^R \sim$ magnetic induction $|\mathscr{B}_y^b|$) with that of Quak, shown in the inset. The highest value in this plot is normalized to 0 dB. Corresponding curves show equal results. Curves 2 and 3 also show that a permeability $\mu_2 = 100$ and thickness $t_2 = 0.1$ μm of the backlayer suffice for optimal playback of a 2 μm wavelength perpendicular recording.

6.3.6 General double-layer (DL) media

In a recent paper [6.5] Quak investigated the influence of the backlayer permeability and thickness on the reproduce voltage in perpendicular recording. He assumed a uniform permanently magnetized coating with unit permeability. Relative to other papers these calculations are rather general since the permeability of the backlayer is variable, although not anisotropic. Therefore it seems useful to check our analytical results for the case where there is no gap-smear ($t_5 = 0$), unit tape permeability ($\mu_3 = \beta_3 = 1$), isotropic backlayer ($\beta_2 = 1$) and uniform perpendicular magnetization ($M_{px} = 0$) with his results. In Fig.

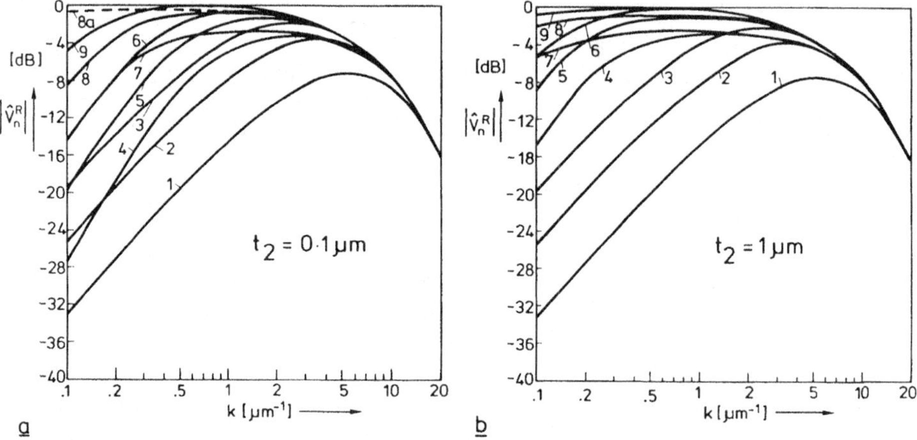

Fig. 6.7. The influence of a backlayer on the playback voltage from perpendicularly uniformly recorded media as a function of k.
The head is assumed ideal, i.e. has zero gap and infinite permeability, hence $\left|\hat{V}^R(k)\right| \sim \left|\mathcal{B}_y^b(k)\right|$.
The highest value in both plots a) and b) is normalized to 0 dB.
Fixed parameters: $d_4 = 0.1$ μm, $\mu_3 = \beta_3 = 1$.
Variable parameters:
1, 2 and 3:   SL medium with coating thickness $t_3 = 0.2$, 0.5 and 1 μm
4, 5 and 6:   DL medium with coating thickness $t_3 = 0.2$, 0.5 and 1 μm and backlayer with $t_2 = 0.1$ μm, $\beta_2 = 1$ and $\mu_2 = 100$
7, 8 and 9:   DL medium as 4, 5 and 6 but with $\mu_2 = 1000$
8a: DL medium as 5 but with $\mu_2 = 10^6$
a) backlayer thickness $t_2 = 0.1$ μm
b) backlayer thickness $t_2 = 1$ μm.
Clearly *a backlayer is advantageous for perpendicular recordings*. The permeability and thickness of the backlayer are of less importance. At very low frequencies the backlayer is no longer quite so effective in short-circuiting the flux, especially when the permeability and thickness of the backlayer are small; see curves 4, 5 and 6 in a). The results then tend to those of the SL medium; see curves 1, 2 and 3. The thinner the coating the greater the interaction with the backlayer and the greater the advantage of using a backlayer in the region of moderate wavelengths. But the demagnetizing fields in a perpendicularly magnetized medium when $d_4 \neq 0$ are larger in the case of a thinner coating, so that, assuming uniform magnetization, the thicker coating is still advantageous. At the highest frequencies, where all curves come together, the backlayer is ineffective because of the too large interaction distance ($2kt_3 \gg 1$). Curve 8a is added to show the influence of an infinite permeability backlayer ($\mu_2 = 10^6$ and other parameters as for curve 8).

6.6, for example, the output is plotted as a function of the coating thickness t_3 with the backlayer thickness t_2 and permeability μ_2 as parameters. In the inset the result of Quak is shown on a $\log(d)$ scale. The gap length of the assumed ideal Karlqvist head plays no role in this plot. The results are exactly the same.

The influence of the coating thickness t_3, backlayer permeability μ_2 and backlayer thickness t_2 on the normalized output is shown in Fig. 6.7 for uniform perpendicular recording and in Fig. 6.8 for uniform longitudinal recording. Fig. 6.7 corresponds to the results in Fig. 3 of ref. [6.5], but we did not include the gap loss of an ideal Karlqvist head with a gap length of 0.2 μm. The main conclusion of Quak's investiga-

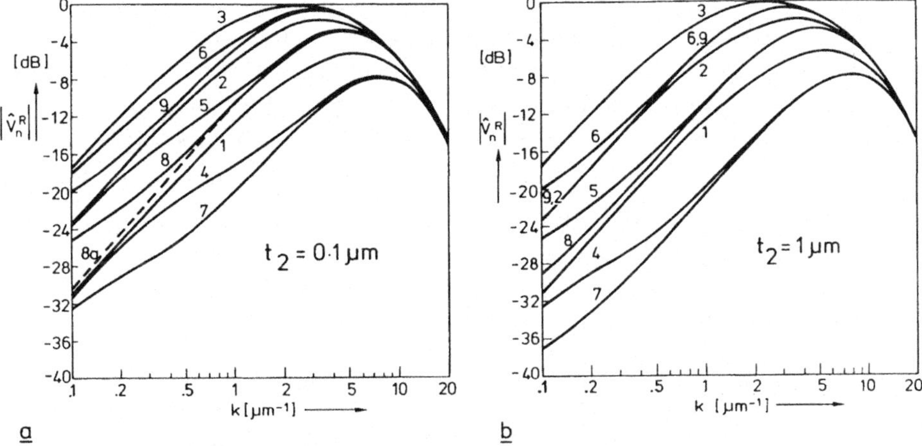

Fig. 6.8a and b. The same curves as in Fig. 6.7a and b but *for playback of longitudinal recordings*. A *backlayer* is now clearly *disadvantageous*. Again the permeability and thickness of the backlayer are of less importance. At very low frequencies the backlayer is no longer quite so effective, and the results tend to the results of the SL medium. The lower the permeability and thickness of the backlayer the sooner the backlayer becomes ineffective, when going to longer wavelengths. The thinner the coating the greater the interaction with the backlayer at the higher frequencies and the greater the disadvantage there. At the highest frequencies all curves come together because of the too large interaction distance ($2kt_3 \gg 1$). The results in the case of the SL medium are exactly the same as in Fig. 6.7. The seeming difference in the absolute value is a consequence of the normalization. Curve 8a is added to show the influence of an infinite permeability backlayer ($\mu_2 = 10^6$ and other parameters as for curve 8 and curve 5).

tions is that the thickness of the backlayer in the case of a high permeability has little influence on the output. From his Fig. 3 or our Fig. 6.7 it follows in more detail that only in the case of a backlayer thickness of only 0.1 μm and a coating thickness of only 0.2 μm is there a clear drawback (some dBs) for wavelengths longer than some μms when a backlayer with a relative permeability of only 100 is used instead of 1000. The only noticeable influence is from the coating thickness. This is not surprising since even for infinite backlayer permeability and thickness the remagnetization due to the reflection in the backlayer of especially long wavelength perpendicular magnetization patterns can only compensate a (small) part of the demagnetization fields; although the charges (including virtual mirror charges) at the interface coating-backlayer have vanished almost completely (even for $\mu_2 = 100$ and $t_2 = 0.2$ μm) the virtual mirror charges of the charges at the head side of the coating (front-surface charges) are not yet far enough away ($2t_3 \not\gg \lambda/2\pi$) in the case of the smaller coating thickness to be negligible; see also Fig. 2.17 curves 1 and 2 for t_3 smaller than say 0.5 μm. The smaller the field in the coating the more the flux density entering the head approaches the permanent magnetization in the tape, since div B = div M + div H. When the front surface charge density is reduced as well ($d_4 \rightarrow 0$) all 'permanent' magnetization is reproduced.

Our interpretation is in seeming contradiction with that of Quak [6.5] who attributed the influence of the thickness of the coating on the relative output to the not completely annihilated surface-charge density at the interface coating-backlayer, even in the case of the highest permeability (1000) and thickness (1 μm) of the backlayer. However, Quak speaks in terms of 'real' charges (div $M \neq 0$) whereas we used the concept of virtual (mirror) charges in the physical interpretation. In the 'real' charge concept a backlayer with infinite permeability can only have charges on its front surface, in the mirror concept virtual charges may even lie behind the backlayer itself when the coating thickness exceeds the backlayer thickness (see also Chapter 2).

Since our results coincide with Quak's results, our results too are in contradiction with the results of Yamamori et al. [6.6]. See also [6.5], who assumed that the real surface-charge density on the interface between coating and ideal backlayer is zero, instead of assuming that only the surface charge at the rear side of the coating plus its mirror charge is zero.

For longitudinal recordings, see Fig. 6.8, the influence of the coating thickness is precisely opposite. A backlayer is disadvantageous for playback of longitudinal recordings, since the reflections in the backlayer decrease the fields emanating from the tape, in contrast to perpendicular recordings, especially at the lower frequencies, where losses due to distance (to the reflection) are small. The smaller the coating thickness the clearer the disadvantage.

From the above results we can concluded that even a thin backlayer approaches the ideal reflector in the wavelength range of interest. Therefore in most of the following sections a backlayer with infinite permeability will be assumed.

6.3.7 DL media with $\mu_3 = 1$ and $\mu_2 = \infty$

With the following assumptions:

$\mu_2 = \infty$ (ideal soft-magnetic backlayer)

$\mu_3 = \beta_3 = 1$ (ideal square MH curve of the coating)

$t_5 = 0\ \mu\mathrm{m}$ (no gap smear)

it can easily be calculated that $\alpha_3 = \infty$, $\mu_{32} = 0$ and $\alpha'_{32} = 0$ such that $R = e^{-2kt_3}$, i.e. remagnetization occurs by way of the backlayer. Further it is assumed that the tape is uniformly magnetized over the full coating thickness, i.e. $t_x = t_y = t_3$ in (6.11). Hence the complex output voltage $\hat{V}^R_{DL} = -j\omega\hat{\Phi}^R_{DL}$ (see also (5.6)) equals

$$\hat{V}^R_{DL}(\omega) = v\,2\mu_0 W\, \frac{e^{kt_3}}{2\sinh\big(k(d_4 + t_3)\big)}\, h^{DL}_x(y_p; k, \omega)$$

$$\times \left[-j\hat{M}^*_{px}(t = 0; k)\, \frac{\cosh(kt_3) - 1}{e^{kt_3}} + \hat{M}^*_{py}(t = 0; k)\, \frac{\sinh(kt_3)}{e^{kt_3}} \right]$$

$$(6.24)$$

where the plane $y = y_p$ is situated at a distance d_4 from the tape.

The superscript 'a' in h^a_x in (6.11) (originally defined in chapter 3 as the 'writing state' with the magnetic tape, without permanent magnetization, present) is changed into the superscript 'DL', indicating that h_x is calculated in the presence of the soft-magnetic backlayer. The perpendicular term (6.24) is the same as that used by Lopez [6.7]. The expression between square brackets represents the complex conjugate of the free-space field $\hat{H}_f$. At low wavenumbers the contribution of the perpendicular magnetization component to $\hat{H}_t^*$ is considerably larger than the contribution of the longitudinal component, but at high frequencies the difference vanishes.

The distance loss factor that follows from (6.24), but is essentially independent of the assumed magnetization pattern (uniform in (6.24)), equals

$$DL = \frac{(1 - e^{-2kt_3})e^{-kd_4}}{1 - e^{-2k(t_3 + d_4)}} = \frac{\sinh(kt_3)}{\sinh(kt_3 + kd_4)}, \qquad (6.25)$$

assuming that the sensitivity function does not alter with d_4.

The distance loss is large when t_3 is small, because demagnetizing effects in the tape (backlayer) when d_4 differs from zero are stronger then; see Fig. 6.9. The curves are qualitatively the same as for an SL medium when instead of a decreasing t_3 the permeability μ_3 increases (see chapter 2, fig. 2.6). The reason is also the same. *An increasing permeability of the tape, no matter if it concerns the backlayer or the coating, causes larger demagnetization and so larger distance losses. The*

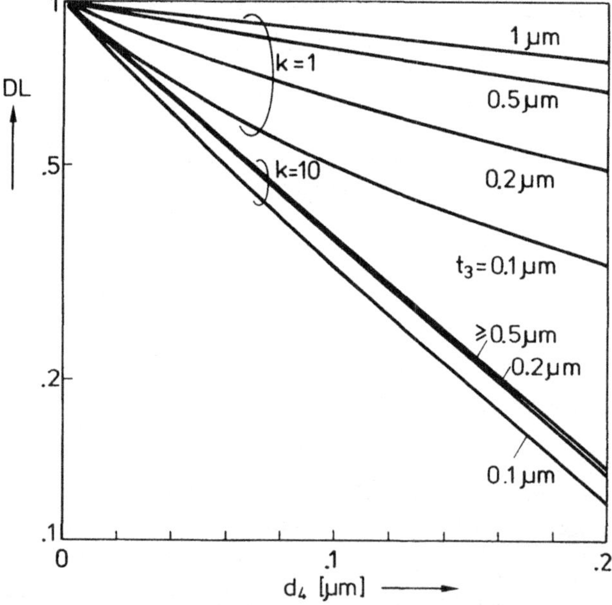

Fig. 6.9. The distance loss DL in the case of a double-layer medium with $\mu_2 = \infty$ and $\mu_3 = 1$ according to (6.25). A decreasing coating thickness t_3 increases the distance loss because of increasing demagnetizing effects in the backlayer. This is mainly visible at moderate wavelengths ($k \simeq 1$) where the influence of the backlayer is strongest.

demagnetization-remagnetization effects are also stronger according as the layer with $\mu_r > 1$ is located nearer to the head, no matter whether it concerns a backlayer or a permeable coating.

6.3.8 Influence of backlayer on factorization

In the case of the DL medium, see (6.24), the influence of thickness t_3 and head-tape distance d_4 could not be separated, unless the normalized interaction distance $k(d_4 + t_3)$ between head and permeable backlayer was larger than about 1.

This must be expected since Westmijze's [6.8] results for a coating with permeability $\mu_3 \neq 1$ already made a separation in thickness and distance loss impossible, because of non-zero interactions between head and tape that change with distance d_4. In addition the contributions from both magnetization components in the coating of the DL medium are not equal, as is also the case for an SL medium if $\mu_3 \neq 1$.

6.3.9 DL versus SL and ring head versus probe head

The ratio of the read-back voltages of the idealized DL and SL media from perpendicular recordings, assuming equal M_{py} and t_y, equals

$$\frac{\hat{V}^R_{DL_y}}{\hat{V}^R_{SL_y}} = \frac{\sinh(kt_3)}{\sinh(k(d_4 + t_3))(1 - e^{-kt_3})e^{-kd_4}} \frac{h^{DL}_x(y_p; k, \omega)}{h^{SL}_x(y_p; k, \omega)} . \qquad (6.26)$$

When $h^{DL}_x = h^{SL}_x$ is assumed, the result (6.26) equals the result of Lopez [6.7] and the ratio in (6.26) then equals the first factor. This factor approaches 1 at short wavelengths, is slightly higher for moderate wavelengths, and approaches infinity for very large wavelengths which are not of interest.

The corresponding ratio for the longitudinal magnetization components reads

$$\frac{\hat{V}^R_{DL_x}}{\hat{V}^R_{SL_x}} = \frac{\cosh(kt_3) - 1}{\sinh(k(d_4 + t_3))(1 - e^{-kt_3})e^{-kd_4}} \frac{h^{DL}_x(y_p; k, \omega)}{h^{SL}_x(y_p; k, \omega)} . \qquad (6.27)$$

This first factor in this equation again approaches 1 at short wavelengths, but is slightly lower for moderate wavelengths and approaches $(t_3/2)/(t_3 + d_4)$ at long wavelengths ($k(d_4 + t_3) << 1$) outside the region of interest.

In the frequency range of interest usually

$$h_x^{DL} < h_x^{SL} \qquad \text{for a ring head (RH)} \qquad (6.28)$$

$$h_x^{DL} > h_x^{SL} \qquad \text{for a single-pole head (PH)} \qquad (6.29)$$

due to concentrations in the field when a DL medium is used.

The result (6.28) is clear from what was said in section 5.6: the higher wavenumber components of the sense field decrease when a high permeability layer (coating and/or backlayer) approaches the head. The efficiency, see section 5.3, of the ring head can be defined as the integral of the normalized sense field h_x divided by N. Since the backlayer shunts the head, this integral decreases and so too therefore does the efficiency. These effects may compensate the advance of the first factor so that the sensitivity of a ring head may be reduced by the presence of a backlayer. Applied to the longitudinal case described by (6.27) these effects reduce the output even further.

For probe-type heads a split of the sensitivity function into an efficiency (5.17) and gap-loss function (5.18) seems more difficult because of (5.20) etc. or even impossible, so that the sensitivity function must be used in comparisons.

The possible differences in sensitivity functions of the ring head (RH) and probe-type head are sketched in Fig. 6.10.

For well-defined structures like probe-type heads of infinite-permeability material in combination with double-layer media with unit permeability coating and infinite-permeability backlayer, Minuhin [6.31-6.35] calculated analytically, by way of the Schwarz-Christoffel transformation, the wavelength response ($\simeq$ sensitivity function) for many ratios of probe thickness to head-backlayer distance. In an article about cross-talk, Luitjens and v. Herk [6.9] calculated the sensitivity functions of three-dimensional ring- and probe-type heads numerically. The gap-null wavelength λ_0 we distil from their plots for a probe-type head with thickness T at distances 0.5 T and $>>>$ T from the high-permeability backlayer are 1.3 T and 1.5 T respectively. From plots in a paper by Bertram [6.10] it follow that $\lambda_0 \simeq 1.4$ T for a distance 2/3 T. Other papers where sensitivity functions of ring, and probe-type heads are reported, or in which statements concerning sensitivity functions can be found, are references [6.11] and [6.12]. The latter by Minuhin only describes sensitivity functions of zero-gap ring heads and zero-thickness probe heads of infinite-permeability material on ideal single or double-layer media in the x-domain. Information about the sensitivity function of asymmetrical thin-film heads (without the influence of tape permeabilities) can be found in a paper by Baker [6.13]. Baker used a boundary integral method to solve the potential problem. He calculated the corresponding frequency response by using a minimum number of grids, depending on the frequency intervals required in the frequency response. The effect of a sublayer of infinite permeability of a similar thin-film head has also been calculated numerically by Baker and Koziol [6.14]. For idealized symmetric, shielded magnetoresistive heads, Heim [6.15] applied the Schwarz-Christoffel transformation to obtain the sensitivity function in the x domain. Finegan [6.16] discussed an algorithm for evaluating conformal transformations on a personal computer.

The results of non-ideal or practical heads are more difficult to describe. Fig. 6.10 can therefore only be seen as a very rough comparison of sensitivity functions.

In Fig. 6.11 the playback signals of different tapes as calculated in this section and normalized to $M_p h_x^a v \mu_0 W$ are plotted as a function of k. Multiplication of the curves in Fig. 6.10 by the curves in Fig. 6.11 gives some idea of the playback behaviour of all the combinations. The result (6.29) for a PH is precisely the opposite of that for an RH; see also Fig. 6.10. For a probe-type head (PH) the concentration of the field at the (main) pole leads to two peaks in the sense field that are shifted towards each other (the edges of the main pole) and are steeper, i.e. contain higher wavenumber components.

For a ring-head peaks are formed at the outer corner of the gap instead of a more or less uniform distribution of H_x over the gap, giving an increase of the gap-null wavelength and a decrease of the high wavenumber content of h_x, except for the wavenumbers beyond the first gap null, which are not of interest. When the efficiency of the PH does not decrease dramatically in the case of a DL medium, which seems a reasonable assumption, the read sensitivity of a probe-type head is increased by the presence of a backlayer, in contrast with the ring head as sketched in Fig. 6.10. Combining the rough sketches in

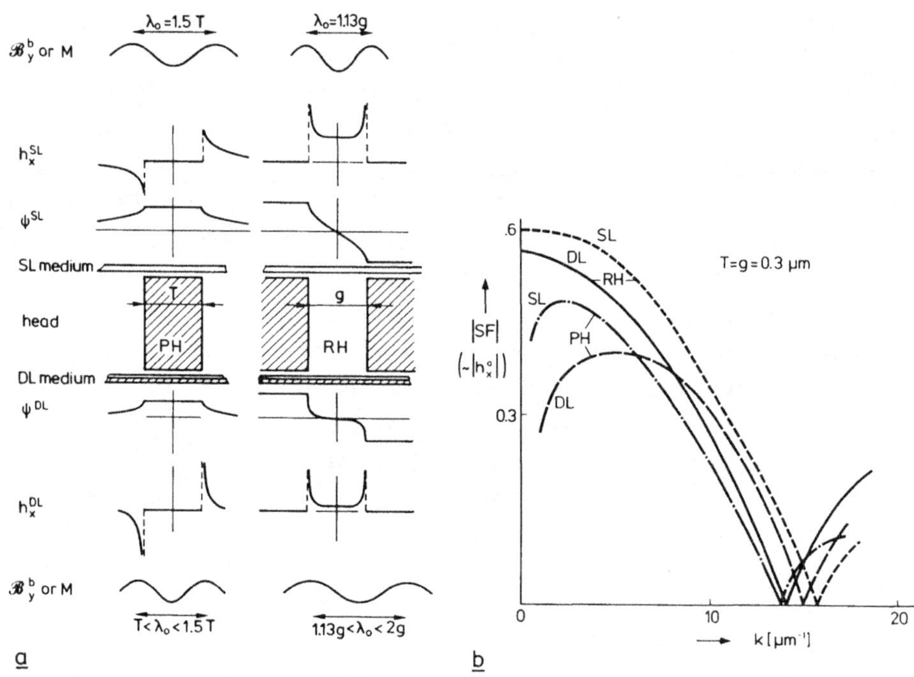

Fig. 6.10. Qualitative sketch of sense fields and sensitivity functions of the probe type head (PH) and ring head (RH). The situations with a unit-permeability tape (SL) and highly permeable tape (*DL*) are sketched.
a) Sketch of the dimensionless potential ψ and normalized field component h_x at pole-surface level. Each depicted sinusoidal magnetization M or magnetic induction component $\mathscr{B}_y$ has such a wavelength (the gap-null wavelength λ_0) that, weighted with the field or potential, the result is zero. The absolute value of the potential of the PH is assumed to decrease more than that of the RH when a high permeability DL medium is placed in front of the head which 'short-circuits' the head.
b) Rough sketch of the corresponding sensitivity functions. On a DL medium the SF of the PH tends to $\sin(kT/2)$ and the SF of the RH tends to $\cos(kg/2)$ because of the Fourier transforms of the symmetric and antisymmetric $h_x^{DL}(x)$ respectively.

Fig. 6.10 with the calculated curves in Fig. 6.11 leads to the conclusions that for playback a probe-type head (PH) with a perpendicularly- recorded double-layer medium (DL_y) forms a good combination but a ring head with a DL_y or SL_y surely cannot be excluded. For longitudinal recording the single-layer medium (SL_x) in combination with the ring head (RH) is undoubtedly the best combination.

A disadvantage of the sensitive combination of a PH with a DL_y medium is that the combination of a PH with a DL medium is most sensitive to distance variations. In this respect head and medium act together:

— The distance loss (assuming that the head sensitivity function is not

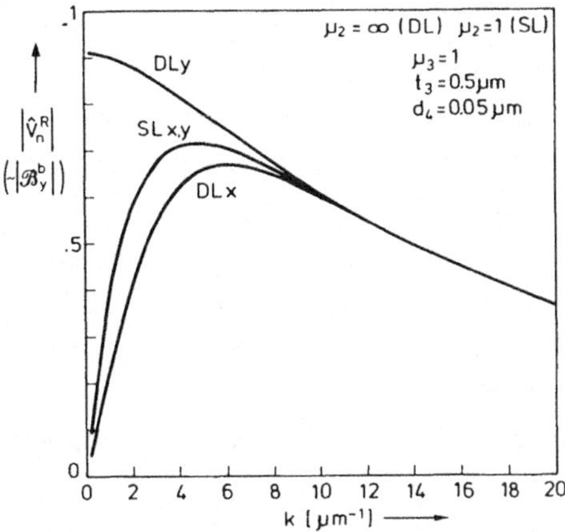

Fig. 6.11. The normalized output voltage $|\hat V^R_n| = |\hat V^R|/\left(|\hat M_p h^a_x|\nu\mu_0 W\right) \sim |\mathscr{B}^b_y|$ according to equations (6.26) and (6.27).
The playback of the perpendicular magnetization component is favoured by the use of a DL medium, while the playback voltage of the longitudinal component is reduced. For $k \to 0$, $|\hat V^R_{nDLx}|/|\hat V^R_{nSLx,y}| = 0.5\,t_3/(t_3 + d_4) \simeq 1/2$ and $|V^R_{nDLy}| = t_3/(t_3 + d_4) \simeq 1$.

altered, see (6.18) and (6.25)), is larger for a DL than for an SL medium. This factor is tape-dependent but head- and magnetization component-independent.

– The sensitivity function $|SF|^{DL}$ of a PH decreases with increasing distances in the frequency range of interest; see the sketch in Fig. 6.10 and calculations in chapter 9 on PHs.

The sensitivity function of the ring head, on the other hand, increases with increasing head-type distances (higher efficiency *and* higher gap-null wavenumber) which compensates partly the tape-dependent distance-loss factor. Together with the reasonable playback sensitivity of this combination the ring-head looks favourable for the playback of vertically recorded DL media, especially when reproducible results are required.

Experimental studies of spacing loss in perpendicular recording can be found in papers by Iwasaki et al. [6.17] and Yamamoto et al. [6.18].

Theoretical and experimental studies of spacing loss in both longitudinal and perpendicular recording, including the write process, can be found in [6.19], but the model assumptions underlying the theoretical analysis are not well documented in that paper.

6.3.10 Use of both sense-field components

In section 5.2.4, see equation (5.11), it was shown that the sensitivity ratio for perpendicular and longitudinal magnetization components is given by the ratio of the Fourier transforms of the corresponding sense-field components in the considered infinitely thin layer. (We worked with the alternative reciprocity theorem using only the Fourier transform of the longitudinal field at a plane $y = y_p$, i.e. the 'sensitivity function' $h_x^a(y_p; k)$ determined the response to both the perpendicular and longitudinal magnetization components once $\hat{\mathcal{B}}_y^b$ was known.) In working with these Fourier transforms in the common reciprocity theorem, see (5.11), and dividing them into the part corresponding to the head and the part corresponding to the image of the head in the ideal backlayer we are in fact using the ideas of Tsuboi and Fushida as employed in geophysics [6.20]; see a recent article about this method by Minuhin [6.21]. For certain qualitative descriptions, e.g. comparison of playback of perpendicularly and horizontally stored information, this method (in fact the use of (5.11) as a general expression for $\hat{\mathcal{B}}_y^{b*}$) is perhaps more suited than a description based on the very detailed but less general equation (2.63) we used in the previous sections. Minuhin also concludes that *a probe-type head in combination with a double layer medium* (perpendicularly magnetized, permeability one) *causes most severe separation losses*. Another conclusion of Minuhin that coincides with our results is that these *separation losses are exactly the same for perpendicularly and longitudinally magnetized media* (where we must add 'when the same read head is used'). See (6.24), where both read-back voltages are influenced in the same way by the term(s) containing d_4. Evidently the same is true for SL media, see (6.17) and (6.18). This statement is generally valid, see (2.64); it is not necessary to distinguish between different magnetizations, whether or not the tape has a unit permeability, etc.

6.3.11 DL versus SL, including gap smear

Many of the previous statements can be generalized for DL and SL media with coatings of permeable material, even when gap smear (i.e. $t_5 \neq 0$) is included. For instance, with the aid of (6.11) the two different modes, perpendicular and longitudinal permanent magnetization, can easily be compared with each other in the case of DL as well as SL media, especially when it is assumed for a moment, as before, that $t_x = t_y$, and for simplicity that M_{px} and M_{py} are uniform over the coating thickness. From the structure of (6.11) it is obvious that the following relation is very useful in this comparison:

$$\frac{\sinh(\beta_3 k(t_3 - t_x) + \alpha'_{32}) - \sinh(\beta_3 k t_3 + \alpha'_{32})}{\cosh(\beta_3 k(t_3 - t_x) + \alpha'_{32}) - \cosh(\beta_3 k t_3 + \alpha'_{32})} =$$
$$= \coth(\beta_3 k(t_3 - \tfrac{1}{2}t_x) + \alpha'_{32}) \geqslant 1. \tag{6.30}$$

Realizing that the upper functions in (6.11) are valid when $\mu_{32} > 1$ and the lower functions when $\mu_{32} < 1$ (see definitions in (6.1)) relation (6.30) applied to (6.11) states that for

1) $\boldsymbol{\mu_{32} > 1}$ **at all wavelengths**, i.e. $\mu_3 > \mu_2$, e.g. an **SL medium** with $\mu_3 \neq 1$

$$\boxed{\frac{|\mathcal{V}^R_{SL_x}|}{|\mathcal{V}^R_{SL_y}|} \geqslant 1} \tag{6.31}$$

for all ω if $|\hat{M}_{px}|/\beta_3$ at least equals $|\hat{M}_{py}|$.

This factor approaches $\mu_3/\beta_3 = \mu_{3y}$ at lower frequencies (since in (6.30) $\coth \to \mu_{32} \to \mu_3$ when $k \to 0$) and approaches $1/\beta_3$ for high frequencies ($\beta_3 k(t_3 - \tfrac{1}{2}t_{x,\,y}) \gg 1$, i.e. $\lambda \ll 2\pi\beta_3(t_3 - \tfrac{1}{2}t_{x,\,y})$) if $|\hat{M}_{px}| = |\hat{M}_{py}|$ is assumed. Thus the playback of the longitudinally recorded information is further favoured when the medium is longitudinally oriented, i.e. $\beta_3 < 1$.

2) $\boldsymbol{\mu_{32} < 1}$, i.e. $\mu_3 < \mu_2/\coth(\beta_2 k t_2 + \alpha_2)$, e.g. a **DL medium**

$$\boxed{\frac{|\mathcal{V}^R_{DL_x}|}{|\mathcal{V}^R_{DL_y}|} < 1} \tag{6.32}$$

for moderate and high frequencies where $\lambda \leqslant 100$ µm if $|\hat{M}_{\mathrm{py}}|$ at least equals $|\hat{M}_{\mathrm{px}}|/\beta_3$. (The maximum wavelength follows for the 'worst case' practical circumstances $t_2 = 0.16$ µm, $\beta_2\mu_2 = 1000$ and $\mu_3 = 10$ from $\mu_3 = \mu_2/\coth(\beta_2 k t_2 + \alpha_2)$ which, because of $\mu_2 \gg \mu_3 \geqslant 1$, can be rewritten as $\beta_2 k t_2 + \alpha_2 = \mu_3/\mu_2$, so that $\beta_2 k t_2 > \mu_3/\mu_2$ or $\lambda < 2\pi\beta_2 t_2\mu_2/\mu_3 = 100$ µm suffices for (6.32) to hold.) For a hypothetical backlayer of infinite permeability (6.32) would hold for all frequencies and an extreme $|\hat{V}^R_{\mathrm{DL}_y}|$ would occur for $k \to 0$, as was shown in Fig. 6.10 for an idealized DL medium. However, for a practical backlayer is μ_{32} larger than 1 at (very) long wavelengths ($\lambda \gg 100$ µm) so that '(6.31)' holds for the long wavelength region and therefore also for a DL medium the ratio $|\hat{V}^R_{\mathrm{DL}_x}|/|\hat{V}^R_{\mathrm{DL}_y}|$ approaches μ_{3y} when $k \to 0$. A large μ_{3y} causes a large demagnetization of the long-wavelength perpendicular components. However this region is far away from the region of interest. At short wavelengths $\lambda \ll \pi\beta_3 t_3$ the factor approaches $1/\beta_3$ if $|\hat{M}_{\mathrm{px}}| = |\hat{M}_{\mathrm{py}}|$ is assumed. *The practical backlayer is ineffective at (very) long wavelengths* since it is too thin, *and at short wavelengths* since it is too far away from the head-side of the coating.

3) $\boldsymbol{\mu_{32} = 1}$ **for all wavelengths**, i.e. $\mu_3 = \mu_2/\coth(\beta_2 k t_2 + \alpha_2)$ for all k, e.g. $\mu_3 = \mu_2 = 1$ or **ideal tape without a backlayer (ISL)**

$$\boxed{|\hat{V}_{\mathrm{ISL}_x}| = |\hat{V}_{\mathrm{ISL}_y}|} \tag{6.33}$$

for all ω if $|\hat{M}_{\mathrm{px}}| = |\hat{M}_{\mathrm{py}}|$.

6.3.12 DL versus SL, including gap smear and non-uniform magnetization

Equation (6.30) was specially suited for application to (6.11) which described uniform permanent magnetization over the coating thickness. The results (6.31), (6.32) and (6.33) are also valid for non-uniform permanent magnetization distributions. This follows easily from inspection of the general expression for non-uniform magnetization distributions (6.1). The relation that replaces (6.30) obviously is

$$\frac{\cosh(\beta_3 k(t_3 - y) + \alpha'_{32})}{\sinh(\beta_3 k(t_3 - y) + \alpha'_{32})} = \coth(\beta_3 k(t_3 - y) + \alpha'_{32}) \geqslant 1. \tag{6.34}$$

Compared to (6.30) only $\frac{1}{2}t_x$ ($= \frac{1}{2}t_y$) is replaced by the variable y, which means, for example, that *when $M_{px}(y)$ and $M_{py}(y)$ have the same y dependence all previous results hold.*

6.4 Coating; constraints on μ_3 and β_3

6.4.1 Introductory remarks

It cannot be said in general which easy-axis orientation is the best. This depends on a lot of factors whose influence may change with orientation and permeability
- writability (depends on g and B_s of write head, H_c, M_r and square-ness of the tape, as explained in chapter 7)
- noise (particulate noise, modulation noise, rubbing noise)
- irreversible self-demagnetization and reversible demagnetization after leaving the write head and reversible remagnetization during playback.

Only the last two reversible processes have been described here in the context of the playback process. These processes depend on β_3 and μ_3 i.e. orientation and mean permeability. In an idealized particulate medium β_3 depends on the geometric length/width ratios of the individual particles, their crystal anisotropy and their orientational distribution and to a lesser extent on the volumetric packing density v if not $v \rightarrow 1$. μ_3 mainly depends on v.

When we consider the playback process only, interesting questions are:
1) What combinations of μ_3, β_3, M_r are physically allowable?
2) Which of these allowable combinations gives the highest playback voltage?

The second question can be answered by calculations using the expressions given in this and preceding chapters.

The first basic question however is difficult to answer.

A start will be made in this and following sections.

The physical model for the particulate tape used here is approximately equal to the model of Stoner and Wohlfarth [6.22]. Results apply to assemblies of idealized non-interacting particles, i.e. $v \ll 1$. Results also apply to uniaxial continuous media ($v = 1$) with e.g. distributed easy-axis orientation (dispersion).

6.4.2 Model assumptions and simplifications

For simplicity the particles are assumed not to interact 'directly' with each other. Only 'indirect' interaction, i.e. self-demagnetization via the magnetization pattern, is then possible. This can be a rough approximation of reality when v is not very small [6.36].

Within each ellipsoidal particle the magnetization is assumed to change coherently (the applied field is approximately uniform) by rotation; domain walls are excluded. This is justified for small particles by the effect that the exchange energy A ($\simeq 1.5\ 10^{-12}$ J/m) has on the ordering of the spins in a ferromagnetic body. Above certain critical sizes of the particle mechanisms other than coherent rotation become favourable. For details concerning the critical size see Tebble and Craik [6.23]. They also summarized the influences that non-uniform rotation processes such as curling and buckling have on reducing the critical field of isolated particles and the influence that fanning in chains of (spherical) particles has on increasing the critical field. Fanning within an elongated particle reduces the critical field.

When the particles are not ellipsoidal, an exact analytical treatment becomes impossible. Then they can better be treated as being of ellipsoidal shape and uniformly magnetized rather than assuming uniform magnetization (i.e. coherent rotation) in the actual non-ellipsoidal particle. The assumption of uniform magnetization in a non-uniform particle would lead to over-estimated free-energy densities E_f at sides of the particle normal to the magnetization direction. This would give rise to torques and thus change the magnetization direction at these sides so that finally a more homogeneous energy density distribution is obtained. Only in very small particles ($< 0.1\ \mu$m) where exchange dominates, is a uniform magnetization possible in a non-ellipsoidal particle.

6.4.3 Susceptibility tensor of ellipsoidal particle at small fields

In Fig. 6.12 the ellipsoid corresponding to a non-ellipsoidal particle, for simplicity a rectangle, is defined. For the general ellipsoid the reversible susceptibilities in the direction of the three axes are solved from the two simultaneous torque equations for this three-dimensional ferromagnetic particle, i.e. with constraint $M_x^2 + M_y^2 + M_z^2 = M_{si}^2$:

$$\partial E_f/\partial\theta' = \partial E_f/\partial\varphi' = 0, \qquad (6.35)$$

where the free energy E_f equals the sum of magnetostatic and anisotropy energy, θ' is the polar angle and φ' the azimuthal angle of the magnetization relative to the major axes x' respectively the $x'y'$ plane and M_{si} is the saturation magnetization of an individual particle.

After straightforward calculations, slightly deviating from the calculations of Stoner and Wohlfarth [6.22], the following (implicit) expressions are obtained for the magnetization components,

$$M_{x'} = \sqrt{M_{si}^2 - (M_{y'}^2 + M_{z'}^2)} \tag{6.36a}$$

$$\frac{M_{y'}}{H_{y'}} = \frac{1}{H_k/M_{si} + (N_{y'} - N_{x'}) + H_{x'}/M_{x'}} \tag{6.36b}$$

$$\frac{M_{z'}}{H_{z'}} = \frac{1}{H_k/M_{si} + (N_{z'} - N_{x'}) + H_{x'}/M_{x'}}, \tag{6.36c}$$

where
-    H_k is the uniaxial crystal anisotropy field in the x' direction defined as $2K/(\mu_0 M_{si})$
-    $\boldsymbol{H} = (H_{x'}, H_{y'}, H_{z'})$ is the homogeneous field applied to the particle.
-    $N_{x'}$, $N_{y'}$, and $N_{z'}$ are the demagnetizing factors defined as $N_{x'} \equiv -H_{dx'}/M_{x'}$ with $\boldsymbol{H}_d$ the demagnetizing field within the particle.

$N_{x'} + N_{y'} + N_{z'} = 1$, see e.g. [6.24] in which handbook explicit expressions for ellipsoids of revolution ($N_{y'} = N_{z'}$) can also be found.

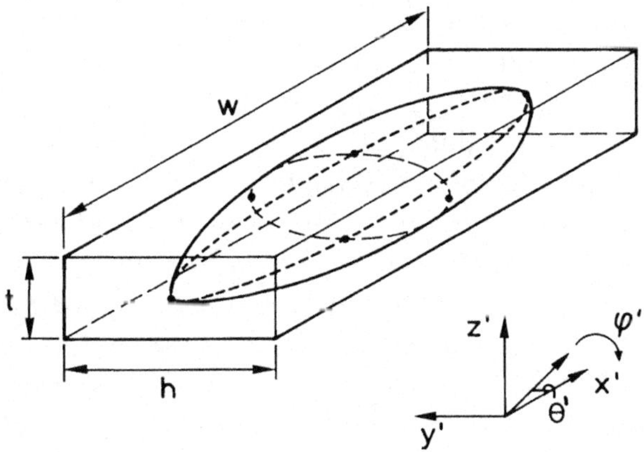

Fig. 6.12. A non-uniformly magnetized rectangular box approximated by a homogeneously magnetized ellipsoid. The magnetization direction is denoted by the polar and azimuthal angles θ' and φ'.

An (elliptic) integral expression for the demagnetizing factors for general ellipsoids was first given in 1881 by Maxwell [6.25] in his famous 'Treatise on Electricity and Magnetism'. He used the results derived by Dirichlet in 1839 for a similar problem in geophysics, the gravitational potential in an ellipsoid with uniform density. Maxwell's result reads:

$$N_v = \frac{abc}{2} \int_0^\infty \frac{ds}{(v^2 + s)R_s},$$
(6.37)

where $v = x'$, y', z' and a, b and c are the major axis and minor semi-axes and $R_s \equiv \sqrt{(a^2 + s)(b^2 + s)(c^2 + s)}$.

Stoner [6.26] expressed this result in terms of other elliptic integrals (Legendre's normal integrals of the first and second kinds), since these could be found in tables. Comprehensive tables and a graph for the demagnetizing factors of general ellipsoids have been given by Osborn [6.27]. Today a numerical evaluation of the simpler form (6.37) is easy.

The results (6.36a), (6.36b) and (6.36c) applied to ellipsoids of revolution, i.e. prolate ($t = h < W$) and oblate ($t = h > W$) spheroids, are consistent with those of Stoner and Wohlfarth. They also examined the general ellipsoid, but only for fields that lie in a principal plane of the ellipsoidal particle, while we are interested in all directions because of application to media with distributed orientations. On the other hand, they considered the stability of the solutions, which involves most of the work. We will consider mainly elongated particles, with or without a subsidiary (crystal) easy axis in the length (x') direction and not too large fields for which the above solutions are undoubtedly stable (energy minima $\partial^2 E_f/(\partial\theta')^2 > 0$). It is only for larger fields with components in the reversed direction that the above equations lead to unstable ($\partial^2 E_f/(\partial\theta')^2 = 0$) or invalid ($\partial^2 E_f/(\partial\theta')^2 < 0$) solutions and consequently to irreversible (switching) effects. Effects such as coercivity and associated power consumption are then revealed. Stoner and Wohlfarth calculated the MH curves for prolate and oblate spheroids and for an assembly of such particles oriented at random. For playback we are only interested in demagnetization-remagnetization processes when the playback head is in the vicinity of the tape for which the small-field behaviour, i.e. linear susceptibilities without switching effects, suffices. Then (6.36) relaxes to

$$\boxed{\chi_{x'} \equiv \left(\frac{M_{x'}}{H_{x'}}\right)_{H \to 0} = 0} \qquad (6.38a)$$

$$\boxed{\chi_{y'} \equiv \left(\frac{M_{y'}}{H_{y'}}\right)_{H \to 0} = \frac{1}{(N_{y'} - N_{x'}) + H_k/M_{si}}} \qquad (6.38b)$$

$$\boxed{\chi_{z'} \equiv \left(\frac{M_{z'}}{H_{y'}}\right)_{H \to 0} = \frac{1}{(N_{z'} - N_{x'}) + H_k/M_{si}}} \; . \qquad (6.38c)$$

Thus the tensor susceptibility reads

$$\boldsymbol{\chi'} = \begin{pmatrix} 0 & & \\ & \chi_{y'} & \\ & & \chi_{z'} \end{pmatrix} . \qquad (6.39)$$

For an assembly of non-interacting particles, volumetric packing density ν and equal orientation

$$\boldsymbol{\chi'} = \begin{pmatrix} 0 & & \\ & \nu\chi_{y'} & \\ & & \nu\chi_{z'} \end{pmatrix} . \qquad (6.40)$$

The elements of the relative permeability tensor $\boldsymbol{\mu'}$ follow easily by adding 1 to the three diagonal elements in (6.39) and (6.40).

6.4.4 Orthonormal-transformation matrices

In Fig. 6.13a the orientation of an elongated particle in a particulate medium is defined by the polar angle θ relative to the longitudinal (x) direction and the azimuthal angle φ relative to the xy plane. The principal axes of the ellipsoid with unit vector $\boldsymbol{e}_{x'}$, $\boldsymbol{e}_{y'}$ and $\boldsymbol{e}_{z'}$ coincide with the unit vectors $\boldsymbol{e}_x$, $\boldsymbol{e}_y$ and $\boldsymbol{e}_z$ when $\theta = \varphi = 0$. The coordinate-transformation matrix $\boldsymbol{T}$ defined in

$$\begin{pmatrix} \boldsymbol{e}_{x'} \\ \boldsymbol{e}_{y'} \\ \boldsymbol{e}_{z'} \end{pmatrix} = \boldsymbol{T} \begin{pmatrix} \boldsymbol{e}_x \\ \boldsymbol{e}_y \\ \boldsymbol{e}_z \end{pmatrix} \qquad (6.41)$$

is found by differentiation of the vector $\boldsymbol{r} = (x, y, z)$ to $|\boldsymbol{r}|$, θ and φ and normalizing the resulting vectors to unit length. The matrix $\boldsymbol{T}$ can be used to transform any vector from the (x, y, z) to the (x', y', z') space and reads

$$\boldsymbol{T} = \begin{pmatrix} \cos\theta & \sin\theta\cos\varphi & \sin\theta\sin\varphi \\ -\sin\theta & \cos\theta\cos\varphi & \cos\theta\sin\varphi \\ 0 & -\sin\varphi & \cos\varphi \end{pmatrix}. \tag{6.42}$$

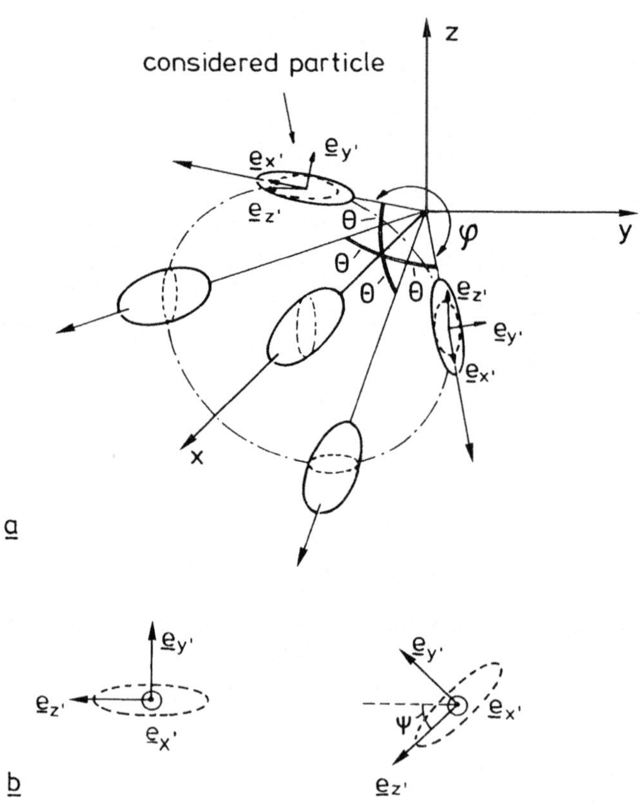

Fig. 6.13. Particles with different directions of their major and minor axis. The coordinate system x', y', z' is fixed to the particles. $\boldsymbol{e}_{x'}$ is the unit vector in the major axis direction with positive x component. $\boldsymbol{e}_{y'}$ and $\boldsymbol{e}_{z'}$ are unit vectors fixed to the two minor-axes directions.

a) Characterization of the considered particle direction and orientation by the polar angle θ with respect to the x axis and the azimuthal angle φ with respect to the xy plane ($\approx 90°$ for the considered particle). Note that by the definitions chosen in this figure the minor axis direction $\boldsymbol{e}_{z'}$ is always alongside the circumference of the dashed-dotted circle around the x axis and hence cannot yet be chosen freely.

b) Definition of the rotation angle ψ of the minor axes directions of the actual particle *with respect to the minor axes directions obtained in fig. 6.13a for the considered particle.*

Any particle direction and orientation can be described by the three variables θ, φ and ψ. Note that for $\theta = 0$ the angles φ and ψ have an equal effect and that the *order* is chosen to be θ, φ, ψ and not ψ, θ, φ.

In the inverse matrix T^{-1} the non-diagonal elements are interchanged. T^{-1} thus equals the transposed matrix T^{T}.

For $\theta = 0$, choice of φ makes every orientation of the minor axis of the particle possible (i.e. every 'rotation' of the particle's minor axes around the x' axis).

However for $\theta \neq 0$ this is not possible, see Fig. 6.13a; the above-mentioned rotation is fixed by the choice of the direction of the major axis of the particle. To make this rotation still possible, the rotation angle ψ is introduced, see Fig. 6.13b. For $\theta = 0$, changes in ψ are equivalent to changes in φ. The matrix that describes the rotation over the angle ψ in the clockwise direction around the x' axis equals according to (6.42) by taking $\theta = 0$ and replacing φ by ψ:

$$T_{\mathrm{rot}} = \begin{pmatrix} 1 & 0 & 0 \\ 0 & \cos\psi & \sin\psi \\ 0 & -\sin\psi & \cos\psi \end{pmatrix}. \tag{6.43}$$

This rotation must be carried out *after* the definition of the direction of the major axis of the particle by θ and φ according to Fig. 6.13a. Hence, for the most general cases, T in (6.42) has to be replaced by the matrix product $T_{\mathrm{rot}}T$ (and not TT_{rot}). This replacement leads to the new T:

$$T = \begin{pmatrix} \cos\theta & \sin\theta\cos\varphi & \sin\theta\sin\varphi \\ -\cos\psi\sin\theta & \cos\psi\cos\theta\cos\varphi - \sin\psi\sin\varphi & \cos\psi\cos\theta\sin\varphi + \sin\psi\cos\varphi \\ \sin\psi\sin\theta & -\sin\psi\cos\theta\cos\varphi - \cos\psi\sin\varphi & -\sin\psi\cos\theta\sin\varphi + \cos\psi\cos\varphi \end{pmatrix}$$

$$\tag{6.44}$$

For $\psi = 0$ this equals T in (6.42) and for $\theta = 0$ the effect of ψ is equivalent to the effect of φ, as was expected. Also $T_{\mathrm{rot}}^{-1} = T_{\mathrm{rot}}^{\mathrm{T}}$ and $T^{-1} = T^{\mathrm{T}}$.

6.4.5 Transformed susceptibility tensor

When a superscript ' (accent) refers to quantities in the reference frame x', y', z' of the particle and no superscript to quantities in reference frame x, y, z of the tape, then:

$$M' = TM \qquad \text{and} \qquad H' = TH \tag{6.45a}$$

such that

$$M = T^{-1}\chi' T H \quad \text{i.e.} \quad \chi = T^{-1}\chi' T. \qquad (6.45b)$$

Substitution of (6.39) and (6.44) in (6.45b) leads to the general expression in the x, y, z coordinate system for the susceptibility tensor of an individual particle χ, with tensor elements χ_{ij} (i = row, j = column):

$$
\begin{aligned}
\chi_{11} &= T_{21}^2\chi_{y'} + T_{31}^2\chi_{z'} \\
\chi_{21} &= \chi_{12} = T_{21}T_{22}\chi_{y'} + T_{31}T_{32}\chi_{z'} \\
\chi_{31} &= \chi_{13} = T_{21}T_{23}\chi_{y'} + T_{31}T_{33}\chi_{z'} \\
\chi_{22} &= T_{22}^2\chi_{y'} + T_{32}^2\chi_{z'} \\
\chi_{32} &= \chi_{23} = T_{22}T_{23}\chi_{y'} + T_{32}T_{33}\chi_{z'} \\
\chi_{33} &= T_{23}^2\chi_{y'} + T_{33}^2\chi_{z'}
\end{aligned}
\qquad (6.46)
$$

The tensor χ is symmetric and the sum of the diagonal elements is unchanged by the orthonormal coordinate transformation, i.e. $\chi_{11} + \chi_{22} + \chi_{33} = \chi'_{11} + \chi'_{22} + \chi'_{33} = \chi_{y'} + \chi_{z'}$, as can easily be checked.

6.4.6 Permeability tensor of arbitrary assembly of particles

The average $\langle \chi \rangle$ of an assembly of particles is obtained by integration over all possible θ (between 0 and $\pi/2$), all possible φ (between $-\pi$ and $+\pi$) and over all possible ψ (between $-\pi$ and $+\pi$), of the elements of the susceptibility tensor multiplied by the probability density function $p(\theta, \varphi, \psi)$ and the volumetric packing density.

When θ, φ and ψ are statistically independent, then $p(\theta, \varphi, \psi) = p(\theta)p(\varphi)p(\psi)$. For a random distribution $p(\theta) = \sin\theta$ and $p(\varphi) = p(\psi) = 1/2\pi$.

The relative permeability tensor follows generally by adding 1 to the diagonal elements of the susceptibility tensor.

Since the sum of the diagonal elements of the susceptibility tensor is invariant, this is also valid for the permeability tensor, i.e.

$$\mu_{11} + \mu_{22} + \mu_{33} = \chi_{11} + \chi_{22} + \chi_{33} + 3 = \chi_{y'} + \chi_{z'} + 3. \qquad (6.47)$$

6.4.7 Squareness vector of an arbitrary assembly of particles

The squareness, S, for a certain field direction is defined as the inverse of the ratio between the magnetization in that field direction when the field is infinite, M_{si}, and the magnetization component in the same direction when the field is reduced to zero after this premagnetization, M_r. This component is always positive because of the assumed premagnetization. For the three field directions x, y and z a squareness 'vector' is thus defined by

$$S \equiv \begin{pmatrix} S_x \\ S_y \\ S_z \end{pmatrix} = \begin{pmatrix} |M_x|/M_{si} \\ |M_y|/M_{si} \\ |M_z|/M_{si} \end{pmatrix}_{H=0} = \left| T^{-1} \begin{pmatrix} 1 \\ 0 \\ 0 \end{pmatrix} \right| = \begin{pmatrix} |\cos\theta| \\ |\sin\theta\cos\varphi| \\ |\sin\theta\sin\varphi| \end{pmatrix}.$$

$$(6.48)$$

The absolute signs are essential, and a result of the assumed premagnetization in the considered direction. $H = 0$ means that the magnetization is turned back to a major-axis (i.e. $+$ or $-x'$) direction, giving the column vector $(1\,0\,0)$ in the following expression where 1 stands for $M_{x'}/M_{si} = 1$ (or -1, which gives the same result because of the absolute sign).

The expression on the far right, obtained by substituting (6.44), shows that φ has no effect when $\theta = 0$ and that ψ never has an effect on the squareness.

For an assembly of non-interacting particles,

$$S = \begin{pmatrix} \langle |\cos\theta| \rangle \\ \langle |\sin\theta\cos\varphi| \rangle \\ \langle |\sin\theta\sin\varphi| \rangle \end{pmatrix}. \qquad (6.49)$$

6.4.8 Zero non-diagonal elements

In all expressions in the previous chapters we assumed zero non-diagonal elements in the permeability tensor, i.e. the magnetization (change) resulting from a field in the x or y direction must also be in the x or y direction. This is only the case when the probability density function p is symmetrical with respect to the xy and with respect to the zx plane.

Expression (6.48) applied to an assembly of particles with the above symmetry properties thus leads to a permeability tensor with the following non-zero tensor elements ($k = 1$, 2 and 3)

$$\mu_{kk} = 1 + \nu \iiint p(\theta, \varphi, \psi)(T^2_{2k}\chi_{y'} + T^2_{3k}\chi_{z'})\,d\theta d\varphi d\psi. \qquad (6.50)$$

Thus the main permeability

$$\mu \equiv \sqrt{\mu_x \mu_y} = \sqrt{\mu_{11}\mu_{22}} \qquad (6.51)$$

and the anisotropy constant

$$\beta \equiv \sqrt{\mu_x/\mu_y} = \sqrt{\mu_{11}/\mu_{22}}. \qquad (6.52)$$

6.4.9 Examples

$\psi = 0$ and uniform $p(\varphi)$

Assume for simplicity that $\psi = 0$ for all particles. In this case (6.46) reduces to

$$\mathbf{X} =$$

$$\begin{pmatrix} \chi_{y}' \sin^2\theta & -\chi_{y'} \sin\theta \cos\theta \cos\varphi & -\chi_{y'} \sin\theta \cos\theta \sin\varphi \\ -\chi_{y'} \sin\theta \cos\theta \cos\varphi & \chi_{y'} \cos^2\theta \cos^2\varphi + \chi_{z'} \sin^2\varphi & \chi_{y'} \cos^2\theta \sin\varphi \cos\varphi - \chi_{z'} \sin\varphi \cos\varphi \\ -\chi_{y'} \sin\theta \cos\theta \sin\varphi & \chi_{y'} \cos^2\theta \sin\varphi \cos\varphi - \chi_{z'} \sin\varphi \cos\varphi & \chi_{y'} \cos^2\theta \sin^2\varphi + \chi_{z'} \cos^2\varphi \end{pmatrix}$$

$$(6.53)$$

The sum of the diagonal elements is $\chi_{y'} + \chi_{z'}$.

It is further assumed that φ is statistically independent from θ and that there is rotational symmetry around the x axis, i.e. $p(\varphi) = 1/2\pi$. From (6.53) it then easily follows that the non-diagonal elements are zero as is expected, and we read for the assembly of particles:

$$\mathbf{X} = \begin{pmatrix} \nu\chi_{y'}\langle \sin^2\theta \rangle & & \\ & \tfrac{1}{2}\nu(\chi_{y'}\langle \cos^2\theta \rangle + \chi_{z'}) & \\ & & \tfrac{1}{2}\nu(\chi_{y'}\langle \cos^2\theta \rangle + \chi_{z'}) \end{pmatrix}$$

$$(6.54)$$

and

$$S = \begin{pmatrix} \langle |\cos\theta| \rangle \\ \frac{2}{\pi} \langle |\sin\theta| \rangle \\ \frac{2}{\pi} \langle |\sin\theta| \rangle \end{pmatrix}. \tag{6.55}$$

Randomly distributed minor axes

For $\theta = 0$ the particles of the previous example have randomly distributed minor axes $e_{y'}$ and $e_{z'}$ because of the assumption concerning $p(\varphi)$, thus leading to

$$\chi' = \begin{pmatrix} 0 \\ & \nu\chi' \\ & & \nu\chi' \end{pmatrix} \tag{6.56}$$

with scalar $\chi' \equiv \frac{1}{2}(\chi_{y'} + \chi_{z'})$.

For $\theta \neq 0$ and the assumed $p(\varphi) = 1/2\pi$, the minor axes are not randomly distributed, see Fig. 6.13a.

When the minor axes are randomly distributed, the susceptibility tensor can, instead of using the general but complicated expression (6.46) or using (6.53), be found by first averaging for a fixed θ over the randomly distributed minor axes. This simply means that $\chi_{y'}$, and $\chi_{z'}$, in (6.53) must be replaced by the average value χ' as in (6.56). Subsequently assuming rotational symmetry around the x axis, $p(\varphi) = 1/2\pi$, then leads to (6.54) with $\chi' \equiv \chi_{y'} = \chi_{z'}$:

$$\chi = \begin{pmatrix} \nu\chi' \langle \sin^2\theta \rangle \\ & \frac{1}{2}\nu\chi'(\langle\cos^2\theta\rangle + 1) \\ & & \frac{1}{2}\nu\chi'(\langle\cos^2\theta\rangle + 1) \end{pmatrix} \tag{6.57}$$

and to a squareness vector equal to (6.55).

In contrast to the previous example, χ_x is now determined by both $\chi_{y'}$ and $\chi_{z'}$. This was to be expected from Fig. 6.13a.

Randomly distributed major and minor axes

When the particles are distributed at random, then $p(\theta) = \sin\theta$ and $p(\varphi) = p(\psi) = 1/2\pi$. This is equivalent to assuming $\psi = 0$ and replacing $\chi_{y'}$ and $\chi_{z'}$ by $\chi' = \frac{1}{2}(\chi_{y'} + \chi_{z'})$ as in the above example. For a random distribution, $\boldsymbol{\chi}$ in (6.57) is thus averaged with the aid of $p(\theta) = \sin\theta$, leading to:

$$\boldsymbol{\chi} = \begin{pmatrix} \frac{2}{3}v\chi' & & \\ & \frac{2}{3}v\chi' & \\ & & \frac{2}{3}v\chi' \end{pmatrix} \tag{6.58}$$

The squareness vector equals:

$$\boldsymbol{S} = \begin{pmatrix} \frac{1}{2} \\ \frac{1}{2} \\ \frac{1}{2} \end{pmatrix} \tag{6.59}$$

6.4.10 MH loops

The results of the last example are consistent with the MH loop derived 'numerically' by Stoner and Wohlfarth (their fig. 7 in [6.22]) for randomly-oriented uniaxial crystals and ellipsoidal particles of revolution.

It was possible to derive explicit expressions here for the squareness and susceptibility or permeability because in the definitions of these quantities $\boldsymbol{H} \to 0$.

Other parts of the MH curves of individual particles follow from the simultaneous equations (6.36a, b and c). From the two solutions the one with the $M_{x'}$ component in the premagnetization direction ($+x'$ when $H_{x'p} > 0$ and $-x'$ when $H_{x'p} < 0$ during premagnetization) must be chosen, unless $\boldsymbol{H}$ has a component in the direction opposite to the premagnetization direction and the absolute value of this field $|\boldsymbol{H}|$ exceeds the critical value H_{cr} for which $\partial^2 E_f/(\partial\theta')^2 = 0$. We found for H_{cr} the following very accurate expression (better than one percent when compared to the numerically obtained values in Table 4 in [6.22]) valid for all angles θ'' between field and major-axis direction x' of a prolate spheroid:

$$H_{cr}(\theta'') \simeq \frac{H_{sw}}{1 + \left(\left|\sin(2\,\theta'')\right|\right)^{0.7}} \, , \qquad (6.60a)$$

where the switching field of the particle is defined as

$$H_{sw} \equiv H_{cr}(\theta'' = 0) = H_k + (N - N_{x'})M_{si} = H_k + 3N - 1$$

$$(6.60b)$$

with $N \equiv N_{y'} = N_{z'}$.

When $\boldsymbol{H} = (H_x, 0, 0)$, $\theta'' = \theta$, otherwise $\theta'' = f(\theta, \varphi, field\ direction)$.

The results from equation (6.36) and the above approximation (6.60) fit the data in Fig. 6.14. This figure is from Tebble and Craik (Fig. 11.5 in [6.23]) and was based on the results of Stoner and Wohlfarth, see also Fig. 6 in [6.22]. The abciss and ordinate are adapted to our purpose.

When $H_k = 0$ and the length/width ratio is very large, $N \rightarrow \tfrac{1}{2}$ and

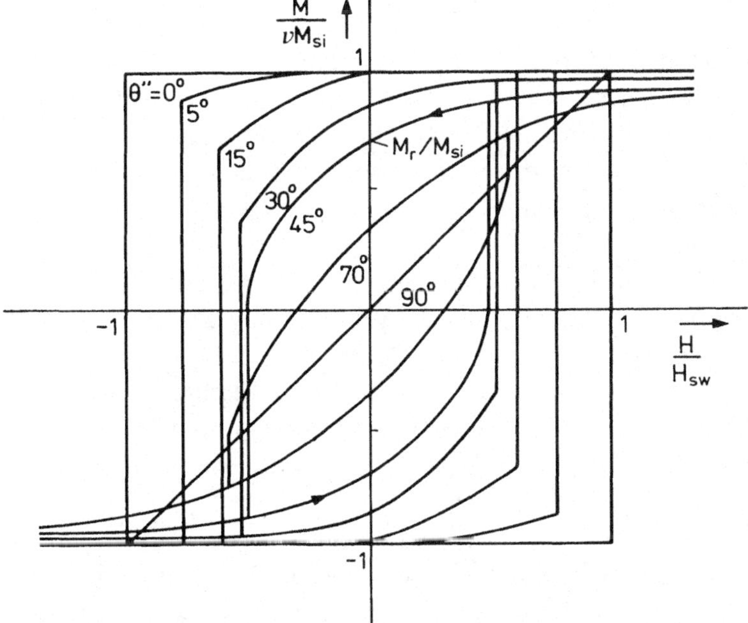

Fig. 6.14. MH loops of assemblies of non-interacting particles with volumetric packing density v $\ll 1$ and sharp orientational distributions. The angle between the orientation direction and the field is θ''. Uniaxial crystal anisotropy with easy axis in the major axis direction of the particles is allowed, see (6.36). For $v = 1$ the results apply exactly to uniaxial crystal anisotropy.

consequently the switching field H_{sw} approaches $\frac{1}{2}M_{si}$. Relative to the remanent magnetization of the assembly as a whole, M_r, we then find for the coercivity when $\theta'' = 0$ (and v is small), $H_c/M_r = 1/(2v)$. A broadening of the particle distribution away from the direction of the field, $\theta'' = 0$, will reduce the coercivity for that field direction.

This happens rather quickly, see Fig. 6.14. At larger v the H_c is smaller, due to interactions between parallel particles (effectively like thicker particles). Also the other possible rotation mechanisms such as curling and buckling, if relevant, can only reduce this coercivity further. The opposite effect due to fanning is not relevant here, since it was assumed that the particles were already strongly elongated. When fanning occurs within individual elongated particles, i.e. individual particles are split up into some 'domains', rotating in alternate directions so as to reduce the demagnetizing field, the coercivity is also reduced. This means that $H_c = M_r/(2v)$ determines the ultimate coercivity for shape anisotropy. Usually lower values are found.

The magnetization $\boldsymbol{M}$ of an assembly of non-interacting prolate spheroids in a non-vanishing field $\boldsymbol{H}$ (neglecting the demagnetizing field of the *assembly* of particles) can now be calculated in the following straightforward, although time-consuming, way:

- Transform $\boldsymbol{H}$ and $\boldsymbol{M}$ to $\boldsymbol{H}'$ and $\boldsymbol{M}'$ with the aid of $\boldsymbol{T}$, see (6.45a), and substitute in (6.36) with $N_{y'} = N_{z'} \equiv N$.
- Substitute the values of the components of $\boldsymbol{H}'$ in (6.36b) and (6.36c) and solve the simultaneous equations for $M_{x'}$, $M_{y'}$ and $M_{z'}$, giving two solutions.
- Calculate the critical field H_{cr} and choose from the two possible solutions for $\boldsymbol{M}'$ the right one as previously described and transform back $\boldsymbol{M}'$ with the aid of $\boldsymbol{T}^{-1}$ to $\boldsymbol{M}$, see (6.45a).
- Repeat this sequence of calculations for all particle directions, average the results by using the actual probability density $p(\theta, \varphi)$ $\left(\equiv p(\theta) \times p(\varphi)\right.$ when θ and φ are statistically independent$\left.\right)$ and multiply the result with v.

The MH loops in Fig. 6.14 are obtained for the simple case of a particulate medium with well-oriented prolate ellipsoidal particles with crystal anisotropy in the major-axis direction, premagnetized by a strong field.

For (vector) fields which gradually change, the direction of the component of $M_{x'}$ of a particle after the preceding step replaces the premagnetization direction in the choice of the right solution.

6.4.11 Interim discussion

From (6.49) it follows for particles oriented in the z direction ($\theta = 90°$, $\varphi = 90°$ and $\psi = 0°$, i.e. z' coincides with z) that the squareness ratios $S_x = S_y = 0$. This means that the maximum attainable amplitudes of the written information in the two possible head-field directions, M_{rx} and M_{ry}, are zero. Thus particles with their major axis x' in the z direction must be avoided.

In all playback analyses, see chapter 2 equation (2.1), the permeability tensors were restricted to those with zero non-diagonal elements, see also section 6.4.8. This is, for example, obtained for $p(\theta, \varphi, \psi)$ distributions that lead to minor and major orientational directions of the particle assembly as a whole, that correspond with the three orthonormal directions x, y and z. We will restrict ourselves to those situations here. Thus two major-axis assembly orientations remain when z is excluded for the above reason, the x or the y orientation. See in this context also particles 1 and 4 or 2 and 5 in Fig. 6.15.

The elongated particles oriented in the x direction may be flat in either the z or y direction, see particles 1 and 4. The elongated particles oriented in the y direction may be flat in the z or x direction, see particles 2 and 5.

Assume that the assembly of particles is well-oriented in the x or y direction (i.e. $x' = x$ or $x' = y$) and z' coincides with z. Then equation (6.50) or the much simpler equation (6.38) and equations (6.51), (6.52) and (6.49) lead directly to:

orientation of x' in x direction	orientation of x' in y direction
$\mu_x = 1$	$\mu_y = 1$
$\mu_y = 1 + \nu\chi_{y'}$	$\mu_x = 1 + \nu\chi_{y'}$
$\mu \equiv \sqrt{\mu_x\mu_y} = \sqrt{1 + \nu\chi_{y'}}$	$\mu = \sqrt{1 + \nu\chi_{y'}}$
$\beta \equiv \sqrt{\mu_x/\mu_y} = \sqrt{1/(1 + \nu\chi_{y'})} = 1/\mu$	$\beta = \sqrt{1 + \nu\chi_{y'}} = \mu$
$S_x = 1$	$S_y = 1$

Thus, in the major-axis (easy axis) direction of a particle the (small-field) reversible permeability is always 1, see also (6.38a). In the direc-

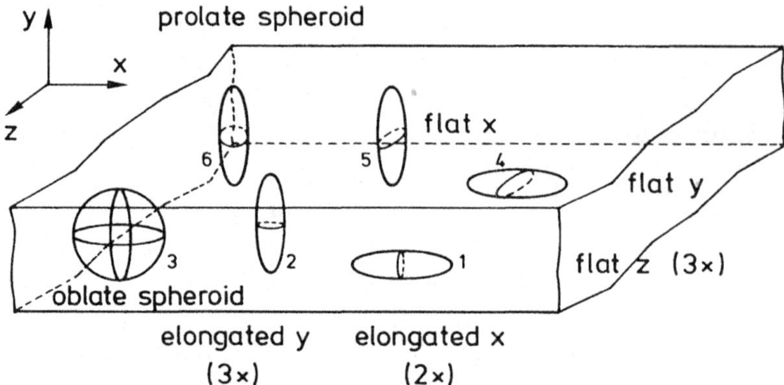

Fig. 6.15. Ellipsoidal particles with different orientations and shapes in a particulate tape. The two particles on the left side are ellipsoids of *revolution* and so-called oblate and prolate *spheroid*.

tions perpendicular to the major axis the reversible permeability is greater than 1, see also (6.38b) and (6.38c).

A large μ_y (or μ_x), i.e. a large $\chi_{y'} = 1/(N_{y'} - N_{x'} + H_k/M_{si})$, is obtained when $N_{y'} \to N_{x'} + H_k/M_{si}$. For a particle with negligible uniaxial crystal anisotropy this would be obtained when the cross section of the particle in the $x'y'$ plane is almost circular ($N_{y'} \to N_{x'}$). Rotation from the x' to the y' axis is then very easy. Flattening the particle in the z direction, see particle 1 in Fig. 6.15, will facilitate this rotation further, due to a further reduction of the self-demagnetizing fields in the $x'y'$ plane within the particle. Consequently, $N_{y'}$ and $N_{x'}$ and hence $N_{y'} - N_{x'}$ are reduced at the cost of an increase of $N_{z'}$ (since $N_{x'} + N_{y'} + N_{z'} = 1$).

However, the disadvantage of the above measures to obtain a more extreme β is that they reduce the coercivity. For the well-oriented case described here, the coercivity equals the switching field, see (6.60b), i.e. $H_c = H_k + (N_{y'} - N_{x'})M_{si} = M_{si}/\chi_{y'}$, (and not $H_k + (N_{z'} - N_{x'})M_{si}$, because we assumed $N_{z'} > N_{y'}$). A large $\chi_{y'}$ causes a small H_c.

An extreme value of β ($\ll 1$ for playback of longitudinal and $\gg 1$ for playback of perpendicular permanent magnetizations) gives good results at small head-to-tape distances. All internal longitudinal or perpendicular permanent flux is reproduced when the head-to-tape distance is zero, see Sections 6.2.1, 6.2.2 and 6.2.3, due to virtual shortening or virtual lengthening of the coating thickness, respectively. However, the necessarily high value of μ (at least $\mu = 1/\beta$ or $\mu = \beta$) causes strong demagnetization (in free space) and strong remagnetization (during playback). Hence the distance losses are considerably higher at small head-to-tape distances than usual, see (6.8) and (6.6).

As the orientation of the assembly becomes less favourable, in both the x and the y direction, then the reversible permeability in the main orientation direction will increase and the other will fall, such that β will become less extreme. This must be avoided in longitudinal or perpendicular recording; the disadvantage of a high permeability causing the large distance loss then would remain, while the advantage of reproducing a larger amount of all the available flux in the tape at small head-to-tape distances would be lost. In addition, the squareness S_x or S_y would fall (for a random distribution $p(\theta) = \sin(\theta)$ to $\frac{1}{2}$).

The low H_c with respect to M_{si}, associated with the extreme value of β is disastrous for small wavelength recordings. This is explained in chapter 7 and is due to the large 'erasing' effects during the recording and self-demagnetizing effects when H_c/M_s is small.

It is difficult to conclude here the optimum choice of β. This depends on the attainable head-tape contact, the relevant wavelengths and the chosen type of recording. Also if these values and choice are fixed, a precise determination of the most favourable β within the present model does not mean that this would be the best choice in practice. The restrictions of the models and the complexity of all the factors that determine the final signal-to-noise ratio are still too great.

Within the context of the present work, it is not possible to give much more than the above general recommendations concerning the orientation of the particles in the tape.

6.5 Concluding remarks

From the considerations concerning demagnetization and subsequent remagnetization it can be concluded that:

a) For a fixed permeability of the coating, μ_3, and wavelengths for which $\beta_3 k t_3 \gg 1$ and for each head-to-tape distance d_4,

   – maximal possible orientation in the longitudinal direction (β_3 minimal) is favourable for the reproduction of the longitudinal magnetizations.

   – maximum possible orientation in the perpendicular direction (β_3 maximal) is favourable for the reproduction of the perpendicular magnetizations.

b) For a fixed anisotropy β_3 the lowest possible μ_3 is favourable. When $\beta_3 kt_3 \gg 1$ the remagnetization-demagnetization process is completely dominated by μ_3 and no longer β_3, μ_2 and β_2.

c) Demagnetization is completely cancelled by remagnetization when $kd_4 \ll 1$ and the head is ideal, i.e. has a zero gap and infinite permeability. Then, as a consequence of the absence of net demagnetization, only anisotropy effects and no permeability effects affect the output.

d) All internal longitudinal flux is reproduced when $d_4 \rightarrow 0$, head is ideal *and*

1) $t_3 \rightarrow 0$ (no thickness loss)

or 2) $\beta_3 \rightarrow 0$ (no thickness loss).

e) All internal perpendicular flux is reproduced when $d_4 \rightarrow 0$, head is ideal *and*

1) $t_3 \rightarrow \infty$ and the coating is uniformly magnetized (no thinness loss)

or 2) $\beta_3 \rightarrow \infty$ (no thinness loss)

or 3) coating is uniformly magnetized and a backlayer with infinite permeability is used (no thinness loss).

f) The advantage of a reduced reproduction loss at $d_4 = 0$, obtained by 'artificial' means, i.e. by d2, e2 or e3, resulted in greater distance losses as more effort was made to reduce the reproduction loss.

g) Quantitatively the distance loss is different for the 5 cases mentioned above:

d1) Usual distance loss.

d2) Enhanced distance loss, especially when $\mu_{3x}kt_3$ is large (this is in conflict with the desirability of a small β_3 when kt_3 is large). $DL \simeq \left(1/(1 + \mu_{3x}kt_3/2)\right)\mathrm{e}^{-kd_4}$ when $kd_4 \gg 1$.

e1) Usual distance loss.

e2) Enhanced distance loss, especially when μ_3 is large which is inseparable from the intended high β_3 value. $DL \simeq \left(2/(1 + \mu_3)\right)\mathrm{e}^{-kd_4}$ when $kd_4 \gg 1$.

e3) Enhanced distance losses occur, especially when kt_3 is small (this is in conflict with the desirability of a backlayer when kt_3 is small). The usual distance loss is only approached in the wavelength region where the backlayer is ineffective, i.e. where $kt_3 > 1$.

h) For circular magnetization, an ideal zero-gap head and at non-zero head-to-tape distance d_4, an orientation of the medium leads to an output that is larger than that of an isotropic unit relative permeability medium in a certain wavelength band and is smaller at other wavelengths. Qualitatively it seems that:

1) a longitudinally oriented medium is more advantageous than an isotropic (unit relative permeability) medium when no backlayer is used for almost all wavelengths of interest,

2) a perpendicularly oriented medium is disadvantageous to an isotropic (unit relative permeability) medium for many wavelengths of interest, but improves when a backlayer is used.

i) A backlayer is advantageous for the playback of all perpendicularly stored information in the case of an ideal zero-gap head and has an unfavourable effect on the playback of all longitudinally stored information.

j) The influence of the thickness t_3 and the head-to-tape distance d_4 cannot be separated exactly in the case of a non-unit relative permeability of the coating or the backlayer.

For an ideal head all playback results, including those stated above, can be made quantitative by using the expressions in this chapter. For non-ideal heads a (numerical) calculation of the head's potential or sensitivity function, including the permeability effect of the tape, is first required.

From considerations concerning head fields, types of head and media can be concluded, independent of recordings being longitudinal or perpendicular, that:

k) The GLF changes with the head-to-tape distance d_4 and coating thickness t_3 when the tape has a non-unit relative permeability. This change effects the distance loss.

l) Use of a backlayer is advantageous for the playback sensitivity in the case of a probe-type head; the high-wavenumber content of the GLF is then larger.

m) Use of a backlayer is disadvantageous for the playback sensitivity of ring heads at moderate and high frequencies. Only for playback at frequencies beyond the first gap-null frequency is the backlayer advantageous.

n) The first gap-null frequency of a probe-type head is 2-times higher than of a ring head when $T = g$ and the distance to the infinite permeability backlayer approaches zero, and is 0.7 times lower at large head-to-tape distances or in the case of a unit permeability tape.

o) The combination of a probe-type head and a DL medium is most sensitive to head-to-tape distance variations, since the tape's $\mathscr{B}_y^b$ and the head's ψ^a are very sensitive to distance variations and moreover they act in this respect in the same direction.

From the considerations concerning longitudinally and perpendicularly stored information with equal y-dependences it can be concluded that:

p) Equal playback results are obtained for longitudinally and perpendicularly stored information if an ideal ($\mu_{3x} = \mu_{3y} = 1$) single-layer (ISL) medium is used.

q) Longitudinally stored information is benefited relative to perpendicularly stored information during playback when an SL medium with $\mu_{3y} > \mu_{3x}$ is used, especially when $\mu_{3y} \gg \mu_{3x}$, as would be the case for extremely-well oriented longitudinal media.

r) Perpendicularly stored information is benefited relative to longitudinally stored information during playback when a DL medium with $\mu_{3x} \geqslant \mu_{3y}$ is used, especially when $\mu_{3x} \gg \mu_{3y}$, as would be the case for extremely-well oriented perpendicular media.

s) The above-mentioned benefits are most pronounced at moderate wavelengths.

t) At short and (very) long wavelengths the above-mentioned benefits vanish when the coating is not permeable.

u) When the coating has a non-unit relative permeability, the readback of a horizontal component of (very) long wavelength is favoured by a factor μ_{3y} with respect to the readback of the vertical component and a short-wavelength horizontal component is favoured (or deteriorated) by the factor $\sqrt{\mu_{3y}/\mu_{3x}} = 1/\beta_3$ relative to the vertical component.

v) A backlayer is ineffective at short and very long wavelengths.

With regard to the constraints on tape parameters it can be concluded that:

w) Orientation of particles in the z direction must be avoided, since maximum attainable permanent-magnetization amplitudes are reduced or vanish.

x) Orientation in the xy plane is favourable, e.g. longitudinal or perpendicular orientation of the particles.

y) Flattening of the particles in the z-direction enhances the playback of the written information but reduces the coercivity (switching field) and thus may reduce the amplitude of the written permanent-magnetization pattern at small wavelength where self-demagnetizing fields are largest.

z) Almost oblate spheroids with negligible crystal anisotropy, i.e. particles with an easy xy plane, will have a mean permeability μ and an extreme anisotropy constant β, i.e. $\beta \ll 1$ or $\beta \gg 1$, but still must be avoided because of an almost vanishing coercivity-to-saturation magnetization ratio.

The question whether a DL medium is better than an SL or ISL ($\equiv$ SL with $\mu_3 = 1$) medium is not answered, neither is the question about the writeability of the coating. The assumed permanent magnetization distributions, however, are not unphysical, but, as in the case of the uniform permanent magnetization distribution, they are not always practically realizable. It is only assumed that the permanent states (or remanent states if initially saturation was reached during the write process) for perpendicular and longitudinal magnetization are equal and the BH curves are such that in both cases the demagnetization-remagnetization process is a reversible and linear process. Moreover, using a BH curve implicitly assumes that micro-magnetic effects do not play a

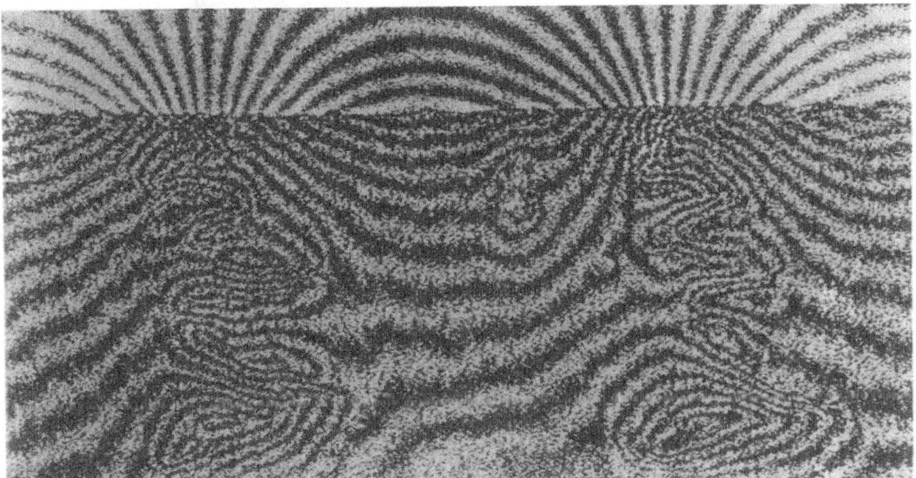

Fig. 6.16. Electron-holography picture of the edge of a magnetized tape (top view) after Yoshida et al. The regular distribution of stray-flux lines is visible in the free space at the top of the picture. Between every two 'flux lines' flows a fixed amount, 4.1 fWb, of flux. An irregular sawtooth-like domain wall characterizes the transition in the 45 nm thick Co film (longitudinally oriented since no Cr is added). The distance between the two transitions is 5 µm.

role on the scale of the magnetization reversals ($\simeq \lambda$). Especially in the case of perpendicular (continuous) media it is doubtful whether (at high bit densities) on the scale of the magnetization reversals it is allowed to use a BH curve as a representative of the micro-magnetic (domain walls?) state of the coating (and backlayer). As an illustration we mention that Yoshida et al. [6.28] observed effects due to sawtooth-like domain walls in (continuous) films of sputtered Co of 45 nm thickness and with longitudinal orientation (see Fig. 6.16). The electron holography method they used to observe the flux lines is proposed by Tonomura et al. [6.29]. Between every two lines flows a magnetic flux $h/e = 4.1 \times 10^{-15}$ Wb. For a realistic flux density of ca. 25 nW/m this means ca. 6 flux lines per μm track width. For a (longitudinal) particulate medium with particles that are short compared with the bit lengths, the use of a BH curve seems less contradictory since the finite distance between the particles (always more than some atomic distances) prevents the formation of domain walls via exchange coupling, which is a pure atomic short-range interaction [6.30].

References

[6.1]   Bertram, H. Neal., *Anisotropic reversible permeability effects in the magnetic reproduce process*, IEEE Trans. Magn., Vol. Mag-14, 111-118 (1978).

[6.2]   Siakkou, M., *Zur Wiedergabe beliebiger Magnetisierungsstrukturen mit einem Magnetkopf endlicher Spaltweite und einem Magnetband beliebiger Permeabilität*, Z. elektr. Inform.- und Energietechnik, 4 (6), 311-316 (1974).

[6.3]   Fan, George J., *A study of the playback process of a magnetic ring head*, IBM Journal, Oct. (1961).

[6.4]   Jorgensen, Finn., *The Complete Handbook of Magnetic Recording*, Tab Books Inc.: Blue Ridge Summit, P.A., 1980, p. 135.

[6.5]   Quak, Dirk., *Influence of the layer thicknesses of a double-layer medium on the reproduced signal in perpendicular recording*, IEEE Trans. Magn., Vol. Mag-19, 1502-1505 (1983).

[6.6]   Yamamori, K., Nishikawa, R., Asana, T. and Fujiwara, T., *Perpendicular magnetic recording performance of double-layer media*, IEEE Trans. Magn., Vol. Mag-17, 2538-2540 (1981).

[6.7]   Lopez, Orlando., *Reproducing vertically recorded information-double layer media*, IEEE Trans. Magn., Vol. Mag-19, 1614-1616 (1983).

[6.8]   Westmijze, W.K., *Calculation of the fields in and around the tape*, Philips Res. Rep. 8, 255-269 (1953).

[6.9]   Luitjens, S.B. and Herk, A. van., *A discussion on the crosstalk in longitudinal and perpendicular recording*, IEEE Trans. Magn., Vol. Mag-18, 1804-1812 (1982).

[6.10] Bertram, H. Neal., *Interpretation of spectral response in perpendicular recording*, IEEE Trans. Magn., Vol. Mag-21, September (1985).

[6.11]   Satake, S., Honda, N. and Hokkyo, J., *Formulation of readout process in perpendicular magnetic recording*, Journal of the Magnetics Society of Japan, Vol. 8, 81-85 (1984).

[6.12]   Minuhin, Vadim B., *Comparison of sensitivity functions for ideal probe and ring-type heads*, IEEE Trans. Magn., Vol. Mag-20, 488-494 (1984).

[6.13]   Baker, Bill R., *Effects of a sublayer on readback from a vertical film*, J. Appl. Phys. 55(6), 2217-2219 (1984).

[6.14]   Baker, Bill R. and Koziol. Mark R., *Response of asymmetrical thin film heads to vertical and horizontal magnetization*, IEEE Trans. Magn., Vol. Mag-17, 3123-3125 (1981).

[6.15]   Heim, David E., *The sensitivity function for shielded magnetoresistive heads by conformal mapping*, IEEE Trans. Magn., Vol. Mag-19, 1620-1622 (1983).

[6.16]   Finegan, J.D., *Algorithm for evaluating conformal transformations*, IEEE Trans. Magn., Vol. Mag-19, 2177-2179 (1983).

[6.17]   Iwasaki, S., Speliotis, D.E., and Yamamoto, S., *Head-to-media spacing losses in perpendicular recording*, IEEE Trans. Magn., Vol. Mag-19, 1626-1628 (1983).

[6.18]   Yamamoto, S., Nakamura, Y. and Iwasaki, S., *Read spacing loss in perpendicular magnetic recording*, Journal of the Magnetics Society of Japan, Vol. 9, 73-78 (1985).

[6.19]   Kugiya, F., Koizumi, M., Okuwaki, T., Shinagawa, K., Uesaka, Y. and Tamura, T., *Studies of spacing loss in longitudinal and perpendicular recording*, J. Appl. Phys. 55, 2220-2222 (1984).

[6.20]   Tsuboi, Chuji, and Fushida, Takato, *Relations between gravity values and corresponding subterranean mass distribution*, Bull. Earthquake Res. Institute, Vol. XV-XVII, Tokyo, 636-649 (1937).

[6.21]   Minuhin, Vadim B., *Theory of playback process with soft magnetic underlayer*, IEEE Trans. Magn., Vol. Mag-21, 28-35 (1985).

[6.22]   Stoner, F.R.S. and Wohlfarth, E.P., *A mechanism of magnetic hysteresis in heterogeneous alloys*, Phil. Trans. Roy. Soc. London, Vol. 240, 599-642 (1948).

[6.23]   Tebble, R.S. and Craik, D.J., *Magnetic materials*, Wiley-Interscience, London etc., p. 374.

[6.24]   Chikazumi, S., *Physics of magnetism*, Robert E. Krieger Publ. Comp., Huntington, New York, p. 21 (1978).

[6.25]   Maxwell, J.C., *A treatise on electricity and magnetism*, Vol. ii 2nd ed., Clarendon Press Oxford (1881).

[6.26]   Stoner, E.C., *The demagnetizing factors for ellipsoids*, Phil. Mag. [7], Vol. 36, 803-821 (1945), equations (2.6) and (3.5).

[6.27]   Osborn, J.A., *Demagnetizing factors of the general ellipsoid*, Phys. Rev., Vol. 67, 351-357 (1945).

[6.28]   Yoshida, K., Okuwaki, T., Osakabe, N., Tanabe, H., Horinchi, Y., Matsuda, T., Shinagawa, K., Tonumura, A. and Fujiwara, H., *Observation of recorded magnetization patterns by electron beam holography*, IEEE Trans. Magn., Vol. Mag-19, 1600-1604 (1983).

[6.29]   Tonomura, A., Matsuda, T. and Endo, J., *Direct observation of fine structure of magnetic domain walls by electron holography*, Phys. Rev. Lett. 44, 1430-1433 (1980).

[6.30]   Heisenberg, W., *Zur Theorie des Ferromagnetismus*, Z. Physik 49, 619-636 (1928).

[6.31]   Minuhin, V.B., *Characteristics of ideal probe heads with ideal windings in the presence of a permeable media underlayer*, IEEE Trans. Magn., Vol. Mag-21, 1289-1294 (1985).

[6.32]   Minuhin, V.B., *Dependence of readback output on medium thickness in the presence of a permeable underlayer – Part 1: General theory*, IEEE Trans. Magn., Vol. Mag-21, 2595-2606 (1985).

[6.33]   Minuhin, V.B. and Steinback, M., *Dependence of readback output on medium thickness in the presence of a permeable underlayer – Part 2: wavelength responses of probe heads in the presence of a permeable underlayer*, IEEE Trans. Magn., Vol. Mag-21, 2607-2612 (1985).

[6.34]   Minuhin, V.B., *Theoretical comparison of readback harmonic responses for longitudinal recording and perpendicular recording with probe head over a medium with permeable underlayer*, IEEE Trans. Magn., Vol. Mag-22, 388-390 (1986).

[6.35]   Minuhin, V.B., *Wavelength responses of probe heads with ideal windings in the presence of a permeable media underlayer*, IEEE Trans. Magn., vol. Mag-22, 1315-1320 (1986).

[6.36]   Schwab, E., and Veitch, R.J., *Aging phenomena in Cobalt-modified iron oxides: influence of dilution and magnetic interactions*, to be published in IEEE Trans. Magn., Vol. Mag-25, Intermag issue (1989).

Chapter 7

A simple model for unbiased sine-wave recording

7.1 Introduction

In the history of magnetic recording there have been many attempts to describe the write process. Because of the strong non-linearity of the tape, especially during the write process, and because of the influence of the past on the response of the tape to a field, the write process is extremely difficult to describe. The most dramatic expression of the past is the existence of the coercive field, without which a tape could not be permanently magnetized. As a consequence the BH curve is at least double valued which complicates calculations. The use of simple partly-linearized hysteresis loops in analytical calculations of the write process has had some success, at least for the better understanding of certain effects, such as the exponential write-separation loss.

An analytical model of the write (and read) process of magnetization transitions is described by van Herk and Wesseling [7.1] under several simplifying assumptions. They obtained an expression for the read signal in which the write and read separations appeared as a sum in the exponent of an exponential loss factor.

This leads to the suggestion that the write-separation loss is of the same exponential character as the read-separation loss, e^{-kd_p}, which resulted in the well-known loss of 55 d_p/λ [dB] with d_p the head-to-tape distance during playback. From large-scale models of Hersener [7.2] it was already known that the write-separation loss was about 40 d_r/λ [dB] for optimally written sinusoidal signals, with d_r the head-to-tape distance during writing. Lopez [7.3] used the model of van Herk and Wesseling in the analytical calculation of the write-induced separation loss for sinusoidal components. By disregarding the zero-frequency magnetization component due to H_c, Lopez found for the Fourier transform of the magnetization

$$M(k, y) = \frac{\dfrac{M_s}{\Delta H} H_{\text{appl}}(k, y)}{1 + \dfrac{M_s}{\Delta H} N_d(k, y)} \tag{7.1}$$

where ΔH, M_s and H_c are defined in Fig. 7.1 and N_d represents the demagnetizing factor for the longitudinal or perpendicular magnetization pattern in an SL or DL medium. (7.1) results easily from Fourier transforming

$$M = \frac{M_s}{\Delta H} H \qquad \text{with } H = H_{\text{appl}} + H_{\text{dem}} \tag{7.2}$$

and the relation

$$H_{\text{dem}}(k, y) \equiv -N_d(k, y)\, M(k, y). \tag{7.3}$$

The influence of head-medium separation in N_d is not taken into account when image charges are neglected. In this particular case N_d can be written as $N_{0d}(k, y')$ where $y' = y - d_r$. Hence, for variations of d_r, (7.1) reads $M(k, y' + d_r) \sim H_{\text{appl}}(k, y' + d_r)$ such that $M(k, y'$

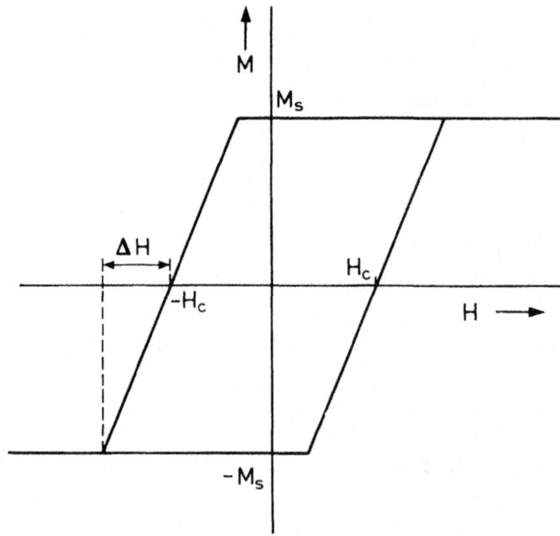

Fig. 7.1 The *M-H* loop model of the medium of Lopez [7.3].

+ d_r) must have the same dependence on d_r as $H_{appl}(k, y' + d_r)$, i.e. decreases by 55 d_r/λ [dB] (this is not valid for variations in y'), resulting in a write-separation loss of 55 d_r/λ [dB]. This is in essence the message of Lopez. He assumed a uniform magnetization also in (7.1), where we still assumed an arbitrary y dependence. This simplified the calculation of the demagnetizing factor N_d in [7.3].

Three things must be realized:

1) Taking into account the head-medium interaction would change the write-separation loss.

2) Although not explicitly stated in Lopez' paper, he obtained a sinusoidal magnetization pattern in the tape by assuming a sinusoidal $H_{appl}(x)$ in space. This is valid for e.g. contact duplication, but is not realizable during the write process with a head. How the write process is thought to progress in time is not discussed by Lopez.

3) Without making assumptions concerning details of the MH-curve, including influences of 'history', and without iterative numerical calculations, the write process cannot be described accurately.

Lopez' description is focussed on a simple theoretical calculation of the write-separation loss and cannot be emphasized as a complete simplified description of writing effects.

Examples of accurate (and complete) models can be found between those which use the Preisach diagram [7.4] or models that use the (irreversible) switching field distribution to describe the magnetization changes inside the tape. They lead to time-consuming iterative calculations.

A simplified use of this irreversible switching field distribution in two dimensions is made in the recently published two-dimensional model of Eiling [7.5]. In a rough approximated way this model takes account of demagnetization during writing. It still requires iterative computations. With Eiling's model effects in different types of magnetic recording (direct or bias recording, sine-wave or square-wave recording, one or more signals recorded simultaneously or after each other, longitudinal and perpendicular recording) can be described. It is hard to say, however, if Eiling's description can be called quantitative, because of a few rough simplifications in his model.

For the write-read model calculations we want to carry out, we do *not* require that the write model describes the magnetization changes in the tape during the write process. A calculation scheme for computing the

'effects' of the write process on the output is sufficient. Instead we require that computations resulting from the calculation scheme are:
1) so fast that output voltage versus frequency curves can be displayed almost interactively with the computer,
2) the most important write effects in the output curves in the case of unbiased longitudinal (optimal [7.6]) sine-wave recording are, at least qualitatively, well described.

The second requirement includes the effects of the gap length and the saturation magnetization of the head material and the coercivity and remanence of the tape.

Because of the first requirement, models derived from Preisach's concept are not appropriate. Simplified analytical models of the write process, such as those of Middleton [7.27], Williams and Comstock [7.15] and Middleton and Wiseley [7.23] seem more appropriate for our purpose. We did not use one of these models, although some elements of these models can be found in our write model, described below. The main reason for not using those models is that they do not completely satisfy our second requirement. These models are not applicable to our relatively thick media and high bit densities or do not take account of the finite saturation magnetization of the write head. This is explained in appendix 7.2 and not in this introduction, since it requires an extensive discussion of those models in the light of (recently) published experimental results, our current experimental recording conditions and published and own theoretical considerations, which would possibly disturb the reader in the introduction. The shift of this discussion to an appendix also enables us to comment on our own write model.

Our model describes the effects of gap length and saturation magnetization of the head core, in contrast to the previously mentioned analytical models. It does not have the accuracy and generality of the description of the read process in the previous 5 chapters. Our model describes the write process analytically, is mathematically simple and needs no time-consuming iterative computations.

7.2 Theory and verification

In the write process three things are of importance:
1) The absolute value of the write field.

2) The gradient of the write field.

3) The steepness of the MH curve.

As to point 1, during the write process the demagnetizing field within the tape reduces the write field. This means that it is not sufficient in general to apply a head field equal to the coercive field, H_c, not even if the intrinsic MH curve of the tape is perfectly square. The demagnetizing field depends on the head-to-medium distance during the write process and the wavelength to be written.

Long wavelength

The video heads we will describe with our model have a very small gap length because they are meant to write *and* read. The fringe field of these heads is only able to write several tens of micrometers deep into the coating of a CrO_2 tape or metal-powder tape with higher H_c. The thickness of the magnetization pattern is therefore small compared to half the wavelength. Therefore during the write process the demagnetizing fields are small. In turn, half the wavelength is large compared to the head-to-tape distance, so that the small demagnetizing field is even reduced by the imaging of the magnetization pattern in the soft-magnetic surface of the head. Hence the demagnetizing fields can safely be neglected in the long-wavelength write process.

For the same reasons (small thickness of the magnetization pattern to be written and small head-to-tape distance relative to the half wavelength) the finite gradients of the fringe field of the write head always suffice and therefore their influence can be neglected.

Hence the long-wavelength write behaviour of the head is determined by the (maximum) *absolute* value of the *applied* field and hardly by gradients of the write field or steepness of the MH curve. At the longest wavelengths the maximum writing depth possible with the head, given its g and M_s and the threshold field H_c, is also the optimal writing depth.

Short wavelength

At short wavelengths the optimal writing depth is less than the maximum writing depth of the head. In this case one expects that the optimum writing depth is determined by the wavelength λ, the gradient of the applied field, and by the M_r, H_c and steepness of the MH curve of the tape. The gradient in the applied field is of importance, since the demagnetizing fields shear the MH_{appl} loop, even if the intrinsic MH loop is perfectly square. The length of a magnetization reversal (the transition length) or the gradients in magnetization $M(x)$ thus decrease with a reduction of the gradients. For a sinusoidal signal this means a decrease of the amplitude of $M(x)$ as well. What can be expected for this decrease with the head-to-tape distance? In the introduction there followed a decrease of 55 d_r/λ [dB] from the simplified model of Lopez. The exact write-distance loss depends on details concerning the MH loop and the progress of the applied field in time.

We will apply a phenomenological write-distance loss factor of $e^{-0.7\,kd_r}$ in our model for tapes with a 'demagnetization ratio' $M_r/H_c \approx 3$. This is also applied at long wavelengths, where this factor approaches 1. The choice is mainly based on the results of model experiments of Hersener [7.2]. He used a coating with $M_r/H_c = 3.7$. From his results concerning optimal recording we deduced this factor.

Bertram and Niedermeyer [7.4] used a constant write current in their work on the effect of spacing during the writing on the output for demagnetization ratios ranging from 1 to 5. To create well-defined head-to-tape spacings during writing, they sputtered thin layers of a non-magnetic ceramic on the head surface. The range of increasing M_r/H_c values was obtained by increasing the volumetric packing density in the tape and thus M_r, while H_c only reduced slightly. With some inaccuracy we deduced from their curves a write-distance loss factor:

$$\boxed{DL_r = e^{-\alpha k d_r}}\,, \tag{7.4}$$

with $\alpha(M_r/H_c, S^* = const) \simeq 0.55 + 0.07\,M_r/H_c$.

The write distance loss, resulting from (7.4), decreases at decreasing demagnetization ratios. The reason for this must be the reduction of the demagnetizing field relative to H_c (given the same writing depth), since the coercivity squareness, S^* (defined in Fig. 7.4.I), was kept constant in the experiments.

We may also expect a decrease of the write spacing loss when the *influence* of the demagnetizing field is reduced by way of an increase of the coercivity squareness, S^*.

For $M_r/H_c = 3.7$, expression (7.4) results in a slightly higher write-distance loss as deduced from Hersener's work. (The non-optimal write current at the larger wavelengths in Bertram and Niedermeyer's experiment cannot be the cause of this.)

A note must be made in the case of applying relation (7.4) to model calculations at (extremely) short wavelengths, since this leads to physical absurdities as will now be explained.

Physically one must expect that a tape with a reduced M_r (and the same H_c and steepness) has a higher free-space field at the head side of the coating after optimal writing. This is explained by our model in the following paragraphs in more detail. Hence the output must always be larger. Now assume that the output at large wavelengths is indeed larger. Then, because of the larger exponential write-distance loss, the output versus frequency curves of the tape with the higher M_r falls more strongly than the curve of the tape with the lower M_r. Hence the curves always cross each other at a certain short wavelength, and at still shorter wavelengths a lower output results from the best tape, i.e. the one with the highest M_r. Since this is nonphysical, the phenomenological relation can only hold for a limited range of not too short wavelengths. In most of the experiments of Bertram and Niedermeyer the shortest wavelength was 1.6 µm. Obviously it is not quite correct to assume an exponential write-distance loss over too large a range of wavelengths.

7.2.1 Maximum writing depth

The definition of the applied field used above is not the complete field from the head during the write process, since the mirror charges in the head due to the already written magnetization distribution in the tape are included in the demagnetizing field instead of the applied head field. H_{appl} is defined as the write field of the head when the tape with its magnetization was absent. The H_{appl} defined in this way can be calculated separately and has the advantage that at long wavelengths it can be reasonably well approximated by the (total) field H during the write process because the demagnetizing field then vanishes due to the opposite field of the mirror charges, as outlined in chapter 4. Thus H_{appl} is not influenced by the magnetization pattern, but it is still influenced by the relative permeability of the tape $\mu_r \equiv B/(\mu_0 H)$. During the write process, however, this permeability changes constantly when the medium is not premagnetized, see Fig. 7.1. This makes an analytical calculation of H_{appl} impossible. Therefore we will *approximate* the permeability in the coating during the write process by 1. To include DL media the calculation of H_{appl} will still be set up rather generally.

The field (or potential) on a distance y from the head can be calculated by multiplying each wave number component of the field (or potential) at the head surface at $y = 0$ by the proper distance factor f. This leads to

$$H(x, y) = \frac{1}{2\pi} \int_{-\infty}^{\infty} f(k, y, 0)\, H(k, 0)\, e^{jkx} dk \quad . \tag{7.5}$$

With the aid of (4.14) expressions for f follow immediately from the results in chapter 2. For an SL medium with the approximation for the relative permeability during writing $\mu_r = 1$, f equals simply $e^{-|k|y}$. Furthermore, we use the simple Karlqvist (i.e. deep-gap field) approximation for $H(x, 0)$. This is a very good approximation for the present purpose since the distance y from the head where $H_{x,y} \simeq 1.25\, H_c$, in the case that the magnetization in the head equals about M_s, will be much larger than $g/2$. At a distance $y \geqslant g/2$ the accuracy of the Karlqvist approximation for the amplitude of the H_x field component is better than 10%, see e.g. reference [7.7]. The influence of a permeability $\mu_r \neq 1$ during the write process will change the field shape at $y = 0$ slightly, but this will not change the accuracy of the Karlqvist approximation drastically.

The Fourier transform of the x component of the Karlqvist field at $y = 0$ gives the well-known gap-loss function, $\text{sinc}(g/\lambda)$. The x component of the applied field is maximal in the middle of the gap ($x = 0$) in the Karlqvist approximation. Substitution of $x = 0$ and the symmetric functions $H_x(k, 0) = \eta NI\, \text{sinc}(g/\lambda)$ and $f = e^{-|k|y}$ in (7.5) leads to a standard integral, the solution of which can be found in e.g. [7.8] and reads

$$\frac{g \cdot H_x(x = 0, y)}{\eta NI} = \frac{2}{\pi} \arctan (g/2y) \tag{7.6}$$

The Fourier transform of the y component of the Karlqvist field at $y = 0^+$ leads to the *gap-loss function* $j\,\text{sgn}(k)\,\text{sinc}(g/\lambda)$. The 90° phase difference and sign-change with k are a result of the Hilbert-transform relationship between the x and y components of two-dimensional fields, see e.g. Minuhin [7.9] and [7.10] or appendix 5.1. For not too large values of y the y component is maximal at $x = \pm g/2$. Substitution of $x = \pm g/2$, $H_y(k, 0) = \eta NI\, j\,\text{sgn}(k)\,\text{sinc}(g/\lambda)$ and $f = e^{-|k|y}$ in (7.4) gives after some manipulations a standard integral, see e.g. [7.11]. The result is

$$\frac{g\left|H_y(x = \pm g/2, y)\right|}{\eta NI} = \frac{1}{2\pi} \ln \left[\frac{y^2 + g^2}{y^2}\right]. \qquad (7.7)$$

The maximum field components H_{xm} and H_{ym} are reached when the deep-gap field H_{dg} (in the Karlqvist approximation equal to $H_x(0, 0)$) equals about M_s. The corresponding write current $I_s = g M_s/(N\eta)$. Pole-tip saturation then occurs only very locally. Larger fields, at currents $I > I_s$, give rise to larger pole tip saturation and hence to a reduction of the gradients. This leads to larger transitions in pulse recording and to smaller amplitudes in sine-wave recording. Thus we assume $H_{dg} = M_s$ and, with the aid of (7.6) and (7.7), arrive at the following maximum field components:

$$H_{xm} = \frac{2M_s}{\pi} \arctan\left(\frac{g}{2y}\right) \xrightarrow{y > g} \frac{M_s}{\pi}\left(\frac{g}{y}\right) \qquad (7.8)$$

$$H_{ym} = \frac{M_s}{2\pi} \ln\left(\frac{y^2 + g^2}{y^2}\right) \xrightarrow{y > 2g} \frac{M_s}{2\pi}\left(\frac{g}{y}\right)^2. \qquad (7.9)$$

For the regions indicated the accuracy of the right-hand approximations is better than 10% (also when compared to the exact result of Westmijze, see p. 171 in [7.12], instead of the Karlqvist results right of the equal signs).

It is now possible to estimate the maximum depth in the tape where the magnetization can be reversed, i.e. where the field reaches H_c during the write process.

Let $C_d \geq 1$ be a phenomenological constant that takes into account the not-quite compensated demagnetizing field. Then a field H_c and an applied field C_dH_c can be reached at a maximum distance

$$y_{xm} = \frac{g}{2 \tan \dfrac{\pi C_d H_c}{2M_s}} \xrightarrow{C_d H_c < \frac{M_s}{\pi}} \frac{g M_s}{\pi C_d H_c} \qquad (7.10)$$

for longitudinal recording on SL media, supposing that $C_d H_c \leq M_s$, and from

$$y_{ym} = \frac{g}{\sqrt{e^{2\pi C_d H_c/M_s} - 1}} \xrightarrow{C_d H_c < \frac{M_s}{5\pi}} \frac{g\sqrt{M_s}}{\sqrt{2\pi C_d H_c}} \qquad (7.11)$$

for perpendicular recording* on SL media, according to (7.8) and (7.9). (For the regions indicated the accuracy of the right-hand approximations is better than 10%, also compared to Westmijze's exact results.) The field H_c can be reached at larger distances only at the cost of a reduction of the gradients.

7.2.2 Optimal magnetization distribution

In order to find the maximal possible output, it is necessary to know the optimal magnetization distribution in the tape. Important here are the tape and head parameters and the question of whether an optimal writing depth exists.

Tape parameters that play a role in the simple model are the remanent magnetization M_r and the coercive field H_c.

For the head the gap length g and the saturation magnetization M_s are important, or if the head is not saturated the ratio I/I_s.

Since we do not intend to calculate the magnetization distribution into the depth of the medium, we must assume one. In the following sections the question concerning the writing depth is therefore answered for 2 simple assumed magnetization distributions. One of them is then chosen for the write-read model calculations.

We must warn beforehand that the concept of writing depth is an oversimplification of the actual writing process.

7.2.2.1 Uniform magnetization model (UMM)

Assumptions:
1) $M_{max} = M_r$ if $0 < y < t_x$ otherwise $M_{max} = 0$.
2) t_x is determined by the distance from the head during the recording where the applied field was $C_d H_c$ according to (7.10) if $t_x < t_3$, otherwise $t_x = t_3$.

---

* Equation (7.9) does not prescribe the condition $C_d H_c \leq M_s$ in (7.10). Physically however, (when the relative permeability above the head is 1, and not zero as implicitly assumed in the Karlqvist model) the field cannot exceed the saturation magnetization anywhere, except when a homogeneous field is added to the highly inhomogeneous write field emanating from the gap edges. The homogeneous field, however, would deteriorate the write field (gradient).

3) $\mu_3 = 1$ (during writing).

4) M demagnetizes irreversibly and uniformly after the contact with the head to a state where the free-space demagnetizing fields do not exceed H_c.

The demagnetizing field can easily be calculated for a sinusoidal magnetization. First the free-space field amplitude H_f at $y = 0$ is calculated with (2.43):

$$H_f = \left| -\frac{k}{2} M_r \int_0^{t_x} e^{-ky} dy \right| = (1 - e^{-kt_x}) \frac{M_r}{2}. \qquad (7.12)$$

In the middle of t_x where the field is maximal (amplitude H_m):

$$H_m = \left| -\frac{k}{2} M_r \left(2 \int_0^{t_x/2} e^{-ky} dy \right) \right| = (1 - e^{-kt_x/2}) M_r. \qquad (7.13)$$

Hence

$$\boxed{H_f = \left(\frac{1 + e^{-kt_x/2}}{2} \right) H_m = \left(1 - \frac{H_m}{2M_r} \right) H_m} . \qquad (7.14)$$

Since $H_m \leq M_r$ and the output voltage $V \sim H_f$ (according to (5.6) and (2.46) with $\mathcal{B}_y = \mu_0 H_y$), the highest output is achieved when

– H_m is maximal, i.e. $H_m = H_c$ (if $H_c \leq M_r$, else $H_m = M_r$) and

– M_r is maximal.

Thus higher M_r and/or H_c values must lead to higher outputs, assuming that they can be recorded optimally. This is possible when the maximal writing depth of the head-tape combination considered is sufficiently large. The writing depth t_x follows for $I \leq I_s = g M_s / N\eta$ with the aid of (7.10). Hence

$$\boxed{t_x = \frac{g}{2 \tan \dfrac{\pi C_d H_c g}{2IN\eta}} - d_r} \qquad (IN\eta \geq C_d H_c g) \qquad (7.15)$$

with d_r the effective distance between recording head and tape.

When $I = I_s$, the maximum writing depth requiring maximum gradients is obtained. For very long wavelengths, where write-field gradients are of less importance, it may be advantageous to use currents higher than I_s. This possibility will not be taken into account in the model.

From (7.12) it follows that an increase of t_x increases the output as long as M_r is constant. For short wavelengths it can easily reach the value at which H_m equals H_c. A further increase of t_x introduces demagnetization of M_r, at least when the tape comes in free space directly after the recording, and demagnetizes to a lower remanent state according to the assumptions 3) and 4). Equation (7.14) with the substitutions $H_m = H_c$ and $M_r = M_p$ then describes the decrease of the output and the permanent magnetization, M_p. The optimal writing depth, $t_{x\,opt}$, thus follows from (7.13) or from equalizing (7.12) with (7.14) and taking $H_m = H_c$. Hence

$$\boxed{t_{x\,opt} = -\frac{\lambda}{\pi} \ln\left(1 - \frac{H_c}{M_r}\right)} . \qquad (7.16)$$

To obtain this optimal situation a current with an amplitude of about

$$I_{opt} = \frac{\pi\, C_d H_c g}{2N\eta\, \arctan \dfrac{g}{2(t_{x\,opt} + d_r)}} \qquad (7.17)$$

is necessary, according to (7.15).

Fig. 7.2 gives H_f, proportional to the playback voltage, versus the recording depth, t_x, and the recording current, I.

7.2.2.2 Discussion and verification of the UMM

Qualitatively the above model is satisfactory since, in spite of its simplicity, it describes several phenomena that can be observed in practical circumstances. These include:
- the existence of an optimal write current,
- the existence of a threshold in the $V(I)$ curve,
- the strong influence of coercivity H_c at small wavelengths [7.18]

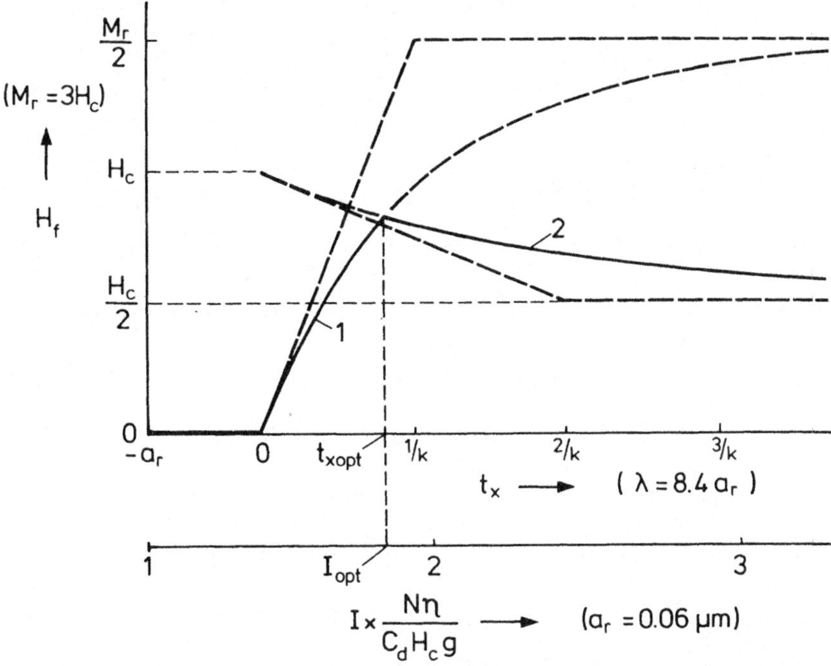

Fig. 7.2 The free-space field, H_f, (the output voltage is proportional to this field) as a function of the write depth for longitudinal recording, t_x, according to Eq. 7.12 (curve 1) and Eq. 7.14 (curve 2), sketched for $M_r = 3 H_c$ and $\lambda = 8.4 a_r$. The I scale is added for the case that $a_r = 0.06$ μm and thus $\lambda = 8.4 d_r = 0.5$ μm, by using relation 7.15, i.e. assuming no head saturation. The optimal current in the case of a metal-powder tape ($H_c = 115$ kA/m) would according to the curve equal 9 mA (if $N = 16$, $\eta = 0.5$, $C_d = 1.25$ and $g = 0.3$ μm), which is in agreement with experimental results for small wavelengths.

where the tape determines the optimal writing depth,
– the strong influence of the remanence M_r at long wavelengths [7.18] where the head determines the maximum writing depth ($<$ optimal writing depth).

In order to arrive at the output voltage, one has to add the read-distance losses and the remagnetization factor to the free-space field, H_f, as outlined in the chapters 2 to 6. In appendix 7.1 this is carried out for the optimal write condition ($t_x = t_{x\,opt}$), while taking permeability effects in the tape into account. The result is expression (a7.1) with $\alpha = 0$ (no write-distance loss, see below) and $\Delta g = 0$ (sharp write- and sharp read-gap edges, see also section 7.3.1). With the aid of this expression the shape and the absolute value of the optimized output voltage are described reasonably well, if one were free to choose the best-fitting

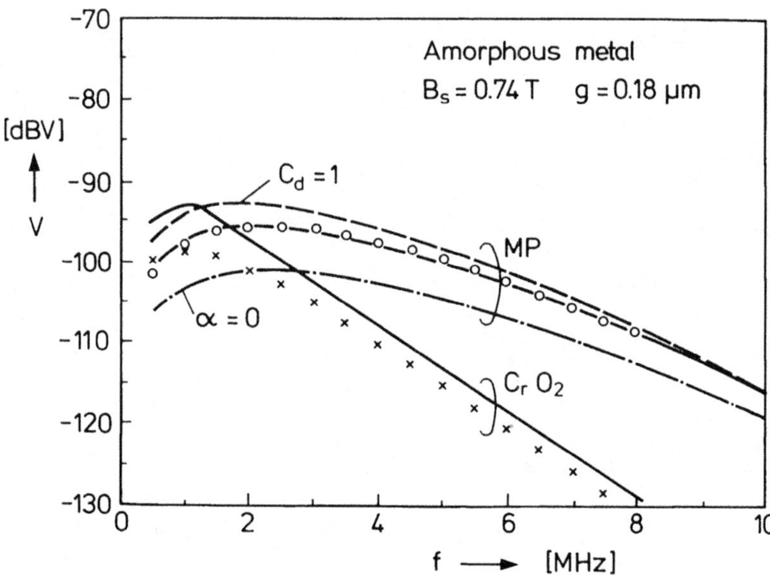

Fig. 7.3.1 Results of the UMM versus experimental results for an amorphous head on a chromium-dioxide (CrO_2) tape and on a metal-powder (MP) tape.
For the rougher tape (CrO_2) a 30 nm larger head-to-tape distance is obtained, see below.
The parameters used in the model are as follows.
CrO_2 tape: H_c = 49 kA/m, M_r = 130 kA/m, t = 4 µm and roughness about 12 nm.
MP tape: H_c = 116 kA/m, M_r = 159 kA/m, t = 3.6 µm and roughness about 8 nm which obviously results in a much lower head-to-tape distance, see below.
Amorphous head: 2-fold laminated sandwich of the construction given in Fig. 8.1c with trackwidth W = 2 × 6 µm = 12 µm, M_s = 585 kA/m (B_s = 0.74 T), g = 0.18 µm and N = 17. η is assumed 1 for simplicity.
Tape permeabilities in playback model: μ = 1.4 and β = 0.9 for both tapes is chosen because of the measured data for comparable tapes in Table 7.3 and Fig. 7.4.
— Theory: C_d = 1.25; α = 0.7; d = 120 nm (CrO_2), or 90 nm (MP)
-- Theory: C_d = 1; α = 0.7; d= 90 nm (MP).
-·- Theory: C_d = 1.25; α = 0; d = 1.7 × 90 nm (MP).
o and + *Experiment:* optimal unbiased sinewave recording; recording and playback with the same head; head-to-tape velocity v = 3.14 m/s.

head-to-tape distance, d, during recording and playback. (In appendix 7.1 the symbol d_{4r} is used for the distance during recording and d_{4p} for the distance during playback).

Write-distance losses have not yet been taken into account. However, large-scale experiments of Hersener [7.2] and measurements by Bertram and Niedermeyer [7.4] and also our hollow-out measurements (see later on in this discussion) show that the calculated distance is then about twice the actual distance. See also the theoretical result of Lopez, mentioned in the introduction of this chapter. According to the experiments one has to add an exponential write spacing-loss factor $e^{-k\alpha d_r}$ with $\alpha \approx 0.7$ in order to obtain reliable values for the actual head-to-

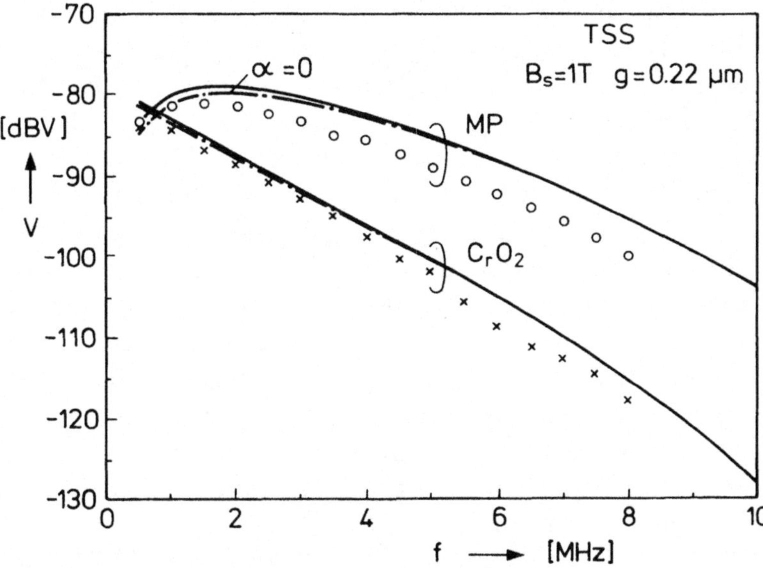

Fig. 7.3.II Results of the UMM versus experimentral results for a tilted sputtered sendust (TSS) head.
The difference in head-to-tape distance is again 30 nm for the two tapes. The absolute values of d however decreased. The parameters used in the model are as follows.
TSS head: $W = 25$ μm, $M_s = 790$ kA/m (1T), $g = 0.22$ μm and $N = 20$. η is assumed 1 for simplicity. Other parameters are as given in the caption of Fig. 7.3.I.
—— Theory: $C_d = 1$, $\alpha = 0.7$, $d = 90$ nm (CrO_2) or 60 nm (MP).
–·– Theory: $C_d = 1.25$; $\alpha = 0$, $d = 1.7 \times 90$ nm (CrO_2) or 1.7×60 nm (MP).
o and + *Experiment:* see also caption of Fig. 7.3.I.

tape distance. The total distance loss then equals about $e^{-1.7kd}$ or about 93 d/λ [dB] (neglecting permeability effects in the tape).

Often it is not possible to measure d or changes in d, in which case d might be chosen 1.7 times larger than the actual value in combination with $\alpha = 0$. Write plus read-spacing losses are then approximately the same as when the actual distance and $\alpha = 0.7$ are substituted in the equations. However, in the region where the head is not able to write optimally, the output decreases, and the maximum in the frequency curve shifts to higher frequencies, see the curve for $\alpha = 0$ in Fig. 7.3.I. The TSS head in Fig. 7.3.II is less sensitive to this choice because of its better writing capabilities (higher M_s and larger gap length). On the low-coercive tape (CrO_2), the effect is negligible.

An increase of the phenomenological constant C_d has about the same effect, see curves for $C_d = 1$ and $C_d = 1.25$ (full curve) in Fig. 7.3.I. Completely cancelling the demagnetizing field, i.e. choosing this constant 1, leads to values which are slightly too high at the long-wavelength side of the curve, where signals are far from being recorded optimally. In our experience the best choice for C_d is always somewhere between 1 and 1.3. We will use $C_d = 1.25$ and $\alpha = 0.7$ in the UMM unless stated otherwise.

In Fig. 7.3 we assumed for simplicity that $\eta = 1$. We also measured the actual efficiency of the heads using the method that will be described in chapter 8; the efficiency of the amorphous sandwich head of Fig. 7.3.I reduced gradually from 0.88 to 0.57 and the efficiency of the tilted sputtered sendust (TSS) head of Fig. 7.3.II reduced from 0.7 to 0.45 between 0.2 MHz and 6 MHz. Hence both efficiencies were reduced by 3.8 dB when the wavelength was about 0.5 μm (at 6 MHz and $v = 3.14$ m/s in this measurement). This is equivalent to a head-to-tape distance increase of 3.8×0.5 μm$/93 = 20$ nm since the distance loss is about 93 d/λ [dB]. Thus the actual distances are about 20 nm smaller than mentioned in the captions of Fig. 7.3.I and II.

Fig. 7.3.I and II show that the present UMM is satisfactory in optimal unbiased sinewave recording. The absolute value and the shape of the frequency curve are predicted reasonably by the UMM for various heads on different tapes. The absolute value is usually within 3 dB of the experimental value when the (frequency-dependent) efficiency is taken into account.

In spite of the simplicity of this model, it helped us several times in understanding observed phenomena qualitatively as well as quantitatively. An example is the accurate determination of the hollow-out value due to differential wear in a metal-in-gap (MIG) head.

Example: hollow-out in a MIG head

The hollow-out phenomenon is typical for a MIG head and also observed in bandpass heads, and therefore discussed in chapter 12 and chapter 13. Hollow-out of the gap region results in an increase of the head-to-tape distance that can be measured accurately either mechanically or optically. This phenomenon thus provides a very simple, direct and accurate way of verifying almost the complete write model. Therefore we will also discuss it here, while referring to section 12.2.3 and the figures 12.10, 12.13 and 12.14. Here it is discussed with greater emphasis on the model. Details concerning the electrical measurements (chapters 10 and 12) and the principles of the mechanical or optical hollow-out determination (chapter 13) are not discussed.

In fig. 12.10 the level difference between the gap edges and the ferrite core in a MIG head is sketched. This 'hollow-out' is due to the greater wear of the metal layers on both sides of the gap with respect to the wear of the MnZn ferrite core. Since the wear is different on different tapes, the hollow-out values are also different. Therefore it takes some time, usually a few hours at a head-to-tape velocity of 3.14 m/s, before the final hollow-out value is stabilized on another tape. Correspondingly one observes the following in electrical measurements of the recording (R) and the playback (P) behaviour:

When first the MIG head 'runs-in' against a 'high-abrasive' CrO_2 tape and subsequently is measured (repeatedly writing and reading) against a 'low-abrasive' metal-powder (MP) tape, one observes a gradual decrease of the output. After a quarter of an hour and up to a few hours at most, the output stabilizes at a lower level.

It is strange that when the gap length is small (≈ 0.2 μm) the reduction of the recording behaviour at large wavelength (4 μm) is about as large as at small wavelength (0.5 μm). This is not as expected from exponential write-spacing losses. Is this peculiar effect correctly calculated by the simple model? The peculiar effect vanishes at larger gap length ($g \gtrsim 0.3$ μm) and also when a tape with a lower coercivity (for instance CrO_2 tape with $H_c = 52$ kA/m instead of the MP tape with 115 kA/m) is used. Then one observes smaller effects at larger wavelengths. This is in agreement with previously mentioned investigations on the effects of head-to-tape distance (in our case hollow-out) variations; write-spacing loss has an exponential character, as does read-spacing loss.

It is not surprising that in this case qualitatively as well as quantitavely the effects are well described by the simple model.

At a small gap length and high-coercive medium the situation is more complicated because the write fields of the MIG head are not sufficient anymore for optimal writing of the large wavelengths. This effect becomes stronger as the wavelength increases. For the actual gap length of 0.2 μm and coercivity of 115 kA/m the increasing loss due to the insufficient write depth at increasing wavelength adds to the decreasing exponential write-spacing loss. The result is an almost uniform write-spacing loss between 0.5 MHz and 7.5 MHz when the head-to-tape velocity is 3.14 m/s; see Fig. 12.14. To our surprise this peculiar behaviour is quite accurately described by the simple model without any modification (see Fig. 12.13); the write-spacing loss is about 10 dB for a 120 nm increase of the spacing, i.e. 3.3 dB for 'every' 40 nm increase in the spacing, in agreement with the 30% worsening of the recording behaviour, see the reduction of the recording figure R in Fig. 12.14. The degradation of the playback behaviour is also accurately described, but this is not surprising.

Qualitatively (and often quantitatively) we observed that the effects of changes in gap length, head-to-tape distance, saturation magnetization of the head and coercivity and remanence of the tape on the shape and on the absolute value of the frequency curve are reasonably predicted by the UMM.

A disadvantage of the UMM is that the introduction of e^{-kad_r} *after* estimating the optimum writing depth $t_{x\,opt}$ makes it a physically-inconsistent model; a lower signal *from* the tape must be caused by a lower magnetization *in* the tape and hence lower demagnetizing fields in the tape than those used for calculating $t_{x\,opt}$. Therefore an attempt will be made in the following section to arrive at a physically more consistent model, or at least to explain why the UMM still seems to be adequate.

7.2.2.3 Exponentially-decaying magnetization model (EMM)

Two changes are necessary with respect to the uniform-magnetization model (UMM), in order to obtain a consistent and more physical model for optimized recording:

a) The magnetization is reduced *beforehand* by the write-distance loss factor experimentally observed in optimized recording, e^{-kad_r}.

b) In the tape, i.e. $y > 0$, the magnetization is not uniform but has about the same dependence of y as the overall magnetization in the tape would have of a head-to-tape distance variation, i.e. $M(y) \approx M(0)\, e^{-kay}$.

When there was no magnetic interaction between different layers within the coating during the write process, then the magnetization within the tape must decay according to the write-distance loss factor, i.e. the write-distance loss factor as determined from an experiment with a fixed current at all d_r values. This loss will differ from the loss in the case that the write current is optimized at each d_r. Hence b) follows exactly from a) only if there is no magnetic interaction between layers in the coating and the write-distance loss is the same for optimized and fixed currents at different write distances. In fact this is assumed in the exponential-decaying magnetization model (EMM), defined by the following four assumptions:

1) $M = M_\mathrm{r}\, e^{-ka(d_\mathrm{r} + y)}$ if $0 < y \leqslant t_\mathrm{x}$, else $M = 0$.
2) I is such that just H_c is reached at one 'characteristic distance' $\tau = 1/ka$ from the head.
3) $\mu_3 = 1$ (during writing).
4) M demagnetizes irreversibly after the contact with the head to a state where the free-space demagnetizing fields do not exceed H_c.

Assumption *1* contains the essential changes *a* and *b* with respect to the EMM, while a writing depth t_x is introduced above which the magnetization changes are assumed to vanish abruptly. Assumption *2* is not an essential part of the UMM. The choice is rather arbitrary, in contrast to the current in the UMM. Another choice for I would be one in which $M_\mathrm{r}/H_\mathrm{c}$ has more influence than the weak influence via α solely, see the $\alpha(M_\mathrm{r}/H_\mathrm{c})$ dependence in e.g. (7.4). Assumption *3* is equal to that in the UMM. Assumption *4* is only of importance when a free-space demagnetizing field follows from the magnetization distribution defined by *1* for a proper value of α, that exceeds the coercive field H_c. When this field turns out to be considerably smaller than H_c, it *seems* unlikely that self-demagnetizing effects played a role in the observed distance loss.

When this field is about the same as H_c or even higher, then it is sure that the write process was influenced by self-demagnetization or was self-demagnetization limited.

For the understanding of the write process these are very important questions, see e.g. the confusing results and (hidden) opinions about this subject in the literature, as outlined in appendix 7.2.

In order to reduce the confusion we will focus our attention on this subject. Also of importance in this section and the subsequent discussion will be the question whether the EMM is more convenient for write-read model calculations than the UMM.

In order to answer these questions, the amplitude of the demagnetizing field in free space, H, and its maximum, H_m, must be calculated for the assumed magnetization distribution in *1*.

(Demagnetizing) field calculation

In 'analogy' with (7.12) and (7.13) the field amplitude at a distance y from the head-facing side of the coating follows from

$$H = - \frac{k}{2} M_r \left[\int_y^{t_x} e^{-k(y'-y)} e^{-k\alpha(d_r + y')} dy' \right.$$

$$\left. + \int_0^y e^{k(y'-y)} e^{-k\alpha(d_r + y')} dy' \right]$$

$$= - \frac{M_r}{2} e^{-k\alpha d_r} \left[\chi^{-\alpha} \left(\frac{2}{1-\alpha^2} \right) - \chi \frac{e^{-k(\alpha+1)t_x}}{1+\alpha} - \chi^{-1} \left(\frac{1}{1-\alpha} \right) \right] ,$$

$$(7.18)$$

where $\chi \equiv e^{ky}$ and t_x is the writing depth.

The maximum value of this field, H_m, is reached for χ_m following from $\partial/\partial\chi\{H(\chi)\} = 0$ which leads to the following implicit expression for χ_m:

$$t_x = - \frac{1}{k(\alpha+1)} \ln \left\{ \frac{1+\alpha}{1-\alpha} \chi_m^{-2} - \frac{2\alpha}{1-\alpha} \chi_m^{-(\alpha+1)} \right\}. \qquad (7.19)$$

This expression describes the relation between t_x/λ and y_m/λ with $y_m = (1/k) \ln \chi_m$ the coordinate where the demagnetizing field is maximal. In Table 7.1, y_m/t_x is calculated for different values of α when $t_x/\lambda =$

Table 7.1   The maximum field position y_m relative to the writing depth t_x.

α	y_m/t_x	
	$t_x/\lambda = 1/2\pi$	$t_x/\lambda = 1/\pi$
0	0.500	0.500
0.1	0.485	0.464
0.2	0.470	0.430
0.3	0.456	0.397
0.4	0.441	0.366
0.5	0.427	0.339
0.6	0.413	0.314
0.7	0.399	0.292
0.8	0.386	0.272
0.9	0.373	0.254
$\rightarrow$ 1.0	0.362	0.240
4.0	0.150	0.078

$1/2\pi$ or $1/\pi$. Because of the decrease of M with y, y_m/t_x is smaller than 0.5 especially when t_x/λ is large and α is large.

The optimal writing depth $t_{x\,\text{opt}}$ is reached when the maximum field in the tape $H_m \equiv H(\chi_m)$ equals H_c. Substitution of H_c and χ_m and expressing t_x as a function of χ_m in (7.18) with the aid of (7.19) leads to the following simple, although implicit, expression for χ_m:

$$H_m = \frac{M_r\, e^{-kad_r}}{1 - \alpha}\,\left(\chi_m^{-\alpha} - \chi_m^{-1}\right) = H_c. \qquad (7.20)$$

Next $t_{x\,\text{opt}}$ is derived from (7.19).

The free-space field $H_f \equiv H(y = 0) = H(\chi = 1)$ can be expressed in relation to the maximum field H_m by

$$H_f = \frac{1 - \alpha - (1 + \alpha)\,\chi_m^{-2} + 2\alpha\,\chi_m^{-(\alpha + 1)}}{2(\alpha + 1)\,(\chi_m^{-\alpha} - \chi_m^{-1})}\,H_m. \qquad (7.21)$$

When α is large $H_f \to H_m$, but then $H_m \to 0$ as well. The maximum output is reached when H_m equals the maximum possible value, i.e. H_c. Increase of M_r/H_c reduces χ_m according to (7.20) (and next $t_{x\,\text{opt}}$ according to (7.19)) and next increases H_f according to (7.21). Thus again as must be expected, higher H_c and M_r values lead to a higher output.

With the optimal write current an applied field H_c must be created at one characteristic distance $\tau = 1/k\alpha$ from the head instead of an applied field $C_d H_c$ at a distance $d_r + t_{x\,\text{opt}}$ from the head in the UMM. (7.17) is thus replaced by

$$I_{\text{opt}} = \frac{\pi\, H_c\, g}{2N\eta\,\arctan\dfrac{g}{2\tau}} \qquad (7.22)$$

In Table 7.2 this I_{opt} is compared to typical experimental values and to values obtained with the UMM. The strongly frequency-dependent efficiency is measured (using the method described in sect. 8.2) and taken into account. The optimum at low frequencies is very unclear since the amorphous head with $M_s = 630\,\text{kA/m}$ (8000 Gauss) saturates at 18 mA. Only when (very) long wavelengths are being recorded, does the optimal current clearly exceed the saturation current, because gradients are less important there.

Table 7.2 Optimal currents as obtained from the two models compared to typical experimental values. The highly frequency-dependent efficiency of the amorphous head (non-laminated, no ferrite cladding and small trackwidth) is taken into account. Data:

tape: $H_c = 115\,\text{kA/m}$ (1460 Oe), $M_r = 200\,\text{kA/m}$ (2560 Gauss) $\alpha = 0.7$
head: $g = 0.25\,\mu\text{m}$, $N = 26$
tape + head: $d_r = 0.09\,\mu\text{m}$ (0.05 μm), $C_d = 1.25$

λ	f	$t_{\text{xopt}} + d_r$	τ	η	I_{opt} [mA]		
[μm]	[MHz]	[μm]	[μm]		UMM	EMM	EXP
0.5	6.28	0.136 + 0.09 (0.05)	0.114	0.36	11.95 (10.20)	5.8 (5.8)	5-9
1.0	3.14	0.272 + 0.09 (0.05)	0.227	0.51	12.82 (11.50)	6.8 (6.8)	8-12
2.0	1.57	0.545 + 0.09 (0.05)	0.455	0.61	18.30 (17.18)	10.6 (10.6)	15-20
5.0	0.63	1.362 + 0.09 (0.05)	1.136	0.73	34.63 * (33.68)*	21.7 * (21.7)*	20-40*

* Head already saturates at this level.

Much more important than knowing the optimal write current, is knowing how $t_{x\,\text{opt}}$ is influenced by the early introduction of the write-separation loss by the value of α. This applies more particularly in the 'worst case' situation $t_x = \infty$; for which values of α can the maximum field H_m still exceed H_c? The χ_m value from (7.19) for $t_x = \infty$ substituted in (7.18) with $t_x = \infty$ or in (7.20) yields

$$\frac{H_m(\alpha, t_x = \infty)}{M_r} = \frac{e^{-kad_r}}{1 - \alpha}\left[\left(\frac{2\alpha}{1 + \alpha}\right)^{\frac{\alpha}{1 - \alpha}} - \left(\frac{2\alpha}{1 + \alpha}\right)^{\frac{1}{1 - \alpha}}\right]. \quad (7.23)$$

The maximum value of this ratio, obtained when kd_r approaches zero, is printed in Table 7.3 for different values of α. For $\alpha = 0.7$ this ratio is still smaller than the H_c/M_r ratio of most tapes;

Metal powder (MP): $H_c/M_r = 0.7$-0.5,
Chromium-dioxyde (CrO_2): $H_c/M_r = 0.5$-0.3,
Metal evaporated (ME): $H_c/M_r = 0.3$-0.2.

Table 7.3 The maximum of the ratio H_m/M_r obtained when $t_x \to \infty$ and $kd_r \to 0$.

α	$(H_m/M_r)_{max}$
0	1
0.1	0.75
0.2	0.63
0.3	0.55
0.4	0.49
0.5	0.44
0.6	0.41
0.7	0.37
0.8	0.35
0.9	0.32
1.0	0.30
1.5	0.23
2	0.19
3	0.14
4	0.11
10	0.05

Hence self-demagnetizing effects in the MP tape and in the CrO_2 tape will be weak in the worst case situation and thus seem negligible in practical circumstances. The thickness of the ME tape $(0.1 \sim 1 \, \mu m)$ is such that self-demagnetizing effects will only occur at small wavelengths and hence only at very small write distances.

7.3 Discussion

In the UMM the magnetization used in the calculation of the (final) self-demagnetizing field was, inconsistently, not reduced by the distance loss. In other words, the magnetization or writing depth that corresponds to the actual output is (much) smaller than the one corresponding to the calculated (final) self-demagnetizing field.

In the modified model, the EMM, the experimentally observed write-distance loss was introduced before the calculation of the (final) self-demagnetizing field. This more consistent model, however, pointed to a self-demagnetization-free write process for most tapes, even at the smallest head-to-tape distances. This seems to be in contradiction with experiments on all kinds of tapes, showing a clear H_c dependence. Compare, for instance, the hf results on the CrO_2 and MP tapes with comparable M_r and quite different H_c values in Fig. 7.3.I and II. However, the finding of a demagnetization-free write process from the EMM ($H_m/H_c \lesssim 0.5$), due to the exponential write-distance loss factor (characterized by the phenomenological constant $\alpha \approx 0.7$), is also supported by a calculation in appendix 7.3 of the maximum H_m/H_c ratio from the best output values we measured.

Is there perhaps another mechanism that causes the H_c dependence?

The UMM did describe the H_c dependence much better. However, the introduction of the write-distance loss afterwards made the UMM non-physical. In fact it took away from the model its previous arguments to come to its satisfying predictions:

a) M_r influences the long-wavelength results,
b) H_c influences mainly the short-wavelength results.

Besides it is doubtful if the mechanism behind these effects can be self-demagnetization.

This doubt is also supported by experiments of Bertram and Niedermeyer [7.13]. They observed:

1) *remanence-limited* results for *intimate-contact* recording till the smallest wavelength (1.6 or 0.8 μm) used in their experiments,
2) *coercivity-limited* results for *out-of-contact* recording for the shorter wavelength in their experiments.

There is some contradiction between observation 1 and our experiments. We observe already a coercivity dependence at wavelengths as large as 1.6 and 0.8 μm in our intimate-contact recordings, see e.g. Fig. 7.3.

In the appendix 7.2 it is explained that 'erasing' by the reversed write field due to the finite gradient is probably responsible for the H_c-dependent effects! When the demagnetizing field is neglected (supported by the results of the EMM and above experiments), then the following very rough rule of thumb, see Eq. (a7.4) in the appendix, follows for the ratio λ/d_4 for which no H_c dependence is expected:

$$\boxed{\lambda/d_4 \gtrsim 2\pi\,(1 - c_1 c_2)} \,, \qquad\qquad (7.24)$$

where $c_1 c_2$ is about $\tfrac{1}{2}$.

This was derived for recording with a current optimized at a short wavelength, as in the experiments of Bertram and Niedermeyer. Results with this rule of thumb agree fairly well with their observations, as can easily be verified, see also the appendix. For optimal recording, as we applied in our experiments, the λ/d_4 ratio becomes larger, see appendix. This explains part of the discrepancy between our and their observations.

We must conclude that the H_c-dependence is mainly caused by 'demagnetization' or erasing effects of the alternating write field ('snow shoe' effect).

The UMM describes the resulting H_c dependences quite well, at least in our intimate-contact and optimized recordings. However the H_c dependence is contributed to the wrong cause: self-demagnetization instead of 'erasing' by the reversed write field.

For output calculations after write and read the UMM suffices. The EMM is not suited because of the absence of the strong H_c dependence at shorter wavelengths.

In a first attempt to declare the H_c dependence at larger distances (where it was sure that the self-demagnetizing fields are low), we focussed our attention on the possibility that the threshold field H_{thr} (which replaces in our definition, the coercive force of the major hysteresis loop, see Fig. 7.4) decreases quickly with the (applied-) field amplitude and thus with the permanent magnetization.

In order to investigate this possibility, we measured major *and* minor loops of various tape samples in longitudinal and transversal fields in a vibrating-sample magnetometer. (No measurements in perpendicular fields are carried out because of the strong influence of self-demagnetizing fields in this situation. See also the comment in the caption of Table 7.4). The measurements in the transversal applied field, one of which is shown in Fig. 7.4.II, show that H_{thr} decreases proportionally to M_p. For the relevant longitudinal applied field the decrease is less: $H_{thr}/M_p \sim 1/\sqrt{M_p}$. This means that self-demagnetization is also plausible for weak transitions if these were written by weak write fields. Hence in (7.16) one 'might' replace H_c/M_p by a more or less constant H_{thr}/M_p ratio. The decrease of the write field amplitude with the distance, however, is not so large that this choice, and hence a large H_c dependence due to self-demagnetization, can really be justified. Also an *increased* H_c dependence at larger distances cannot be explained in this way, and

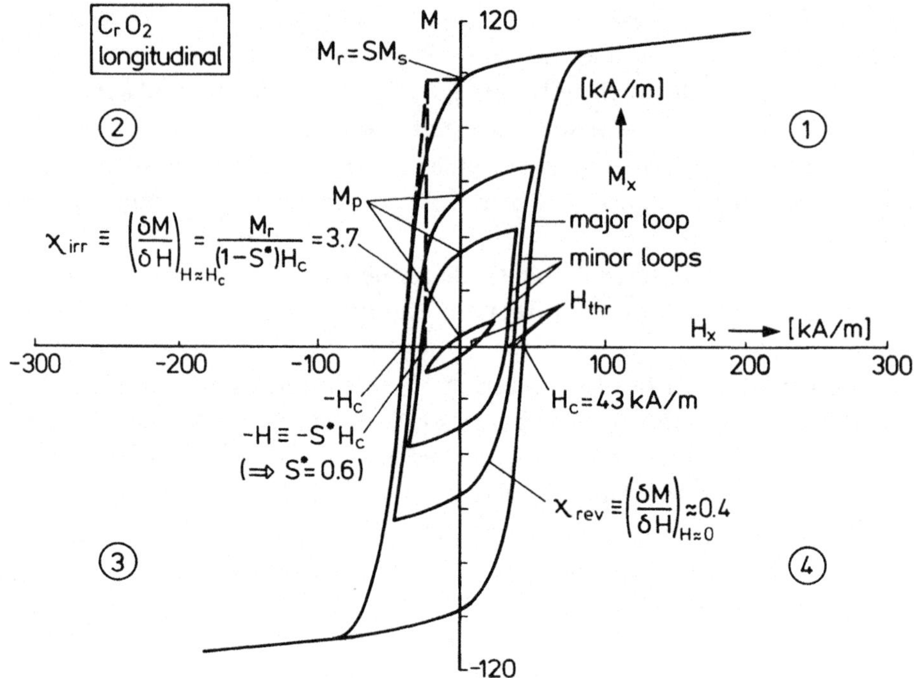

Fig. 7.4.I M-H loops of a CrO$_2$ tape for longitudinal applied fields measured by a VSM.
M_s = saturation magnetization
M_r = remanent magnetization (of major loop)
$S \equiv M_r/M_s$ is the squareness
S^* = coercivity squareness, defined in the figure
χ_{irr} = susceptibility of the steep (irreversible) part of the M_x-H_x loops around $H_x = H_c$ or H_{thr}
$\chi_{x\,rev}$ = susceptibility of the reversible part of the M-H loops around slightly negative values of H
in the second quadrant or slightly positive values in the third quadrant, where the point (H, M)
is located after a recording.
M_p = the 'remanent' magnetization of a minor loop, i.e. at $H = 0$.
H_{thr} = the 'coercive' field of a minor loop, i.e. at $M = 0$.
(x, y, z) = (longitudinal, perpendicular, transversal).

is also not described by the UMM.

The characteristic parameters defined in Fig. 7.4 of three measured tapes are collected in Table 7.4 for reference. The permeability data which are necessary for an accurate description of the playback process are hardly found in the literature and are never given by the tape manufacturers.

The relation between H_{thr} and the amplitude of the applied field is of importance for an accurate description of the write process. This relation is also used in the rough calculations concerning the 'erasing' effects by the reversed write field in appendix 7.2, leading to (7.24).

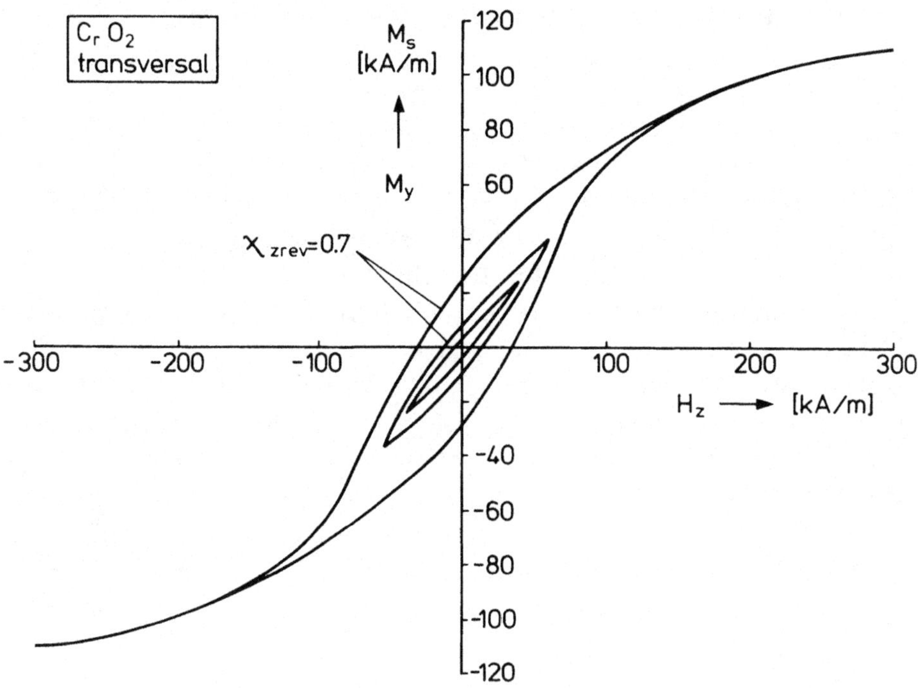

Fig. 7.4.II M-H loops of a CrO$_2$ tape for transversal applied fields.
$\chi_{z\,rev}$ = 'reversible' susceptibility for transversal fields, see also the definition of $\chi_{x\,rev}$.

Table 7.4 Parameters of a chromium-dioxide (CrO$_2$), a metal-evaporated (ME) and a metal-powder (MP) tape. The magnetic values for the CrO$_2$ tape, are determined from the M-H loops in Fig. 7.4 and for the ME and MP tape from similar loops obtained by the VSM.

tape	t [μm]	M_r [kA/m]	H_c [kA/m]	S	S^*	μ_{irr}	$\mu_{x\,rev}$	$\mu_{z\,rev}$	μ	β
CrO$_2$	4.6	95	43	0.80	0.60	4.7	1.4	1.7	1.5	0.9
ME	0.18	240	83	0.67	0.45	6.3	2.4	2.8	2.6	0.9
MP	3.6	160	118	0.70	0.66	5.0	1.5	1.5	1.5	1.0

In the calculation of the relative permeability and anisotropy, $\mu \equiv \sqrt{\mu_{x\,rev} \cdot \mu_{y\,rev}}$ and $\beta \equiv \sqrt{\mu_{x\,rev}/\mu_{y\,rev}}$ according to the definitions in chapter 2, we assumed that $\mu_y = \mu_z$. For the particulate tapes CrO$_2$ and MP with rotational-symmetric elongated particles directed along the x-axis this will be a good assumption. It is impossible to determine the M_y-H_y loop from the measured M_y-$H_{y\,applied}$ loop accurately, because of the highly dominating influence of the demagnetizing field on $H_y = H_{y\,applied} + H_{dem}$. In the x and y direction the demagnetizing field is negligible during the VSM measurement.

7.4 Influence of low permeability around the gap

7.4.1 Gradual gap

The consequences of a gradual gap on the playback signal are extensively discussed in Sect. 5.7 and other sections. The 'arctan' approach, see Sect. 5.7.9, will be used for further application, especially in the write-read channel described in this chapter.

The great advantage of this 'arctan' approach is that the write fields of this head are completely equivalent to those of an ideal head on a $\Delta g/\pi$ larger distance. This is proven by relation (5.72). Δg can be interpreted as the gap widening due to the poor magnetic gap-core transition with permeability profile $\mu_r(x) = \pi/(\arctan (x + g/2)/(\Delta g/\pi) - \arctan (x - g/2)/(\Delta g/\pi))$.

Consequently the losses in optimal recording are described by the phenomenologically derived write-spacing loss, i.e. the gradual gap recording-loss, GGL_r, equals

$$\boxed{GGL_r = e^{-\alpha k\, \Delta g/\pi}} \, , \qquad (7.25)$$

where $\alpha \simeq 0.7$ for most tapes in the region of present head-to-tape distances $(0.1 g - g)$.

This factor is therefore incorporated in the expressions for the output voltage in appendix 7.1.

As explained in chapter 4, it is noted that the efficiency of this head is equal to that of a head with a sharp gap with length g. It is important to realize this in write-read model calculations.

7.4.2 Uniform low permeability around the gap

When the permeability around the gap is about constant but relatively low, the gradients of the 'write' field will be low too. This will deteriorate the write process.

This deteriorated (indicated by a prime) deep-gap field, H'_{dg}, is sketched, for a relative permeability $\mu_r = 5$, in the left side of Fig. 7.5. It is essential that a 'ground level' or 'dc level' of $1/\mu_r$ times the maximum deep-gap field (H'_{max} at $y \leqslant 0$) exists in the write field. At

$y = 0$ the field is assumed to be equal to the deep-gap field. Karlqvist made this approximation for an ideal gap.

At $y > 0$ the field is then given by:

$$H' = \frac{H'_{dg}}{\mu_r} + \frac{\mu_r - 1}{\mu_r} H'_{dg} \frac{1}{\pi} \left(\arctan \frac{x + g/2}{y} - \arctan \frac{x - g/2}{y} \right).$$

$$(7.26)$$

This is sketched in Fig. 7.5. The dashed line represents the ideal Karlqvist head. The ratio of the slopes of the ideal ($\mu_r = \infty$) and non-ideal (μ_r finite) head equals $(\mu_r + 1)/(\mu_r - 1)$. (It is assumed that the proper value of the write field at the gap edges is unaffected by the choice of μ_r. In practice this will be about the coercive field, H_c, of the tape, obtained by applying the proper write current.) This means that *the low-permeability head acts approximately like an ideal one on a* $(\mu_r + 1)/(\mu_r - 1)$ *times larger distance, y'', i.e.*

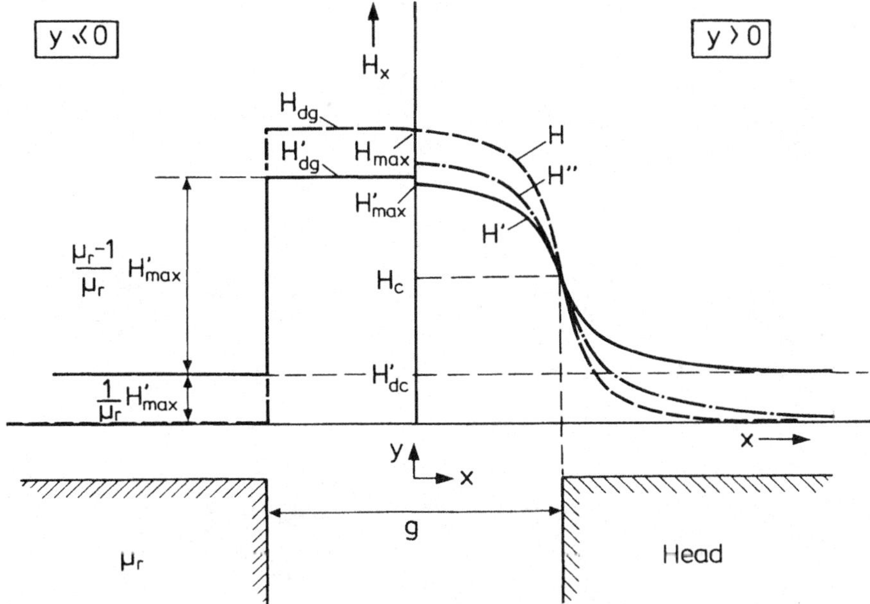

Fig. 7.5 Arctan approximation of the write field above an ideal and non-ideal head at $y = 0$ (sketched for negative x values) and at $y/g = 0.07$ (sketched for positive x values).
----: ideal head; $\mu_r = \infty$
——: non-ideal head; $\mu_r = 5$
—·—: ideal head at a $(\mu_r + 1)/(\mu_r - 1)$ larger distance i.e. at $y''/g = 0.11$.
The slopes of H'' and H' coincide at $H' = H'' = H_c$.

$$y'' = y(\mu_r + 1)/(\mu_r - 1) \tag{7.27}$$

Steele and Mallinson [7.14] calculated the write fields of low-permeability heads (down to $\mu_r = 10$) numerically. Our results are consistent with their results for $y = g/3$ (Fig. 3 in [7.14]); their slope reduces roughly by 25% when μ_r reduces from 10^6 to 10, while in our approximation this reduction is 20%. The discrepancy with their results is larger for $y = 2g/3$ (fig. 4 in [7.14]), which may be due to inaccuracies in their line drawings.

What is the influence of the permeability-dependent increase of the virtual head-to-tape distance, y'', on the playback signal?

For this assume, as in the EMM, that the magnetization in the tape is proportional to $e^{-\alpha k y_f}$, because field gradients and consequently transition widths (here in the pulse-crowded situation such that $M(x) \to M \sin kx$) are proportional to $1/y$ (for large enough g). The proportionality factor α in the exponent is expected to decrease with the steepness of the MH curve and thus decreases when the coercivity squareness S^* and the M_r/H_c ratio increase [7.15]. (α will be a little lower than in the UMM model, where α is only applied to the head-to-tape spacing.) Substitution of $e^{-\alpha k y_f}$ in the EMM, while neglecting reversible-permeability effects, leads to a loss after writing with the low-permeability head relative to the loss after writing with a high-permeability head (LPL) given by

$$LPL = \frac{\displaystyle\int_{d_r}^{d_r + t_3} e^{-ky}\, e^{-\alpha k y''}\, dy}{\displaystyle\int_{d_r}^{d_r + t_3} e^{-ky}\, e^{-\alpha k y}\, dy} = e^{-(k_2 - k_1)d_r}\;\frac{k_1}{k_2}\,\frac{1 - e^{-k_2 t_3}}{1 - e^{-k_1 t_3}}, \tag{7.28}$$

where $k_1 \equiv (1 + \alpha)k$ and $k_2 \equiv (1 + \alpha(\mu_r + 1)/(\mu_r - 1))k$.

At short wavelengths where the loss may become relevant, the last factor can be neglected when the coating is 'thick' ($t_3 > \lambda/2\pi$), giving

$$LPL = \frac{1 + \alpha}{1 + \alpha(\mu_r + 1)/(\mu_r - 1)}\; e^{-\frac{2\alpha}{\mu_r - 1} k\, d_r}. \tag{7.29}$$

For ultimate-contact recording ($k\,d_r \to 0$) only the first factor remains. Application of the UMM would have given only the last exponential term and hence no effect in the case of ultimate contact, as was to be expected. It is interesting to note that the low-permeability recording loss increases with the distance to the head exponentially $\left(55 \times 2\alpha d_r /((\mu_r - 1)\lambda)\ [\mathrm{d}B]\right)$, besides the first factor that depends on μ_r. Equation (7.28) gives for example $LPL = 0.7$ for $\lambda = 0.7\,\mu\mathrm{m}$, $\mu_r = 5$ and $\alpha = 0.5$.

The actual LPL is underestimated by (7.28) when the μ_r is very small, say $\mu_r < 5$, because of the alternating 'ground level' (in the rhythm of the alternating write signal) which deteriorates the write process further. This effect becomes stronger at larger head-to-tape distances, because the maximum field at the centre of the gap, H'_{max}, decreases with respect to the distance-independent dc level. (E.g. at $y/g = 0.11$, H_{max} decreases by about 10% with respect to the case $y = 0$, see the curves H'' and H_{dg} in Fig. 7.5, due to the increasing overlap of the arctan functions at both gap edges in the Karlqvist approximation. See also our numerical results in Fig. 7.6 compared to those in Fig. 3 of Steele and Mallinson [7.14].)

During playback, a low permeability of the head core material works out in a low efficiency. This is described numerically in e.g. [7.14] and analytically in chapter 8. When the low permeability is restricted to a small area around the gap, e.g. restricted to the metal layers in a metal-in-gap (MIG) head, then it changes both the efficiency as well as the gap loss function, as will be explained in chapter 12.

In conclusion, a low permeability ($\mu_r < 10$) of the core near the gap gives a write-distance loss that is considerably larger than that of a high-permeability head.

The α in the exponent of the exponential write-distance loss factor e^{-kad_r} is then replaced by $\alpha(1 - 2/(\mu_r - 1))$.

It depends on the model (EMM or UMM) whether or not an extra distance-*in*dependent loss is predicted. The actual loss is expected to be larger because of a further deterioration of the write process due to the alternating ground level that is not taken into account.

7.5 Conclusions

A simple model for longitudinal recording was presented. This uniform-magnetization model (UMM) fits for head-to-tape distance variations only when a phenomenological write-spacing loss term $e^{0.7\,kd_r}$ was added. The same factor incorporated in a physically more consistent exponential-decaying magnetization model (EMM) led to almost self-demagnetization-free write processes for practical 'demagnetization ratios', M_r/H_c. However, our experimental observations and also those of others showed clearly an increase of the short-wavelength output of tapes with increasing H_c values, which must be due to a reduction of the demagnetization. In the UMM and EMM the characteristics of the major loop, H_c and M_r, were used for simplicity. In more accurate analysis it seems necessary to use the varying threshold field H_{thr} and permanent magnetization M_p of the minor loops. VSM measurements show an only slowly decreasing 'demagnetization ratio' M_p/H_{thr}, when the amplitude of the applied field is reduced. Thus 'demagnetizing' effects (such as erasure by the reversed write field and self-demagnetization) are stronger at lower applied fields as might be expected. A 'weakness' of the model (but an important simplification for the calculations) is that nearly all the H_c-dependent effects are attributed to self-demagnetization. The complete 'demagnetizing' effects are described by this 'self-demagnetizing' term and a phenomenological exponential write-distance loss factor. The self-demagnetization calculation is carried out before the introduction of the phenomenological distance loss factor, and is thus overestimated. This will have compensated for the underestimated H_c-dependence due to the other 'demagnetizing' effect caused by the finite write-field gradient (erasing). The remaining part of the demagnetizing effects is put into the phenomenological write-distance loss factor.

With the aid of this phenomenological factor, e^{-kad_r}, it was also possible to estimate the signal loss in the write process due to a gradual gap and due to a low permeability of the head core near the gap.

The model is not suitable for a detailed description of the write effects in the magnetic coating, it only describes the effects on the output voltage after playback on the basis of simple physical and experimental arguments. For a detailed local description of the write processes in the tape one needs self-consistent numerical calculations that include minor loops in the longitudinal and perpendicular direction. Especially the theoretical and experimental investigation of the (self-)demagnetization process needs further attention, also because of some discrepancy in the literature concerning the possible absence of (self-)demagnetization effects in intimate (ultimate) contact recording. See also the conclusions at the end of appendix 7.2.

Appendix 7.1 Playback voltage in the case of the UMM

The remanent-magnetization distribution that must be substituted in the expression for the output voltage (6.11) is approximated by the permanent magnetization distribution that emanates from the UMM. (A different approach would be to approximate the free-space field H_f in the playback situation by the 'free-space' field calculated in the UMM under the (temporary) assumption $\mu_r = 1$).

As a result, the effective output voltage, $V_{rms} = \omega|\Phi^R(\omega)|/\sqrt{2}$, after optimal unbiased sinewave recording and subsequent playback, reads for an SL medium, using the equations (6.11), (5.7) and (5.19):

$$V_{rms} = \frac{1}{\beta_3 \cosh(\hat{\beta}_5^* k t_5)}$$

$$\sqrt{2}\, \nu N W \mu_0 M_r \mathrm{e}^{-\alpha k(d_{4r} + \Delta g/\pi)}\, \eta^a |GLF_i^a|$$

$$\times \left[\left(1 + \frac{\tanh(\hat{\beta}_5^* k t_5)}{\hat{\mu}_5^*}\right) \mathrm{e}^{k(d_{4p} + \Delta g/\pi)} - R\left(1 - \frac{\tanh(\hat{\beta}_5^* k t_5)}{\hat{\mu}_5^*}\right) \mathrm{e}^{-k(d_{4p} + \Delta g/\pi)} \right]$$

$$\times \frac{\sinh \alpha_3}{\sinh(\beta_3 k t_3 + 2\alpha_3)} \left\{ \sinh(\beta_3 k t_3 + \alpha_3) - \sinh(\beta_3 k(t_3 - t_{x\,opt}) + \alpha_3) \right\},$$

$$(a7.1)$$

with

$$\alpha_3 = \tfrac{1}{2} \ln \frac{\mu_3 + 1}{\mu_3 - 1},$$

$$R = \frac{\sinh(\beta_3 k t_3)}{\sinh(\beta_3 k t_3 + 2\alpha_3)},\ \text{the remagnetization parameter,}$$

η^a efficiency including the influence of the gap smear (reduces the efficiency, see Sect. 8.1.6.1) and of the coating (usually negligible reduction),

GLF_i^a gap-loss function of ideally-sharp gap including the influence of gap smear (makes the Karlqvist approximation more accurate) and non-unit relative permeability coating (see Sect. 5.6 and reference [5.12]); in the case of no gap smear ($t_5 = 0$), the accurate approximation GLF_0 given further on is used in our calculations,

μ_3, β_3 mean permeability ($\sqrt{\mu_{3x}\mu_{3y}}$) and anisotropy ($\sqrt{\mu_{3x}/\mu_{3y}}$) of coating around $H = 0$ in a permanently magnetized state,

$\hat{\mu}_5, \hat{\beta}_5$ as μ_3, β_3 but complex time-independent quantities, $*$ denotes complex conjugated,

t_3, t_5 thickness of coating and gap-smear layer in [m],

d_{4r}, d_{4p} head-tape distance (between gap-smear layer and tape) during recording or playback in [m],

v relative head-tape velocity during playback in [m/s],

N number of windings through the coil chamber of the playback head,

Δg gap widening due to poor magnetic transition, see sections 5.7.9 and 7.3.1, in [m],

W trackwidth in [m],

μ_0 $= 4\pi \, 10^{-7}$ [Vs/Am] is the permeability of vacuum,

M_r remanent magnetization of the tape in [A/m],

H_c coercivity of the tape in [A/m], while $H_c > M_r$,

M_s saturation magnetization of the head in [A/m],

k $= 2\pi/\lambda$ is the wavenumber in [m^{-1}],

λ wavelength in [m],

$t_{x\,opt}$ $= -\dfrac{\lambda}{\pi} \ln \left(1 - \dfrac{H_c}{M_r}\right)$ if $< t_{x\,max}$ else

$t_{x\,opt}$ $t_{x\,max}$ if $t_{x\,max} > 0$ and $t_{x\,opt} = 0$ if $t_{x\,max} < 0$, is the optimal writing depth according to the UMM, in [m],

$t_{x\,max}$ $= \dfrac{g_r}{2 \tan \dfrac{\pi \, C_d \, H_c}{2 M_s}} - d_{4r}$ if $< t_3$ else $t_{x\,max} = t_3$, is the maximum writing depth of the recording head with gap length g_r, in [m],

C_d ≈ 1.25 is the 'demagnetization' constant and

α ≈ 0.7 is write-spacing loss coefficient.

This expression can be greatly simplified without much loss of accuracy, when using the factorized form proposed in section 6.3.5, valid for $\beta_3 kt_3 \gtrsim 1$ and no gap smear:

$$V_{\rm rms} = \frac{kvN}{\sqrt{2}} \, |\hat{\Phi}_{\rm ix}| \, RL_{\rm x}$$

(a7.2)

where

$|\hat{\Phi}_{\rm ix}|$ $\quad = Wt_{\rm x}\mu_0 M_{\rm r}\, e^{-ak(d_{4\rm r} + \Delta g/\pi)}$ is the longitudinal permanent flux in the coating according to the UMM and including the recording effect of the gradual gap,

$RL_{\rm x}$ $\quad = TL \times GDL \times \eta_0 \times GLF_0$ is the reproduction loss factor for longitudinal recording,

TL $\quad = \dfrac{1 - e^{-\beta_3 kt_{\rm x}}}{\beta_3 kt_{\rm x}}$ is the thickness loss factor,

GDL $\quad = \dfrac{2}{\mu_3 + 1} \dfrac{1}{1 - \dfrac{\mu_3 - 1}{\mu_3 + 1}\, e^{-2k(d_{4\rm p} + \Delta g/\pi)}}\, e^{-k(d_{4\rm p} + \Delta g/\pi)}$

is the generalised distance loss factor, including the playback effect of the gradual gap,

η_0 $\quad$ efficiency of the head (without the influence of the tape),

GLF_0 $\quad = \frac{1}{2}\{J_0(\pi g/\lambda) + \operatorname{sinc}(g/\lambda)\}$ is the gap loss function of the idealized (i.e. sharp) gap, disregarding the influence of the coating permeability, and

g $\quad \equiv \displaystyle\int_{\text{gap area}} \frac{1}{\mu_{\rm r}(x)}\, dx$ the definition of the gap length according to (5.57).

The above expression does not describe losses accurately for $\beta_3 kt_3 \lesssim 1$, but this is not important for e.g. CrO_2 and MP tapes. These tapes have a coating thickness that is much larger than the head-to-tape distance, so that even when $\beta_3 kt_3$ is as large as 1 the total reproduction loss is still small and hence irrelevant for the description. For some ME tapes with a very small coating thickness the above expression may not be appropriate.

In the above expression we chose then to add $\Delta g/\pi$ to every head-to-tape distance, d. The effect of a uniform low permeability around the gap, as described in Sect. 7.3.2, is not incorporated in the expressions (a7.1) and (a7.2), since the effect is usually negligible.

Appendix 7.2 Survey of literature and comments

This appendix will mainly be focussed on the role of self-demagnetization in the write process, or more accurately formulated: on the effects and observations attributed *in the literature* to self-demagnetization. It is noted beforehand that the opinions seem divided and the observations not quite consistent. Further it is often unclear whether demagnetization was meant to be *self*-demagnetization or has any more general meaning. Perhaps the following discussion will reduce the confusion to some extent.

The discussion of the literature and comments is divided into four parts:

a) A discussion of the observations and suggestions of Bertram and Niedermeyer [7.13].

b) A discussion of the observations and conclusions of Lee and George [7.16] concerning the role of self-demagnetization. In this context a lot of other 'underlying' papers will be discussed.

c) A correction on a statement of Mallinson concerning the limits of magnetic recording and a subsequent own estimation of the highest self-demagnetizing field to coercivity field ratio obtained today.

d) Concluding remarks.

Bertram and Niedermeyer

Recently Bertram and Niedermeyer [7.13] published a new experimental result. They observed, also with the smallest wavelength in their experiments being 0.8 µm, that the output of metal heads in close contact with various tapes with almost equal H_c but quite different M_r and vice versa is proportional to the remanent magnetization and not to the coercivity. This result is fairly unexpected at this 'small' wavelength.

The slope of the output versus k curves, defined by the well-known and simple approximation expression $e^{-k \cdot slope}$, was 0.34 µm (the slope being the result of both write and read distance losses). According to our read-write model, the distance-loss factor points to a head-to-tape distance during reading and writing of about 0.2 µm, which is more than two times larger than follows from our measurements on video tapes.

After sputtering the write head with a non-magnetic layer of 0.3 µm the unexpected result vanished, i.e. the M_r dependence at the shorter wavelengths vanished and the output became coercivity dependent, i.e. demagnetization limited.

Hence the above unexpected results seem typical for intimate contact recording and would therefore surely be visible in our experiments with heads in contact with video tapes. However, in spite of the considerably smaller head-to-tape distances in our video experiments, we did experience a clear H_c dependence at wavelengths smaller than say 1 μm, see e.g. Fig. 7.3.I and the description at the end of section 7.2.2.3.

The most well-known mechanism that explains a clear H_c dependence is self-demagnetization. For this reason Bertram and Niedermeyer suggest that the write process is (self-)demagnetization free in ultimate-contact recording. With the aid of the demagnetizing-field calculations in one of the proposed models, the EMM, it was shown that even at very small head-to-tape distances the self-demagnetization fields of the final magnetization do not approach too closely the 'threshold field' of the major loop, i.e. the coercive field H_c. This may explain why no clear H_c dependence is observed till rather short wavelengths in ultimate-contact recording. However, after writing at a larger head-to-tape distance

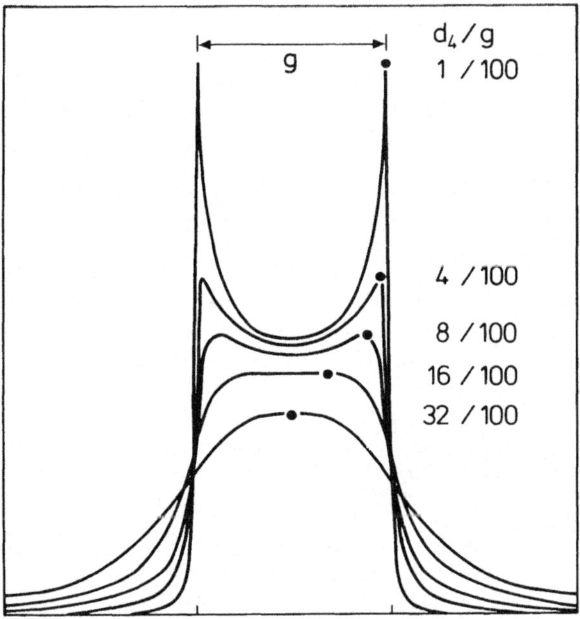

Fig. 7.6 Accurate shape of the field of an ideal ring head at small head-to-tape distances where the Karlqvist approximation fails. (The curves are obtained with the aid of the simple expression (5.45), see also the text.)

the self-demagnetizing fields are still smaller and a possible H_c dependence would be reduced further. Just the opposite is experienced by Bertram and Niedermeyer.

In the discussion in Section 7.3 we also reported on measurements of minor hysteresis loops on various tape samples that may have a clarifying effect. These measurements showed that the 'demagnetization' ratio of a minor hysteresis loop, M_p/H_{thr} (which replaced the demagnetization ratio of the major hysteresis loop, M_r/H_c), only decreases slowly with the applied-field amplitude. Besides, at larger head-to-tape distances the *optimal* write field amplitude is smaller. This follows from the experimental observation that at larger head-to-tape distances the optimal write current is hardly increased and the theoretical result (shown by the dots in Fig. 7.6) that the maximum of the x component of the write field (at constant write current) decreases considerably with the head-to-tape distance.* Hence the (optimal) write field at larger distances decreases and so H_{thr} decreases such that in spite of a lower level of the written magnetization self-demagnetization may still have some influence on the write process. A *larger* self-demagnetization effect at larger distances (read 'lower write fields') can however never be explained in this way.

Before proposing a possible explanation of the observed increasing H_c dependence at increasing write distances, it is useful to return to the observation of a decreasing optimal writing *field* in the tape at increasing write distances. The only possible mechanism behind this effect is 'erasing' by the reversed write field, H_{rev}. This happens approximately half a period after the magnetization was written and half a wavelength away from the gap (centre). This 'snow-shoe' effect increases drastically at larger head-to-tape distances if one applies the same write field amplitude in the tape. One must expect that the compromise between

---

*The strong decrease of the peak and gradient of the x component of the write field with increasing d_4 at this short head-to-tape distances in Fig. 7.6 would not be visible in the Karlqvist approximation. This approximation has been used in all analytical and also self-consistent numerical computations of the recording process until now, see for example the work of Potter and Beardsley (Ortenburger) [7.28 + 7.29].

The accurate result in Fig. 7.6 is obtained by carrying out the (fast) Fourier transform of the product of the accurate but simple expression (5.45) for the *GLF* of an ideal ring head with the distance-loss factor e^{-kd_4}. Use of the exact result of Westmijze, e.g. the *GLF* expression in (5.37), is too complicated. The results in Fig. 7.6 indicate that it is necessary to use one of the more accurate field expressions and not the Karlqvist expression when effects in intimate-contact recording (say $d_4 < \frac{1}{3}g$) are to be described accurately. One such an effect is the possible absence of a H_c dependence at short head-to-tape distances.

field amplitude and erasing effects leads at lower gradients (read 'larger distances') to lower field amplitudes. These erasing effects are stronger when the threshold field, which is correlated to H_c, is lower. Thus, it is quite feasible that the observed H_c dependence is a result of erasing or 'demagnetization' (in this context not *self*-demagnetization by the demagnetizing field) under influence of the sum of the reversed write field and the self-demagnetizing field. The strong increase of the reversed write field (and not the self-demagnetization field which decreases) at larger distances, possibly in combination with a reduced threshold field, thus may explain the stronger H_c dependence at larger distances with respect to small distances. The EMM indicated that the demagnetizing field is completely negligible at larger head-to-tape distances and may have some influence (but not large in Bertram and Niedermeyer's experiment) at small head-to-tape distances and small wavelengths. In two recent papers [7.30 + 7.31] Bertram neglected the demagnetizing field in a simplified description of the write process for in-contact recording ($d_4 = g = 0\,\mu\text{m}$).

When we make the same simplification, but for large head-to-tape distances too, then we can (very roughly) estimate the ratios λ/d_4, for which no H_c dependence will be observed.

Estimation of H_c-independent region

H_c-dependent effects are not expected when:

$$H_{\text{dem}} + H_{\text{rev}} \lesssim c_1 H_{\text{thr}}, \tag{a7.3}$$

where c_1 is of the order of 0.5.

When demagnetizing fields are neglected, then

$$H_{\text{rev}} \lesssim c_1 H_{\text{thr}}. \tag{a7.4}$$

The total write field now equals the applied field and will be approximated, also for small y/g ratios (y is the distance from the head's surface) by the trapezoidal depicted in Fig. 7.7 with the same gradient (for $g > \pi y$) as the Karlqvist field. (In fact the Karlqvist approach only becomes inaccurate for $g > \pi y$. For the present purpose of a rough approximation we are not concerned about this.)

Hence:

$$H_{\text{rev}} \approx H_{\text{appl}} \left(1 - \frac{\lambda}{2\pi y}\right). \tag{a7.5}$$

From our hysteresis-loop measurements, see section 7.3 and Fig. 7.4, we estimated:

$$H_{\text{thr}} \approx c_2 H_{\text{c}}^{c_3} \cdot H_{\text{appl}}^{1-c_3}, \tag{a7.6}$$

with $c_2 \approx 1$ and $c_3 \approx \frac{1}{2}$.

Substitution of (a7.5) and (a7.6) in the criterion (a7.4) yields:

$$\lambda/y \gtrsim 2\pi\left[1 - c_1 c_2 (H_{\text{c}}/H_{\text{appl}})^{c_3}\right]. \tag{a7.7}$$

Thus at large λ or small y no H_{c} dependence will be observed!

Optimizing the write current at the shortest wavelength (as Bertram and Niedermeyer did in their experiments) gives $H_{\text{appl}} \approx H_{\text{c}}$ at $y \approx d_4$. Hence (a7.7) reduces in this case for the surface layer of the magnetized medium to the following very simple expression, independent of the value of c_3 (!):

$$\boxed{\lambda/d_4 \gtrsim 2\pi(1 - c_1 c_2)} \quad . \tag{a7.8}$$

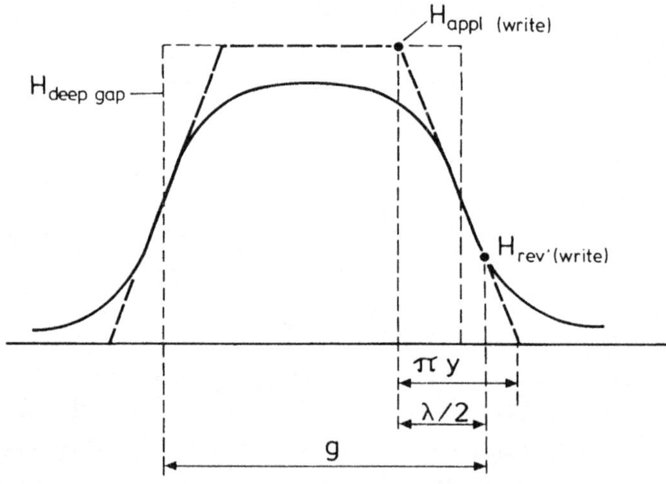

Fig. 7.7 Approximation of the (Karlqvist) field of a ring head (——) by a trapezoidal (---). H_{appl} is the amplitude of the applied write field in this approximation and H_{rev} is the amplitude of the reversed write field as experienced by the magnetization just written after a half period ($\lambda/2$).

Deeper in the tape there may be some H_c dependence because y is larger (partly compensated by a lower applied field). However, the effect on the output will be small, because of larger read-distance losses, and will therefore be neglected. The (very rough) rule of thumb (a7.8) then predicts the following observations (when the least arbitrary choices $c_1 = \frac{1}{2}$ and $c_2 = 1$ have been made):

$\lambda = 1.6$ μm $d_4 = 0.2$ μm   definitely no H_c dependence
$\lambda = 0.8$ μm $d_4 = 0.2$ μm   just no H_c dependence
$\lambda = 1.6$ μm $d_4 = 0.5$ μm   just a H_c dependence
$\lambda = 0.8$ μm $d_4 = 0.5$ μm   definitely a H_c dependence.

These results correspond fairly well with the observations of Bertram and Niedermeyer.

When the current is optimized at each wavelength separately (as we did in our experiments, see e.g. Fig. 7.3), then the H_c dependence will remain visible up to *longer* wavelengths, since then $H_{appl} = H_c$ at $y > d_4$ (or $H_{appl} > H_c$ at $y = d_4$). Perhaps this may explain the difference between our observations and those of Bertram and Niedermeyer.

Lee and George and others.

Recently Lee and George [7.16] concluded from their intimate-contact recording experiments that self-demagnetization in thin metallic films with thickness t_3 occurs when roughly both $t_3 \geq 0.1$ μm and the recording distance $d_{4r} \lesssim 0.1$ μm. They based their conclusions partly on a theoretical result of Williams and Comstock concerning transition widths.

Williams and Comstock [7.15] state that self-demagnetization can occur in media with high demagnetization ratio, M_r/H_c, and low spacing, which gives rise to transition widths, a_t, just after writing that are comparable to or larger than the transition width a_d (rewritten in SI units) 'defined' by:

$$a_d = \frac{M_r}{2\pi H_c}\, t_3\,(3.3\text{-}2.3\,S^*) \quad , \qquad \text{(a7.9)}$$

with S^* the coercivity squareness defined in Fig. 7.4.I.

Williams and Comstock calculated this expression in an alternative way, by minimizing the free energy ($\sim \frac{1}{2}\,\boldsymbol{H}_{\mathrm{d}} \cdot \boldsymbol{M}$) of an arctan transition (constraint) until the work necessary to do this ($\sim \int \boldsymbol{H} \cdot \mathrm{d}\boldsymbol{M}$) equaled the energy saving. For a loop (or coercivity) squareness $S^* = 1$ this coincides with the result obtained when the maximum demagnetizing field equals H_{c}. It must be realized that this expression is derived for a single transition in a thin medium and applied by Lee and George to 'sinusoidal' magnetization patterns resulting from very high bit densities! It is discussed further on whether this was allowed.

Mallinson and Steele [7.17] discussed why linear superposition of isolated pulses holds reasonably well even when there is overlap of pulses (pulse crowding) in the coating, in spite of the fact that the magnetic recording process is quite non-linear.

Kosters and Speliotis [7.25] subsequently showed experimentally that linear superposition of isolated pulses with shape $f(x) = 1/\bigl(1 + (x/PW_{50})^2\bigr)$ leads to an accurate determination of the amplitude as a function of the bit density, but peak shift is not accurately predicted.

Tjaden [7.22] finally investigated the validity of superposition theoretically (Preisach-model calculations and self-consistent calculations) as well as experimentally (large-scale measurements on aligned iron-oxide particles). He arrives at the following statements:
Superposition is not consistent with the Preisach model (unless no particle interaction is assumed and demagnetizing fields are zero).
For three or more transitions a weaker form of superposition applies.

In particular demagnetizing effects reduce the validity of the superposition principle. This has been shown by Tjaden's self-consistent computations as well as by his large-scale measurements.

Williams and Comstock also calculated analytically the *transition length a_{t}* of an isolated transition in a thin coating ($t_3 \ll \lambda$) when it is not self-demagnetization but the head-field gradient (which we called 'erasing' by the reversed write field) which limits the steepness of the transition. The use of the arctan model, proposed by Miyata and Hartel [7.21], for the shape of the transition made the analytical calculation possible. William and Comstock took account of the shunting (or shielding) effect of the soft-magnetic head. Due to this effect the transition width decreases usually less than 10%, see examples in reference [7.16]. Their transition length a_{t} corresponded reasonably well with self-consistent iterative numerical computations from Curland and Speliotis [7.19].

Maller and Middleton [7.20] tried to extend the analysis of Williams and Comstock to thick media. For greater analytical-flexibility they introduced another mathematical model for the hysteresis loop, whose characteristics however are essentially the same as those of the loop proposed by Williams and Comstock. The self-demagnetizing limit was not calculated by energy considerations as did Williams and Comstock, but by taking the demagnetizing field equal to (about) the coercive field, as has been used in all earlier publications, e.g. in the earlier paper of Middleton [7.27]. The calculations of the demagnetizing field were carried out for the case $t_3 \ll a_t$. Furthermore, the determination of the transition length, when due to field gradients rather than self-demagnetization, was carried out in the mid plane of the medium. Thus, since no y-dependences were assumed, thick media had not been described realistically.

In a later paper Middleton and Wisely [7.23] accounted for this by introducing a transition length that changed linearly from a_t at the 'top' of the coating (head side) to $a_b = a_t + \alpha t_3$ at the bottom. a_t, a_b and α depend on the squareness $S \equiv M_r/M_s$ of the theoretical hysteresis loop (it is noted that this squareness in the above work is determined rather peculiarly, by comparing the characteristics in the second quadrant of the theoretical and actual loop rather than by the squareness of the actual loop!), the demagnetizing ratio M_r/H_c, the reversible susceptibility χ_{rev}, the spacing during recording d_{4r} and the thickness of the coating t_3. Implicit expressions have been found such that a_t, a_b and α follow after an iterative process. In contrast to Williams and Comstock, non-optimal recording is included. In addition, linear superposition of isolated pulses is used to describe the effects of pulse crowding at high bit densities where the magnetization becomes almost sinusoidal. Calculated pulse widths agreed quite well with theory. The same is true for the output as a function of bit density *except for currents smaller than the optimal current*. It was experimentally verified for a relatively thin coating ($t_3 = 2\,\mu m$) with respect to the recording gap (11 μm!), spacing (1 μm) and wavelengths ($> 6\,\mu m$) that according to the theory the slope of the frequency characteristic at high frequencies approached the sum of transition width at the head side plus spacing during playback, $a_t + d_{4p}$. In the case of a thick coating (10 μm) the predicted values were too high, especially at high current levels. The satisfying correspondence for the 2 μm coating cannot be expected for video recording situations ($g = 0.3\,\mu m$, $t_3 = 3\,\mu m$, $d_{4r} \leqslant 0.1\,\mu m$) since t_3/g, t_3/d_{4r} and t_3/λ

are much larger than for the thick tape in Middleton and Wisely's experiments. The assumption that the transition length increases linearly with depth into the medium will of course be less accurate for larger values of the ratios t_3/d_{4r}, t_3/g and t_3/λ.

In addition, for ultimate-contact recording head-field gradients can be so high that transition lengths are not gradient limited anymore but are dominated by self-demagnetization (or only dominated by the remanent magnetization supposing that the coercive field is not yet reached by the strong self-demagnetizing field in ultimate-contact recording). In the paper of Middleton and Wisely this was not taken into account (in contrast to the earlier paper of Maller and Middleton where in fact thin media ($t_3 \ll a_t$) were assumed), possibly because of the complexity of this phenomenon in a thick coating. Another weakness in Middleton and Wisely's model is the assumption of either complete or no remagnetization. They chose the last in their further analysis, which is probably better than the first for the shorter wavelengths. However it is possible to take remagnetization into account accurately, as described in chapters 2 and 4. For this purpose the permanent magnetization pattern $M_{px}(x, y)$ as follows for $H_x = 0$ must be used. This means that just the solution of Middleton and Wisely for complete remagnetization must be substituted in the various expressions, e.g. (2.33) or (2.43) etc.

The actual remagnetization is now easily taken into account by multiplying the output by the remagnetization factor F_R, see (2.46). F_R is only dependent on the reversible permeabilities of the minor loops of the magnetic layer(s) of the tape and their thicknesses and not on the magnetization pattern.

Lee and George [7.16] considered the self-demagnetization transition-length parameter a_d given in (a7.9) for thin coatings in their discussion on experimental results at very high bit densities. But (a7.9) was calculated by Williams and Comstock for single arctan transitions. As already indicated, linear superposition of isolated pulses *seems* justified for pulse trains and these lead to almost sinusoidal magnetization patterns [7.23] with a slope of the log $V(\omega)$ curve proportional to $a_t + d_{4p}$, where a_t is the width of the single arctan transition and d_{4p} the head-to-tape distance during playback. When however demagnetization effects are important, then the validity of the superposition principle was strongly reduced, as followed from the self-consistent computations and also from the large-scale measurements of Tjaden. Also from the following two arguments we can argue that *linear superposition of single arctan transitions limited by self-demagnetization* ($a_t = a_d$) will be *not allowed* when bit densities are too high and the magnetization pattern is almost sinusoidal:

–   Firstly, it is doubtful if the constraint of an arctan transition holds for transitions limited by self-demagnetization, because *the highest demagnetizing fields H_d do not occur at the centre $x = 0$ of the transition.* For example in a uniformly-magnetized thin coating ($t_3 \ll a_t$) for simplicity H_d of a positive arctan transition in free space equals [7.24]

$$H_d = - \frac{M_r t_3}{\pi} \frac{x}{x^2 + a_t^2}, \tag{a7.10}$$

which has its maximum at $x = \pm a_t$.

–   Secondly, *it is not possible to define an a_d in the pulse-crowded situation that is independent of the bit density or wavelength.*

This second argument needs further explanation. Let us first show that a_d decreases at increasing bit densities by an (not necessarily practical) example. Consider the demagnetizing field of an infinite number of alternating arctan transitions separated by a distance b and with a positive arctan transition at $x = 0$:

$$H_d = - \frac{M_r t_3}{\pi} \sum_{n=-\infty}^{\infty} (-1)^n \frac{x + nb}{(x + nb)^2 + a_t^2}$$

$$= - \frac{M_r t_3}{b} \frac{\cosh(\pi a_t / b)}{\sinh^2(\pi a_t / b) + \sin^2(\pi x / b)} \sin(\pi x / b). \tag{a7.11}$$

Hence, in the pulse-crowded situation ($b < a_t$) the demagnetizing field becomes sinusoidal:

$$H_d = - \frac{2}{\pi} k M_r t_3\, e^{-k a_t} \sin(kx). \tag{a7.12}$$

Since within a sinusoidally-magnetized thin coating $|\hat{H}_d| = k\, t_3 |\hat{M}|/2$, as follows from (7.13), obviously

$$\frac{|\hat{M}|}{M_\mathrm{r}} \simeq \mathrm{e}^{-ka_\mathrm{t}}. \qquad (a7.13)$$

(From (7.13) we also know that the validity region of (a7.12) and hence of (a7.11) is $\lambda > 2\pi t_3$. This region is for small wavelengths, $\lambda \equiv 2b \ll a_\mathrm{t}$, much smaller than the validity region of the individual terms in (a7.11). This latter validity region reads $a_\mathrm{t} > 2\pi t_3$.)

When self-demagnetization limits the above pulse-crowded magnetization pattern, then $|\hat{H}_\mathrm{d}| = H_\mathrm{c}$ and $a_\mathrm{t} \equiv a_\mathrm{d}$. (This last definition implies that the sinusoidal pattern is thought to be built up from, linearly superponated, fictitious arctan pulses with width a_d!) Then (a7.12) leads to

$$a_\mathrm{d} = \frac{1}{k} \ln \frac{2k\,M_\mathrm{r}\,t_3}{\pi\,H_\mathrm{c}}. \qquad (a7.14)$$

For $2kM_\mathrm{r}t_3/\pi H_\mathrm{c} < 1$ no solution exists for $H_\mathrm{d} = H_\mathrm{c}$ and then the occurrence of self-demagnetization is impossible if the write head-field gradients is infinite.

According to Middleton and Wisely's equation (23) [7.23], the output

$$|\hat{V}| \sim \mathrm{e}^{-k(d_{4\mathrm{p}} + a_\mathrm{t})}. \qquad (a7.15)$$

This equation was derived for linearly-superponated (arctan) transitions with width a_t independent of bit density. It also holds for the fictitious bit-density-dependent transitions with width a_d. Substitution of (a7.14) into (a7.15) leads to

$$|\hat{V}| \sim \frac{2kM_\mathrm{r}t_3}{\pi H_\mathrm{c}}\, \mathrm{e}^{-kd_{4\mathrm{p}}}. \qquad (a7.16)$$

Obviously a_d is such a strongly-decreasing function of k in (a7.14), due to the decrease of the demagnetizing field in this model, that the first factor in (a7.16) even turned over in a linearly-*increasing* function of k. According to (a7.16) the slope of the Log $\hat{V}(\omega)$ curve for $a_\mathrm{t} \neq 0$ would even be *reduced* with respect to the slope valid when a_t were 0, $d_{4\mathrm{p}}$! A note must be made on the validity of the above expression. The coating was assumed to be thin compared to the wavelength and pulse crowding was assumed, see (a7.12). If we accept some inaccuracy by taking $t_3 < \lambda/4$ and $a_\mathrm{d} > \lambda/4$, i.e.

$$\frac{1}{k} \ln \frac{2kM_r t_3}{\pi H_c} > \frac{\pi}{2k} > \frac{\pi t_3}{4}, \qquad (a7.17)$$

then $M_r/H_c \geqslant \pi\, e^{\pi/2}/4 = 3.8$ is required to give (a7.16) validity over a certain frequency range. Thus the validity of the expression (a7.16) is usually doubtful. But the example shows that the demagnetizing field reduces in the pulse-crowded situation such that a_d *is a decreasing function of k* and not a constant, as was assumed by Lee and George in their interpretation of experimental results. *Their statement that the self-demagnetization limit was reached for some of their thicker tape samples is therefore doubtful.* In either case their arguments are invalid. On the other hand, their statement would support our model (UMM).

We must conclude that the theoretical and experimental investigation of the self-demagnetization effects in case of ultimate-contact recording, including thicker media, need further work.

Maximum flux density at the head's surface in ultimate-contact recording

An ideally-recorded tape, with sufficiently thick coating and $M_r \gg H_c$, will give rise to a free-space field H_f that approaches H_c. It can not be larger because in the tape on the head side the tangential field H_x equals H_f (since $\nabla \times \boldsymbol{H} = 0$, i.e. $\oint \boldsymbol{H} \cdot d\boldsymbol{r} = 0$) and would cause irreversible self-demagnetization if larger than H_c. According to (2.46) a $2F_R$-times larger perpendicular field results at the surface of an ideally soft-magnetic zero-gap playback head. The factor 2 which is essential in the following discussion was due to the mirror effect of the head. The factor 2 is easily visualized by the factor two increase of the drawn perpendicular fields just outside the medium when in Fig. 2.3 the opposing mirror charges in the head are placed at a small distance from the coating. F_R is the remagnetization factor which was 1 when the tape has a unit relative permeability ($\mu_3 = \mu_2 = 1$), see (2.47), or too far away from the head to be effectively remagnetized by the opposite mirror image. In near contact, see (2.48), F_R is large when the coating permeability is large, i.e. $\mu_3 \gg 1$ (this is not a practical situation). On the other hand, when $\mu_3 \gg 1$, then the magnetization of the tape in free space $M_f \ll M_p$, because of strong reversible demagnetizing effects, while $H_f \leqslant M_f$ and $M_p \leqslant M_r$ are always valid. Consequently $H_f \leqslant M_f \ll M_p \leqslant M_r$ such that an extra large 'demagnetization ratio' M_r/H_c is needed if one requires a free-space field that approaches the coercive field in the case of a high permeable medium.

In either case, at least a factor of 2 higher head-surface field can be reached in ultimate contact recording. This is in contradiction to the opinion that the ultimate flux density at the head's surface equals $\mu_0 H_c$ as stated by Mallinson [7.26]. Thus, his conclusion that the best published data (of Ampex, Hitachi and Fujitsu on longitudinal and perpendicular metallic media of ~ 71 kA/m (900 Oe)) already approach the ultimate limit is untrue. There is still a possible gain of at least a factor 2.

We can show this too by our own measurements. For this purpose an expression that relates the free-space demagnetizing field (just outside the coating at the head side of the tape) to the output voltage is needed first. From the expressions (2.46) and (5.6) and the definitions (5.7) and (5.19) it follows that the peak value of this free-space field relative to the coercive field of the tape equals:

$$\frac{|\hat{H}_f|}{H_c} = \frac{V_{rms}\sqrt{2}}{2\,\mu_0 v\,WN\eta(\omega)\,GLF(k)\,F_R\,e^{-kd_4}\,H_c}\,. \qquad (a7.18)$$

In our experiments the largest value of this ratio is obtained with the rather smooth and high-coercive MP tape ($H_c = 116$ kA/m) and a TSS or MIG head at about $\lambda = 0.7\,\mu$m ($f = 4.5$ MHz since $v = 3.14$ m/s). For the very best MIG heads, with $\eta \simeq 0.8$ and $d_4 \simeq 70$ nm, (hence $e^{-kd_4} \simeq 0.53$), the $V_{rms} \simeq 40\,\mu$V when $N = 20$, $W = 20\,\mu$m and $g = 0.20\,\mu$m (hence $GLF \simeq 0.87$). With $F_R \simeq 1$ then follows a ratio $|\hat{H}_f|/H_c \simeq 0.4$. The same ratio is obtained for the TSS head in Fig. 7.3.II, with its little lower efficiency $\eta \simeq 0.7$ and lower head-to-tape distance of 60 nm (see experimental data in Fig. 7.3.II and head and tape data in the captions of Figs. 7.3.I and 7.3.II).

The maximum demagnetizing field in the tape will be a little higher than the peak value of the free-space field just outside the tape. Thus there may be a factor of say 2 still to win, before the recording process is completely self-demagnetizing limited.

Concluding remarks

The observations of Bertram and Niedermeyer of no H_c dependence in intimate-contact recording at moderate and longer wavelengths and strong H_c dependence at large head-to-tape distances during writing must be explained by negligible self-demagnetization and 'erasing' effects (due to the finite write-field gradient) at small distances and dominating write-field gradient effects at larger distances for the considered wavelengths.

We measured an H_c dependence up to larger wavelengths in even more intimate contact recordings than the above mentioned. This discrepancy is only partly declared by the use of optimized currents in our recording.

Linear superposition of isolated pulses in the pulse-crowded situation is not allowed when self-demagnetization effects dominated the width of the isolated pulse. The transition width (or pulse width) is then to be considered as a decreasing function of the bit density or of the wavenumber k.

Existing analytical models are derived for thin media and single transitions (Williams and Comstock) and do not take account of a finite saturation magnetization of the head, of importance for the optimal recording of long wavelengths 'sinusoidal' signals. Another analytical model derived for thicker media (Middleton and Wisely) has doubtful accuracy in the case of our current video conditions (relatively large coating thickness) and is mathematically more difficult to handle than our UMM.

In ultimate-contact recording a perpendicular field of *two times* H_c (and not H_c as stated in [7.26]) is theoretically possible at the boundary surface between an ideal replay head and a unit relative permeability medium. A higher field would have caused irreversible self-demagnetization in free space after the recording in the idealized square-loop material of the medium.

Therefore in our best recordings and those of others the self-demagnetization limit is not yet reached, but approached by roughly a factor 0.5.

Detailed knowledge of the role of self-demagnetization, theoretical as well as experimental, is not yet available.

For self-consistent calculations to be accurate the actual field instead of the Karlqvist field used until now needs to be applied for intimate contact recording. In addition, particle interaction and switching-field distributions need to be taken into account.

In continuous films a major difficulty arises because of zigzag domains present in these films (see the end of chapter 6 and [7.32]), which probably necessitates a three-dimensional and statistical description.

References

[7.1]   Herk, A. van and Wesseling, P., *An analytical model of the write and read process in digital recording with a 'linear' recording medium*, IEEE Trans. Magn., Vol. Mag-10, 761-764 (1974).

[7.2]   Hersener, Jurgen, *Modelluntersuchungen zur Magnetisierungsverteilung im Magnetband bei sinusvörmigen Signalen*, Wiss. Ber. AEG-Telefunken, Vol. 46, 15-24 (1973).

[7.3]   Lopez, Orlando, *Analytic calculation of write induced separation loss*, IEEE Trans. Magn., Vol. Mag-20, 715-717 (1984).

[7.4]   Schwantke, G., *The magnetic tape recording process in terms of the Preisach representation*, Jour. AES, Vol. 9, 37-47 (1961).

[7.5]   Eiling, A., *High-density magnetic recording: Theory and practical considerations*, J. Appl. Phys., Vol. 62, 2404-2418 (1987).

[7.6]   Optimal unbiased sinewave recording is characterized by the use of that write current at each wavelength that gives maximum output during playback.

[7.7]   Middleton, B.K. and Davies, A.V., *Gap effects in head field distributions and the replay process in longitudinal recording*, Journal of the Institution of Electronic and Radio Engineers, Vol. 55, 111-118 (1985).

[7.8]   Gradshteyn, I.S. and Ryzhik, I.M., *Table of integrals, series, and products*, Academic Press Inc., New York. See equation 3.941.

[7.9]   Minuhin, Vadim B., *Theory of playback process with soft magnetic underlayer*, IEEE Trans. Magn., Vol. Mag-21, 28-35 (1985).

[7.10]   Minukhin, V.B., *Hilbert transform and phase distortions of signals*, Radio Eng. Electronic Physics, Vol. 18, no. 8, 1189-1193 (1973).

[7.11]   See e.g. 3.946(2) in [7.8]. It is noted that the last p in the numerator of 3.946(2) is wrong and must be changed into p^2.

[7.12]   Westmijze, W.K., *Field configuration around the gap and the gap-length formula*, Philips Res. Rep. 8, 161-183 (1953).

[7.13]   Bertram, H. Neal and Niedermeyer, Rex, *The effect of spacing on demagnetization in magnetic recording*, IEEE Trans. Magn., Vol. Mag-18, 1206-1208 (1982).

[7.14]   Steele, C.W. and Mallinson, J.C., *Theory of low permeability heads*, IEEE Trans. Magn., Vol. Mag-8, 503-505 (1972).

[7.15]   Williams, M.L. and Comstock, R.L. *Analytical model of the write process in digital magnetic recording*, Proc. AIP Conf. Magnetism and Magnetic Materials, Vol. 6, 738-742 (1971).

[7.16]   Lee, Jae I. and George, Peter K., *Demagnetization-free longitudinal recording on flexible thin film metal media*, IEEE Trans. Magn., Vol. Mag-21, 1221-1227 (1985).

[7.17]   Mallinson, J.C. and Steele, Charles W., *Theory of linear superposition in tape recording*, IEEE Trans. Magn., Vol. Mag-5, 886-890 (1969).

[7.18]   Köster, E. and Pfefferkorn, D., *The effect of remanence and coercivity on short wavelength recording*, IEEE Trans. Magn., Vol. Mag-16, 56-58 (1980).

[7.19]   Curland, N. and Speliotis, D.E., *A theoretical study of an isolated transition using an iterative hysteretic model*, IEEE Trans. Magn., Vol. Mag-6, 640-646 (1970).

[7.20]   Maller, V.A.J. and Middleton, B.K., *A simplified model of the writing process in saturation magnetic recording*, The Radio and Electronic Engineer, Vol. 44, 281-285 (1974).

[7.21] Miyata, J.J. and Hartel, R.R., *The recording and reproduction of signals on magnetic medium using saturation-type recording*, IRE Trans. Electron. Comput., Vol. 8, 159-169 (1959).

[7.22] Tjaden, D.L.A., *Some notes on 'superposition' in digital magnetic recording*, IEEE Trans. Magn., Vol. Mag-9, 331-335 (1973).

[7.23] Middleton, Barry K. and Wisely, Paul L., *The development and application of a simple model of digital magnetic recording to thick oxide media*, IERE Conf. Proc. *35* (1976).

[7.24] Tjaden, D.L.A. and Tercic, E.J., *Theoretical and experimental investigations of digital magnetic recording on thin media*, Philips Res. Repts *30*, 120-161 (1975).

[7.25] Kosters, A.J. and Speliotis, Dennis E., *Predicting magnetic recording performance by using single pulse superposition*, IEEE Trans. Magn., Vol. Mag-7, 544 (1971).

[7.26] Mallinson, John C., *The next decade in magnetic recording*, IEEE Trans. Magn., Vol. Mag-21, 1217-1220 (1985).

[7.27] Middleton, B.K., *The dependence of recording characteristics of thin metal tapes on their magnetic properties and on the replay head*, IEEE Trans. Magn., Vol. Mag-2, 225-229 (1966).

[7.28] Ortenburger, Irene B. and Potter, Robert I., *A self-consistent calculation of the transition zone in thick particulate media*, J. Appl. Phys., Vol. 50, 2393-2395 (1979).

[7.29] Potter, Robert I. and Beardsley, Irene A., *Self-consistent computer calculations for perpendicular magnetic recording*, IEEE Trans. Magn., Vol. Mag-16, 967-972 (1980).

[7.30] Bertram, H. Neal, *Geometric effects in the magnetic recording process*, IEEE Trans. Magn., Vol. Mag-20, 468-478 (1984).

[7.31] Bertram, H. Neal, *The effect of the angular dependence on the particle nucleation field on the magnetic recording processes*, IEEE Trans. Magn., Vol. Mag-20, 2094-2104 (1984).

[7.32] Arnoldussen, T.C. and Tong, H.C., *Zigzag transition profiles, noise, and correlation statistics in highly oriented longitudinal film media*, IEEE Trans. Magn., Vol. Mag-22, 889-891 (1986).

Chapter 8

Video-head parameters: model and measurements

In this chapter two main subjects are described. The first subject is an analytical model for ring heads. Equivalent schemes, known from network analysis, are derived for the magnetic circuit of the head and applied to various problems. Several ring heads for video applications to which the above model applies are shown in Fig. 8.1.

Based on results with this model a quick method for measuring the efficiency is developed: this is the second main subject.

The results of the model are then compared with the results of the efficiency measurement and other head-parameter measurements. In the meantime different modern types of heads and the applicability of physical models for permeability spectra are discussed.

Finally, for the quasi-static case (for μ is real), a comparison is made with results of finite-element method computations.

It should be noted that the reading of the description of the saturation process in a video head in section 8.1.7 is not necessary for a rough understanding of the quick efficiency measurement described in section 8.2. This concerns especially the subsections 8.1.7.4-7 which appear in small print.

Because of the extensive nature of this chapter each main subject has its own introduction.

8.1 Analytical approach to 3-dimensional video head

8.1.1 Introduction

In this section a reasonably accurate analytical model for a video head is developed. Compared with calculations with the aid of numerical packages using the finite-element method (FEM), calculations can be done very quickly. For a manageable number of finite elements, the FEM is about four times more accurate; however, the computations in

Fig. 8.1. Various video heads.
a Ferrite video head with a lasered track-narrowing and ground asymmetric coil chamber. Our calculations relate mainly to this configuration when the coil chamber is symmetric or symmetrized.
b Ferrite video head with laser-cut symmetric coil chamber and reduced coil size. With a slight modification of the back of the head in the model, results will also apply to this configuration. Scale 1 mm/div.
c Amorphous head-core consisting of a magnetic amorphous ribbon between non-magnetic ceramic substrates. Calculations also apply, with a symmetric or symmetrized coil chamber, to this configuration. Scale: 100 μm/div.

the analytical model are carried out about a million times faster. Hence, results are immediately available. The analytical model describes heads with *complex* permeabilities in contrast to the usual FEM packages. At the frequencies of interest in video recording this aspect is very important and makes these FEM packages unsuitable for serious calculations of the head's efficiency and its electric impedance. These FEMs are then only applicable for calculations of the shape of local fields. It is even doubtful whether saturation effects (if not local) are well described at high frequencies by those FEM calculations.

In the analytical model the head and its surroundings are divided into a small number of sections; on the order of 10 for a symmetrical and 20 for an asymmetrical video head. See for the sake of clarity Figs. 8.2 and 8.15. For the video head with its non-negligible stray flux, its extreme ratio of maximum to minimum cross section of the core and maximum to minimum flux density and field, these are very low numbers of elements. Calculations with these numbers of elements with the usual FEM (which is set up for universal use) would be extremely bad. The results obtained with these small numbers of elements can only be reasonably accurate when:

1. The sections are chosen so that their beginning and end surfaces are reasonably approximated by equipotential surfaces.
2. The decay of the magnetic potential over sections, from which a considerable part of the total flux leaves the core as stray flux, is relatively small.
3. In the calculations of the reluctance of sections one takes account of the expected direction of the flux flow or the location of equipotential surfaces.

In the description of the model details will become clear.

Shortly after the development of the model (1980) described in the first sections of this chapter, Katz [8.1], Jorgenson [8.2] and also Berglund [8.3], who referred to Enger's thesis on permeances in magnetic circuits [8.4], proposed comparable analytical models. Recently Corcoran and Pope [8.5] proposed such a model in which they used a transmission line description (see also chapters 9 and 11) to describe the front end of the ring head. This fits in better with the distributed character of the stray and core reluctances. In contrast to those models, the present model is more accurate and is worked out in much more detail. The reason for the higher accuracy, particularly of the head's impedance, is the introduction of a phenomenological stray flux term in the model.

With the aid of Thevenin's and Norton's theorems of circuit analysis it is possible to characterize the 'type' of head by internal efficiencies and impedances. Once these 'internal' head parameters are known (by the analytical head model, FEM computations or measurements) various kinds of problems can be solved accurately by hand. Examples are changes in the head's efficiency and impedance due to:

1) gap length changes,
2) stresses near the gap,
3) insufficient permeability of gap cladding and
4) saturation effects in the tapered part of the head.

The analytical model, especially when extended with the equivalent schemes, gives much greater insight into the strengths and weaknesses of a head design than FEM calculations can do. The proposed analytical head model has proven to be very useful in designing ring heads.

8.1.2 Definition of the configuration

In Fig. 8.2 a video head is sketched, which consists of a core, a (front) gap, a back gap and a coil. The core is divided into a number of parts. Clockwise these are A', B', C', D', E', F', G', G, F, E, D, C, B and A. The dashed sections A', A, B', B, C' and C are smaller than the corresponding parts of the core, indicating that a negligible amount of flux flows through the corners of the head. Other choices for these sections will not alter the results noticeably but complicate the calculations. The dashed sections in Fig. 8.2 have two boundary surfaces with the neighbouring sections which are both assumed to be equipotential surfaces. Consequently the reluctances of the sections C' and C are zero in the present model. The other sections determine the reluctances in the core. All parts of the head that correspond to a dashed section have four boundary surfaces with the surrounding of the head; these are indicated by 1, 2, 3 and 4 starting at the outer surface of the head and again going clockwise. Exceptions are parts C' and C which have one boundary surface less and parts A' and A which have one more. From (most of) these surfaces stray flux leaves the core in the model; see the diagram in Fig. 8.3a.

All reluctances (complex Z or real R) with single subscripts A', B' etc. are core or gap reluctances, while all reluctances with a double subscript denote stray reluctances between sections corresponding to the two (letter) subscripts. All stray reluctances, except $R_{CC'}$ are com-

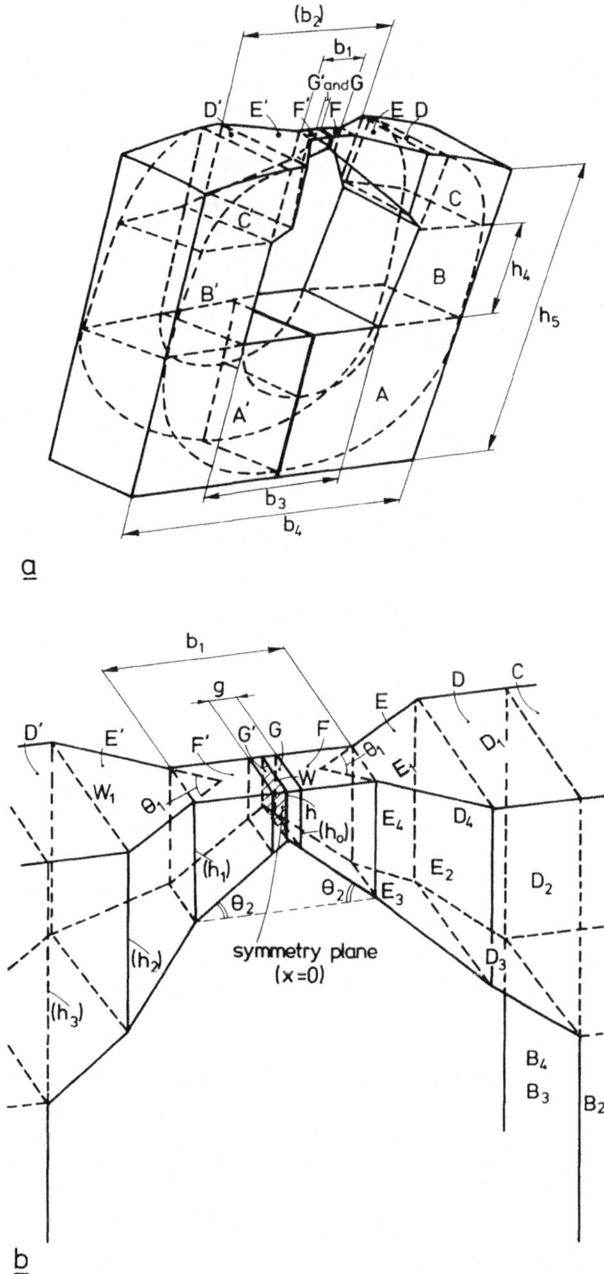

Fig. 8.2. Subdivision of the head in sections denoted by dashed lines. Not all sections fill the corresponding part of the head completely. Dimensions derived from the basic dimensions and angles are placed between brackets.
a Video core with greatly exaggerated coil chamber and bridge dimensions.
b Enlargement of the bridge.

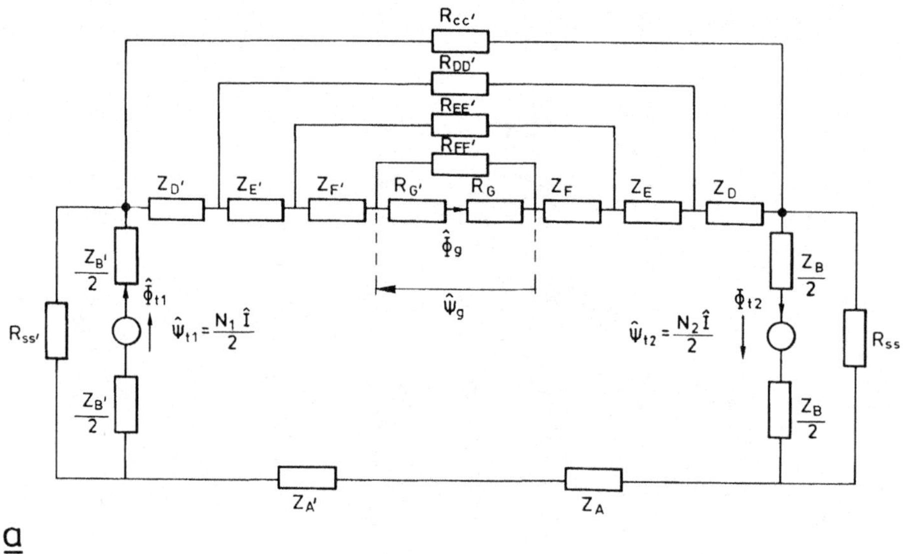

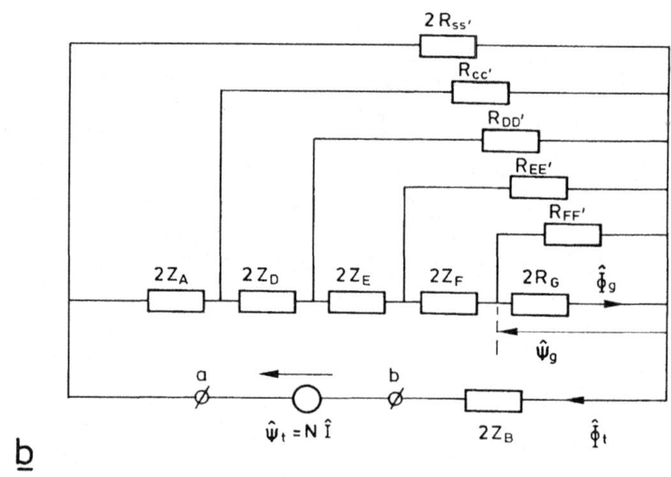

Fig. 8.3. Magnetic circuit representations of the video head. $N_1 + N_2 = N$, Ψ = magnetic potential and Φ = flux. c = core and g = gap. Magnetic-circuit representations of the video head.
a General circuit for asymmetric and symmetric heads.
b Reduced equivalent circuit for symmetric video heads as used in the calculations, which implies $N_1 = N_2$.

posed of the contributions of four pairs of surfaces of the corresponding two selections and are indicated by a third (number) subscript. This number subscript reads 1, 2, 3 or 4 and corresponds to the number of the surface in Fig. 8.2. In Fig. 8.4 the simplified flux tubes corresponding to $R_{DD'1}$ and $R_{DD'2}$ have been drawn. In the calculation of these reluctances the possibility (and reality) of a spread of flux, especially at the edges of the surfaces, which causes a certain reduction of the reluctance, is not taken into account. For satisfactory agreement with experimental data it will turn out to be necessary to take this effect into account. Since this effect cannot be described analytically, a phenomenological term with number subscript 5 is added for this purpose, which further reduces the permeance. For example:

$$\frac{1}{R_{DD'}} = \sum_{i=1}^{5} \frac{1}{R_{DD'i}}$$

In the circuit representation of the head, see Fig. 8.3a, all stray flux reluctances are connected to those sides of the corresponding pair of sections from which usually most of the stray flux can be expected to leave the core and to come back into the core. These can be the high potential-difference sides and/or the sides that are nearest together.

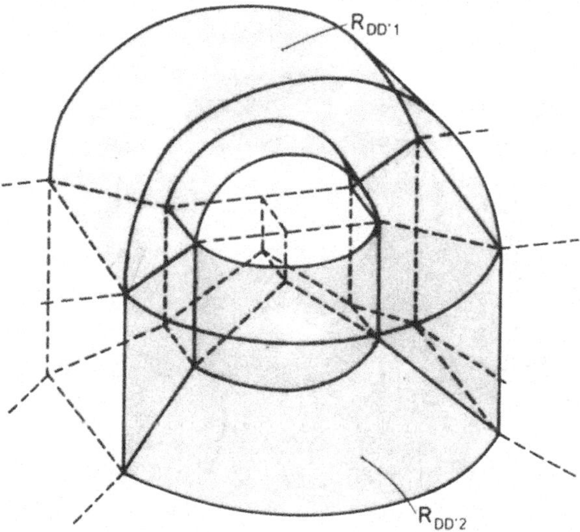

Fig. 8.4. Sketch of two of the four stray-flux tubes between the sections D' and D, which are assumed to be circular for simplicity. The numbers of the surfaces of each section that correspond to the number subscript of the reluctances R increase from 1 (outer surface) clockwise; see also Fig. 8.26.

For stray flux over the coil this leads clearly to the coil side of sections A and C; these are the high potential-difference ($N/2I$) sides as well as the sides that are nearest together. The high-potential difference sides differ from the minimum-distance sides in the case of flux tubes that cross the symmetry plane of the head. Usually the best choice is the gap side, especially for stray flux from F to F', since, in spite of the fact that the potential difference between the other sides of the sections may be $1/\eta$ times higher, the gap is so small relative to b_1 that most of the stray flux still leaves and comes back at the gap side. In more detail this will become clear from expressions for the stray reluctances that have yet to be derived. This is not a good choice, however, when the permeability is very low around the gap, i.e. strong saturation.

To the stray fluxes from C to A and from C' to A' are also added stray fluxes that leave the core through the turns of the coil. The corresponding 'total' stray reluctances are called $R_{SS'}$ and R_{SS} respectively; see fig. 8.3. They are calculated in the following section.

A simplified, but equivalent, circuit representation is possible because of assumed symmetry of the head; see the result in Fig. 8.3b.

The present symmetric video head configuration is completely defined by the location of the windings around sections B' and B and the number of turns N, a permeability tensor with non-zero diagonal elements $\mu_0\mu_x$ and $\mu_0\mu_y$, and the following two angles and nine basic dimensions:
– Track-narrowing angle θ_1.
– Height-reducing angle θ_2.
– Gap length g.
– Bridge length b_1.
– Coil-chamber length b_3.
– Core length b_4.
– Track width W.
– Core width W_1.
– Gap-centre height h.
– Coil-chamber height h_4.
– Core height h_5.

Dimensions between brackets in Fig. 8.2 are derived from the above basic dimensions and angles which are also defined in Fig. 8.2. They are listed in Table 8.1.

Table 8.1. Quantities derived from the basic quantities in the text.

$$b_2 = b_1 + \frac{W_1 - W}{\tan(\theta_1/2)} \quad \text{if } b_2 < b_3$$

$$h_0 = h + (g/2)\tan\theta_2$$

$$h_1 = h + (b_1/2)\tan\theta_2$$

$$h_2 = h + (b_2/2)\tan\theta_2$$

$$h_3 = h + (b_3/2)\tan\theta_2$$

8.1.3 Reluctance expressions

Around a cylindrical and infinitely long conductor, the flux lines are circles because of circular symmetry and consequently equipotential surfaces are radially oriented. Thus the flux lines between two inclined surfaces, with magnetic potentials Ψ_1 and Ψ_2 respectively, must be circular arcs when the surface is elongated in the z direction ($l \gg r_{max}$) and when the other direction is large compared to the gap ($r_{max} \gg r_{min}$); see Fig. 8.5. Only near the edges will this not hold.

A differential section of this configuration has a permeance

$$dP \equiv d\Phi/(\Psi_2 - \Psi_1) = \frac{\mu_0\hat{\mu}_r l dr}{\theta r} \tag{8.1}$$

which leads after integration to a reluctance $Z = 1/P$ that reads

$$Z = \frac{\theta}{\mu_0\hat{\mu}_r l \ln(r_{max}/r_{min})} . \tag{8.2}$$

The reluctances to which this expression applies are listed in Table 8.2.

When the potential difference $\Psi_1 - \Psi_2$ is expected to be too strong a function of r, then it is necessary to divide the section into subsections. It is advisable to subdivide in such a way that r_{max}/r_{min} is about the same for each subsection if the potential drop between so-defined areas in the core are about comparable, i.e. $\Psi_1 - \Psi_2 \simeq \ln r$, and to let r_{max}/r_{min} decrease more with r according as the potential difference increases more strongly with r.

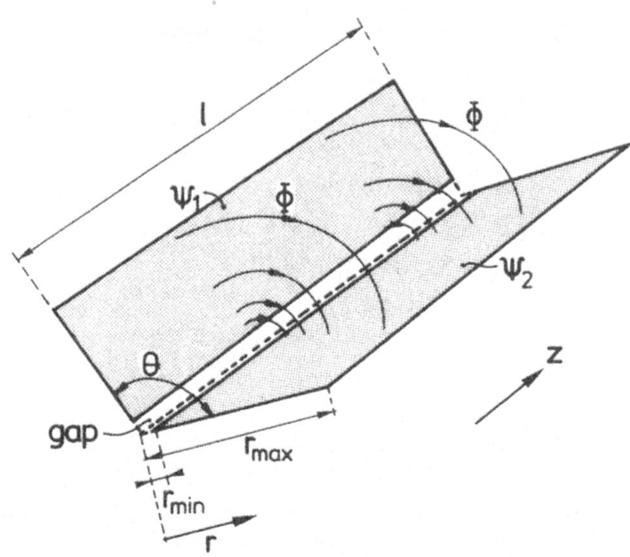

Fig. 8.5. Flux (Φ) lines between inclined equipotential (Ψ_1 and Ψ_2) surfaces approaching circular arcs. The reluctance of this configuration is expressed by (8.2) and applies to many stray-flux tubes; see also Table 8.2.

Table 8.2.   Reluctances expressed by (8.2), valid for, the situation sketched in Fig. 8.5.

Substitutions in (8.2)				
Z	θ	l	r_{max}/r_{min}	μ_r
$R_{DD'1}$	π	W_1	b_3/b_2	1
$R_{DD'3}$	$\pi - 2\theta_2$	W_1	b_3/b_2	1
$R_{FF'1}$	π	W	b_1/g	1
$R_{FF'3}$	$\pi - 2\theta_2$	W	b_1/g	1
$Z_A, Z_{A'}$	$\pi/2$	W_1	b_4/b_3	$\sqrt{\mu_x\mu_y}$*
$R_{CC'}$***	π	$W_1 + 2h_3$	b_4/b_3	1
$R_{SS}, R_{SS'}$***	π	$b_4 - b_3 + 2W_1$	$h_5/(h_4/\sqrt{e})$**	1

* This, not exact, expression does not influence the results noticeably, since Z_A and $Z_{A'}$ are almost negligible.
** When the region within r_{min} is filled with uniformly distributed turns, as is the case for R_{SS} and $R_{SS'}$, then Fig. 8.9 and expression (8.11) apply to this region. Adding the reluctance expressed in (8.11) to the one in (8.2) is equivalent to reducing r_{min} in (8.2) by a factor $\sqrt{e}$; see later on.
*** Except the phenomenological terms $R_{CC'5}$, R_{SS5} and $R_{SS'5}$, which are given in Table 8.5b.

For flux travel in the direction perpendicular to that in Fig. 8.5, see cross-sectional view in Fig. 8.6a, the reluctance follows easily from integrating the reluctance of differential sections (instead of the permeance in (8.1)), hence

$$Z = \frac{\ln\left(r_{max}/r_{min}\right)}{\mu_0 \hat{\mu}_r l \theta}.$$

(8.3a)

Compared to flux travel in the perpendicular direction, see (8.2), only $\ln(r_{max}/r_{min})$ and θ are interchanged.

Reluctances to which (8.3a) applies are given in Table 8.3. It is easily verified that for small r_{max}/r_{min} the reluctance of structure c after transformation into structure a has been increased exactly by a factor $\tan\theta/\theta$. Thus for small r_{max}/r_{min} and θ not small, it is advisable not to transform c (or b) into a. However, it is not totally correct to divide the structures b and c into differential sections that are bounded by equipotential surfaces that are parallel to the upper and lower (equipotential) surfaces (indicated by Ψ_1 and Ψ_2). (This means that in fact the reluctances of the structures b and c are not uniquely defined.) Still doing so leads to the approximations

$$Z \simeq \frac{1}{\mu_0 \hat{\mu}_r} \int_{r_{min}}^{r_{max}} \frac{1}{A(r)}\, \mathrm{d}r = \frac{\ln\left(r_{max}/r_{min}\right)}{\mu_0 \hat{\mu}_r l\, 2 \tan\left(\theta/2\right)}$$

(8.3b)

and

$$Z = \frac{\ln\left(r_{max}/r_{min}\right)}{\mu_0 \hat{\mu}_r l \tan\theta}.$$

(8.3c)

for the configurations b and c respectively, which are valid when θ is not too large. When, however, many of the flux lines are expected to traverse the section under an angle comparable to θ (i.e. the end faces are no longer both equipotential surfaces), then, for small r_{max}/r_{min} too, the transformation and hence (8.3a) is chosen, because a higher 'effective' reluctance must then be expected. For this reason (8.3a) was chosen for the section FF', DD', B and B'. In the description of the core saturation, however, use is made of (8.3c) in order to simplify calculations there.

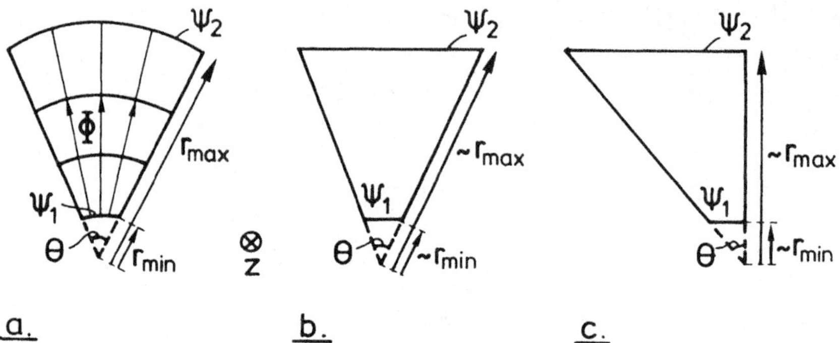

Fig. 8.6. *a.* Cross-sectional view of a configuration similar to the one in Fig. 8.5, but with flux in the direction perpendicular to the flux in the configuration of Fig. 8.5.
b and *c.* Configurations to which, for not too large θ, expression (8.3a), valid for configuration a, also applies. For larger θ's and small r_{max}/r_{min}, the approximations (8.3b) and (8.3c) are more suited for the configurations b and c respectively.
The configurations apply to various core sections; see Table 8.3.

Table 8.3. Reluctances expressed by (8.3a), valid for the situation sketched in Fig. 8.6a.

Substitutions in (8.3a)				
Z	θ	l	r_{max}/r_{min}	μ_r
$Z_F, Z_{F'}$	θ_2	W	h_1/h_0	μ_x
$Z_D, Z_{D'}$	θ_2	W_1	h_3/h_2	μ_x
$Z_B, Z_{B'}$	$\arctan\left[\dfrac{b_4 - b_3 - 2h_3}{2h_4}\right]$	W_1	$(b_4 - b_3)/(2h_3)$ if > 1	μ_y

In the present video head the track width W is obtained from a core of width W_1 by laser cutting perpendicularly to the tape-facing surface of the head; see also Fig. 8.1a and b. This leads to sections E' and E, as shown in Fig. 8.2b. This structure is first transformed into the structure depicted in Fig. 8.7, from which the reluctance can easily be approximated. The outer planes are assumed to be equipotential surfaces (Ψ_1 and Ψ_2). The flux lines do not meet (virtually) at one point but in two lines, except when $\theta' = \theta''$. The side planes are curved, except when $\theta' = \theta''$ or when at least one of them approaches zero. The flux lines are approximately perpendicular to the planes indicated by Ψ_1 and Ψ_2 and all planes in between. Two such planes, bounding a subsection of length dr and surface of about $r'\theta'r''\theta''$, are indicated by dashed lines. Thus these planes are also equipotential surfaces. This means that each differential section contributes a reluctance of about

$$dZ \simeq \frac{1}{\mu_0\hat{\mu}_r}\;\frac{dr}{r'\theta'r''\theta''} \qquad (8.4)$$

to the total reluctance R. Hence,

$$Z \simeq \frac{1}{\mu_0\hat{\mu}_r\theta'\theta''}\;\frac{\ln(r'_{max}/r'_{min}) - \ln(r''_{max}/r''_{min})}{r''-r'}. \qquad (8.5)$$

This expression applies to Z_E and $Z_{E'}$ which are given in Table 8.4. In the limit $\theta' \to 0$, this expression relaxes to (8.3a) when $\theta \equiv \theta''$, $r_{max}/r_{min} \equiv r''_{max}/r''_{min}$, and $l \equiv \theta'r'_{min}$ or $l \equiv \theta'r'_{max}$ is kept constant.

The reluctance between any two other pair of surfaces of the structure in Fig. 8.7 follows in an analogous way from integrating the permeance of a differential section, while referring to Fig. 8.8:

$$dP \simeq \mu_0\hat{\mu}_r\;\frac{r'\theta'dr}{r''\theta''} \qquad (8.6)$$

and reads

$$Z \simeq \frac{\theta''}{\mu_0\hat{\mu}_r\theta'}\;\frac{1}{(r''_{max}-r''_{min})-(r''-r')\ln\left(\dfrac{r''_{max}}{r''_{min}}\right)}. \qquad (8.7)$$

This expression will be used for most stray-flux tubes that cross the symmetry plane of the head (see Table 8.5), except for those with θ'' or $\theta' = 0$, which are described by (8.3) and given in Table 8.3.

In the neighbourhood of the coil the largest contributions to the stray flux must be expected, because there the magnetic potentials are high and the areas are largest. For this reason this stray flux cannot be approximated by only the stray flux from the surfaces of C to those of A (leading to R_{CA}) and the surfaces of C' to those of A' (leading to $R_{C'A'}$) as reasonably described by (8.2). The stray flux emanating from the core through the turns of the coil must be taken into account but can not be expressed exactly. An accurate expression that is easy to

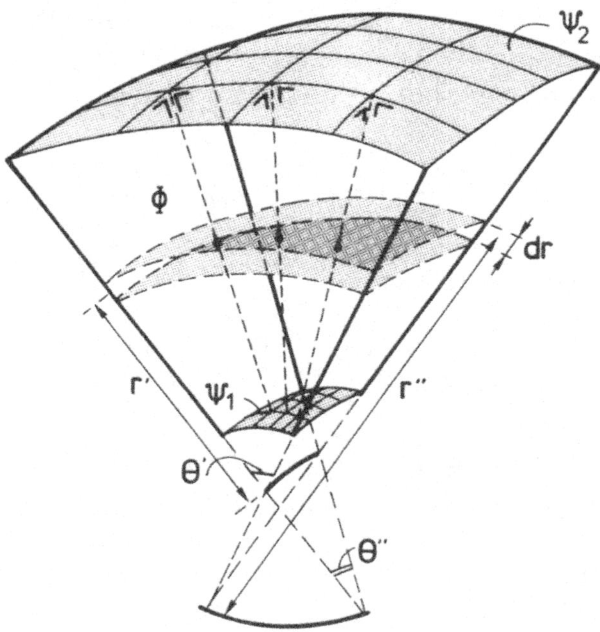

Fig. 8.7. The circular-symmetric structure that leads to expression (8.5). The sections E' and E, see Fig. 8.2b, can be approximated by this configuration; see also Table 8.4.

Table 8.4. Reluctances expressed by (8.5), valid for the situation sketched in Fig. 8.7.

Substitutions in (8.5)						
Z	θ'	θ''	r'_{max}/r'_{min}	r''_{max}/r''_{min}	$r'' - r'$	μ_r
$Z_E, Z_{E'}$	θ_1	θ_2	W_1/W	h_2/h_1	$\dfrac{h_1}{\tan\theta_2} - \dfrac{W}{2\tan(\theta_1/2)}$	μ_x

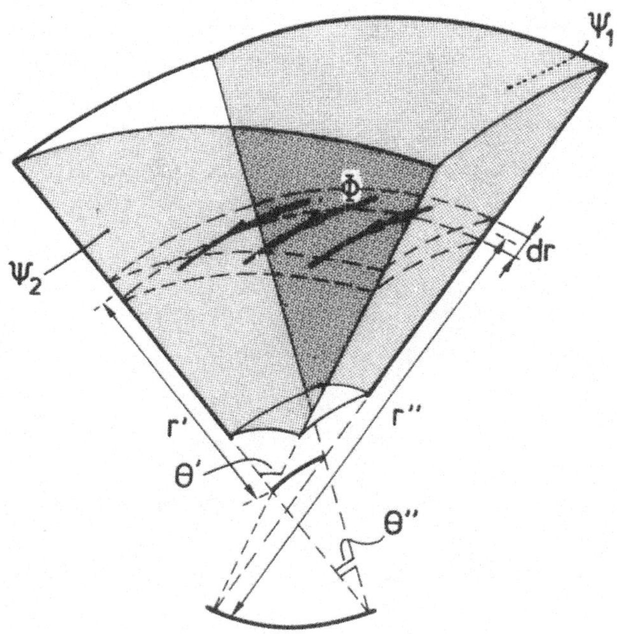

Fig. 8.8. The same configuration as in Fig. 8.7, but with flux travel in the perpendicular direction. The resulting reluctance, expressed in (8.7), is applied to the stray-flux tubes that cross the symmetry plane of the head and obey θ_1, $\theta_2 \neq 0$; see also Table 8.5.

Table 8.5. Reluctances expressed by (8.7), valid for the situation sketched in Fig. 8.8.

Substitutions in (8.7)						
Z	θ'	θ''	r''_{min}	r''_{max}	$r'' - r'$	μ_r
$R_{DD'2}, R_{DD'4}$	θ_2	π	$\dfrac{b_2}{2}$	$\dfrac{b_3}{2}$	$\dfrac{b_2}{2} - \dfrac{h_2}{\tan\theta_2}$	1
$R_{EE'1}$	θ_1	π	$\dfrac{b_1}{2}$	$\dfrac{b_2}{2}$	$\dfrac{b_1}{2} - \dfrac{W}{2\tan(\theta_1/2)}$	1
$R_{EE'2}, R_{EE'4}$	$\arctan\left[\cos(\theta_1/2)\tan\theta_2\right]$	$\pi - \theta_1$	$\dfrac{b_1}{2\cos(\theta_1/2)}$	$\dfrac{b_2}{2\cos(\theta_1/2)}$	$r''_{min} - \dfrac{h_1}{\tan\theta'}$	1
$R_{EE'3}$	$2\arctan\left[\cos\theta_2 \tan(\theta_1/2)\right]$	$\pi - 2\theta_2$	$\dfrac{b_1}{2\cos\theta_2}$	$\dfrac{b_2}{2\cos\theta_2}$	$r''_{min} - \dfrac{W}{2\tan(\theta'/2)}$	1
$R_{FF'2}, R_{FF'4}$	θ_2	π	$\dfrac{g}{2}$	$\dfrac{b_1}{2}$	$\dfrac{b_1}{2} - \dfrac{h_1}{\tan\theta_2}$	1

calculate is found when, instead of the uniformly distributed current (I)-carrying multiturn winding, one flat turn with height $2r_{max}$ and with a uniform surface-current density $\sigma = I/(2r_{max})$ is assumed. The situation is sketched in Fig. 8.9. The permeance of a differential section (differential stray flux tube) at distance r from the centre of the turn is

$$dP = \frac{\mu_0 l\,dr}{r\theta} \tag{8.8}$$

and the magnetic potential over this section equals

$$\Psi(r) = 2r\sigma. \tag{8.9}$$

The flux through this differential flux tube ΨdP also flows through a part r/r_{max} of the flat turn, and 'thus' the effective permeance that can be added to P_{CA} ($= 1/R_{CA}$) or to $P_{C'A'}$ ($= 1/R_{C'A'}$) follows from

$$P \equiv \frac{\Phi}{I} = \frac{1}{\sigma 2r_{max}} \int_0^{r_{max}} \left(\frac{r}{r_{max}}\right)\left(2r\sigma\right) dP \tag{8.10}$$

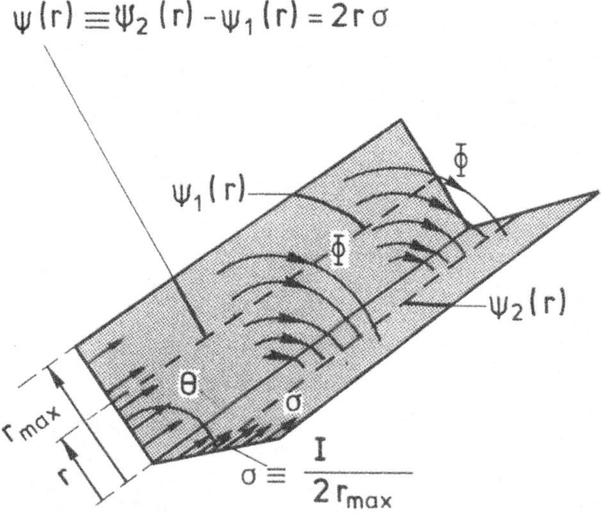

Fig. 8.9. Model for the stray flux that emanates from the core through the turns of a coil. The configuration leads to the reluctance in (8.11), which applies, for $\theta = \pi$, to the stray flux through the windings of the head.

and leads to an effective reluctance

$$Z = \frac{2\theta}{\mu_0 l} = \frac{\theta}{\mu_0\, l\, \ln\sqrt{e}} \tag{8.11}$$

between both sides of the coil.

For the present purpose $\theta = \pi$. (8.11) looks like (8.2); only r_{max}/r_{min} is replaced by $\sqrt{e}$. Added to R_{CA} and $R_{C'A'}$ this leads to reluctances R_{ss} and $R_{ss'}$ of the form (8.2) with $r_{max} = h_5/2$ but $r_{min} = (h_4/2)/\sqrt{e}$ reduced by a factor $\sqrt{e}$. Thus especially when h_5/h_4 is small, as is the case in video heads, the contribution of the stray flux that emanates from the core through the turns is important. Table 8.2 gives the substitutions necessary for R_{SS} and $R_{S'S'}$ when (8.2) is applied.

Finally the expression for the gap reluctances

$$R_G = R_{G'} = \frac{1}{\mu_0 W h_0/(g/2) + \mu_0 2(W + h_0)/(\pi/2)} \tag{8.12}$$

completes the list of expressions used in the model. The last term in the denominator is the permeance that accounts for the stray flux that emanates from the gap, i.e. the fringes of the gap itself. It is calculated analogously to expression (8.11) by assuming a linearly decreasing potential over the gap (Karlqvist approach) (and of course without the first factor in the integral in (8.10) that accounted for the enclosed part of the winding). Usually this term is negligible ($g \ll h$) and h_0 approaches h.

8.1.4 Head parameters

The electric impedance Z and complex efficiency $\eta = |\eta| e^{j\varphi_\eta}$ are calculated in this section. The electric impedance Z is proportional to the permeance P_{ab} and inversely proportional to the reluctance Z_{ab} between the points a and b of the magnetic circuit in Fig. 8.3b. This follows after application of Faraday's law $V = N d\Phi_t/dt$, while neglecting capacitances between the turns of the coil and the resistance of the wire:

$$Z = j\omega N^2 P_{ab} = j\omega N^2/Z_{ab}. \tag{8.13}$$

Use of the definition $Z \equiv R + j\omega L$ thus leads to

$$R = \omega N^2(-\mathrm{Im}P_{ab}) \tag{8.14}$$

and

$$L = N^2 \mathrm{Re}P_{ab} \tag{8.15}$$

It is convenient to define a quality factor, Q, which expresses the imaginary part of the electric impedance of a coil relative to the real part:

$$Q \equiv \frac{\omega L}{R} = \frac{\mathrm{Re}P_{ab}}{-\mathrm{Im}P_{ab}} \tag{8.16}$$

To the resistance R, due to losses in the core $(-\mu'')$, the losses in the wire can be added, described by the resistance R_w of the wire which includes the skin effect. The thermal-noise voltage generated by the head is proportional to the square-root of this total resistance. Compared with other noise sources, the contribution from R_w can be neglected in video applications. The influence of the capacitance can be neglected for a quite different, practical, reason. If the head's impedance were adjusted for optimum noise behaviour, then the number of turns would be so large that the resulting high inductance L and capacitance C would cause resonance in the video-frequency band. This would complicate the equalization. Therefore the number of turns is reduced at the expense of a lower signal-to-noise ratio S/N. As a consequence, C is more or less negligible at video frequencies.

The field efficiency η_H is, by definition,

$$\eta_H \equiv \frac{\hat{\Psi}_g}{\hat{\Psi}_t} \tag{8.17}$$

This is what we usually call the efficiency η.

With the definitions $\eta_H \equiv |\eta_H| \, e^{j\varphi_{\eta H}}$, $\hat{\Psi}_g \equiv |\hat{\Psi}_g| \, e^{j\varphi_g}$ and $\hat{\Psi}_t \equiv |\hat{\Psi}_t| \, e^{j\varphi_t}$ follow:

$$|\eta_{\mathrm{H}}| = |\hat{\Psi}_{\mathrm{g}}|/|\hat{\Psi}_{\mathrm{t}}| \tag{8.18}$$

$$\text{and } \varphi_{\eta\mathrm{H}} = \varphi_{\mathrm{g}} - \varphi_{\mathrm{t}}. \tag{8.19}$$

In an analogous way a flux efficiency can be defined:

$$\boxed{\eta_{\Phi} \equiv \frac{\hat{\Phi}_{\mathrm{g}}}{\hat{\Phi}_{\mathrm{t}}} = \frac{Z_{\mathrm{ab}}}{2R_{\mathrm{G}}}\,\eta_{\mathrm{H}}} \tag{8.20}$$

and be split into an amplitude and phase factor.

Expressions for all the above quantities, as a function of the reluctances of the circuit elements, follow in a straightforward way from the circuit representation in Fig. 8.3b. However the expressions are lengthy and are therefore not reproduced here. The expressions in Tables 8.2 to 8.5 and expression (8.12) are calculated by simple routines and subsequently substituted in the above-mentioned expressions in a relatively simple computer program for symmetric video heads; see Fig. 8.3b. The eleven basic head quantities, the number of turns around sections B and B' and the permeability components can freely be chosen. The requirements $b_2 < b_3$ in Table 8.1, $(b_4\text{-}b_3)/2h_3 < 1$ in Table 8.3 and $h_5 - h_4 - h_3 \geq b_4/2$ are nearly always fulfilled in video heads.

8.1.5 Enhancement of the accuracy of the model

The model as described in the previous sections was used for some years to fit the experimentally obtained head inductance value. Often the permeability had to be chosen higher than expected. The reason for this became apparent when we were able to measure absolute values of head efficiency more accurately than by recording/playback experiments and found that the permeability necessary to fit this efficiency was substantially lower than the permeability necessary to fit the inductance value. The latter permeability values were more realistic. In cases where the permeability values were known fairly accurately, a reasonable fit was obtained to the experimentally obtained efficiency, but the predicted inductance value was usually 20-40% too low. Since the inductance is proportional to the total flux, which contains a large amount of stray flux (for a video head 50 to 70%) and the efficiency is mainly determined by core reluctances and usually not much by the stray flux, the inaccuracy had to be caused by a too low estimate of the stray permeance. Possible reasons for this too low estimate are:

- the assumption of circular-shaped stray flux lines;
- the assumption that all stray-flux lines are parallel with each other, accounting for no spread.

It is important to realize that deviation of the above assumptions from reality always leads to lower reluctances (from minimum energy considerations).

The first assumption seems fairly reasonable. The second is expected to give rise to more inaccuracies, see Fig. 8.10, especially far away from the coil (or gap) where the spread is relatively large. The earlier derived results were proportional to about $\ln(r_{max}/r_{min})$, i.e. the main contributions to the stray permeances were from areas close to the coil or gap. So, far away ($> b$) from the gap or coil, where the flux density is low, the influence of the spread of flux is largest. This suggests that permeance terms, P_{spr}, which are more linear with the length of the various segments of the head, must be added to the formerly derived expressions.

Attempts to derive analytical expressions that take account of an assumed spread of flux fail because of lack of symmetry. Even when a cylinder is used to describe the effects of spread, integrals remain which do not have analytical solutions. Therefore the extra term had to be determined experimentally.

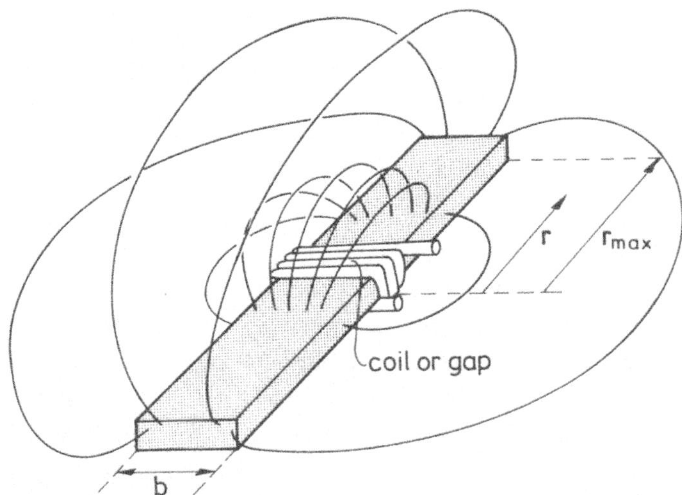

Fig. 8.10. Sketch of the growing influence of the spread of flux with r, which is not accounted for in the formerly derived expressions.

In the first experiment, a video head was broken into two equal core halves, as shown in Fig. 8.11. We measured an inductance $L = 0.32\,\mu$H, whereas the value calculated using equation (8.2) with $l = 3100$ μm, $N = 8$, $r_{max} = 1500$ μm and $r_{min} = 250/\sqrt{e}$ μm is $L = 0.18\,\mu$H. The difference would have been a little larger if the coil had been placed at the centre of the core, as assumed in the calculation. (On the other hand, the contribution of the end surfaces is not included in the calculation.) Assuming that $P_{spr} = L_{spr}/N^2 \leqslant (0.32\text{-}0.18)10^{-6}/N^2 = 2.2\ 10^{-9}$ H equals a constant C times half the core length (r_{max}) gives $C < 1.46\ 10^{-6}$, which equals roughly μ_0. (The $<$ sign is used since it is not intended to include the contribution from the end surfaces in $P_{spr} = Cr_{max}$.)

The second experiment verified the above assumption that P_{spr} is about proportional to r_{max}. For this purpose different numbers of ferrite bars, with dimensions given in Fig. 8.13, were placed in series and in parallel. One ribbon with a thickness of 0.1 mm and a width of 16 mm was tightly wound around the middle of the bar obtained in this way. The inductance values and changes for 2, 3 and 4 bars in series, and for 2 bars in parallel, or one single bar, were measured at 100 kHz and are given in Table 8.6. It is noticed that for these 'open' configurations the tiny gaps between the bars do not influence the results and the assembly of bars can be very well approximated by one larger bar with infinite permeability, unless the frequency is much higher than 1 MHz. Expression (8.2) can thus be applied, results of which are also tabulated. For dimensional ratios like those in the head, i.e. the lower n values, adding $L_{spr} = P_{spr} = \mu_0 r_{max}$ leads to reasonable agreement with the experimental data, realizing that in the experimental data the con-

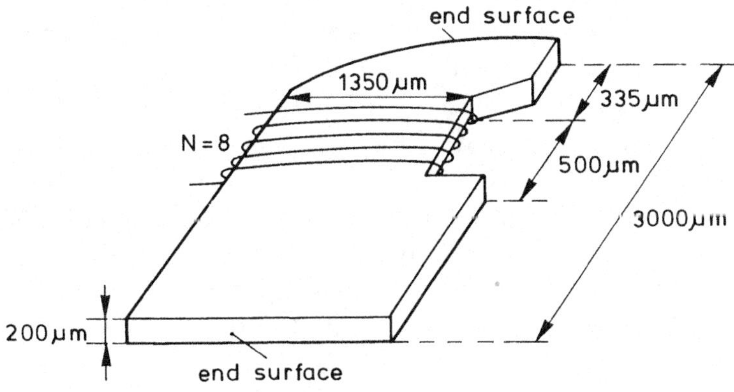

Fig. 8.11. Convenient geometry of one half of a (V2000) ferrite video head.

Table 8.6. The inductance values, L_n, and changes, $L_n - L_{n-1}$, of n single or double bars with dimensions given in Fig. 8.13 in series.

Single bars in series ($8 \times 18 \times 100n$) [mm^3]:

	measured		calculated			
n	L_n	$L_n - L_{n-1}$	expr. (8.2)	$L_n - L_{n-1}$	$\mu_0 r_{max}$	(8.2) incl. $\mu_0 r_{max}$
			[nH]			
1	125		49		63	112
2	185	60	63	14.5	126	199
3	235	50	71	12	188	259
4	265	30	77	6	251	328

Double bars in series ($8 \times 36 \times 100n$) [mm^3]:

n	L_n	$L_n - L_{n-1}$	expr. (8.2)	$L_n - L_{n-1}$	$\mu_0 r_{max}$	(8.2) incl. $\mu_0 r_{max}$
1	182		82		63	145
2	246	64	107	24	126	233
3	300	54	121	13.5	188	309
4	340	40	131	10	251	382

tribution from the end surfaces has not been subtracted. For larger bar lengths the effect of the spread of flux is then overestimated.

In the third experiment, three bars are placed in parallel, see Fig. 8.12, and again a copper ribbon of 16 mm width is used, such that the ratios resemble those of one half of a head. Fig. 8.12 shows the effect of placing the ribbon out of the centre, $x = 50$ mm, of the bar. At $x = 20$ mm, corresponding to the location of the coils in a video head, see Fig. 8.11, the inductance dropped to 84%. We will not take this effect into account, because the effect of the end surfaces has to be added to R_{ss}, which compensates most of this effect (in this way $116 + 63 = 179$ nH is calculated instead of the experimental value $L = 187$ nH at $x = 20$ mm).

Subsequently it was verified that the inductance of equally distributed windings over a length of 16 mm is exactly equal, apart from the proportionality constant N^2, to that of the one-turn coil obtained with the 16 mm copper ribbon. This was assumed in the evaluation of (8.2).

Finally the influence of the width of the ribbon was determined by experiments; see Fig. 8.13. It is easily verified that the differences be-

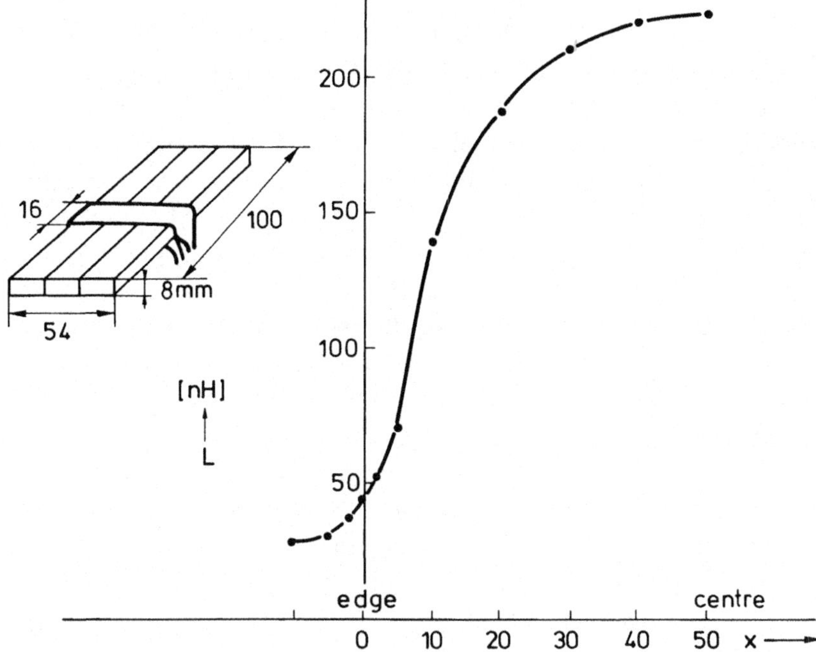

Fig. 8.12. The inductance of a one-turn (thin copper ribbon) coil, tightly wound on a composition of bars and with outer dimensional ratios equal to those of the half video head in Fig. 8.11, as a function of the location of the winding.

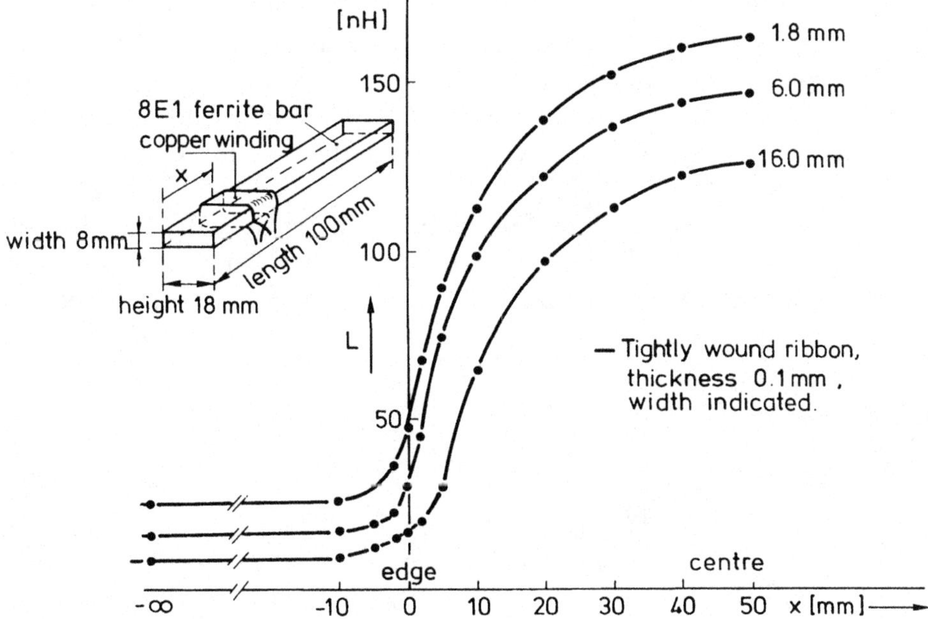

Fig. 8.13. The inductance of one-turn ribbon coils as a function of winding location, x, for different widths of the ribbon.

tween the curves for $x = 50$ mm (centre of the bar) correspond reasonably with values obtained from (8.2). When a single turn of wire is used with a diameter equal to the width of the ribbon, then the inductance of the latter is noticeably higher, unless the 0.1 mm thin ribbon is wound at a spacing comparable to the location of the centre of the wire; see the various 1.8 mm curves. At larger spacings the inductance decreases only slightly. The same displacement from the core in the case of a wider ribbon, see 16 mm curves, has less effect, as expected. So a displacement acts like a tightly wound wider ribbon, which can be well understood. From the examples it follows that both a wider ribbon and a longer coil as well as an increase of the distance between coil and core surface (if not too large), reduces the (stray) inductance of the core; see Fig. 8.14. In a head configuration an increase of this space would soon induce so much extra coupling between the coils at both head sides that via the mutual inductance M the (stray) inductance L would rise instead of decrease. For a head it is therefore recommended to use flat, tightly wound, coils.

Table 8.7 gives the terms chosen to account for the spread of flux. Other authors who used the reluctance model [8.1] did not account for the spread of flux, so that their results must suffer from the same discrepancies as we found when using our preliminary results.

Table 8.8 gives convenient values for permeabilities at 4.5 MHz of ferrite (8X1) and laminated amorphous metal (3×6.5 μm; $Co_{70.3} \cdot Fe_{4.7} \cdot Si_{15} \cdot B_{10}$ at %) and dimensions of video heads made from these materials. The calculated output parameters, impedance and efficiencies are also tabulated.

Many measurements on ferrite heads and on some (laminated) amorphous heads with 0.3 μm gap indicate that the calculated inductance and efficiency agree, on average, with the observations. The cal-

Table 8.7. Extra permeances, $P_{str} = 1/R_{str}$, due to the spread of flux, as determined phenomenologically.

$$
\begin{aligned}
R_{FF'5} &= 1/P_{str\,FF'} = 2/(\mu_0(b_1 - g)) \\
R_{EE'5} &= 1/P_{str\,EE'} = 2/(\mu_0(b_2 - b_1)) \\
R_{DD'5} &= 1/P_{str\,DD'} = 2/(\mu_0(b_3 - b_2)) \\
R_{CC'5} &= 1/P_{str\,CC'} = 4/(\mu_0(b_4 - b_3)) \\
R_{SS'5} &= 1/P_{str\,SS'} = 2/(\mu_0 h_5)
\end{aligned}
$$

Table 8.8. Conventional MnZn ferrite video-head and amorphous (sandwich) video-head dimensions and permeabilities (measured at 4,5 MHz) and resulting head parameters (calculated. The internal efficiencies and reluctances have been added for later purposes.

Input parameters		Ferrite head	Amorphous head
Track-narrowing angle	θ_1	90°	0.0637°*
Height-reducing angle	θ_2	45°	45°
Bridge length	b_1	20 µm	20 µm*
Coil-chamber length	b_3	300 µm	300 µm
Core length	b_4	3000 µm	3000 µm
Track width	W	20 µm	20 µm
Core width	W_1	200 µm	20.1 µm*
Gap height	h	35 µm	35 µm
Coil-chamber height	h_4	500 µm	500 µm
Core height	h_5	3000 µm	3000 µm
Number of turns	N	18	18
Real part of μ_x	μ_x'	550	2225
Minus imaginary part of μ_x	μ_x''	450	1065
Real part of μ_y	μ_y'	550	2225
Minus imaginary part of μ_y	μ_y''	450	1065
Gap length	g	0.3 µm	0.3 µm
Output parameters (for circuit components see Fig. 8.15)			
Inductance	L	1.9 µH	1.48 µH
Quality factor	Q	6.7	6.5
Field-efficiency ampl.	$|\eta_H|$	67%	58%
Field-efficiency phase	$\varphi_{\eta H}$	−15.1°	−12.1°
Flux-efficiency ampl.	$|\eta_\Phi|$	33%	37.5%
Flux-efficiency phase	$\varphi_{\eta\Phi}$	− 6.6°	− 3.4°
Thevenin and Norton representations; see section 8.1.5			
Int. field-efficiency ampl.	$|\eta_{Hint}|$	88%	78%
Int. field-effiency phase	$\varphi_{\eta Hint}$	− 5.7°	− 6°
Int. flux-efficiency ampl.	$|\eta_{\Phi int}|$	73%	79%
Int. flux-efficiency phase	$\varphi_{\eta\Phi int}$	−12°	− 5.8°
Real part of Z_{int}	$\text{Re}\{Z_{int}\}$	97 MA/V.s	117 MA/V.s
Im. part of Z_{int}	$\text{Im}\{Z_{int}\}$	72 MA/V.s	48 MA/V.s
Real part of $Z_{\Phi int}$	$\text{Re}\{Z_{\Phi int}\}$	270 MA/V.s	300 MA/V.s
Im. part of $Z_{\Phi int}$	$\text{Im}\{Z_{\Phi int}\}$	48 MA/V.s	24 MA/V.s

* In reality, $\theta_1 = 0$ and $W_1 = W$ for the amorphous head. This is not allowed in the calculations, because of divisions by zero. To avoid this and to make possible a comparison between the circuit components of both head types, given in Fig. 8.15, θ_1 is chosen so small that the core width increases from W to $W_1 = W + 0.1$ µm over 90 µm, equal to the length of the tapered parts of the ferrite head. Since the bridge length too is chosen equal to that in the ferrite head, all segments have the same length as those in the ferrite head.

The L, Q and η values of the amorphous head are not much influenced by this choice.

Generally, however, in the case of heads without tapered parts, like most amorphous heads, it is best to choose b_1 and θ_1 such that the lengths of the segments D, E and F decrease in this order.

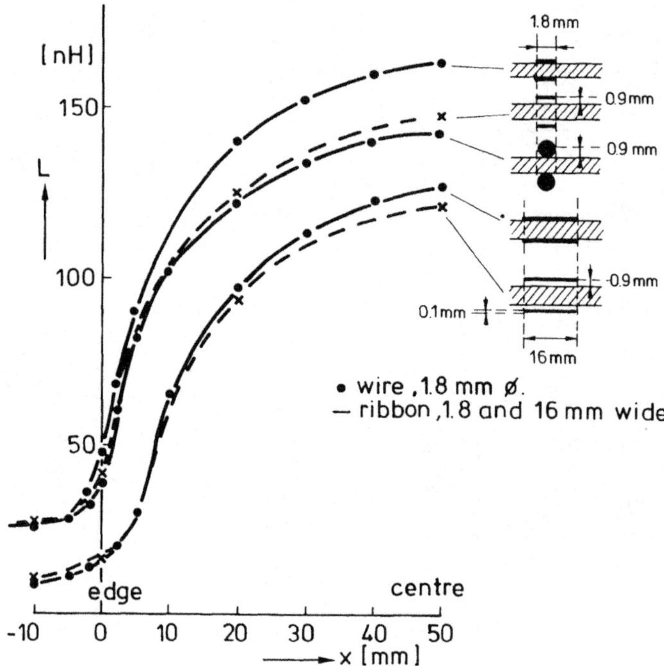

Fig. 8.14. Reduction of (stray) inductance with the distance between the ferrite surface and the centre of the wire or ribbon.

culated quality factor is consequently too high. Measured values for Q are about 5 on average.

Usually we do not ask for the head circuit reluctance values, which have been calculated in the foregoing before the output parameters have been calculated. However, to get insight into the relevance of some components and irrelevance of others, it is sometimes useful to know these values. In Fig. 8.15 they are given for the present ferrite and amorphous head.

From these values we conclude for the ferrite head that:

- The core reluctances of sections A, (C) and D are almost negligible and that of section B plays a minor role.
- The core reluctances of sections E and F play a major role. For section F this is particularly true when the bridge length is large, and results in a low efficiency, η_{H}.
- The stray flux from F to F', E to E' and from D to D' is much smaller than the core flux.
- Stray flux in the neighbourhood of the windings (SS'), and to a

lesser extent from section C to C', determine almost all stray flux.

Due to the higher permeability and much smaller core width of the laminated amorphous head, the core reluctances are higher, except in region E where the reluctance is comparable to that of the tapered section in the ferrite head, and in the 'bridge region' F where the widths of both heads are equal and thus the reluctance of the amorphous head considerably smaller.

Since stray permeances are determined by outer surfaces (more in particular by circumferences, lengths and length ratios) of sections, rather than by volumes, the stray reluctances of the amorphous head

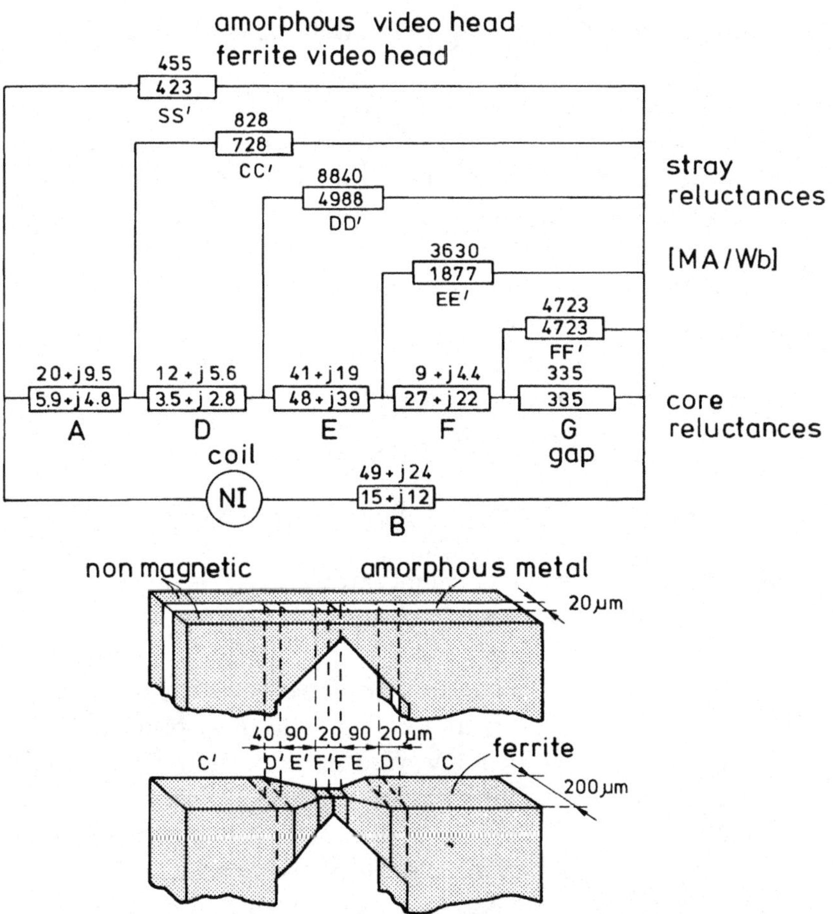

Fig. 8.15. Values of the circuit reluctances in the case of an amorphous head, denoted in mega amperes per weber (MA/Wb) above the circuit components, and of a ferrite head, denoted in the drawn circuit components, for convenient permeabilities at about 4.5 MHz.
For more details concerning the dimensions, permeabilities and the calculated head parameters see Table 8.8.

are comparable to those of the ferrite head. Only stray reluctances from sections *D* and *E* are about twice as high, due to approximately two times smaller circumferences than for the tapered ferrite head.

Corresponding with these differences we read for the laminated amorphous head from the circuit component values in Fig. 8.15:
- in contrast to the case of the ferrite head, the core reluctances of sections *A, B, C* and *D* cannot be neglected, while on the other hand the reluctance of the bridge section *F* is rather small and,
- the same stray-flux reluctances as for the ferrite head are of importance.

8.1.6 Thevenin and Norton representation of a videohead

Effects of any change in the geometry of the head or permeability of the core are described by the previous model. From the values in Fig. 8.15 it follows that the efficiency is mainly determined by the reluctance of the core elements near the gap and the gap length. For this reason efficiency of the head is highly sensitive to faults in this region. The head is also most fragile in this region. Common problems are:
- the gap length is too long or too short;
- the gap definition is bad;
- high saturation metal films around the gap of a metal-in-gap (MIG) head have a poor permeability;
- gap smear;
- stresses in the gap cause, via the magnetostriction constant, a low permeability around the gap;
- the adhesion at the ferrite-metal interface, or the magnetic permeability around this interface is bad;
- short-circuits around the gap.

It also follows from values in Fig. 8.15 that the stray permeance near the gap plays a minor role in the determination of both the efficiency and the inductance of the head. Variations in this stray permeance, due to changes of the permeability near the gap or changes of the gap length, can therefore be neglected. Because of this, the Thevenin and Norton representations, see Fig. 8.16, of the head's equivalent circuit in Fig. 8.3 can be used for predicting effects due to most of the above causes.

The two elements that characterize *Thevenin's* circuit are:

1) the ideal potential source with magnetic potential Ψ_{int} equal to the open-circuit potential Ψ_{load} ($Z_{load} = \infty$) and
2) the source reluctance, Z_{int} which according to Thevenin's circuit equals Ψ_{int}/Φ_{load} ($Z_{load} = 0$).

For the Thevenin circuit $\Psi_t (= NI$, see Fig. 8.3) will be kept constant, e.g. as indicated below the circuit, such that Ψ_{int} is also constant, at different loads. (If Φ_t is kept constant, then Z_{int} will be larger.) This choice is suited for calculations of the (field) efficiency variations due to variations of e.g. the load (gap).

The two elements that characterize *Norton's* circuit are:

1) the ideal flux source with flux Φ_{int} equal to the short-circuit flux Φ_{load} ($Z_{load} = 0$) and
2) the source reluctance, $Z_{\Phi int}$, which according to Norton's circuit equals $\Psi_{load}(Z_{load} = \infty)/\Phi_{int}$.

In contrast, *for Norton's circuit Φ_t will be kept constant.* Because of this *choice*, $Z_{\Phi int} \neq Z_{int}$! ($Z_{\Phi int}$ is the larger value mentioned above.) It depends on the application whether this choice is suited. In the case of equal ampere turns or for *field* efficiency calculations it is obviously not, whereas for calculating the *flux* efficiency variations due to load (gap) variation it is very well suited, since $\hat{\Phi}_{int}$ does not then vary with Z_{load}. Preferably, $Z_{load} = 0$ is obtained by choosing the (fictitious) permeability of the gap, $\mu_{r_{gap}}$, infinite, and zero for $Z_{load} = \infty$.

Since efficiencies η and η_Φ, reluctances Z_{ab} and impedances Z are

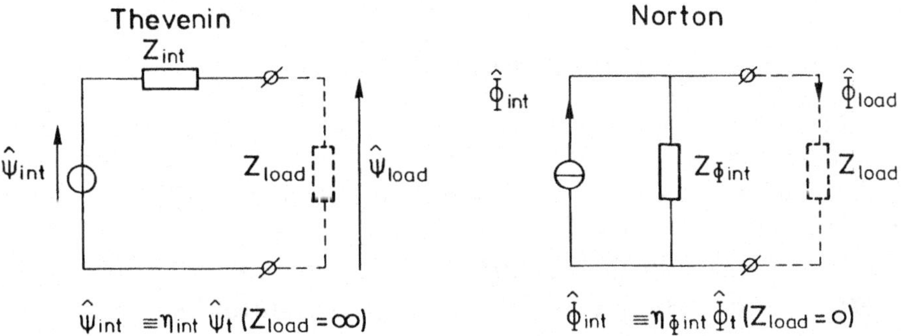

Fig. 8.16. Thevenin and Norton representation of a video head, used for predicting effects due to variations in the neighbourhood of the gap.

calculated from the model rather than potentials, it is more appropriate to characterize the two equivalent circuits by η_{int} and Z_{int}, and $\eta_{\Phi_{int}}$ and $Z_{\Phi_{int}}$, respectively. For this purpose the following relations have been derived:

$$\boxed{\eta_{int} = \eta(\mu_{r_{gap}} = 0)} \tag{8.21}$$

$$\boxed{\eta_{\Phi_{int}} = \eta_\Phi(\mu_{r_{gap}} = \infty)} \tag{8.22}$$

$$\boxed{Z_{int} = \frac{\eta_{int}}{\eta_{\Phi_{int}}} Z_{ab}(\mu_{r_{gap}} = \infty) = \frac{\eta_{int}}{\eta_{\Phi_{int}}} \cdot \frac{j\omega N^2}{Z(\mu_{r_{gap}} = \infty)}} \tag{8.23}$$

$$\boxed{Z_{\Phi_{int}} = \frac{\eta_{int}}{\eta_{\Phi_{int}}} Z_{ab}(\mu_{r_{gap}} = 0) = \frac{\eta_{int}}{\eta_{\Phi_{int}}} \cdot \frac{j\omega N^2}{Z(\mu_{r_{gap}} = 0)}} \tag{8.24}$$

Indeed $Z_{\Phi_{int}} > Z_{int}$, with which we mean $|Z_{\Phi_{int}}| > |Z_{int}|$.

The effects of changes in for instance the gap lengths can now be well described by setting $Z_{load} = R_g$ in Thevenin's and Norton's circuits and subsequently changing g in the expression for R_g, hence

$$\boxed{\eta = \frac{R_g}{R_g + Z_{int}} \eta_{int}} \tag{8.25}$$

$$\boxed{\eta_\Phi = \frac{Z_{\Phi_{int}}}{R_g + Z_{\Phi_{int}}} \eta_{\Phi_{int}}} \tag{8.26}$$

When η and η_Φ are known experimentally for two different gap lengths (R_g), then these equations may serve to determine η_{int}, Z_{int}, $\eta_{\Phi_{int}}$ and $Z_{\Phi_{int}}$.

The permeance of the head $P_{ab} \equiv 1/Z_{ab}$ can easily be derived when information from both the Thevenin and the Norton representation is used. The result is

$$P_{ab} = \frac{\eta}{\eta_\Phi R_g} \tag{8.27}$$

which coincides with (8.20).

Results (8.25) and (8.26) must be substituted in (8.27) in order to calculate the impedance, Z, or inductance, L, and quality factor, Q, according to (8.13) or (8.15) and (8.16), hence:

$$\boxed{Z = \frac{N^2}{Z_{\Phi_{int}}} \cdot \frac{\eta_{int}}{\eta_{\Phi_{int}}} \cdot \frac{R_g + Z_{\Phi_{int}}}{R_g + Z_{int}}} \, , \tag{8.28}$$

which decreases when R_g increases, because $Z_{\Phi_{int}} > Z_{int}$. This electric impedance is determined by the elements of both the Thevenin and the Norton representation, in contrast to the efficiencies.

In (8.25), R_g can be chosen simply $g/(\mu_0 wh)$ up to extreme gap lengths, since at large gap length for which R_g is large, η still approaches η_{int}. In (8.26), on the other hand, this leads to inaccuracies about as large as those on R_g. When the accurate expression (8.12) is used for R_g, then η_Φ is also approximated accurately at the extreme gap lengths, and consequently L and Q can be approximated with the aid of (8.27). This is demonstrated in Fig. 8.17, showing that the agreement is perfect for all practical gap lengths (< 1 μm). The still reasonable agreement at extremely large gap lengths indicates that even when large changes take place outside the gap, the effects can still be described without taking into account the variations in the stray permeance (note that the stray flux may still vary).

This result is of importance for considerations about the bulk saturation process in video heads applied to efficiency determinations further on.

The relation between the efficiencies η' and η of heads with gap length g' and g respectively follows easily from the Thevenin representation and reads for $g, g' \ll W, h$ such that gap edge effects can be neglected and R_g is proportional to g:

$$\boxed{\frac{\eta'}{\eta} = \frac{1}{1 - \dfrac{g' - g}{g'}\left(1 - \dfrac{\eta}{\eta_{int}}\right)}} \, . \tag{8.29}$$

This relation is useful when the gap length of a known head is to be changed. In one of the following figures (Fig. 8.20) η'/η is plotted as a function of $\Delta g/g \equiv (g'-g)/g$ for different values of η/η_{int}. As expected, a high-efficiency head is not sensitive to a gap length reduction, since the core reluctance remains small compared to the gap reluctance. The efficiency of a poor head with $\eta \ll \eta_{int}$, decreases almost proportionally with the gap length, i.e. $\eta'/\eta \rightarrow g'/g$.

The analogous relation between the electric impedance of heads with different gap lengths is derived from (8.28). The most suitable way to

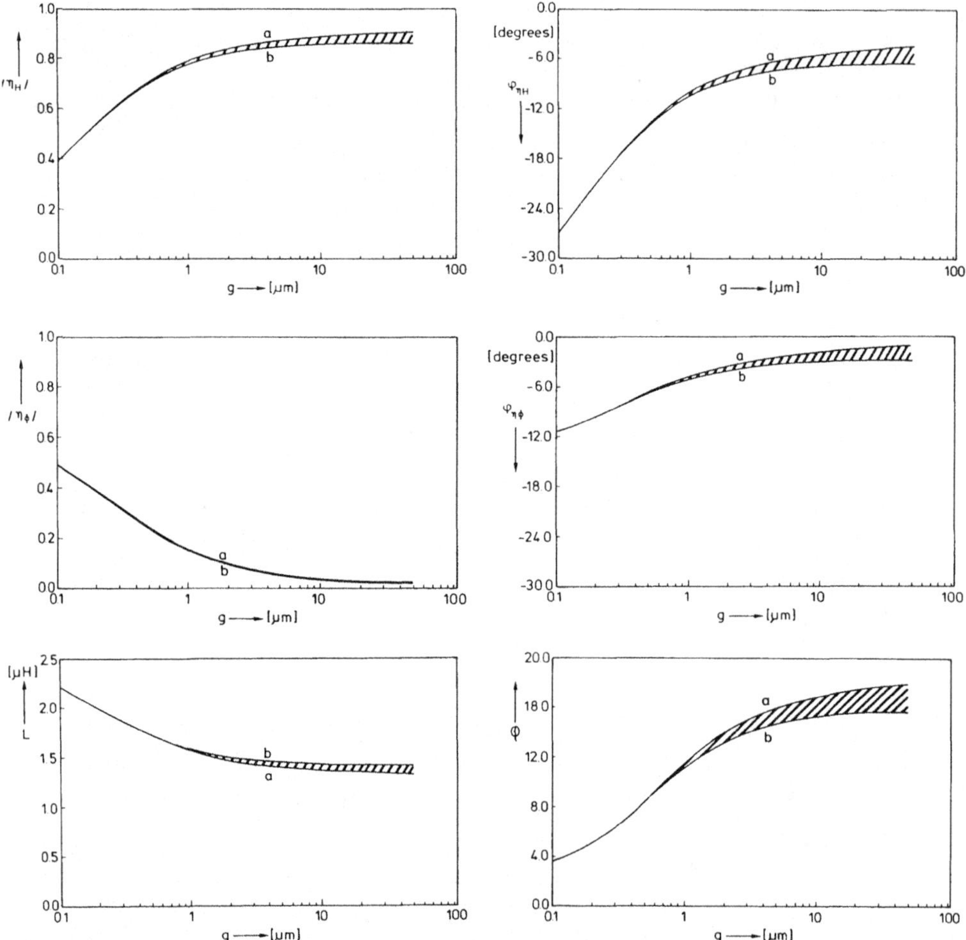

Fig. 8.17. Comparison between results obtained from the complete equivalent circuit in Fig. 8.3, see curves a, and those obtained from the Thevenin-Norton representation, see curves b.
The convenient ferrite, video-head parameters given in Table 8.8 have been used, except for the bridge length, b_1 which is chosen 60 μm in the present example.

express the result reads:

$$\frac{Z'}{Z} = \frac{1 + \dfrac{g' - g}{g}\left(1 - \dfrac{\eta_\Phi}{\eta_{\Phi_{int}}}\right)}{1 + \dfrac{g' - g}{g}\dfrac{\eta}{\eta_{int}}}. \tag{8.30}$$

It follows again from $Z_{\Phi_{int}} > Z_{int}$ with the aid of (8.25) and (8.26) that $1 - \eta_\Phi/\eta_{\Phi_{int}} < H/\eta_{int}$ and hence Z' decreases when g' increases.

Analogously expressions can be derived for changes in the gap permeance $\Delta P_g \equiv P_g' - P_g$, source (core) permeance ΔP_{int} and source (core) reluctance ΔZ_{int}. In addition they can be normalized to Z_{int} or P_{int}, or to R_g or P_g, depending on the application. The parameters in the expressions may be only reluctances or permeances, efficiencies or combinations of both. We will not reproduce them all. In the following sections some of these relations are reproduced and applied. In any case, the results from the most meaningful relations are given in one plot; see Fig. 8.20.

8.1.6.1 Efficiency reduction due to increasing the (gap-)permeance, applied to gap smear and magnetic short-circuits around the gap

In a metal-in-gap (MIG) head part of the metal film may be smeared over the gap by the tape as illustrated in Fig. 8.18a. The thickness of this layer is t_5 and its relative permeability is μ_5. In this type of head hollow-out, h_0, of the metal layers occurs due to a high wear rate of the metal with respect to the ferrite. If the relative permeability μ_5 of the gap-smear layer were 1 and t_5 were larger than h_0, then the effective head-tape distance would increase by the extra geometrical distance $t_5 - h_0$. Since $\mu_5 \neq 1$, it is expected that some shielding of the information on the tape occurs. However in Section 2.9 it was concluded that a gap-smear layer on top of an idealized zero-gap head does not shield the head from information on the tape. Moreover the magnetic layer 'conducted' the information from the tape better than air would do (see Fig. 8.18b as a reminder). In this section it will be shown that practical heads, i.e. with efficiencies $\eta < \eta_{int} < 1$ and non-zero gap length, do indeed experience 'shielding', i.e. the above remarkable conclusion does not always hold.

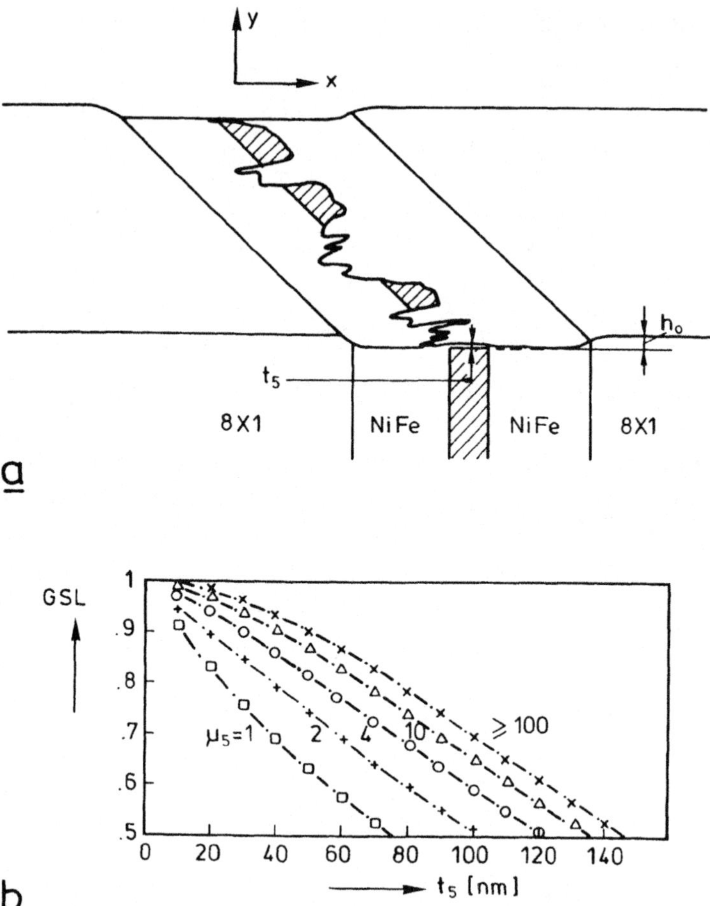

Fig. 8.18. Gap smear
a) A magnetic smear layer with thickness t_5 on the gap, causing a reduction of the efficiency and worse gradients (distance losses).
b) The factor GSL that describes the playback loss for $\lambda = 0.7$ μm 'due to worse gradients', as a function of t_5, for various permeabilities. Tape parameters (coating) are:
$\mu_3 \equiv \sqrt{\mu_{3x}\mu_{3y}} = 2$ and $\beta_3 \equiv \sqrt{\mu_{3x}/\mu_{3y}} = 0.5$; $t_3 = 3$ μm and head (gap-smear layer) to tape distance during playback $d_{4p} = 0.15$ μm.

In fact the shielding takes place via a reduction of the potential at the surface of an actual head, Ψ^a, which for this purpose is defined at the interface gap – gap smear layer; see the alternative expression (3.17) for the read flux. For the wavelength of interest in video recording (see arguments in Sect. 5.3) the effect can be well described after introducing an efficiency $\eta \equiv \hat{\Psi}_g/\hat{\Psi}_t$. The change of the efficiency from η to η' due to a *change of the permeance over the gap*, ΔP_g, is obtained from the Thevenin representation and reads

$$\frac{\eta'}{\eta} = \frac{1}{1 + \dfrac{\Delta P_g}{P_g}\left(1 - \dfrac{\eta}{\eta_{int}}\right)} \qquad (8.31a)$$

or

$$\frac{\eta'}{\eta} = \frac{1}{1 + \dfrac{\Delta P_g}{P_{int}}\dfrac{\eta}{\eta_{int}}}\,, \qquad (8.31b)$$

where $\Delta P_g \equiv P_g' - P_g$.

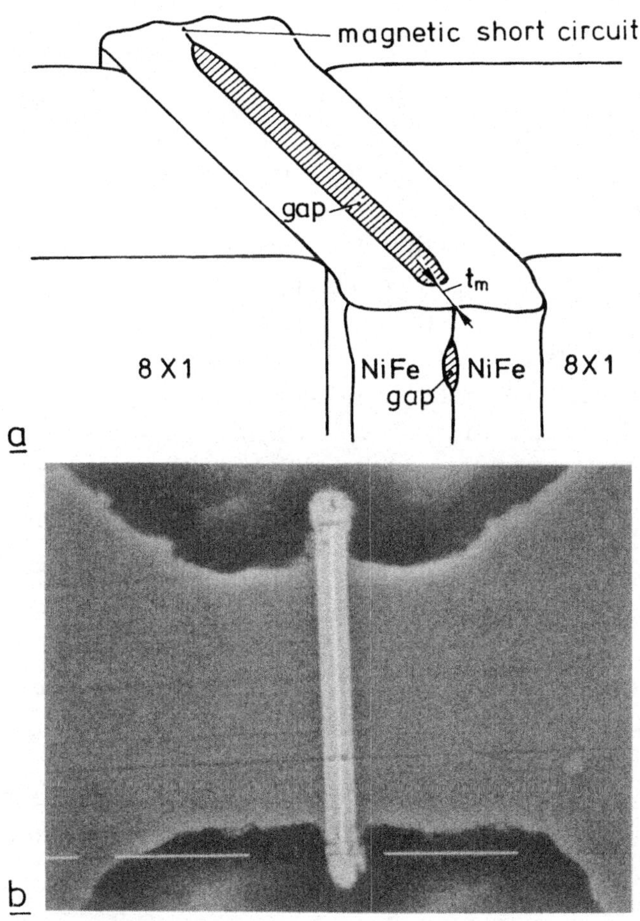

Fig. 8.19. Magnetic short-circuits around the gap.
a) Sketch and definition of t_m.
b) Picture of this phenomenon in a NiFe-MIG head.

The equality $\Delta P_{\mathrm{g}}/P_{\mathrm{g}} = -\Delta Z_{\mathrm{g}}/Z'_{\mathrm{g}}$ shows the correspondence with (8.29), where R_{g} was assumed to be approximately proportional to the gap length, which holds for $g \ll W, h$. When the increase of the permeance over the gap is due to gap-smear or to some other very local change in the permeance, then non-uniformity of the magnetic potential over the gap surface may occur when the reluctance of the core in the neighbourhood of the gap is not very small. This may happen as soon as P_{int} approaches the permeance of the extra flux path, ΔP_{g}. So expression (8.31a) may be inaccurate when $\eta'/\eta \lesssim 1/(1 + \eta/\eta_{\mathrm{int}})$. In the case of only gap smear:

$$\boxed{\frac{\Delta P_{\mathrm{g}}}{P_{\mathrm{g}}} = \frac{\langle \mu_5 t_5 \rangle}{h_0}.} \qquad (8.32)$$

Expression (8.31a) is most appropriate for physical interpretations, since the gap length and gap surface, and consequently P_{g}, are fixed for a given head geometry while P_{int} may vary depending on core material properties. Expression (8.31a) shows that the perturbation ΔP_{g} has more consequences when the efficiency becomes lower or more precisely when the ratio η/η_{int} becomes smaller; see Fig. 8.20, which shows η'/η versus $\Delta P_{\mathrm{g}}/P_{\mathrm{g}}$ for different values of η/η_{int}. When η approaches η_{int}, hardly any effects are observed. This is due to the very low internal reluctance of the head relative to the gap reluctance; see also (8.25). High-efficiency heads are thus less sensitive to gap-smear layers with high permeability, assuming that h_0, see (8.32), is unaltered. When h_0 is larger, the efficiency is allowed to be smaller before effects on η' become noticeable. In contrast to the effect on the efficiency, the 'distance loss', described by the factor GSL, decreases when the permeability of the gap-smear layer increases, see Fig. 8.18b. It is noted that the characteristics of the tape and the head-tape distance play a minor role in this result, as outlined in Sect. 2.9. For completeness the chosen parameters are given in the figure caption.

For example, when $\Delta P_{\mathrm{g}}/P_{\mathrm{g}} = 1.4$, e.g. $\mu_5 = 1000$, $t_5 = 14$ nm and $h_0 = 35$ μm (case 1) or $\mu_5 = 100$, $t_5 = 140$ nm and $h_0 = 35$ μm (case 2) then the reduction in the efficiency is about 1 dB, assuming a ferrite video head with $\eta = 0.6$ and $\eta_{\mathrm{int}} = 0.9$. However when the gap-smear layer completely adds to the geometrical head-tape distance ($h_0 = 0$), then the distance loss effect in only the playback process of case 2 is already 6 dB, according to Fig. 8.18b, whereas this effect is negligible in case 1!

If μ_5 is 1, then the distance-loss effect is maximum, i.e. the well-known distance loss described by e^{-kt_5}, while no reduction in the efficiency is revealed.

Short-circuiting of the gap by a thin magnetic layer around the gap (excluding the short-circuit on the tape-facing surface by gap smear), also reduces the efficiency. The picture in Fig. 8.19 shows this phenomenon. Since this thin flux path already saturates at low flux levels (see Fig. 8.48a in Sect. 8.2.5), it only has consequences during

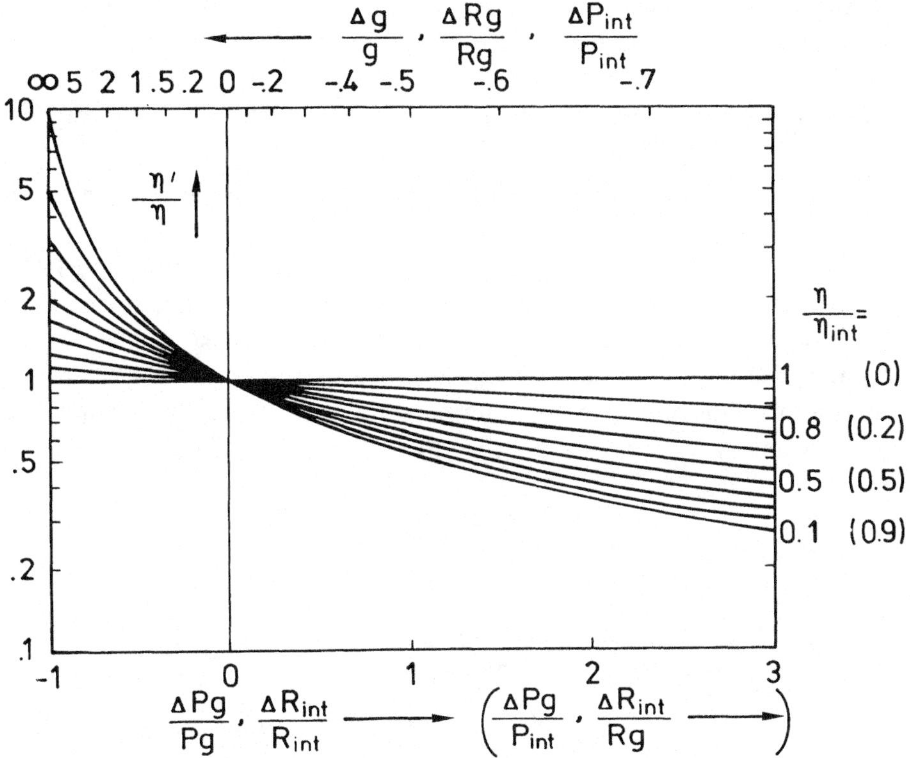

Fig. 8.20. Change of the efficiency from η to η' (in the case of real Z_{int} and P_g) due to
- changing the gap permeance P_g or gap reluctance R_g by an amount ΔP_g or ΔR_g; $\Delta P_g \equiv P_g' - P_g$ while P_g' corresponds to η' and P_g corresponds to η etc.
- changing the internal reluctance R_{int} or permeance P_{int} by an amount ΔR_{int}.
- changing the gap length g by an amount Δg.

For different choices concerning the normalization of the above quantities the curves can be applied. The parameter values, η/η_{int}, between the brackets must be applied for the variables between the brackets. (Not all relations used for the above plot are given in the text. The relations follow immediately from (8.31a) or (8.31b) after it is shown that equal variation of the following quantities has equal result on η'/η:
$\Delta P_g/P_g$, $-\Delta R_g/(R_g + \Delta R_g)$, $\Delta R_{int}/R_{int}$, $-\Delta P_{int}/(P_{int} + \Delta P_{int})$ in (8.31a)
and
$\Delta P_g/P_{int}$, $\Delta R_{int}/R_g$ in (8.31b), where $1 - \eta/\eta_{int} \rightarrow \eta/\eta_{int}$ with respect to (8.31a).)

the playback process. The reduction of the efficiency from η to η' also follows from (8.31a) in which

$$\boxed{\frac{\Delta P_{\mathrm{g}}}{P_{\mathrm{g}}} = \langle \mu_{\mathrm{rm}}\, t_{\mathrm{m}} \rangle \, \frac{W + 2h_0}{Wh_0}} \, , \qquad (8.33)$$

where t_{m} is the thickness and μ_{rm} the relative permeability of the magnetic layer.

A reduction of 1 dB in the efficiency occurs when $\Delta P_{\mathrm{g}}/P_{\mathrm{g}} = 1.4$ (already reached when for instance $\mu_{\mathrm{rm}} = 100$, $t_{\mathrm{m}} = 30$ nm (!), $W = 20$ μm and $h_0 = 35$ μm) and $\eta/\eta_{\mathrm{int}} = 60\%/90\% = 2/3$.

8.1.6.2 Application to ring heads on high permeability media

A permeable layer at a small distance from the head surface can be treated as a transmission line. In chapter 11, the transmission-line concept is applied to the yoke magnetoresistive thin-film head (YMRH). The transmission-line concept is simple when the distance to the head, d_4, and the track width, W, are constant over at least a characteristic distance, λ_{c}, on both sides of the gap, i.e. when $\lambda_{\mathrm{c}} < b_1/2$. Then in a good approximation twice the characteristic reluctance R_{c} of this transmission line can be placed parallel to the gap reluctance, i.e.

$$\Delta P_{\mathrm{g}} = 1/(2R_{\mathrm{c}}). \qquad (8.34)$$

Calculations

The series and parallel reluctance R_{s} and R_{p} determined for a length l are important in the determination of the characteristic length

$$\lambda_{\mathrm{c}} = l\sqrt{R_{\mathrm{p}}/R_{\mathrm{s}}}, \qquad (8.35)$$

and the characteristic impedance

$$R_{\mathrm{c}} = \sqrt{R_{\mathrm{p}}R_{\mathrm{s}}}, \qquad (8.36)$$

see chapter 11. In reference [8.6], some information can be found about the circumstances necessary for the transmission-line concept to hold (in thin-film head applications). R_{s} and R_{p} are given by

$$R_s \;\; = \;\; \frac{l}{\mu_0 \mu_{2x} t_2 W} \qquad (\mu_{2x} t_2 \gg \mu_3 t_3) \tag{8.37}$$

$$R_p \;\; = \;\; \frac{t_3}{\mu_0 \mu_{3y} \, l \, W} \; + \; \frac{d_4}{\mu_0 \, l \, W} . \tag{8.38}$$

Hence:

$$\lambda_c \;\; = \;\; \sqrt{\left(\frac{t_3}{\mu_{3y}} + d_4\right) \mu_{2x} t_2} \qquad [\mathrm{m}], \tag{8.39}$$

$$R_c \;\; = \;\; \frac{1}{\mu_0 W} \sqrt{\left(\frac{t_3}{\mu_{3y}} + d_4\right) \Big/ (\mu_{2x} t_2)} \qquad \left[\frac{\mathrm{A}}{\mathrm{Wb}}\right], \tag{8.40}$$

and so

$$\boxed{\frac{\Delta P_g}{P_g} = \frac{g}{2h_0} \sqrt{\mu_{2x} t_2 \Big/ \left(\frac{t_3}{\mu_{3y}} + d_4\right)} .} \tag{8.41}$$

Results

In the series reluctance expression, (8.37), the coating contribution has been neglected, which is a very good approximation for DL media. Due to this approximation, SL media must be treated separately. By using $\Delta P_g/P_g = (g/2h_0)\sqrt{\mu_{3x} t_3/d_4}$ for an SL medium, it is verified for $\mu_{3_x} = 1.4$, $t_3 \le 3$ µm and $d_4 = 0.1$ µm, and a ring head with $g = 0.3$ µm, $h_0 = 20$ µm, $\eta = 60\%$ and $\eta_{\mathrm{int}} = 90\%$, that the efficiency decreases by less than 1 percent. The effect can thus be neglected.

For high permeability double-layer (DL) media the situation is different. In Fig. 8.21, $\Delta P_g/P_g$ is plotted versus $\mu_{2x} t_2$ with $g/\left(2h_0\sqrt{(t_3/\mu_{3_y} + d_4)}\right)$ as a parameter. This parameter equals $1.68 \; 10^{-2}$ µm$^{-1/2}$ with the following convenient set of parameters for a DL medium and video ring head:
$t_3 = d_4 = 0.1$ µm, $\mu_{3_y} = 1$, $g = 0.3$ µm and $h_0 = 20$ µm. At the higher values of $\mu_{2x} t_2$, the characteristic length may be greater than half the bridge length, which means that the characteristic reluctance is smaller than calculated and consequently the reduction of the efficiency is larger than calculated. With the above convenient parameter values, and further $\mu_{2_x} = 1000$ and $t_2 = 0.5$ µm, we have $\Delta P_g/P_g = 0.375$, which

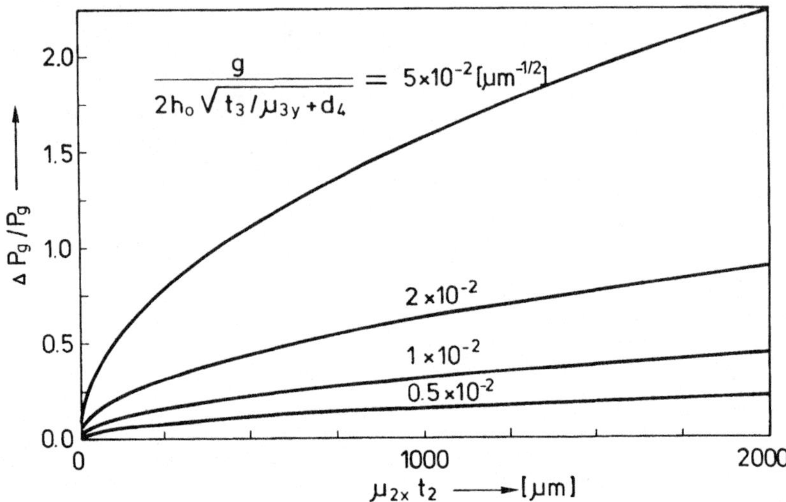

Fig. 8.21. The fictitious increase of the gap permeance of an RH when a DL medium is used.

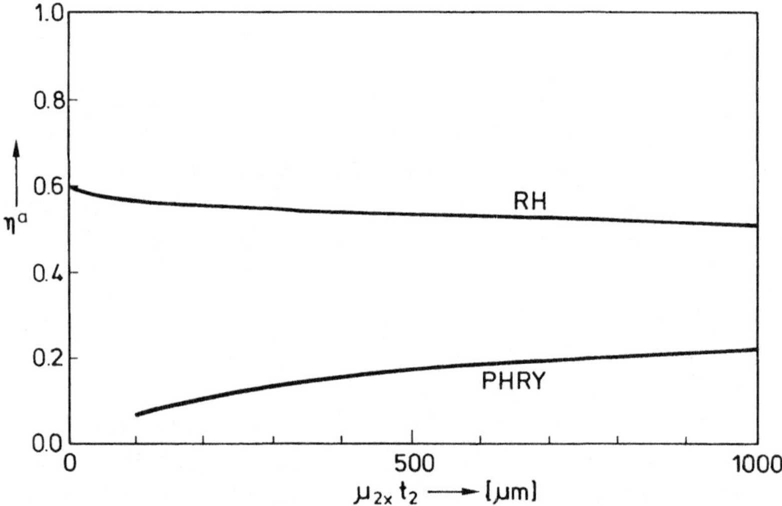

Fig. 8.22. The opposite influence of the permeability of a tape on the efficiency of RHs and PHs.

RH: $\Delta P_g/P_g$ is calculated from (8.41) with $\mu_{3y} = 1$ (valid for perpendicular medium) and substituted in (8.31a) in order to obtain the efficiency for the following parameters: the efficiency in free space $\eta = 0.6$, $\eta_{int} = 0.9$, $g = 0.3$ μm, $h_0 = 20$ μm and $t_3 + d_4 = 0.2$ μm.

PHRY: The 'standard' parameter values have been used as defined in Table 9.1 for the configuration in Fig. 9.2, except for the product $\mu_{2x}\,t_2$ which has been varied. The curve corresponds to the curve indicated $i = 2x$, in Fig. 9.14.

gives a reduction of the efficiency from 60% to 53% when η_{int} equals 90%.

This example shows that a very high permeability and/or thickness of the backlayer may reduce the efficiency of a ring head (RH) noticeably. This is in contrast to the efficiency increase of single-sided probe heads (PH) on media with higher permeability, and is worked out in chapter 9 for various PHs. The reason is simple: the backlayer is an essential part in the series connection of elements that forms the PH's 'core' reluctance, while it shunts the core (or gap) of an RH. In Fig. 8.22 the different character of the efficiency variation of RHs and single-sided PHs with $\mu_{2_x} t_2$ is shown, based on an expression for the efficiency of 2-dimensional PHs with a return yoke (PHRY), derived in chapter 9. For details of the configuration see chapter 9 (Fig. 9.2 and Table 9.1).

8.1.6.3 Efficiency reduction due to increasing the core reluctance, applied to stresses near the gap

Introduction

When the reluctance of the core near the gap is increased by ΔZ_{int}, then the efficiency is reduced from η to η'. Expressions for the effect are easily obtained from the Thevenin representation and read

$$\frac{\eta'}{\eta} = \frac{1}{1 + \dfrac{\Delta Z_{\text{int}}}{Z_{\text{int}}} \left(1 - \dfrac{\eta}{\eta_{\text{inr}}}\right)} \tag{8.42a}$$

or

$$\boxed{\frac{\eta'}{\eta} = \frac{1}{1 + \dfrac{\Delta Z_{\text{int}}}{R_g} \dfrac{\eta}{\eta_{\text{int}}}}} \tag{8.42b}$$

The last expression is most suitable when $\eta \to \eta_{\text{int}}$ and R_g is known.

In Sect. 5.7.10 the exponential gradual gap has been described. There it was noted that, when the gradual gap was caused by stresses from the gap, the effects on the output could best be described by a reduction of the efficiency. The reason is that the balancing stresses in the ferrite relax over a large distance. According to Saint-Venant's principle [8.7] this distance equals about the track width, W.

In evaluating the efficiency reduction due to stresses in the sputtered metal layers in a MIG head, it is also of importance to know the reduction of the permeability in the metal layer itself. Since these metal layers are not thick compared to the longest wavelengths of interest, this effect cannot be described by a reduction of the efficiency only, but must be described by a change of the sensitivity function. This is outlined in chapter 12. The consequences of a very low metal permeability are undulations in the gap loss function and a slight increase of the efficiency.

However these particular undulations have never been observed in frequency characteristics of MIG heads. The type of undulation that is characteristic for (tiny) residual gaps at the metal-ferrite interface has been frequently observed, however, especially in sendust MIG heads; see chapter 12. This means that the permeability component normal to the thin sendust layer is at least 20 (see chapter 12). A direct measurement of this component is impossible because of the stongly dominating demagnetizing field in the normal direction. Hence only in-plane components have been measured in stressed sendust layers on 8X1 tiles (MnZn ferrite) of 8×16 mm^2 surface and 1.5 mm thickness. Because of the small anisotropy in the plane of the sendust layer, only one component in this plane is measured. This is carried out at a temperature of 300°C, which is well above the Curie temperature of the MnZn ferrite (220°C). The mean stress, $\langle \sigma_m \rangle$, in the metal layer has been measured from the difference in curvature of the tiles before (h_b) and after (h_a) sputtering of the sendust layer, according to the expression [8.8]:

$$\langle \sigma_m \rangle = \frac{4Et_s^2(h_a - h_b)}{3L^2(1 - v)p} \tag{8.43}$$

where E is Young's modulus of elasticity in N/m^2, in the x-direction (defined by $\sigma_x = E_x \varepsilon_x$, where ε_x is the strain) when no contractions in the transverse directions (y and z) are present. Further, v is Poisson's ratio, i.e. the ratio between transverse (y and z) and longitudinal (x) strain. Other quantities are defined in Fig. 8.23.

The above expression applies to a beam (small b/d ratio) and not to a plate (large b/d ratio), and in fact isotropy has been assumed. Baratta [8.9] explained that for ratios smaller than, say 20, and strains smaller than, say 10^{-3}, the beam approximation is very accurate for both metal-

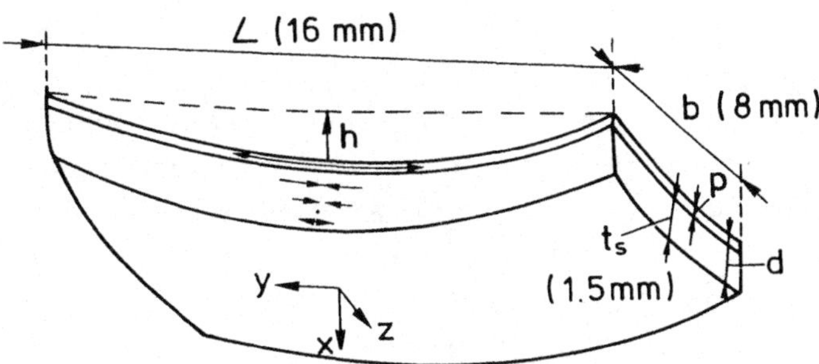

Fig. 8.23. Sketch of an 8X1 (MnZn ferrite) tile used in the experiments and definition of quantities used in the expression for the average stress in the thin layer (thickness p), sputtered on top of the ferrite.

lic materials ($\nu \approx 0.3$) and brittle non-metallic materials ($\nu \approx 0.25$).

Further it is assumed in the above expression that the metal layer is much thinner than the substrate.

In calculating the stresses given in Fig. 8.24a as a function of the substrate heating for different crystal orientations, we took account of the fact that the elasticity modulus depends on this orientation (although in fact (8.43) was deduced for isotropic materials). When no substrate heating is applied, high compressive ($\sigma < 0$) residual stresses build up in the sendust layer, resulting in low (in-plane) permeabilities; see Fig. 8.24b. Annealing (not shown) usually reduces the stresses and enhances the permeability especially at low substrate temperatures. When the normal component of the permeability is of the same order, then the permeability of even the stressed sendust would suffice in an actual MIG head, provided the sendust layer is not too thick ($\lesssim 5\,\mu$m). This corresponds to the experimental observations on MIG heads concerning the undulations, as explained earlier. Hence, for MIG heads too, it is only necessary to take into account the balancing stresses in the ferrite, which is much more sensitive to stresses. Consequently the only effect is a reduction of the efficiency. Since the balancing stress results from the residual stress in the sendust, some understanding of the causes of this stress is desirable.

The final stress in the sendust metal layer is a result of

1) differences in the temperature of metal layer (sendust) and substrate (ferrite) during sputtering;

2) differences between the above temperatures and the final (room) temperature;

3) differences in the thermal expansion coefficient of metal and substrate;

4) bombardment of Ar atoms and ions on the film surface during the sputter process;

5) differences between lattice constants of sendust and ferrite.

For a cool substrate and a hot process, like sputtering, the temperature effects always lead to compressive balancing stress ($\sigma < 0$) in the ferrite, because of shrinking of the surface layer after the hot process is over. Since the expansion coefficient of sendust ($+16 \cdot 10^{-6}°C^{-1}$) is larger than that of ferrite ($+11 \cdot 10^{-6}°C^{-1}$), the substrate temperature needs to be 16/11 times higher than that of the layer during sputtering in order to arrive at a vanishing compressive balancing stress. At least

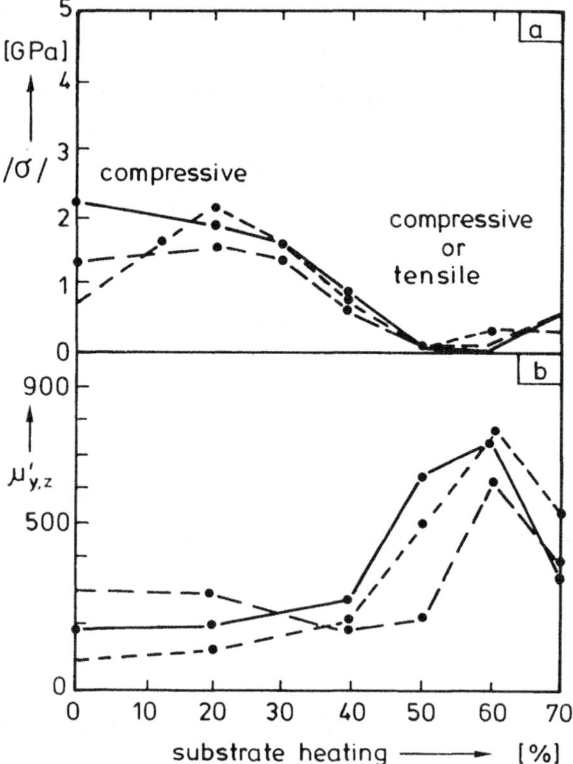

Fig. 8.24. Correlation between the absolute value of the residual stress in sendust layers of about 3 μm thickness (a) and the approximately equal in-plane relative permeability components μ_y and μ_z of the sendust layer (b). Crystallographic orientations (x-axis, see Fig. 8.23) and values of the modulus of elasticity and Poisson's ratio used in the stress determination are:

—— [100] $E_x = 130$, $\nu = 0.3$
--- [211] $E_x = 227$, $\nu = 0.3$
----- [111] $E_x = 189$, $\nu = 0.3$

A substrate heating of 0% corresponds approximately to a substrate and layer temperature of 100°C, 40% to 200°C and 60% to 450°C.
(measurements Sillen/Bode).

at relatively low substrate temperatures when radiation is negligible, this is impossible, because of the kinetic energy of atoms reaching the surface (ranging from some eVs to some tens of eVs) and because of the cohesive energy (some eVs) heating the layer. The observed balancing stresses, see Fig. 8.24, are not compressive, however, but tensile, and most severe without substrate heating. This strong compressive residual stress in the absence of substrate heating is believed to be caused by energetic Ar atoms and ions from the sputtering gas (becoming incorporated in sites with an insufficient volume) as well as from local radiation damage [8.10].

At higher (substrate) temperatures most of the 'Ar atoms' quickly become detached from the surface during the sputter process. In addition, a smaller compressive balancing stress adds to the tensile balancing stress due to the temperature effects. The stress-compensating effect of argon entrapped in sendust is applied by Saito and Mori in a thin-film magnetic head [8.11].

Observed structural changes in the sendust after heating of the substrate may also play a role in the final stress.

Influence of stress on the rotational susceptibility and the efficiency

In the following rough estimate it is assumed that the stresses only vary with x. End effects are thus neglected, and so too are changes of the cross-sectional area over the stressed region. Then, because of equilibrium of forces, the integral of the balancing stress in the ferrite, σ_f, must simply oppose the integral of the residual stress in the metal, σ_m:

$$\int_{\text{ferrite}} |\sigma_f| \, dx = \int_{\text{metal}} |\sigma_m| \, dx. \tag{8.44}$$

We are not interested in very weak effects, hence the stress-induced anisotropy energy $K_\sigma = c\,\sigma\,\lambda$, where λ is the magnetostriction constant and c a constant of the order of 1 (c depends on the crystal symmetry and type of stress), is large compared with the crystal anisotropy energy $K \equiv \frac{1}{2}\mu_0 M_s H_k$, where H_k is the crystal anisotropy field. So the crystal anisotropy field necessary to rotate the magnetization from the easy direction to the hard direction, $H_k = 2K/\mu_0 M_s$, is replaced by the much larger magnetic stress-induced stiffness or anisotropy field $H_{k\sigma} = 2K_\sigma/\mu_0 M_s = 2c\sigma\lambda/\mu_0 M_s$. The rotational susceptibility, χ_r, thus decreases from M_s/H_k to $M_s/H_{k\sigma} = \mu_0 M_s^2/2c\sigma\lambda$.

Since the susceptibility is assumed to be much greater than 1, the expression for the permeability is the same.

We have neglected the tensor character of the stress and the magnetostriction, due to the crystal structure of the ferrite, and also the frequency dependence of the rotational permeability, due to the gyromagnetic ratio, γ, and damping, α. In appendix 8.1, see also [8.12], it is shown that for biaxial tensile stresses in the (001) plane of MnZn ferrite (8X1), one can write

$$\hat{\chi}(\omega) \simeq \chi_{r0} \frac{\{1 - (\omega/\omega_{\text{res}})^2\} - j\alpha\omega/\omega_{\text{res}}\{1 + (\omega/\omega_{\text{res}})^2\}}{\{1 - (\omega/\omega_{\text{res}})^2\}^2 + \{2\alpha\omega/\omega_{\text{res}}\}^2}$$
$$\text{(001)}$$

$$(8.45)$$

where $\chi_{r0} = M_s/H_{k\sigma} = \gamma M_s/\omega_{\text{res}} = \mu_0 M_s^2/3|\lambda_{100}\sigma|$ is the low frequency (dc) rotational susceptibility and $\omega_{\text{res}} = \gamma H_{k\sigma}$ is the stress-dependent (and hence place-dependent) gyromagnetic resonance frequency.

At dc the same result is valid in the case of biaxial compressive stresses in the (001) plane for $\hat{\chi}_{[001]}$, i.e. for the rotational susceptibility in the [001] direction (easy plane (001)) instead of the rotational susceptibility components $\hat{\chi}_{(001)}$ in the (001) plane (easy axis [001]) in expression (8.45). The ac behaviour is much worse since in the extreme case of vanishing anisotropy in the (001) plane, the resonance frequency ω_{res} approaches zero, as explained in Appendix 8.1. The accuracy of this expression in an actual situation with demagnetizing fields is better than that of the corresponding expression for the case where there is no (strong) stress-induced anisotropy, since the latter is much more strongly affected by demagnetizing field components in the direction of the easy axis, which must in the latter case be added to the small H_k instead of the much larger $H_{k\sigma}$ in the present case. In the present case, however, the resonance frequency and hence χ_{r0} is strongly dependent on the stress and thus on the distance to the gap, x. At low frequencies, i.e. $\omega \ll \omega_{\text{res}}$, only the variation of χ_{r0} with x is important. Since at large stresses (before effects on the efficiency become noticeable) ω_{res} is high compared to the video frequencies, this is a realistic assumption.

This simplification makes it very easy to a calculate the increase of the core reluctance, ΔR_{int}, with the assumption that magnetization changes take place only by rotation magnetization. This is the case, for instance, in a MnZn MIG head when the x axis, perpendicular to the gap plane, is taken to be the [001] axis as in the V2000 video heads, and compressive balancing stresses ($\sigma_f < 0$) are assumed in the y-z plane, since $\lambda_{100} < 0$ and consequently the y-z plane is an easy plane (see appendix 8.1), while wall displacement is completely neglected.

Hence, for $\omega \ll \omega_{res}$:

$$\frac{\Delta R_{int}}{R_g} \simeq \frac{\displaystyle\int_{ferrite} \frac{1}{\mu_x(x)}\,dx}{g} \simeq \int_{ferrite} \frac{3|\sigma_f \lambda_{100}|}{\mu_0 M_s^2 g}\,dx$$

$$= \frac{3|\lambda_{100}|\,|\langle\sigma_m\rangle|}{\mu_0 M_s^2 g}\,(p_1 + p_2), \qquad (8.46)$$

where $\langle\sigma_m\rangle$ is the average residual stress in the sendust layers with thicknesses p_1 and p_2, left and right of the gap.

The reason why ΔR_{int} could be evaluated so easily is that ΔR_{int} is proportional to the integral over the stress (assumed $\omega \ll \omega_{res}$ etc).

In the ferrite we have tensile balancing stresses when no substrate heating is applied, instead of the assumed compressive stresses. So the x-axis, perpendicular to the gap plane, is an easy axis and consequently wall displacements dominate the magnetization processes near the gap. Hence effects on the efficiency are less severe than obtained by substituting (8.46) into (8.42b), i.e.

$$\boxed{\frac{\eta'}{\eta} > \frac{1}{1 + \dfrac{3|\lambda_{100}|\,|\langle\sigma_m\rangle|}{\mu_0 M_s^2 g}\,(p_1 + p_2)\,\dfrac{\eta}{\eta_{int}}}} \, . \qquad (8.47)$$

As an example, we assume a biaxial tensile stress $\langle\sigma_m\rangle = 0.2$ GPa $= 0.2 \, 10^9$ N/m^2 in the two sendust layers with 3 μm thickness each, in a MnZn MIG head with $\lambda_{100} = -1. \, 10^{-5}$ and $M_s = 400$ kA/m ($= 0.5$ T). The x axis, perpendicular to the gap, is taken to be the [001] direction, hence the (001) plane is an easy plane. η'/η is then given by the right-hand member of (8.47) when domain wall displacements are neglected. For actual values of g and η/η_{int}, being 0.3 μm and 2/3 respectively, the ratio $\Delta R_{int}/R_g$ then becomes 0.6 and η'/η equals only 0.7. (For simplicity no phase shifts are assumed.)

Although the stress calculations in this section are not accurate, it is shown that stresses in the surrounding of the gap, with integral values larger than 1 GPa μm, may have a considerable influence on the efficiency of MIG ferrite heads. In sendust MIG heads this explains the observed low efficiencies. NiFe MIG heads have efficiencies equal to those of pure MnZn ferrite heads. Consistent with the above description the stresses in NiFe layers are much lower than those in the sendust layers obtained without substrate heating.

Saturation also causes the permeability near the gap to be low, and can therefore be treated analogously to the effects of stress, once $\mu(x)$ has been calculated. This will be carried out in the following sections.

8.1.7 Saturation

8.1.7.1 *Importance of saturation effects for determination of efficiencies*

Important information about the flux efficiency, η_Φ, and (field) efficiency, $\eta_{(H)}$, follows from the inductance change, ΔL, of a head when the head is being saturated in the gap region and from the current at which this happens, I_s.

This saturated state can be achieved by different means, which do not all give such detailed information about the efficiencies. We will restrict ourselves, for application to video heads, to saturation achieved by applying high currents, say five to ten times the saturation current, I_s, to the windings of the head. This method works because of the special shape of a video head as explained later. For this type of head it is usually not necessary to achieve saturation by placing a source of magnetic field, for instance a current-carrying winding, around or in the neighbourhood of the gap region. The already present windings of the coil around the legs of the core can be used.

With respect to currents through the windings, three methods are available; see also Fig. 8.25.

Method a.

A dc current, I_{dc}, is used to saturate the head.

An ac current of frequency f_1 with small amplitude, $|\hat{I}_{ac}(f_1)| \ll |\hat{I}_s(f_1)|$, is used to measure L.

Method b.

An ac current I_{ac}, of frequency f_2 is used to saturate the head.

An ac current of frequency f_1 with small amplitude, $|\hat{I}_{ac}(f_1)| \ll |\hat{I}_s(f_1)|$, is used to measure L.

$L(f_1)$ must be measured 'selectively' because of interference with f_2.

Method c.

The ac current that saturates the head is also used for the $L(f_1)$ measurement, i.e. $f_1 = f_2$.

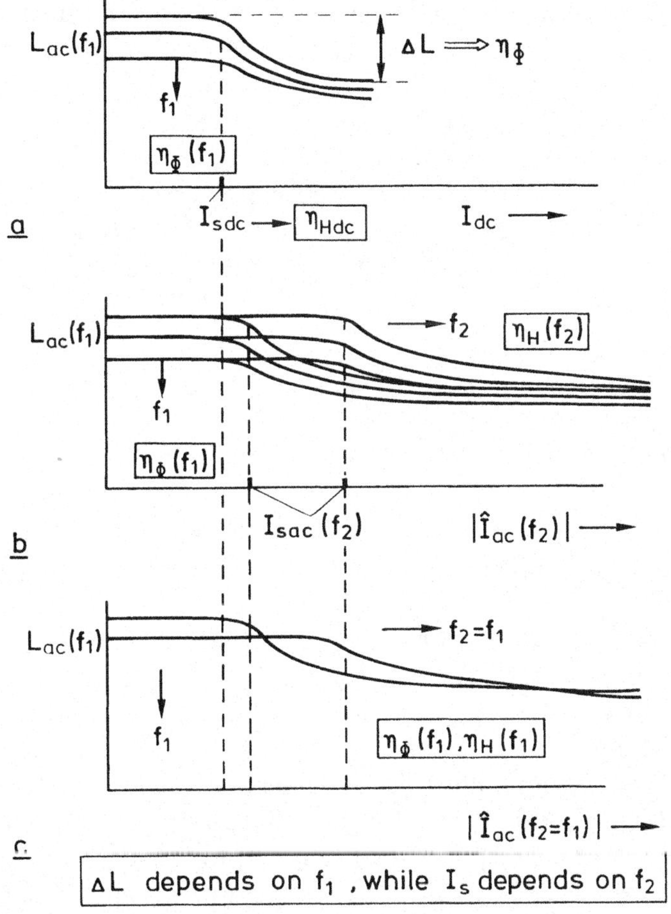

Fig. 8.25. Three methods that give information about the efficiencies of a video head by measuring only the effects of saturation on the head impedance or inductance:
a) Saturation achieved by a direct current I_{dc}.
b) Saturation achieved by an alternating current, $I_{ac}(f_2)$, with a frequency that differs from the frequency f_1 at which the inductance is measured.
c) Like b), however $f_1 = f_2$.

$L(f_1)$ must also be measured 'selectively', but now because of the higher harmonics that occur when $|\hat{I}_{ac}(f_1)|$ exceeds $|\hat{I}_s(f_1)|$.

Clearly, the last method requires fewer measurements to get information about both η_H and η_Φ and is therefore chosen. In section 8.2 this new method of measuring efficiency is implemented and applied to various heads. Instead of L, simply the absolute value of Z is determined there from the ratio $|\hat{V}(f_1)|/|\hat{I}(f_1)|$. This does not influence the results noticeably when $Q \geqslant 3$, since $|Z| = \omega L\sqrt{1 + 1/Q^2}$.

In the following section some theoretical background information is given about the saturation process in a video head, in order to get some insight into the accuracy of the proposed efficiency measurement.

8.1.7.2 General theoretical considerations

When the head saturates near the gap at increasing currents, then the flux Φ_g that traverses the gap will no longer increase linearly but will only slowly exceed the value at the onset of saturation:

$$\Phi_{gs} \equiv B_s h_0 W = \mu_0 M_s \frac{\mu}{\mu - 1} h_0 W, \tag{8.48}$$

where μ is the relative permeability.

Thus, for a sinusoidal current with frequency f_1 and amplitude $|\hat{I}| > 2|\hat{I}_s|$, the gap flux is roughly block-shaped. Therefore, the amplitude of the first harmonic of the distorted flux, $|\hat{\Phi}|$, is approximated by $(4/\pi)\mu_0 M_s h W$ when $|\hat{I}| > 2|\hat{I}_s|$.

In a head, saturation starts at regions with the highest flux density. In a video head these are the regions F and F' near the gap, except when the flux efficiency is very low. In video heads the reluctances Z_F and $Z_{F'}$ become so large when $|\hat{I}| \gg |\hat{I}_s|$ that Φ_g becomes negligible relative to the other, more or less constant, (stray) flux components. The resulting electric impedance is called Z_{fs}, where fs stands for 'full saturation', and corresponds to the asymptotic values in Fig. 8.25. In contrast, the subscript s was used for quantities at the onset of saturation. Full saturation is relative here: it does not mean $|\hat{I}| \to \infty$, but $|\hat{I}|$, say, between 5 and 10 times $|\hat{I}_s|$, in order to prevent the head from reaching saturation at other places than the gap region at those currents. The dominating stray-flux components (through R_{ss}, $R_{s's'}$ and

$R_{cc'}$) are then not influenced too much and the model used to determine the efficiency values from ΔL and $|\hat{I}_s|$ is relatively simple.

These dominating stray-flux components will be influenced at lower $|\hat{I}|/|\hat{I}_s|$ values when the flux efficiency η_Φ is higher, and the ratio of the cross section of the core at the coil and gap, $A_{coil}/A_{gap} = W_1(b_4 - b_3)/(2Wh_0)$ is larger. The above effects oppose each other when $W_1(b_4 - b_3)$ is kept constant and W or h_0 is changed, and may even cancel each other when η_Φ is small.

In contrast, the effects strengthen each other when $W_1(b_4 - b_3)$ is reduced and Wh_0 is kept constant. In that case the reduction of the ratio of the cross-sections also causes a reduction of η_Φ, by way of a reduction of the core permeance. This can be a particular problem with amorphous heads for which hold $W_1 = W$. In Fig. 8.26 the ratio $\eta_\Phi A_{coil}/A_{gap}$ is plotted as a function of the gap length at different track widths and core permeabilities. The other parameters are as usual for an amorphous head (values given in Table 8.8). When the plotted ratio exceeds 5, then at the highest I/I_s values (about 5) during the 'efficiency' measurement, the head may saturate in the neighbourhood of the coil. This

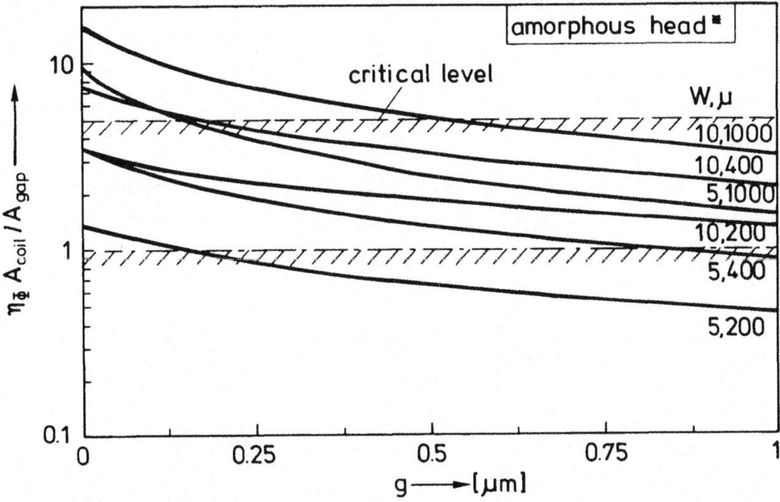

Fig. 8.26. The ratio $\eta_\Phi A_{coil}/A_{gap}$ as a function of the gap length at different track widths and permeabilities for convenient * amorphous head parameters. The ratio of the coil surface to the gap surface equals $(b_4 - b_3)/(2 h_0)$, since $W_1 = W$.
Above the critical level (dashed horizontal line, $y = 5$) the simple saturation model holds for currents to at least 5 times the saturation current, and η_H and η_Φ can be derived from curves like those in Fig. 8.25. Below the chain-dotted line a determination of the saturation current I_s and corresponding efficiency η_H from curves like those depicted in Fig. 8.25 is completely impossible. In between the two levels only η_H can be determined reliably.
* See Table 8.8.

obviously may happen when $W \leqslant 10\ \mu$m and $g \leqslant 0.3\ \mu$m, see Fig. 8.26, especially when the permeability of the amorphous material, μ, is poor. In practice amorphous heads with $W_1 = W \leqslant 10\ \mu$m indeed often show $L(|\hat{I}|)$ or $|Z(|\hat{I}|)|$ curves at higher frequencies, where μ is low, that differ from those sketched in Fig. 8.25, see Fig. 8.27, such that the curves beyond saturation fit less well a hyperbola (that the simple model leads to the hyperbola is explained in the following section) see Fig. 8.27. The model for the saturation process is then too complicated to describe analytically, because the saturation process is strongly dependent on head geometry and permeability (assumed to be unknown before the efficiency measurement).

However, when the ratio is still clearly larger than 1, it is possible to estimate I_s, and consequently η_H, directly from the $L(I)$ curve instead of calculating I_s from the intersection of the (high-current) hyperbola with the (low-current) horizontal, as explained later on. It is not possible anymore to estimate η_Φ from ΔL, because ΔL cannot then be estimated reliably. This is no problem when the gap surface is known, because Wh_0 relates $|\eta_\Phi|$ to $|\eta_H|$, as worked out in detail later on.

If all stray flux of a head were not influenced by changes in the permeance of the core, then the equivalent circuit of the head in Fig. 8.3 could be reduced to an ideal source of magnetic potential NI and two parallel reluctances, R_{stray} and $Z_{\text{core}} + R_{\text{gap}}$. When μ'' is neglected, then we have the simple equivalent circuit of Fig. 8.28. From this circuit it follows for increasing currents above $2|\hat{I}_s|$ that $|\hat{\Phi}_{\text{gap}}|$ does not increase noticeably any further when the material is assumed to be ideal as in Fig. 8.29. Thus, only the part of the head impedance due to Φ_{stray} remains constant and the part due to Φ_{core} is inversely proportional to

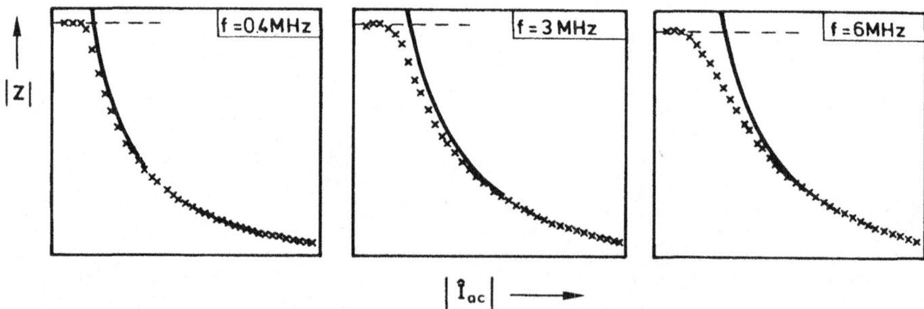

Fig. 8.27. Demonstration, by way of measurements on a narrow track width (10 μm) amorphous-ribbon head, of the effect that the hyperbola fits less well according as the frequency increases (i.e. the permeability decreases), due to saturation far away from the gap region at larger currents.

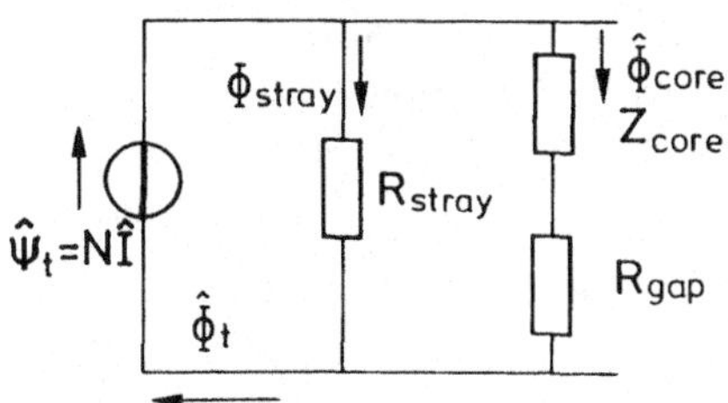

Fig. 8.28. Simplest circuit representation of a head with stray flux. The validity of this representation is usually restricted to high (field) efficiency heads, i.e. $|Z_{core}| \ll R_{gap}$.

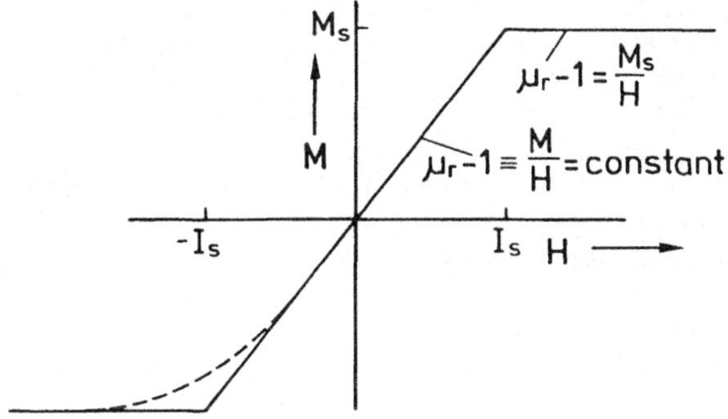

Fig. 8.29. Idealized MH curve as used in the description of geometric effects in the saturation process of a (video) head.

the currents, i.e. the $L(|\hat{I}|)$ for $|\hat{I}| > 2|\hat{I}_s|$, see Fig. 8.25, is part of a hyperbola with a vertical asymptote that coincides with the L axis and with a horizontal asymptote represented by the line $L = L_{fs}$. This derivation is in more detail carried out in Section 8.2. Here it is the aim to check the validity of the hyperbola approach for heads, including those corresponding to the more realistic circuit of Fig. 8.3. For this purpose more detailed knowledge of the saturation process is necessary.

8.1.7.3 Geometric effects in the saturation process

How saturation takes place in the core of a video head can be calculated analytically and rather simply when the following assumptions and approach are introduced:

1) The flux that leaves region F as stray flux, also when this region has been saturated, is neglected during the evaluation of the reluctance of the saturated region and the potential decay in this section.

2) For the highest I/I_s ratios considered it is assumed that the saturation is still restricted to the regions F and F', or, when the bridge length is zero, restricted to the regions E and E'.

3) For the core near the gap, where saturation may take place, the idealized MH curve, depicted in Fig. 8.29, is assumed.

4) The approximation is quasi-static and μ'' is neglected in $\hat{\mu} = \mu' - j\mu''$.

The first assumption is always necessary in an analytical approach where a 3-dimensional configuration is divided into a finite number of components and was already applied in the evaluation of the unsaturated head. After calculation of all core and stray reluctances, the core and stray fluxes can be calculated. Then it can be checked if the stray flux that emanates from a section was indeed negligible with respect to the core flux that traverses the section. If this is not the case, the section can be divided into more subsections.

Since the potential decay over the saturated region is usually large, the stray permeance is calculated by using the potential difference as a weighting function, analogously to the calculation of the stray permeance of the coil in (8.10), such that one permeance instead of a large number of circuit elements suffices. The decay under extreme saturation conditions will even turn out to be so large that the main contribution to the stray flux comes from the high-potential side of the saturated region and not from the gap side. For this reason the stray permeance is placed over the saturated region instead of over only the gap and consequently defined in this way. For less extreme saturation conditions, when the ratio of the stray flux of the saturated region to the gap flux is reduced, this choice is less relevant.

The second assumption is not necessary for obtaining explicit expressions, but simplifies the results. For I/I_s ratios of interest. say $I/I_s < 10$, the assumption will turn out to hold usually.

The third assumption, see BH curve in Fig. 8.29, is accurate at say $|I| < \frac{1}{2}I_s$ and $|I| > 2I_s$. The determination of the value of I_s by asymptotes is therefore not influenced by this idealization.

Application of this BH curve assumes an instantaneously-reacting material, at least in the region that saturates.

The fourth assumption is necessary to keep the description simple. An accurate dynamic description of the first harmonics with the aid of complex quantities would be difficult or impossible, because of the non-linearities. This means that in the following calculations all impedances and efficiencies etc. must be interpreted as real quantities.

Because of assumptions 1 and 3 the flux in the saturated region, $\Phi(x) = B(x)A(x)$, must be constant. ($A(x)$ is the x-dependent cross-sectional area of the core near the gap). When x_s is the distance from the gap centre where saturation starts and μ_m is the relative permeability corresponding to the linearly increasing part of the MH curve, then we can write for $x \leqslant x_s$

$$B(x)A(x) = B_s(x_s)A(x_s) = \mu_0 M_s(x_s) \frac{\mu_m}{\mu_m - 1} A(x_s). \qquad (8.49)$$

Hence the relative permeability, $\mu \equiv B/\mu_0 H$, equals in the saturated region

$$\mu(x) = \frac{B(x)}{B(x) - \mu_0 M_{\mathrm{s}}(x)} = \frac{1}{1 - \dfrac{A(x)}{A(x_{\mathrm{s}})}\dfrac{\mu_{\mathrm{m}} - 1}{\mu_{\mathrm{m}}}} \simeq \frac{1}{1 - \dfrac{A(x)}{A(x_{\mathrm{s}})}} . \qquad (8.50)$$

because of assumption 3 (which includes that M_{s} is x-independent) and (8.49). Since $|\mu_{\mathrm{m}}| \gg 1$, the factor $(\mu_{\mathrm{m}} - 1)/\mu_{\mathrm{m}}$ is approximated by 1. This approximation leads near x_{s} to $\mu(x) \to \infty$, which is of course incorrect, but does not influence the reluctance of the saturated section noticeably. *The permeability $\mu(x)$ in the saturated region is thus completely determined by the geometry of the head in that region (except near x_{s}) and not by the initial permeability!* Analogously to (8.3c) this reluctance is expressed as

$$R_{\mathrm{sat}}(x_{\mathrm{s}}) \simeq \frac{1}{\mu_0} \int_{x'=0}^{x'_{\mathrm{s}}} \frac{\mathrm{d}x'}{\mu(x')A(x')} \simeq \frac{1}{\mu_0} \int_0^{x'_{\mathrm{s}}} \frac{\mathrm{d}x'}{A(x')} - \frac{x'_{\mathrm{s}}}{\mu_0 A(x'_{\mathrm{s}})} ,$$

$$(8.51)$$

where $x' = 0$ corresponds to the gap surface ($x' \equiv x - g/2$) and the approximation of (8.50) is substituted.

In the following sections we will evaluate the integral (8.51) for two practical cases; a linearly-increasing and a 'quadratically'-increasing core cross-section $A(x')$. Even in these 'simple' cases the subsequent derivation of the relation between the gap flux and the current etc. is fairly elaborate. The results, however, are worthwhile. They give an insight into the important 'geometric' effects in the saturation process and into the accuracy of the use of the correction factor $4/\pi$ in the proposed efficiency measurement. A detailed reading of the following sections 8.1.7.4 + 5, that contain the above mentioned calculations, is not necessary for a good understanding of the limitations of the quick efficiency measuring method. Also of secondary importance for that purpose is the discussion in section 8.1.7.6 concerning the application of the results of the calculations to $V_{\mathrm{out}}(I_{\mathrm{write}})$ curves, and the experimental investigation of the applicability of the results to actual situations with the aid of a large-scale head having a more realistic MH-curve than the ideal curve assumed in the calculations in section 8.1.7.7. For a good understanding of the inaccuracies and limitations of the quick efficiency measuring method, the reading of section 8.1.7.8 suffices.

8.1.7.4 Case 1: Constant core width around the gap

When $A(x')$ increases linearly with x', according to $C_2(x_2' + x')$, as is the case for the regions F and F' with $C_2 = W \tan \theta_2$ and $x_2' = h_0/\tan\theta_2$, then it follows from (8.51) that

$$R_{\text{sat}}(x_n') = \frac{1}{\mu_0 C_2} \left[\ln \left(\frac{x_{2n}' + x_n'}{x_{2n}'} \right) - \frac{x_n'}{x_{2n}' + x_{sn}'} \right] \quad (0 < x_n' < x_{sn}') \qquad (8.52)$$

where the upper bound x' is freely chosen between 0 and x_s' instead of the fixed x_s' in (8.51), and dimensionless distances (normalized to half the gap length, i.e. $x_n' = x'/(g/2)$ etc.) have been introduced. The potential at x_n' with respect to the gap centre ($x_n = 0$) equals

$$\Psi(x_n') = \left(R_G + R_{\text{sat}}(x_n') \right) \Phi \qquad (8.53)$$

where $\Phi = \mu_0 M_s \cdot A(x_{sn}')$ is the maximum flux that can enter the saturated region at x_{sn}'. This results in

$$\Psi(x_n') = \frac{g M_s}{2} (x_{2n}' + x_{sn}') \left[\frac{1}{x_{2n}'} + \ln \left(\frac{x_{2n}' + x_n'}{x_{2n}'} \right) - \frac{x_n'}{x_{2n}' + x_{sn}'} \right]. \qquad (8.54)$$

Integration of the field $H(x_n') = M_s/[\mu(x_n')-1]$ with μ given in (8.50), also yields (8.54). The normalized length of the saturated regions, x_{sn}', follows for currents with absolute values that exceed I_s from solving the following equation iteratively:

$$I_n = \frac{2\Psi(x_{sn}')}{M_s g} = \frac{2R_{\text{sat}} + R_g}{R_g} \cdot \frac{A(x_n' = x_{sn}')}{A(x_n' = 0)}$$

$$= (x_{2n}' + x_{sn}') \left[\frac{1}{x_{2n}'} + \ln \left(\frac{x_{2n}' + x_{sn}'}{x_{2n}'} \right) - \frac{x_{sn}'}{x_{2n}' + x_{sn}'} \right], \qquad (8.55)$$

where the positive and dimensionless current $I_n \equiv \eta_s'|I|/\eta I_s$ is introduced and where $\eta_s' \equiv 2\,\Psi(x_{sn}')/NI$. The efficiency η_s' is slightly higher, because of the virtual shortening of the core, than the efficiency η_s of the head when the gap reluctance is replaced by the reluctance of the gap and the saturated regions, R_{gs}. When the reduction of the stray permeance at the saturated region is neglected, then η_s' can easily be calculated from the circuit representation in Fig. 8.3 for any chosen x_{sn}' analogously to η, by replacing R_g by $R_{gs} = 2R_{\text{sat}} + R_g$. Later we give approximations for the stray permeance over the saturated region. A quicker iteration is obtained by approximating η'_s/η by η_s/η, for which, on the basis of the Thevenin representation, it easily can be derived that

$$\frac{\eta_s}{\eta} = \frac{R_{gs}}{R_g} \cdot \frac{R_g + Z_{int}}{R_{gs} + Z_{int}}, \tag{8.56}$$

where $R_{gs} \equiv R_g + 2R_{sat}(x'_{sn})$ and $R_g \equiv 2R_G \simeq g/(\mu_0 W h_0)$ and Z_{int} is the internal impedance in the Thevenin circuit of Fig. 8.16 when $Z_{load} = R_g$. The following useful approximations hold:

$$x'_{sn} = \sqrt{1 + 2x'_{2n}(I_n - 1)} - 1 \qquad \text{for } x'_{2n} \gg x'_{sn} \tag{8.57a}$$

$$= \sqrt{2x'_{2n}(I_n - 1)} \qquad \text{when also } I_n - 1 \gg 1/(2x'_{2n}), \tag{8.57b}$$

where the subscript n stands for normalized as defined earlier. In video heads, $x'_{2n} \ggg 1$, such that in the interesting range, $1 < I_n < 10$, expression (8.57a) always holds.

The flux that traverses the saturated region increases proportionally with the surface at x'_{sn} and so increases proportionally with $x'_{2n} + x'_{sn}$, which leads after application of (8.57) to

$$\Phi_{gn} = \frac{x'_{2n} + x'_{sn}}{x'_{2n}} = 1 + \frac{\sqrt{1 + 2x'_{2n}(I_n - 1)} - 1}{x'_{2n}} \qquad \text{for } x'_{2n} \gg x'_{sn} \tag{8.58a}$$

$$= 1 + \sqrt{\frac{2(I_n - 1)}{x'_{2n}}} \qquad \text{when also } I_n - 1 \gg 1/(2x'_{2n}) \tag{8.58b}$$

where $\Phi_{gn} \equiv |\Phi_g|/\Phi_{gs}$, with $\Phi_{gs} \equiv \mu_0 M_s W h_0$ the gap flux at the onset of saturation.

x'_{sn} and $\Phi_{gn} - 1$ are determined by (8.57a) and (8.58a) respectively with an accuracy better than 7% in practical circumstances, i.e. for $1 < I_n(t) < 10$ and $x'_{2n} > 100$; see also Table 8.9.

Because of the second term between the brackets in (8.54), there turns out to be no way to express the stray reluctance P_{stray} of the saturated region as a finite number of elementary functions. With the practical assumption $x'_{sn} \ll x'_{2n}$ this problem vanishes. Then (8.54) leads to

$$\Psi(x'_n) = \frac{gM_s}{2} (x'_{2n} + x'_{sn}) \left[\frac{1}{x'_{2n}} + \frac{x'_n x'_{sn}}{(x'_{2n})^2} - \frac{1}{2}\left(\frac{x'_n}{x'_{2n}}\right)^2 \right] \qquad \text{for } x'_{sn} \ll x'_{2n}. \tag{8.59}$$

It will turn out that the main contributions to the stray permeance are from the parts of the saturated region that are located further away from the gap, where the potential approaches an I_n times higher value than just at the gap edge. For this reason the stray-permeance will be placed over $R_g + 2R_{sat}$ in the circuit representation and will consequently be defined as

Table 8.9. Case 1: Constant core width around the gap.

Expression used					*		(8.55)	(8.57a)	**	(8.52)		(8.61)$^{-1}$				(8.58a)
g	θ_1	θ_2	W	h_0	(x'_{2n})	I_n	x'_{sn}	$x'_{sn}\approx$	R_g	$R_{sat}(x'_{sn})$		R_{stray}		$\dfrac{R_{stray}}{2R_{sat}+R_g}$		$\Phi_{gn}\approx$
0.3	90°	45°	7.5	15	100	2	13.44	13.18	2.122×10^9	$0.8093 \cdot 10^9$	$0.7807 \cdot 10^9$	$2.5603 \cdot 10^{10}$	$2.6037 \cdot 10^{10}$	6.845	7.069	1.132
						4	24.38	23.52	2.122×10^9	$2.3511 \cdot 10^9$	$2.2089 \cdot 10^9$	$2.7175 \cdot 10^{10}$	$2.8440 \cdot 10^{10}$	3.982	4.349	1.235
						10	44.14	41.44	2.122×10^9	$6.3010 \cdot 10^9$	$5.6998 \cdot 10^9$	$2.8107 \cdot 10^{10}$	$3.1199 \cdot 10^{10}$	1.909	2.307	1.414
			30	60	400	2	27.60	27.30	0.133×10^9	$0.5776 \cdot 10^8$	$0.5655 \cdot 10^8$	$5.1979 \cdot 10^9$	$5.2420 \cdot 10^9$	20.92	21.30	1.068
						4	48.93	48.00	0.133×10^9	$0.1700 \cdot 10^9$	$0.1641 \cdot 10^9$	$5.9691 \cdot 10^9$	$6.1051 \cdot 10^9$	12.62	13.24	1.120
						10	86.70	83.86	0.133×10^9	$0.4787 \cdot 10^9$	$0.4514 \cdot 10^9$	$6.7254 \cdot 10^9$	$7.088 \cdot 10^9$	6.168	6.843	1.210

* $x'_{1n} \equiv W/(g \tan (\theta_1/2))$
$x'_{2n} \equiv 2h_0/(g \tan \theta_2)$

** $R_g = \dfrac{1}{\mu_0} \dfrac{g}{Wh_0}$ has been used

$$P_{\text{stray(sat)}} \equiv \frac{\Phi_{\text{stray}}}{I_n 2\Psi(x_n' = 0)} = \frac{1}{I_n \Phi_g R_g} \int_0^{x_{sn}'} 2\Psi(x_n')\,dP(x_n'). \tag{8.60}$$

Substitution of $\Phi_g R_g = gM_s(x_{2n}' + x_{sn}')/(x_{2n}')$, of the differential permeance $dP(x_n') \simeq \mu_0\,[(W + 2h_0)/\pi + W/(\pi - 2\theta_2)]dx_n'/(x_n' + 1)$ around the saturated region, and of expression (8.59) in the expression (8.60) yields:

$$P_{\text{stray}} = \frac{\mu_0}{I_n}\left(\frac{W + 2h_0}{\pi} + \frac{W}{\pi - 2\theta_2}\right) \times$$

$$\left[\left(1 - \frac{1}{x_{2n}'}\,(\tfrac{1}{2} + x_{sn}')\right)\ln(1 + x_{sn}') + \frac{1}{2x_{2n}'}\,x_{sn}' + \frac{3}{4x_{2n}'}\,x_{sn}'^2\right]. \tag{8.61}$$

Values of $R_{\text{stray}} = P_{\text{stray}}^{-1}$ and R_{sat} are given in Table 8.9 for heads with $W_1 = W$ (e.g. amorphous heads). The ratio of the flux that traverses the gap to the stray flux emanating from the saturated region, $\Phi_g/\Phi_{\text{stray}} = R_{\text{stray}}/(2R_{\text{sat}} + R_g)$, decreases with increasing values of the normalized current, I_n, according to the values in the table. This is as expected, since the outer area of the saturated region increases with increasing currents.

From the values of this ratio equal to $\Phi_g/\Phi_{\text{stray}}$, we may conclude that at least up to $I_n = 5$ it was justified to neglect P_{stray} in the calculation of R_{sat}. Above $I_n = 5$ a simultaneous calculation of the stray and core fluxes in this region is necessary, e.g. by way of a generalized transmission line description as is carried out in chapter 9 for probe-type heads. However these high I_n values are not of interest here, since they hardly limit the application of this model to the proposed efficiency determination which is further worked out in Section 8.2. The limit obtained in narrow-track amorphous heads at higher frequencies (lower permeabilities), due to the low η_Φ and small cross-sectional area ratio $A_{\text{coil}}/A_{\text{gap}}$, as depicted in Fig. 8.26, is more relevant in this context.

8.1.7.5 Case 2: Linearly increasing core width around the gap

Analogous equations can be derived for the case that the bridge length equals the gap length and so all saturation takes place in the regions E and E'. Now the surface increases more rapidly with x', according to $A(x') = C_{12}(x_1' + x')(x_2' + x')$, where $C_{12} \equiv 2\tan(\theta_1/2)\tan\theta_2$, $x_1' \equiv W/2\tan(\theta_1/2)$ and $x_2' \equiv h_0/\tan\theta_2$. The following equations have been derived:

$$R_{\text{sat}}(x_n') = \frac{2}{\mu_0 C_{12} g}\left[\frac{1}{x_{2n}' - x_{1n}'}\ln\frac{1 + x_n'/x_{1n}'}{1 + x_n'/x_{2n}'} - \frac{x_n'}{(x_{1n}' + x_{sn}')(x_{2n}' + x_{sn}')}\right] \tag{8.62}$$

$$\Psi(x_n') = \frac{g}{2}M_s(x_{1n}' + x_{sn}')(x_{2n}' + x_{sn}')\left[\frac{1}{x_{1n}'x_{2n}'} + \frac{1}{x_{2n}' - x_{1n}'}\ln\frac{1 + x_n'/x_{1n}'}{1 + x_n'/x_{2n}'} - \right.$$

$$\left. \frac{x_n'}{(x_{1n}' + x_{sn}')(x_{2n}' + x_{sn}')}\right] \tag{8.63}$$

$$I_n = (x'_{1n} + x'_{sn})(x'_{2n} + x'_{sn}) \left[\frac{1}{x'_{1n}x'_{2n}} + \frac{1}{x'_{2n} - x'_{1n}} \ln \frac{1 + x'_n/x'_{1n}}{1 + x'_n/x'_{2n}} - \frac{x'_{sn}}{(x'_{1n} + x'_{sn})(x'_{2n} + x'_{sn})} \right] \tag{8.64}$$

Because of the second term between brackets, no explicit expression for x'_{sn} exists. A useful approximation follows from (8.64), for $x_s \ll x_1, x_2$:

$$x'_{sn} = \frac{x'_{1n} + x'_{2n}}{2 + x'_{1n} + x'_{2n}} \left(\sqrt{1 + \frac{(4 + 2x'_{2n} + 2x'_{1n})x'_{1n}x'_{2n}}{(x'_{1n} + x'_{2n})^2} (I_n - 1)} - 1 \right)$$
$$\text{for } x'_{sn} \ll x'_{1n}, x'_{2n} \tag{8.65}$$

which leads to (8.57a) when $x'_{1n} \to \infty$.

Also when $x_1 = x_2$ is assumed, an explicit expression results for x'_{sn}:

$$x'_{sn} = \frac{x'_{2n}}{1 + x'_{2n}} \left(\sqrt{1 + (1 + x'_{2n})(I_n - 1)} - 1 \right) \qquad \text{for } x'_{1n} = x'_{2n}. \tag{8.66}$$

This follows from (8.65) as well, but without the knowledge that (8.66) is exact.

The normalized gap flux Φ_{gn} follows for these cases after substitution of (8.65) or (8.66) into

$$\Phi_{gn} = \left(1 + \frac{x'_{sn}}{x'_{1n}} \right) \left(1 + \frac{x'_{sn}}{x'_{2n}} \right), \tag{8.67}$$

see also (8.58).

Again because of the second term in the expression between brackets in (8.63), there turns out to be no way to express the stray reluctance P_{stray} of the saturated region as a finite number of elementary functions. When $x'_{sn} \ll x'_{1n}, x'_{2n}$ or when $x'_1 = x'_2$, this problem vanishes and (8.63) relaxes to

$$\Psi(x'_n) = \frac{g}{2} M_s(x'_{1n} + x'_{sn})(x'_{2n} + x'_{sn}) \left[\frac{1}{x'_{1n}x'_{2n}} + \frac{(x'_{1n} + x'_{2n})x'_{sn}}{x'^2_{1n}x'^2_{2n}} x'_n - \tfrac{1}{2} \frac{(x'_{1n} + x'_{2n})}{x'^2_{1n}x'^2_{2n}} x'^2_n \right]$$
$$\text{for } x'_{sn} \ll x'_{1n}, x'_{2n}, \tag{8.68}$$

which leads for $x_{1n} \to \infty$ to (8.59), and (8.63) relaxes to

$$\Psi(x'_n) = \frac{g M_s}{2} (x'_{2n} + x'_{sn})^2 \left[\frac{1}{x'^2_{2n}} + \frac{x'_n}{x'_{2n}(x'_{2n} + x'_n)} - \frac{x'_n}{(x'_{2n} + x'_{sn})^2} \right]$$
$$\text{for } x'_{1n} = x'_{2n}. \tag{8.69}$$

Especially in these cases it must be expected, because of the strong increase of the surface $A(x'_n)$ with increasing values of x'_n, that the main contribution to the stray flux of the saturated region will be from the part further away from the gap. Thus P_{stray} is again placed over the saturated region and so defined according to (8.60). Substitution of $\Phi_g R_g = g M_s(x'_{1n} + x'_{sn})(x'_{2n} + x'_{sn})/(x'_{1n} x'_{2n})$, $dP(x'_n) = \mu_0[W/\pi + W/(\pi - 2\theta_2) + 2h_0/(\pi - \theta_1)]dx'_n/(x'_n + 1)$ and (8.68) into (8.60) yields:

$$P_{stray} = \frac{\mu_0}{I_n}\left(\frac{W}{\pi} + \frac{W}{\pi - 2\theta_2} + \frac{2h_0}{\pi - \theta_1}\right)\left[(1 - C(\tfrac{1}{2} + x'_{sn}))\ln(1 + x'_{sn}) + \tfrac{1}{2}Cx'_{sn} + \tfrac{3}{4}Cx'^2_{sn}\right]$$

$$\text{for } x'_{sn} \ll x'_{1n}, x'_{2n}, \qquad (8.70)$$

where $C \equiv (x'_{1n} + x'_{2n})/(x'_{1n} x'_{2n})$.

For $x'_{1n} \to \infty$ expression (8.61) remains. Analogously it follows in case $x'_{1n} = x'_{2n}$ from (8.69) etc.:

$$P_{stray} = \frac{\mu_0}{I_n}\left(\frac{W}{\pi} + \frac{W}{\pi - 2\theta_2} + \frac{2h_0}{\pi - \theta_1}\right) \times$$

$$\left[\left\{\left(\frac{x'_{2n}}{x'_{2n} + x'_{sn}}\right)^2 - \frac{1}{x'_{2n} - 1}\right\}\ln(1 + x'_{sn}) + \left\{\frac{x'^2_{2n}}{x'_{2n} - 1}\right\} \times \right.$$

$$\left. \ln\left(\frac{x'_{2n} + x'_{sn}}{x'_{2n}}\right) - \left\{\frac{x'_{2n}}{x'_{2n} + x'_{sn}}\right\}^2 x'_{sn}\right] \quad \text{for } x'_{1n} = x'_{2n}. \qquad (8.71)$$

A comparison between the exact (8.71) and approximation (8.70) is now possible for $x'_{1n} = x'_{2n}$, see results in Table 8.10. For $x'_{sn} > 20$ or $I_n > 5$ the applicability of (8.70), and hence of (8.61), becomes doubtful. For about the same value of I_n, the calculation of P_{stray} subsequent to R_{sat} instead of calculating them simultaneously, becomes arbitrary; see Table 8.11 and Table 8.9. In addition, at $I_n = 10$ the (maximum) of the normalized gap flux, Φ_{gn}, in case 2 is so large that use of the factor $4/\pi$ would lead to extra inaccuracies.

The following differences between case 2 and case 1 follow from comparison of the results in Table 8.11 with those in Table 8.9:
1) The length of the saturated region, x'_{sn}, decreased noticeably. Consequently the potential decrease over the saturated region is steeper, i.e. the fields in and above the saturated region are stronger.
2) The gap flux, Φ_{gn}, increased.

This is to be expected in the case of a more rapidly increasing cross-sectional area around the gap. Qualitatively the same would happen if θ_2 were increased in case 1, i.e. $\theta_1 = 0°$.

What hardly changed is the ratio of the flux that traverses the gap to the stray flux that emanates from the saturated region, $R_{stray}/(2R_{sat} + R_g)$; see Tables 8.11 and 8.9. This means that the justification of calculating R_{sat} and P_{stray} subsequently is hardly influenced by the shape of the regions around the gap.

Table 8.10. Comparison between results obtained when $x'_{1n} = x'_{2n}$ with the approximation (8.70) and exact expression (8.71). Input parameters are: $g = 0.3$ μm, $\theta_1 = 90°$, $\theta_2 = 45°$, $h_0 = 15$ μm and $W = 30$ μm, such that $x'_{2n} = x'_{1n} = 100$.

x'_{sn}	I_n	$R_{stray}\,(8.70)^{-1}$	$R_{stray}\,(8.71)^{-1}$
100	104	11845	37581
40	17,96	11923	19392
20	5.44	11339	13715
10	2.21	10541	11093
4	1.2416	11861	11925
2	1.0804	16025	16038
1	1.0301	24619	24622
0.4	1.0096	49959	49959
0.2	1.0044	91792	91792
0.1	1.0021	175224	175223
0.04	1.0008	425288	425287
0.02	1.0004	841977	841977
0.01	1.0002	1675322	1675321

8.1.7.6 Application to the saturation curves $V_{out}(I_{write})$

In the present analytical calculations, variations of the magnetization over the cross-section of the core are neglected. For a small apex angle (i.e. large θ_2) this is of course untrue. Then the onset of saturation occurs at the apex of the coil chamber. This can only be treated numerically, which has recently been carried out by Valsteyn et al. [8.13]. They used the resulting field in front of the head, calculated for different apex angles and saturation conditions, to explain the effect qualitatively, that (under certain conditions concerning tape coercivity and the saturation magnetization and gap length of the head, etc.), large apex angles (small θ_2) lead to the well-known saturation curves (i.e. output from a tape as a function of the write current) without a maximum, so that no optimum write current is observed. Their numerical results show that a large region, away from the gap, begins to saturate at large apex angles, so that the field gradients at the pole tips hardly decrease.

Table 8.11. Case 2: Increasing core width around the gap.

Expression used				*		*		(8.64)	(8.65)	**	(8.62)		$(8.70)^{-1}$				(8.67)
g	θ_1	θ_2	h_0	x'_{2n}	W	x'_{1n}	I_n	x'_{sn}	$x'_{sn}\approx$	R_g	$R_{sat}(x'_{sn})$		R_{stray}		$\dfrac{R_{stray}}{2R_{sat}+R_g}$		$\Phi_{gn}\approx$
0.3	90°	45°	15	100	7.5	25	2	5.43	5.37	2.122×10^9	$0.5920\cdot10^9$	$0.5806\cdot10^9$	$2.3761\cdot10^{10}$	$2.4014\cdot10^{10}$	7.187	7.314	1.280
							4	10.15	9.93	2.122×10^9	$1.6786\cdot10^9$	$1.6214\cdot10^9$	$2.3112\cdot10^{10}$	$2.3813\cdot10^{10}$	4.218	4.439	1.536
							10	18.54	17.87	2.122×10^9	$4.0783\cdot10^9$	$3.8781\cdot10^9$	$2.2453\cdot10^{10}$	$2.3934\cdot10^{10}$	2.184	2.423	2.021
			60	400	30	100	2	11.76	11.67	0.133×10^9	$0.4900\cdot10^8$	$0.4831\cdot10^8$	$4.2969\cdot10^9$	$4.3257\cdot10^9$	18.60	18.84	1.149
							4	21.18	20.89	0.133×10^9	$0.1416\cdot10^9$	$0.1382\cdot10^9$	$4.5933\cdot10^9$	$4.6745\cdot10^9$	11.04	11.42	1.272
							10	37.74	36.89	0.133×10^9	$0.3736\cdot10^9$	$0.3602\cdot10^9$	$4.8883\cdot10^9$	$5.076\cdot10^9$	5.554	5.948	1.495

* $x'_1 \equiv W/(g\,\tan(\theta_1/2))$
$x'_{2n} \equiv 2h_0/(g\,\tan\theta_2)$

** $R_g = \dfrac{1}{\mu_0}\dfrac{g}{Wh_0}$ has been used.

From our model the flat saturation curve, in the case of a large apex angle, is explained in about the same way: at a large apex angle (e.g. $\theta_2 \to 0$), the saturated region spreads very rapidly into the core, away from the gap, because the increase of the surface with x is so small and consequently, according to (8.50), the permeability is still reasonable. So, the saturation is only weak over this large area (leading to about the same R_{sat} as in case of strong saturation over a small region, as would occur if the apex angle were small). Hence the gradient of the field near the gap and its absolute value are almost unaffected relative to the gradient and field at the onset of saturation. If the apex angle were small or θ_1 were large, then the saturated region would be small and consequently the saturation would be strong and the permeability small in this saturated region, so that the gradient near the gap would be strongly reduced and a clear optimum in the write current would be found. Then the adjustment of the write current is more critical.

8.1.7.7 *Application to quasi-static description and large-scale experiment*

Expression (8.58a) is applied in a quasi-static description of the saturation process, results of which are given in Fig. 8.30. For this purpose the current $I(t)$ changes sinusoidally, but relatively slowly, such that no phase shifts occur. For simplicity η'/η is assumed to be 1 for all instantanous values of $I(t)$, so that $I_n(t)$ is sinusoidal too, as depicted in Fig. 8.30. From the resulting periodic $\Phi_{gn}(t)$ signals the amplitudes of the first harmonics have been calculated. They are shown beside each curve in the figure. For practical circumstances, i.e. $100 \leqslant x'_{2n} \leqslant 400$ and $I_n \geqslant 2$, these values are close to the factor $4/\pi$ that applies to a purely block-shaped flux.

If the increase of η' with I is taken into account, then according to (8.58b), the deviation of the gap flux relative to Φ_{gs} will only be about $\sqrt{\eta'/\eta}$ larger than shown in the figure. In practical circumstances it hence seems allowed to assume a more or less block-shaped gap flux with a first-harmonic amplitude of $(4/\pi) \times (M_s W h_0)$.

In an actual head the shape of the *MH* curve is not as ideal as assumed in the calculations. In many magnetic materials the increase of the flux beyond saturation, due to this non-ideal curve, exceeds the calculated geometric effect.

An effort has been made to measure the gap flux of an actual head by placing a one-turn coil around the gap. Because of the very small dimension of the gap surface a lot of stray flux is linked to this turn, which makes the method unsuited for gap-flux observations. For this reason the same measurement was carried out on a large-scale model; see the set-up in Fig. 8.31. The core was made of 3E1 whose non-ideal magnetization curve can be expected to dominate the saturation process up to at least $I_n = 3$; see ballistic curve of this material, as reproduced in Fig. 8.32 from reference [8.14]. For this reason changing the gap length from 10 to 80 μm did not affect the $\Phi(t)$ curve noticeably; see Fig. 8.33. Subsequently, for a very extreme situation, $g = 170$ μm, i.e. $x'_{2n} = 10$, we checked whether the use of the $4/\pi$ ($= 2.1$ dB) factor still applied. If in this situation, the current is increased from the saturation current to 6 times the saturation current, then the amplitude of the first harmonic increases by 2.6 dB (1.35×). For this extreme situation and non-ideal material this is reasonably close to the factor $4/\pi$.

In order to approach the theoretical model better, large-scale heads have been made out of 8E1 ferrite, which possesses the more perfect *BH* loop, depicted in Fig. 8.34.

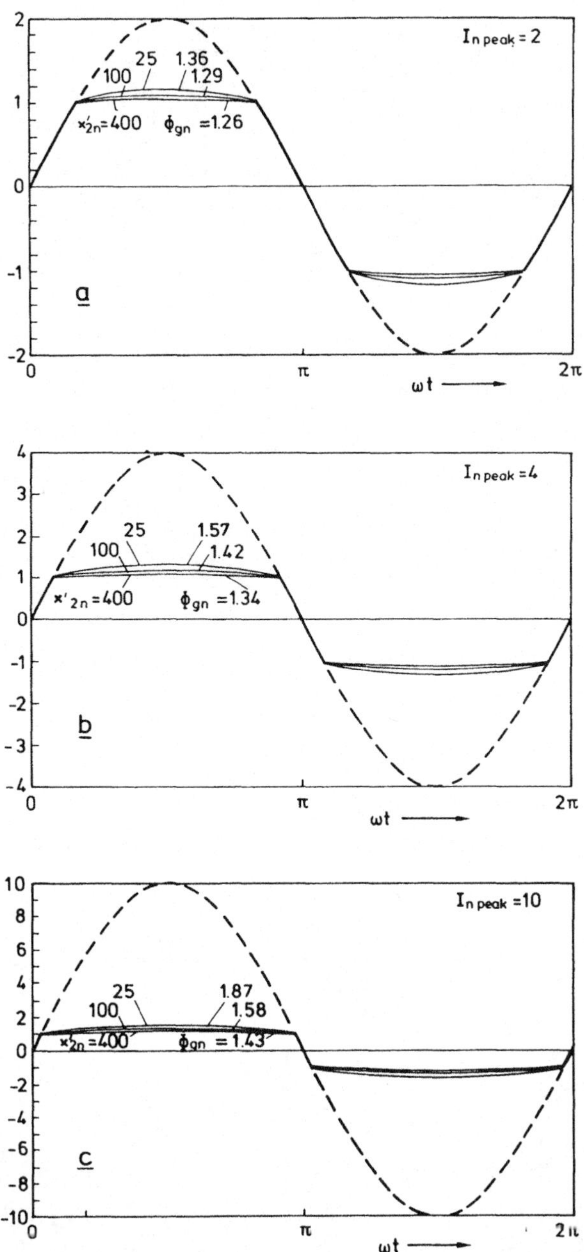

Fig. 8.30. The shape of the gap flux (full curves), $\Phi_g(t)$, due to geometric effects in the saturation process at different 'current levels', I_{peak}, and gap length, gap height and 'height-reducing angle' θ_2. $\Phi_{gn}(t) \equiv \Phi_g(t)/\Phi_{gs}$ where Φ_{gs} is the gap flux at the onset of saturation, $I_n(t) \equiv I(t)/I_s$ (dashed curves) and $x'_{2n} \equiv 2\,h_0/(g\tan\theta_2)$.
a) $I_{n\,peak} = 2$
b) $I_{n\,peak} = 4$
c) $I_{n\,peak} = 10$.

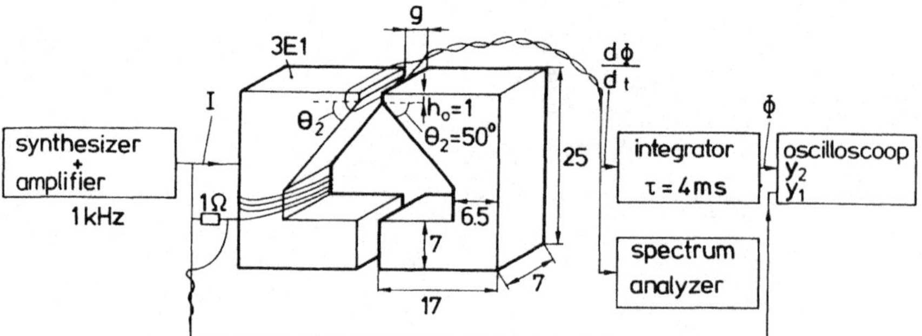

Fig. 8.31. Experimental set-up as used for observation of saturation effects near the gap and for determination of the correction factor in an extreme actual situation. Core dimensions are given in mm.

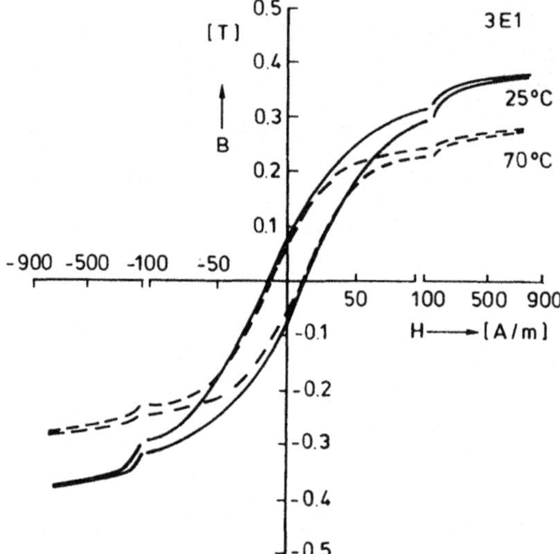

Fig. 8.32. Ballistic curve of the sintered MnZn ferrite, 3E1, after [8.14], used for the large-scale model in Fig. 8.31.

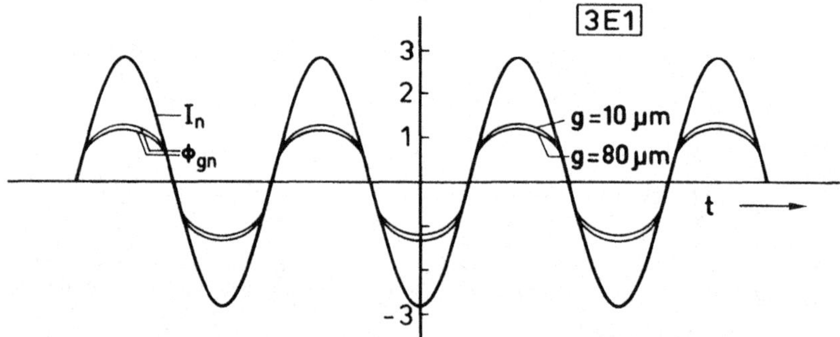

Fig. 8.33. Simultaneously visualized $I(t)$ and $\Phi(t)$ curves. The sensitivity for $I(t)$ is adjusted on the oscilloscope such that $I(t)$ and $\Phi(t)$ coincide in the linear region $I(t) < I_s$.

The dimensions of the tapered version of the large-scale heads are given in the new experimental set-up sketched in Fig. 8.35. The gap surfaces of the heads are reduced in comparison with the gap surface of the head in the previous experiments, in order to reduce the ampere turns and hence the power consumption and heat in the coil at a fixed value of I_n. The gap surface is still 1×1 mm^2, for the tapered head and 3×1 mm^2 when the track width equals the core width of 3 mm. In order during $\Phi_g I$-loop measurements to reduce the power consumption caused by excessive currents (up to $30I_s$!), further alternating pulses with very small duty cycle (about 0.1) are applied. The

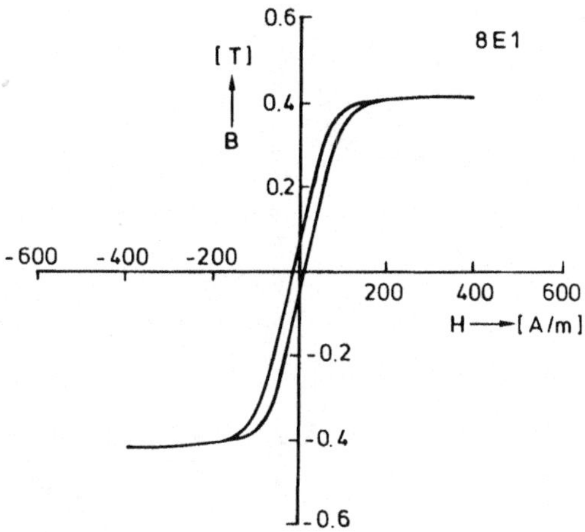

Fig. 8.34. The nice BH-loop of the sintered MnZn ferrite, 8E1, used for the large scale head in the set-up of Fig. 8.35.

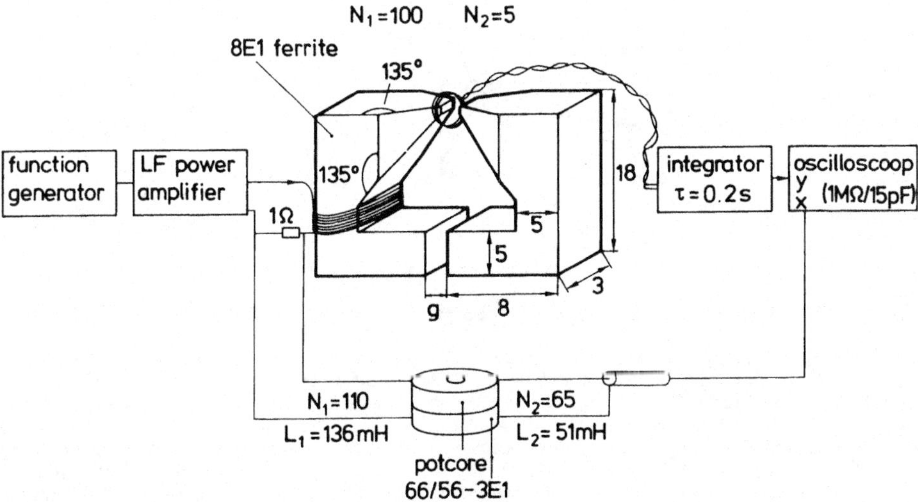

Fig. 8.35. Experimental set-up used for the more serious saturation-model verifications. The large-scale head has a $\theta_1 = 90°$, $\theta_2 = 45°$, $h_0 = 1$ mm, and a track width of 1 mm (tapered head) or 3 mm (= core width). Core dimensions are given in mm.

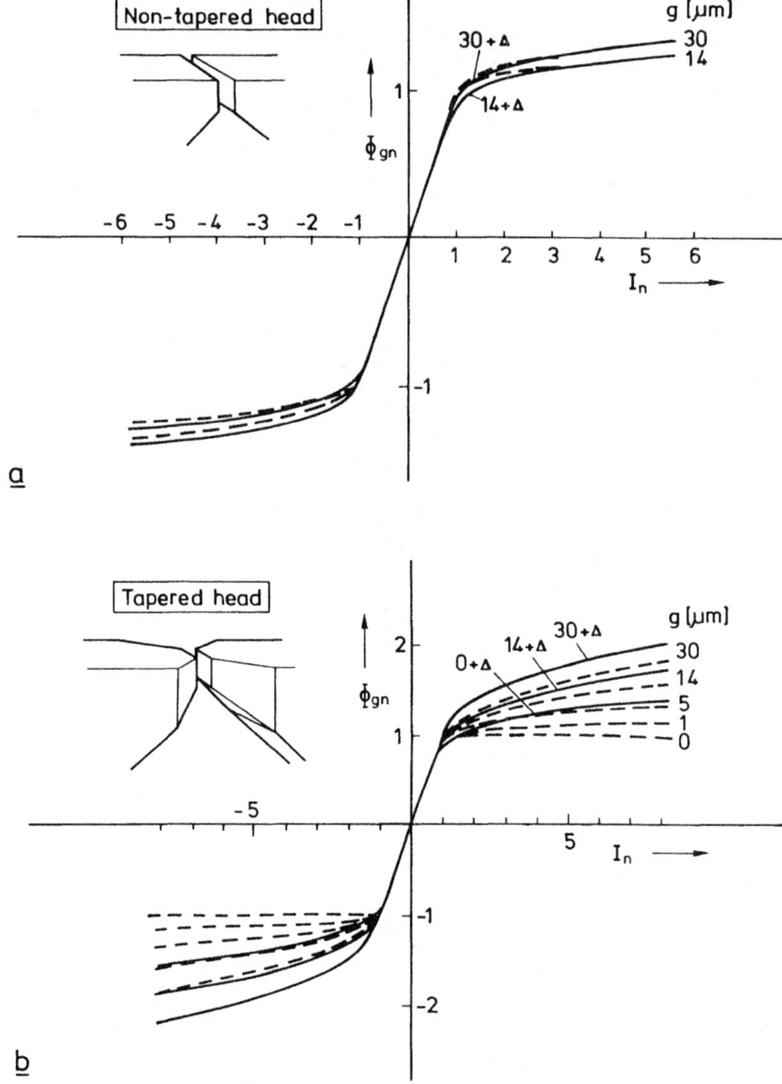

Fig. 8.36. Experimental (full curves) and theoretical (dashed curves) Φ_g-I loops at various gap lengths.
a) Non-tapered head (case 1).
($I_s \simeq C_1$, 3.3 C_1 and 6.6 C_1 for $g = 0 + \Delta$, $14 + \Delta$ and $30 + \Delta$ respectively.)
b) Tapered head (case 2).
($I_s \simeq C_2$, 2.4 C_2 and 3.6 C_2 in the above order.)
The experimentally observed small effect of coercivity at the smallest gap length is not drawn in this plot. However since this effect was relatively small compared to the coercive effect in the BH loop, given in Fig. 8.34, the efficiency must still have been reasonably high (> 50% at least), which means that the minimum gap length Δ at least must have been a few microns (certainly > 1 µm, given the initial relative permeability of ~ 3000 in Fig. 8.34 and the dimensions of the large-scale head).

pot core is used between the current-measuring resistor (1 Ω) and the high-impedance input of the x amplifier of the oscilloscope. This core serves as a necessary galvanic separation between the large-current input circuit and the low-voltage output circuit. The primary and secondary reactances ωL_1 and ωL_2 of the pot core are respectively very large and small with respect to the 1 Ω current-measuring resistor and the x-input impedance, so that hardly any phase shifts are introduced by the core at the main frequency components (between 500 Hz and, say, 30 kHz) of the pulses, which is important for accurate $\Phi_g I$-loop measurements.

Fig. 8.36 shows $\Phi_g I$ loops obtained with this set-up, for various gaps of the large-scale heads (full curves). The gaps are obtained with trapping foils of various thicknesses (14 and 30 μm) between the gap surfaces. Qualitatively the curves are predicted by the model (dashed curves), according to the expressions (8.65) and (8.67) for the tapered head and according to (8.58a) for the head without track width reduction. When no foil is placed between the gap surfaces the agreement seems bad. This has three main causes, which all work in the same direction.

1) The minimal effective gap length is still at least a few microns in the experiments. This is caused by the roughness of the gap surface 5 $\sim$ 10 μm) which is caused by the porosity of the sintered core material and the rough machining of the gap surface (the gap surfaces were not polished).

2) The 'efficiency' η_s', defined following (8.55), increases from the head efficiency η to the internal efficiency η_{int} (which approaches 1 for this configuration) and thus cannot be taken as a constant. This is only allowed at gap lengths larger than about 10 μm, and follows from the observation that I_s did not increase linearly with g for $g < 10$ μm; see also caption of Fig. 8.36.

3) The BH loop of the 8E1 material, given in Fig. 8.34, differs from the idealized loop used in the model.

The first cause can be avoided, partly, by carefully machining and cleaning the gap surfaces. The influence of the efficiency can be taken into account in the way outlined after (8.55) or by substituting (8.56) in the original definition of I_n, which took account of this effect. If this definition were applied to the experimental curves, as ought to be done, then the saturation part of the curve, especially for the lowest gap length, would expand in the x direction. This would flatten the experimental curve beyond saturation, but would only be meaningful if the actual effective gap length were known more accurately. In these respects no further attempts have been carried out.

The last figure in this series, Fig. 8.37, shows the corresponding saturation effects as a function of time when approximately sinusoidal currents are applied to the large-scale head. The peak value of I_n was limited, because of power consumption in the coil and distortion of the applied current in the strongly varying inductive load formed by the head.

8.1.7.8 Application to efficiency measurement

At the end of Sect. 8.1.7.2 it was shown that the impedance of a head, if its magnetic circuit can be modelled by only two parallel branches (Fig. 8.28), depends hyperbolically on applied currents I with ab-

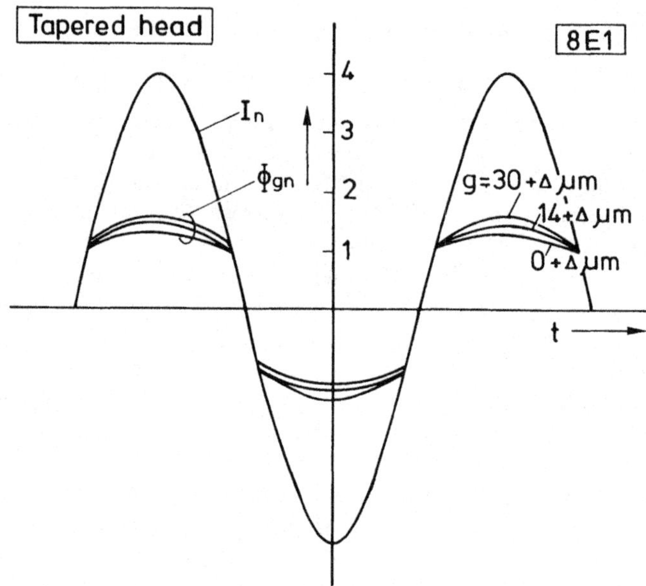

Fig. 8.37. Saturation effects as a function of time in the tapered 8E1-MnZn head, depicted in Fig. 8.35, when sinusoidal currents are applied.

solute values higher than the saturation current I_s.

In this section it will be shown that the hyperbola is a good approximation for actual heads as well, which must be modelled by the complicated circuits shown in Fig. 8.3. Further it will be investigated how the relative decrease of the (absolute) value of the head's electric impedance (or inductance, see Fig. 8.25) after full saturation is related to the (absolute) value of the flux efficiency.

The evaluations can be based on the complete circuit in Fig. 8.3, in which saturated regions near the gap are taken into account in the way outlined next to (8.55).

More suited for a discussion, and hardly less accurate, is an evaluation based on the Thevenin and Norton equivalent circuits (see Sect. 8.1.6) as will be carried out in this section.

Since the impedance of the head, Z, is proportional to the permeance of the head, P_{ab}, and we are interested only in the shape of the function $Z(I)$ for $|I| > I_s$, where the subscript s stands for saturation, i.e. $|I| > I_s$, we will focus on the ratio P_{abs}/P_{ab}. The reluctance in the gap region then increases in a good approximation from R_g to $R_{gs} \equiv R_g + 2\,R_{sat}$, see (8.56). Due to this reluctance change in the neighbourhood of the gap, the efficiencies, η_s and $\eta_{\Phi s}$, corresponding to this 'new gap reluctance', R_{gs}, change. The ratios η/η_s and $\eta_\Phi/\eta_{\Phi s}$, written in terms of fundamental equivalent circuit reluctances, follow easily from the Thevenin and Norton expressions (8.25) and (8.26) respectively. These ratios after being substituted in

$$\frac{P_{abs}}{P_{ab}} = \frac{\eta}{\eta_\Phi R_g} \cdot \frac{\eta_{\Phi s} R_{gs}}{\eta_s}, \tag{8.72}$$

see (8.27), give:

$$\frac{P_{abs}}{P_{ab}} = \frac{R_g + Z_{int}}{R_g + Z_{\Phi int}} \left(1 + \frac{Z_{\Phi int} - Z_{int}}{R_{gs} + Z_{int}} \right). \tag{8.73}$$

Note that Z_{int} was defined as the internal reluctance of the head in the case of an ideal magnetic potential source at the winding (as is the case for a current-driven winding) and $Z_{\Phi int}$ was defined as the internal reluctance in the case of an ideal flux source (i.e. a voltage-driven winding). Hence $Z_{\Phi int} > Z_{int}$ (see also (8.23) and (8.24)), i.e. $P_{abs} > P_{ab}$, as expected.

As a function of R_{gs}, (8.73) represents a hyperbola with the vertical line $R_{gs} = - Z_{int}$ as an asymptote. However it is desired to know P_{abs}/P_{ab} as a function of I/I_s, i.e. to express R_{gs} as a function of I/I_s. In (8.55) this is done:

$$\frac{|I|}{I_s} = \frac{\eta}{\eta_s} \cdot \frac{R_{gs}}{R_g} \cdot \Phi_{gn}, \tag{8.74}$$

where $\Phi_{gn} \equiv A(x'_n = x'_{sn})/A(x'_n = 0)$ is a factor that is slightly larger than 1 when $|I| > I_s$. Again expressing η/η_s with the aid of (8.25) in equivalent circuit reluctances gives

$$R_{gs} + Z_{int} = \frac{R_g + Z_{int}}{\Phi_{gn}} \frac{|I|}{I_s}. \tag{8.75}$$

This means that the total reluctance in the Thevenin equivalent circuit increased by a factor $|I|/I_s$, besides the small reduction due to the increased surface, expressed in Φ_{gn}. Hence,

$$\boxed{\frac{P}{P_{\max}} = \frac{P_{\min}}{P_{\max}}\left(1 + \frac{P_{\max} - P_{\min}}{P_{\min}} \cdot \frac{I_s}{|I|} \cdot \Phi_{gn}\right)} \quad \text{for } |I| \geqslant I_s, \qquad (8.76)$$

where $P \equiv P_{abs}$, $P_{\max} \equiv P_{ab} = P_{abs}\left(|I| = I_s\right)$ and $P_{\min} \equiv P_{abs}\left(|I| \to \infty\right)$,

$$\frac{P_{\min}}{P_{\max}} = \frac{R_g + Z_{int}}{R_g + Z_{\Phi int}} \quad \text{and} \quad \frac{P_{\max} - P_{\min}}{P_{\min}} = \frac{Z_{\Phi int} - Z_{int}}{R_g + Z_{int}}.$$

For $P_{\min}$, $P_{ab}(\mu_{rgap} = 0)$ was preferably chosen for the Norton representation; see Sect. 8.1.6. When the slight increase of Φ_{gn} with $|I|/I_s$ (see Tables 8.9 and 8.11) is neglected, then $P/P_{\max}$ as a function of $|I|/I_s$ (see Fig. 8.38) is an exact hyperbola. In contrast to (8.73), now the vertical asymptote is always the $I = 0$ axis! The horizontal asymptote is the line $P/P_{\max} = P_{\min}/P_{\max}$.

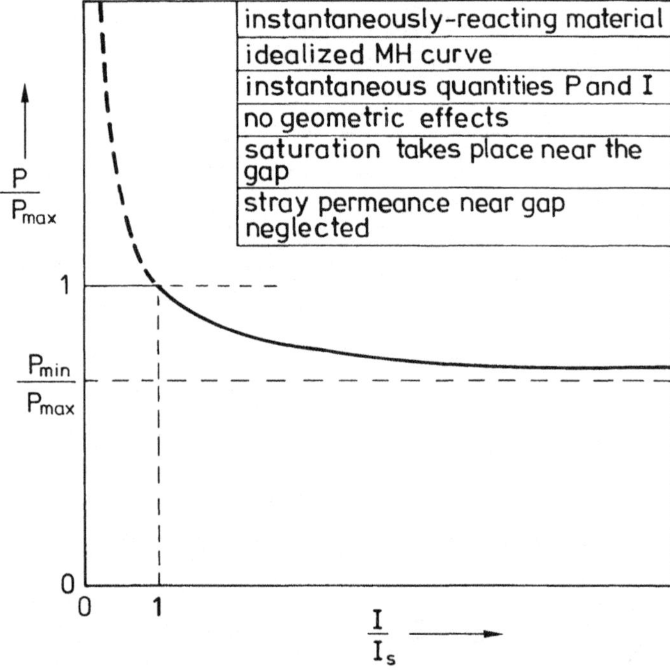

Fig. 8.38. The hyperbola according to (8.76). The result is exact, neglecting the stray permeance near the gap, when:
- the *MH* curve is idealized, see Fig. 8.29, which automatically includes no phase shifts or lags,
- the cross-sectional area of the core near the gap is uniform, i.e. no geometric effects,
- P and I are interpreted as instantaneous quantities, and not as first harmonics (of distorted signals),
- the ratio $\eta_\Phi A_{coil}/A_{gap}$ is large enough to ensure that saturation only takes place near the gap.

Only for heads with small x'_{1n} or small x'_{2n}, i.e. large θ_1, θ_2 or g, or small h_0 or W, may the increase of the surface with $|I|/I_s$, by which the permeance P increased especially at intermediate $|I|$ values, be of importance. This follows from Φ_{gn} values in Tables 8.9 and 8.11.

For $1 < \Phi_{gn} \lesssim 4/\pi$ and I_n larger than about 2 the use of the correction factor $4/\pi$ will be fairly accurate. This means that the inaccuracies of the use of the correction factor in the efficiency determination plays a role for 'tapered' heads (Table 8.11) with g/w and g/h larger than say $1/100$. In a head with a constant core width around the gap (Table 8.9) the inaccuracy arises much later, say above g/w and $g/h = 1/25$. Hence this 'geometric' effect usually plays a minor role.

Flux efficiency, η_Φ, and relation between flux and field efficiency

When the cross-sectional area of the gap and the saturation field are known, then the gap flux at the saturation current I_s is easily calculated. When in addition the impedance Z, or inductance L and quality factor Q, is measured, then in addition the total flux at the saturation current is known. Hence the ratio of gap flux to total flux, i.e. $|\eta_\Phi|$, is known. By using $Z = j\omega N^2 P_{ab}$, see (8.45), and $P_{ab} = \eta/R_g\eta_\Phi$, see (8.20), we arrive at an expression for this flux efficiency:

$$\boxed{\eta_\Phi = \frac{j\omega N^2 \eta}{R_g Z}} \ , \tag{8.77}$$

hence

$$|\eta_\Phi| = \frac{N^2|\eta|}{R_g|Z|/\omega} = \frac{N^2|\eta|}{R_g L \sqrt{1 + 1/Q^2}} \tag{8.78a}$$

or by writing out R_g and $|\eta| = H_{(s)}g/I_{(s)}N$ as a consequence of its definition (8.17):

$$\boxed{|\eta_\Phi| = \frac{\mu_0 H_{(s)} \cdot W h_0 \cdot N}{I_{(s)} L \sqrt{1 + 1/Q^2}}} \ . \tag{8.78b}$$

The last expression also follows directly from writing out $|\eta_\Phi| \equiv |\Phi_g|/|\Phi_t|$ at e.g. saturation (subscript s).

The question arises as to how this η_Φ is related to the relative decrease of the electric impedance:

$$\frac{Z_{max} - Z_{min}}{Z_{max}} = \frac{P_{max} - P_{min}}{P_{max}} = \frac{Z_{\Phi int} - Z_{int}}{R_g + Z_{\Phi int}} \qquad (8.79)$$

which follows from (8.13) and (8.76).

With the aid of (8.26) this is easily written as a function of the flux efficiencies:

$$\frac{P_{max} - P_{min}}{P_{max}} = \frac{\eta_\Phi}{\eta_{\Phi int}} \left(1 - \frac{Z_{int}}{Z_{\Phi int}}\right), \qquad (8.80)$$

so that the question arises when $1 - Z_{int}/Z_{\Phi int} = \eta_{\Phi int}$. In this special case

$$\eta_\Phi = (Z_{max} - Z_{min})/Z_{max} \qquad (8.81)$$

and according to (8.77)

$$\eta = (Z_{max} - Z_{min})R_g/j\omega N^2. \qquad (8.82)$$

This is only the case for a pure parallel circuit as depicted in Fig. 8.28, for which $Z_{int} = Z_{c(ore)}$, $Z_{\Phi int} = R_{str(ay)} + Z_c$ and $\eta_{\Phi int} = R_{str}/(R_{str} + Z_c)$. For more complex circuits, $(P_{max} - P_{min})/P_{max}$ deviates from η_Φ. For instance, when a single series reluctance, Z_s, is placed in the branch of the source, cf. Fig. 8.39, then $Z_{int} = Z_c + Z_s//R_{str}$ is increased, while $Z_{\Phi int}$ and η_Φ are not affected (because of the ideal flux source used in

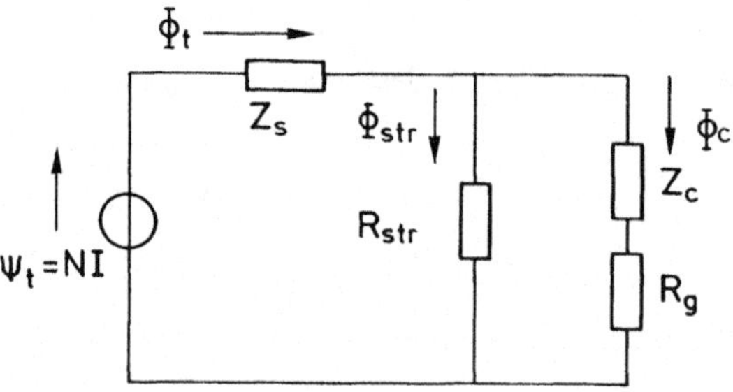

Fig. 8.39. Simplest parallel/series representation of a head with stray reluctance (R_{str}) and 'distributed' core reluctance (Z_c and Z_s). Relative to the parallel circuit in Fig. 8.28, the series (source) reluctance Z_s is added.

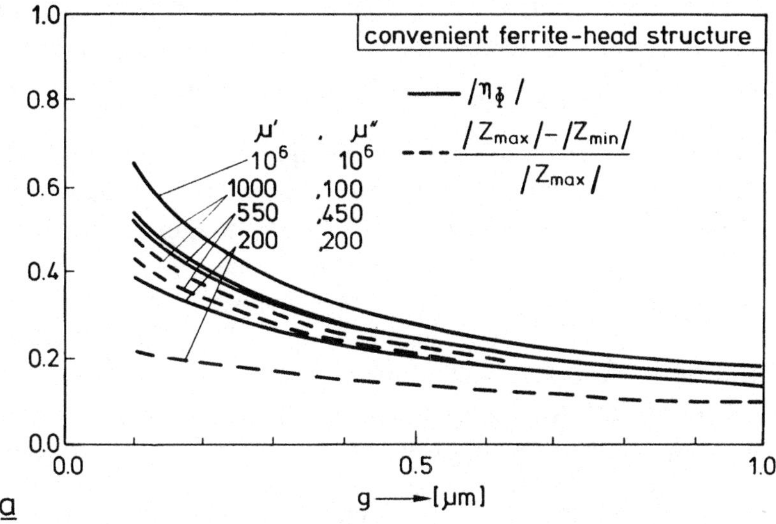

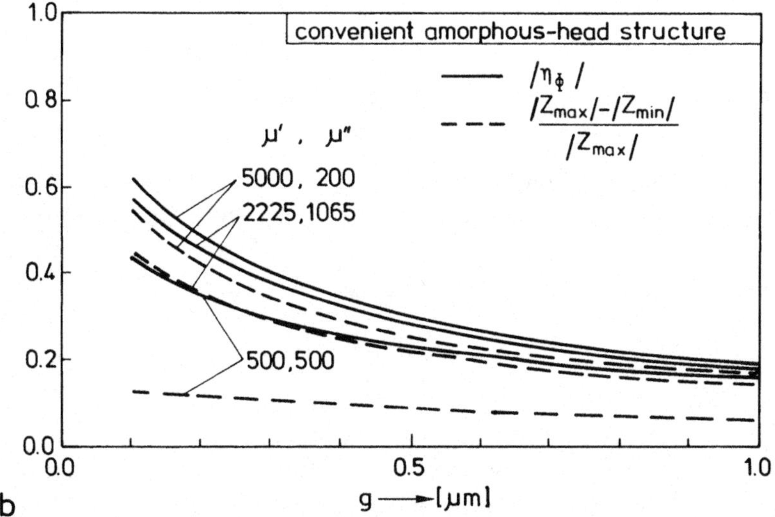

Fig. 8.40. Differences between $|\eta_\Phi|$ (full curves) and $(|Z_{max}| - (Z_{min}))/|Z_{max}|$ (dashed curves), made visible by model calculations using the accurate Thevenin/Norton representations.
a) Carried out for the convenient ferrite-head structure (i.e. with track tapering) given in Table 8.8,
–   at 4.5 MHz ($\mu' = 550$, $\mu'' = 450$)
–   at a considerably lower frequency ($\mu' = 1000$, $\mu'' = 100$)
–   at a considerably higher frequency ($\mu' = 200$, $\mu'' = 200$)
–   with a very high efficiency ($\mu' = 10^6$, $\mu'' = 10^6$) such that the simple parallel circuit of Fig. 8.28 holds and consequently the dashed and full lines coincide (see text).
b) Carried out for the convenient amorphous-head structure (i.e. without track tapering) given in Table 8.8, at 4.5 MHz as in Table 8.8 and at a considerably lower and higher frequency.

the definition of $Z_{\Phi_{int}}$). Hence $(P_{max} - P_{min})/P_{max} < \eta_\Phi$ in this example. So we must conclude that in general it is safer to use expression (8.78) for the $|\eta_\Phi|$ determination. Only when the gap area is not accurately known may $(P_{max} - P_{min})/P_{max}$ serve as an indication for η_Φ.

When the gap area is accurately known and the head approximates to a pure parallel circuit, then the field efficiency can be determined without knowledge of the head's saturation magnetization by using (8.82).

In the actual efficiency measurement we chose to measure the absolute value of the head's impedance, $|Z| = \omega N_2 |P_{ab}|$ see (8.13), and not the phase. Hence, in Fig. 8.40a and b a comparison is made between $|\eta_\Phi|$ and theoretical values of $(|Z_{max}| - |Z_{min}|)/|Z_{max}| = (|P_{max}| - |P_{min}|)/|P_{max}|$. This is carried out for the ferrite and amorphous head structures, given in Table 8.8, as a function of the gap length, g, at different core permeabilities, μ. At higher values of μ or g the agreement is better, because the validity of the pure parallel circuit increases in that case.

8.2 Quick efficiency measurement

A very simple efficiency measurement is described. No external transducer or pick-up coil is needed. The whole measurement takes place via the head-winding terminals. The accuracy seems satisfactory for most video and data-storage heads.

8.2.1 Introduction

Efficiency measurements can be done in many ways. One method makes use of a pick-up coil in front of the head (see for instance references [8.15] and [8.16]) to measure the field in front of the head, with the disadvantage of a moderate resolution and difficult manipulation of the pick-up coil. The magnetoresistive [8.17] or Hall [8.18] element instead of the pick-up coil gives better l.f sensitivity, but the same resolution and manipulation difficulties when used for small gap lengths, e.g. video heads. The above methods were in fact developed for field measurements and not primarily for efficiency measurements. A method specially developed for efficiency measurements is described in reference [8.19], and makes use of an extra coil, usually one winding, around the gap area and through the coil chamber of the head. This winding must be wound rather tightly in order to minimize the stray flux content, although this component is compensated to some extent. Manipulation of this extra winding around the mechanically weak gap-area of a video head through the tiny coil chamber costs time and leads sometimes to destruction of the head. These problems do not occur in the method that will be described in this section. The whole efficiency measurement takes place via the terminals of the head itself where the impedance of the head as a function of the current through the head is measured. It makes use of the assumption that the saturation of the head starts in the neighbourhood of the gap. Thus the method works best when applied to heads with a low cross-section near the gap relative to the cross-sections of the head further away from the gap area. The last (two) methods are in fact extensions of the $L(I_{dc})$ measurement of J.P. Morel [8.20], where the induction L, usually at a high frequency, is determined as a function of the dc current. The new elements in the last method, described in this section, are the quantitative model and evaluation of the results and the extension to ac-biasing currents by which the efficiency can be measured as a function of frequency.

8.2.2 Measuring principle

When the current through the winding of a head increases, the flux through the coil usually starts to increase proportionally to this current. This flux can be thought to consist of two parts, one flowing through the gap of the head the other flowing through the air. The latter is the stray flux and contributes to the total induction and impedance of the head. To a very good approximation the first part cannot exceed a value equal to the cross-sectional area at the gap times the saturation magnetization of the head material. The current at which this first occurs is called the saturation current I_s. At higher currents the area with the highest flux density starts to saturate, i.e. its efficiency becomes 1. This area is usually in the neighbourhood of the gap since the cross-sectional area is relatively (compared to the flux decrease) small there and the flux density thus maximal. The effective length of this saturated region increases proportionally with the current since the flux density remains practically constant. This causes the related impedance part to decrease proportionally with the current. This is worked out in more detail in the following section. This leads to a hyperbolic shape for the related impedance part as a function of $|\hat{I}|$, when $|\hat{I}|$ exceeds I_s. For $|\hat{I}|$ smaller than I_s an approximately constant $|\hat{Z}|(|\hat{I}|)$ relation exists, see Fig. 8.41. The intersection of these two curves defines the intersection

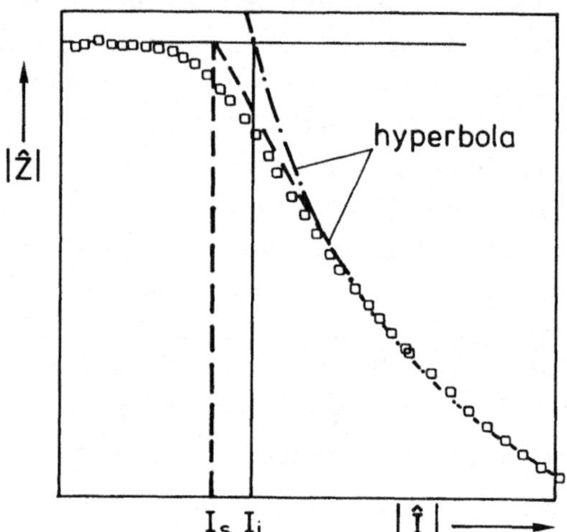

Fig. 8.41. The $\left|\hat{Z}\left(|\hat{I}|\right)\right|$ curve of an actual head and the hyperbola approximation.

current I_i which is related to the saturation current I_s. In practice the $|\hat{Z}|(|\hat{I}|)$ curve is not well desribed well in the neighbourhood of the intersection. One of the reasons is that the MH curve deviates from the 'ideal' curve depicted in Fig. 8.42 in the neighbourhood of the intersection defined by (H_s, M_s). The 'saturation field', H_s corresponds with I_s. (Note that $B_s = \mu_0(M_s + H)$ is not single valued because of H). Usually the ideal curve can realistically be used in the calculation of the asymptotic approximation of the actual curve for both low $\left(|\hat{I}| \ll I_s\right)$ and high $\left(|\hat{I}| \gg I_s\right)$ currents.

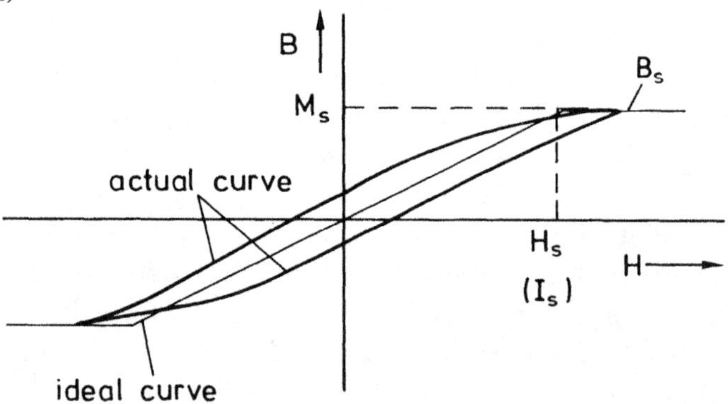

Fig. 8.42. Idealized and actual BH curve of a strongly-saturated head material.

8.2.3 Derivation of asymptotic approximation of Z(I) and determination of the efficiency

For simplicity we will assume $Q \gg 1$ and assume the 'magnetic stray reluctance' $R_{str(ay)}$ parallel to the core- and gap-reluctances R_c and R_g, see Fig. 8.43. For $I(t) < I_s$, the flux $\Phi_g(t)$ is equal to $C_g I(t)$, where C_g is a constant. For $|\hat{I}| > I_s$ the flux is distorted during the time that $I(t) > I_s$. The first harmonic which contributes to the impedance of the head Z then reaches $(4/\pi)\Phi_{gs}$ in a good approximation for a block-shaped gap flux, i.e. $|\hat{I}| \gg I_s$, where Φ_{gs} is the complex amplitude of the first harmonic of the (gap)flux when $|\hat{I}| = I_s$. In principle it is possible to calculate the factor Φ_g/Φ_{gs} as a function of $|\hat{I}|/I_s$ for the ideal MH curve when neglecting geometric effects (in the case of a non-ideal curve this factor cannot be easily derived and in addition will depend more strongly than in the case of the ideal curve on the geometry of the head in the gap region). Since the ideal curve is never the actual curve (see Fig. 8.42), it seems better to take $|\hat{I}| \gg I_s$ anyway such that automatically the block-shaped flux is a reasonable approximation for all kinds of MH curves. Influences of geometric effects on the distortion are then neglected. Geometric effects play a minor role when the cross-sectional area in the neighbourhood of the gap can be treated as a constant over a few times $g\,|\hat{I}_{max}|/I_s$ (where $\hat{I}_{max}$ represents the maximum measuring current). So for $|\hat{I}| \gg I_s$

$$|\Phi_g| \simeq \frac{4}{\pi} C_g I_s \tag{8.83a}$$

$(\text{and } |\Phi_g| \simeq C_g * I_s \text{ for } |\hat{I}| \gtrsim I_s)$

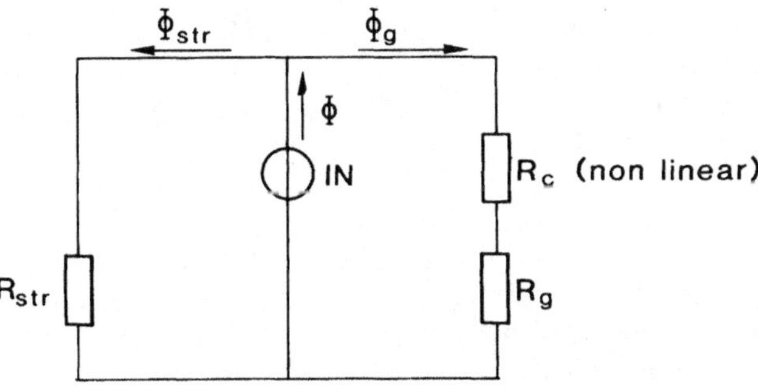

Fig. 8.43. Simplified equivalent magnetic circuit of a magnetic head.

and for $|\hat{I}| \ll I_s$

$$\hat{\Phi}_g = C_g \hat{I} \tag{8.83b}$$

It then easily follows after some manipulations that:

$$|\hat{Z}| = \frac{|\hat{U}|}{|\hat{I}|} \rightarrow \omega N \left(C_{str} + \frac{4}{\pi} C_g \frac{I_s}{|\hat{I}|} \right) \qquad |\hat{I}| \gg I_s \tag{8.84a}$$

$$\rightarrow \omega N (C_{str} + C_g) \qquad |\hat{I}| \ll I_s, \tag{8.84b}$$

where C_{str} accounts for the left branch in Fig. 8.43 ($\hat{\Phi}_{str} = C_{str}\,\hat{I}$), i.e. a horizontal asymptote for $|\hat{I}| \ll I_s$ and a hyperbolic curve as an approximation for $|\hat{I}| \gg I_s$ with the y-axis as asymptote for $|\hat{I}|/I_s \rightarrow 0$. Thus the hyperbola is defined by only two extra measuring points, e.g. at 3 and 5 times I_s. Usually the so defined hyperbola fits well over a wide range of the $Z(I)$ curve. The current at the intersection of both curves (8.84a) and (8.84b), I_i, follows by making (8.84a) equal to (8.84b) and leads to the following relation between the intersection current I_i and the saturation current I_s:

$$I_s = \frac{\pi}{4} I_i. \tag{8.85}$$

When instead of Z at the fundamental frequency the total amplitude of the output voltage is measured and not the amplitude of only the first harmonic, the correction factor vanishes.

The efficiency is defined by the ratio between the magnetic scalar potential over the gap and the magneto motive force, i.e.

$$\eta = \frac{Hg}{NI} \tag{8.86}$$

where:
H is the (deep) gap-field,
g the (magnetic) gap length,
I the current through the winding and
N the number of windings of the coil.

The above defined efficiency at low levels of I is thought to be representative of the efficiency of the head during reading. For a linear head this is true because of the reciprocity principle. For practical heads this is not guaranteed. Nevertheless we believe this principle is applicable to most of our heads. The above definition can be rewritten with the aid of the proposed definitions of the saturation field and current H_s and I_s to:

$$\eta = \frac{M_s g}{N I_{\text{seff}} \sqrt{2}} \tag{8.87}$$

when $B_s \to \mu_0 M_s$ since $\mu_r \gg 1$ is used and a sinusoidal current with effective value $I_{s-\text{eff}}$ is assumed.

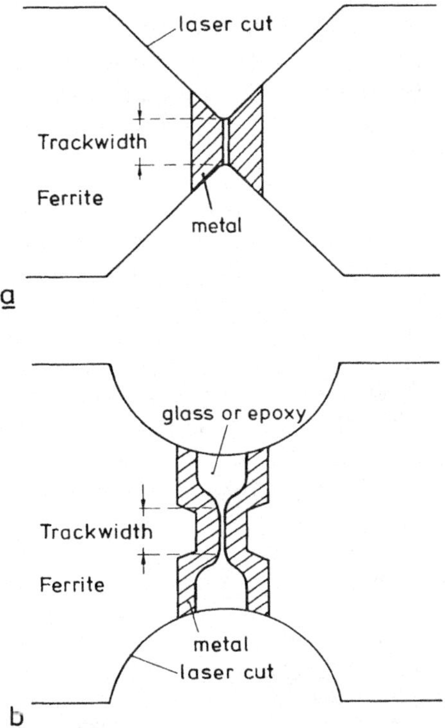

Fig. 8.44. Examples of metal-in-gap heads with thick metal layers (top view).
a. The conic shape and track width are defined by laser cutting after the head is completed. The layer may be sendust (FeSiAl), amorphous metal ($Co_{70} Fe_5 Si_{15} B_{10}$ or CoNb etc.) or permalloy ($Ni_{80} Fe_{20}$).
b. The track width is defined by reactive ion etching or grinding of the ferrite tile. Then the tiles are covered by the high-saturating metal by electroplating or sputtering. After the head is completed, rough laser cutting is usually applied in order to reduce side reading of longer wavelengths.

With the aid of (8.85) the efficiency can then be derived from the current at the intersection I_i when $|\hat{I}| \gg I_s$ is used:

$$\eta = \frac{M_s g}{N I_{\text{ieff}} \sqrt{2}} \cdot \frac{4}{\pi}. \tag{8.88}$$

When $|\hat{I}|$ is not much larger than I_s, the correction factor is smaller than $4/\pi$ and approaches 1, but cannot then be determined reliably.

The magnetic gap length is determined by measuring the gap-null wavelength in a recording experiment. Depending on the recording conditions (permeability of the tape, thickness of the coating, presence of backlayer, head-to-tape distance during playback) this gap-null wavelength must be divided by a correction factor of 1.12-1.2 in practical circumstances. We chose 1.15. For the saturation magnetization in the case of metal-in-gap heads with metal layers that are thin with respect to the track width, the saturation value of the ferrite must be taken into account, since this saturates over a wide area before the high-saturation thin metal film starts to saturate.

When, in contrast, metal layers have been used with thicknesses comparable to (or thicker than) the track width in a configuration as sketched in Fig. 8.44a and b, see also [8.21], the high-saturation metal layers may start to saturate at the same time or before the ferrite does.* The fact that the cross-section of the core widens from the gap further complicates the description of the saturation process for $I(t) \geqslant I_s$; due to the widening of the head near the gap the flux Φ_g may still increase weakly with I when $I(t) > I_s$. The assumption $\Phi_g = $ constant for $I(t) > I_s$ then becomes arbitrary. This 'geometric' effect is extensively discussed in Section 8.1.7, see especially Table 8.11 and the comment at the end of section 8.1.7.8.

---

* This is the case when the product of the saturation magnetization of the metal layer times the minimum cross-sectional area of the metal layer, $M_{sm}A_m$, is equal or smaller than the product of the saturation magnetization of the ferrite times the minimum cross-sectional area of the ferrite, $M_{sf}A_f$ (where $A_f = Wh$), i.e. when

$$M_{sm} \leqslant M_{sf} A_f/A_m.$$

In this case the saturation magnetization of the metal layer M_{sm} need to be substituted for M_s in (8.88). In the opposite case the value on the right side is smallest and consequently $M_{sf}A_f/A_m$ then has to be substituted for M_s in (8.88)

8.2.4 Basic diagram and experimental set-up

The basic diagram is shown in Fig. 8.45 and the experimental set-up is given in Fig. 8.46.

The meaning of the symbols in the black boxes is as follows:

A: Oscillator (part of LCR bridge HP4192LF).

B: Power amplifier (for general-purpose use, including measurements on heads that are difficult to saturate, e.g. ENI 10 kHz-100 MHz amplifier).

B': Amplifier for heads which saturate easily (e.g. PM5175 for currents of max. 30 mA up to frequencies of 5 MHz, although this is actually a 1 MHz amplifier).

C: Protection circuit for the LCR measuring bridge input circuits if amplifier B is used and too high power is accidentally generated.

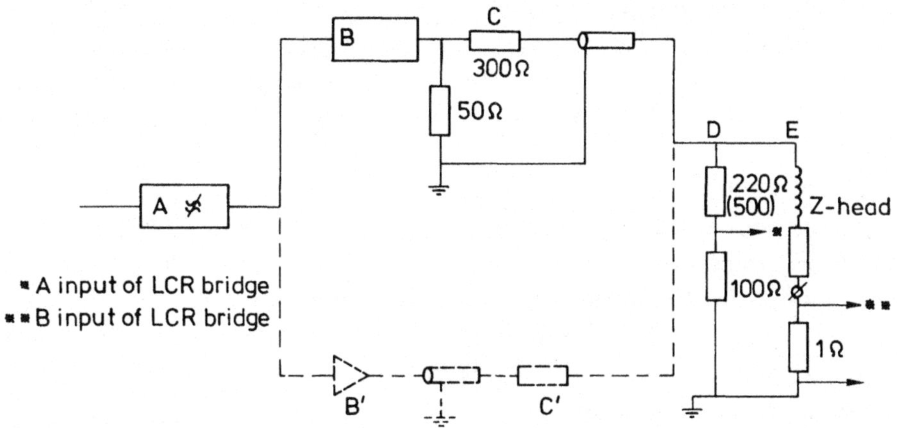

Fig. 8.45. Basic diagram of the experimental set-up of the quick efficiency measurement.

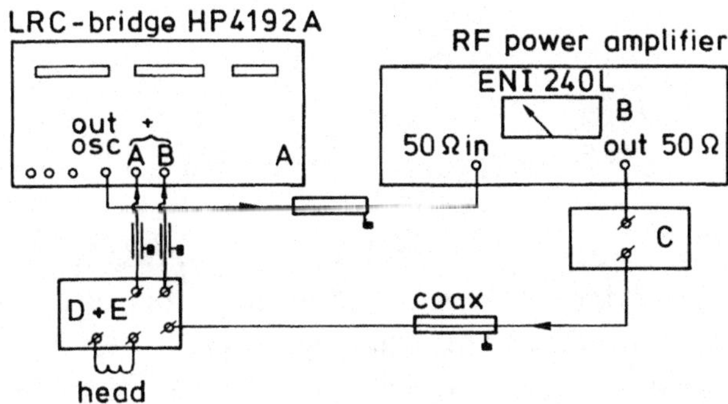

Fig. 8.46. Experimental set-up of the quick efficiency measurement.

C': Current-limiting resistor.

D: Attenuator, to prevent overload at the highest currents and head impedances. In the event of overload the attenuation has to be increased.

E: Head plus measuring resistor. The current through the head is measured with the aid of this resistor. 1 ohm is a convenient value for this resistor, since it does not influence the total impedance to such an extent as to reduce the sensitivity of the measurement. On the other hand the resistance is still so high that soldering of the measuring wires close to the case of the resistor (very important!!) guarantees that the induction of the resistor leads plays no role in the determination of the current. It is therefore important to check this circuit up to the highest frequencies used. The impedances of the (short) wires between the resistor (where the connections have been made by soldering) and the head etc. only play a role in the sensitivity and do not introduce extra inaccuracies in the determination of the efficiency since these impedances add to the impedance part of the head due to leakage flux and coil resistance (including skin effect).

The protection circuit (C) and current-limiting resistor (C') also make better current-sources for the amplifiers. The current through the head coil (and hence the voltage at the B input of the LCR bridge) is therefore almost sinusoidal, as assumed in the model calculations leading to the hyperbola approximation and the $4/\pi$ correction factor. This is particularly important when the PM5175 is used, because of the low source impedance of this amplifier.

When the saturation current is directly estimated from the $|\hat{Z}(|\hat{I}|)|$ curve, i.e. without making use of the intersection current, I_i, then a voltage source works as well.

8.2.5 Discussion

Enhancing the reproducibility and accuracy of the measurement

The relative accuracy of the measuring apparatus must be very good. The absolute accuracy is of less importance, because ratios ($|\hat{Z}| = |\hat{V}|/|\hat{I}|$) are measured. During the measurement the apparatus (HP 4192LF) changes its ranges a number of times for the $|\hat{V}|$ and a number of other times for the $|\hat{I}|$ measurements. When every range has a high absolute accuracy (e.g. 1%) it is still possible for two successive determinations of $|\hat{Z}| = |\hat{V}|/|\hat{I}|$ to change by 2% because of this, giving an irregular $|\hat{Z}|(|\hat{I}|)$ curve with a period that increases with $|\hat{I}|$. When the absolute accuracy is a constant for different ranges and within different ranges $|\hat{Z}|(|\hat{I}|)$ is measured correctly. The HP 4192A fulfils these requirements and delivers smooth $|\hat{Z}|(|\hat{I}|)$ curves. Yet we observed very irregular $|\hat{Z}|(|\hat{I}|)$ curves for different amorphous heads (see Fig. 8.47 top) and assumed that irregular domain wall displacements caused this phenomenon. In order to force the domain walls to form a defined

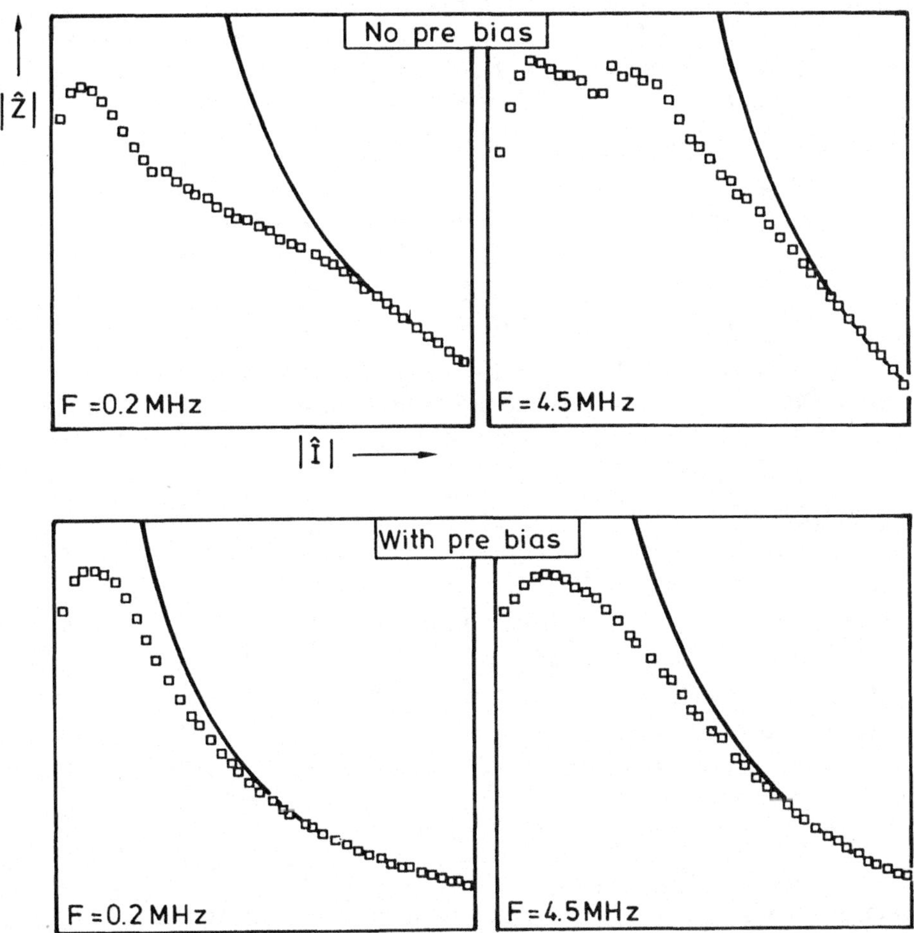

Fig. 8.47. $|\hat{Z}(|\hat{I}|)|$ curve of an unstable amorphous head without (a) and with (b) application of a pre-bias.

pattern we applied a bias to the head in two ways to get rid of irregular domain-wall displacements:

1. before every measuring point,
2. before measuring a new $|\hat{Z}|(|\hat{I}|)$ curve at a new frequency.

Method 2 turned out to be very successful in getting smooth $|\hat{Z}|(|\hat{I}|)$ curves (see fig. 8.47, bottom) and is now standard in the efficiency measurement. The bias-current is obtained by passing a large ac current through the head and decreasing the amplitude with time. We use the following values (note that the exact values are not very essential, only the amplitude has to be large enough):

Amplitude: 150 mA
Time constant: 100 msec
Frequency: 100 kHz

Influence of short circuits and power losses in the core

Short circuits over or in the gap between both core halves, see section 8.1.6.1, cause an increase of the inductance, and analogously in the impedance, at (very) low current levels when the short-circuit paths are not yet saturated. This is visible in Fig. 8.48a for $L(I_{dc})$ and also appears from the extra low Q factor at small I_{dc}. After etching away the (hardly or not visible) short circuits the $L(I_{dc})$ curves – and analogously the $|\hat{Z}|(|\hat{I}_{ac}|)$ curves, not shown – look reliable, see fig 8.48b; a plateau, more or less, is observed at low current levels and a plateau at high current levels.

The main effect of the short circuits is a reduction of the efficiency of the head, as described in Sect. 8.1.6.1. It is noted that the (quick) efficiency measurement determines the efficiency incorrectly, as if there were no short circuits.

The Q factor usually increases with the current, especially when $|\hat{I}|>I_s$, see fig. 8.48, since the relative power losses in the core decrease; the influence of the core with its low quality becomes smaller with respect to the leakage path through the air with its very high quality factor (small μ''). In Fig. 8.49 this is illustrated with the aid of the BH curve; the area of the BH curve relative to the product $B_{max} H_{max}$ characterizes the relative power loss. Thus the Q factor usually influences the $|\hat{Z}|(|\hat{I}|)$ curve in the same direction as the inductance, since $|\hat{Z}| = \omega L\sqrt{1 + 1/Q^2}$. For $Q > 2$ the influence is already smaller than 12% on $|\hat{Z}|$, but on the measured (slight) change in $|\hat{Z}|$ it still plays a role at larger values of Q. Beyond saturation, where the hyperbola is fitted, the Q factor is usually 7-20 and moreover rather constant, and thus changes in $\sqrt{1 + 1/Q^2}$ are completely negligible.

For $|\hat{I}| < I_s$ a small increase of L with $|\hat{I}|$ and decrease of Q with $|\hat{I}|$ is often measured. This leads to an increasing $|\hat{Z}|(|\hat{I}|)$ curve for small values of $|\hat{I}|$; see e.g. Fig. 8.50.

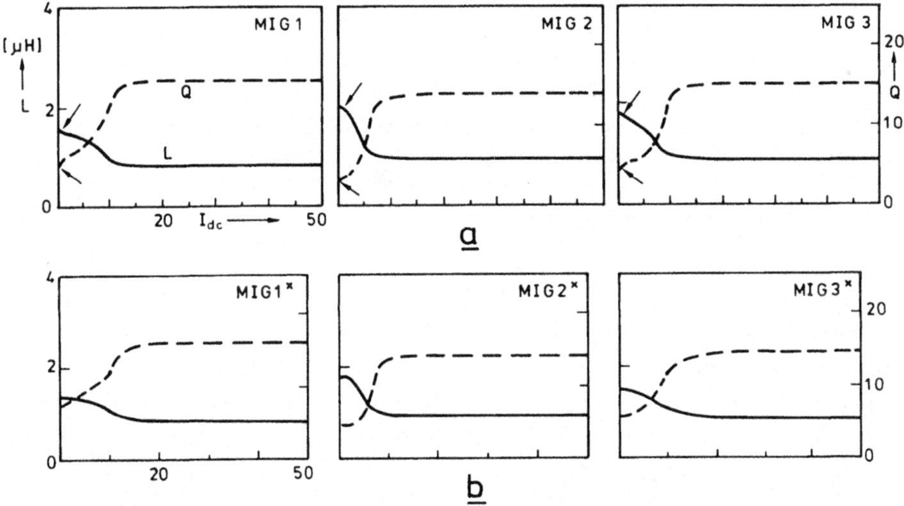

Fig. 8.48. Three NiFe-metal-in-gap (MIG) heads (L and Q measured at 4.5 MHz).
a) Before etching. The arrows indicate the effects of magnetic short-circuits over the gap on the $L(I_{dc})$ and $Q(I_{dc})$ curves.
b) The NiFe-MIG heads after etching. Obviously the short-circuits have disappeared. The effect on the $|\hat{Z}(|\hat{I}_{ac}|)|$ curves (not shown) is alike.

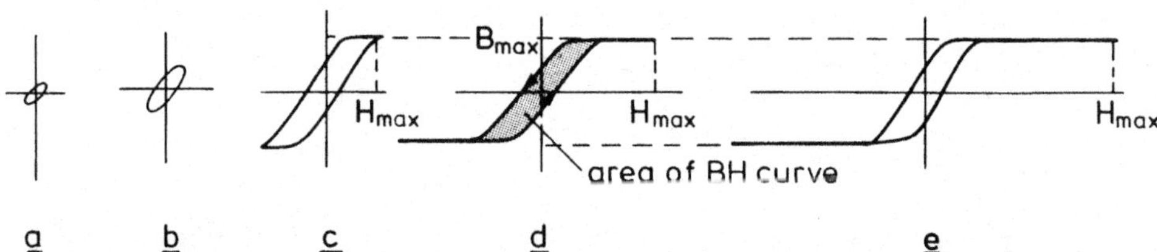

Fig. 8.49. Illustration of the BH curve of a (soft-)magnetic head core (material) at different magnetic-field amplitudes, H_{max}.
The illustrations are consistent with what is usually observed:
1) going from a to c, the μ' (i.e. L) is constant or changes weakly and the ratio μ'/μ'' (i.e. Q) is constant or decreases weakly
2) going from c to e, the μ' (i.e. L) as well as the μ'/μ'' ratio (i.e. Q) decrease.

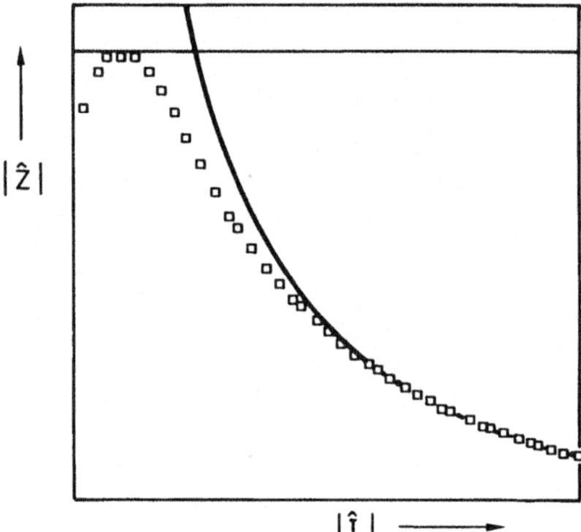

Fig. 8.50. A rather extreme case of what is usually observed; $|\hat{Z}|$ increases with $|\hat{I}|$ at small values of $|\hat{I}|$.

Influence of 1 ohm measuring resistor

The $|\hat{Z}|$ we measure (see basic diagram and experimental set-up) includes the 1 ohm current measuring resistor. Since, in principle, $|\hat{Z}|$ and not the complex $\hat{Z}$ is measured, it is not possible to subtract the 1 ohm resistance in order to get the absolute value of the head-impedance. Nor is it possible to measure the phase of $\hat{Z}$ in the mode of the LCR bridge used here, type HP 4192LF; it would at least require a power amplifier with a defined phase characteristic. Fortunately inclusion of the 1 ohm resistance in $\hat{Z}$ is equivalent to increasing the head impedance by a constant. This does not influence the accuracy of the determination of I_s! The same is true for the induction of the leads to the head, which add to the stray inductance of the head. and the resistance of the head winding (0.1-0.5 ohm) which adds a constant to the head impedance which hardly affects the accuracy of the measurement.

Simple determination of I_s

In the following cases, a simple determination of I_s may be preferable to the one using the hyperbola fit in combination with the $4/\pi$ correction factor.

Narrow-track heads with a core width equal to the track width, at high frequencies.

As explained in Sect. 8.1.7.2, the saturation may then start at a region other than the gap region, see also Fig. 8.27. The correction factor $4/\pi$ is too small in this case.

Heads with *steep permeability roll-off*, e.g. due to eddy currents.

In these heads the high-frequency content of the gap flux may be so strongly suppressed by the core that the correction factor vanishes. Then the hyperbola fits the $|\hat{Z}|(|\hat{I}|)$ curve just beyond I_s better. The right correction factor is then smaller than $4/\pi$. However we never found clear examples of this phenomenon. Possibly the flux efficiency of heads of this type (in our cases these were always narrow-track non-laminated amorphous-ribbon metal heads) was already so small that the above opposite effect dominated.

Heads with a *strongly increasing cross section of the core* around the gap.

An example is the metal-in-gap head geometry sketched in Fig. 8.44, see also Section 8.2.3. Here the gap flux at $I \gg I_s$ may be too much in excess of the gap flux, $\mu_0 M_s hW$, at the onset of saturation. This happens when g/W and g/h are larger than say 1/100, see Table 8.11 and Section 8.1.7.8. In most heads where the track width is constant around the gap and only the gap height increases, this occurs much later, say at g/W and g/h larger than 1/25, see Table 8.9 and Section 8.1.7.8. Hence only in extreme cases is the correction factor $4/\pi$ too small.

In all the above cases it is more accurate to fit the hyperbola (or a straight line) through the first part beyond the 'kink' in the $|\hat{Z}|(|\hat{I}|)$ curve, while disregarding the correction factor (i.e. the current at the intersection point I_i then equals I_s). The computer program opts for this method when the ratio of the currents at the intersection points obtained by both methods exceeds the correction factor $4/\pi$.

Comparison between the 'old' and 'new' efficiency measurement

To check the reliability of our 'new' efficiency measurement we measured several heads using both the old [8.19] and the new measurement. In Fig. 8.51 the results of measurements on three arbitrary heads are presented graphically. Apparently there is good agreement between the two measuring methods so that we may say that the accuracy of both methods is approximately equal. However, the absolute accuracy of both methods in extreme cases is not known yet.

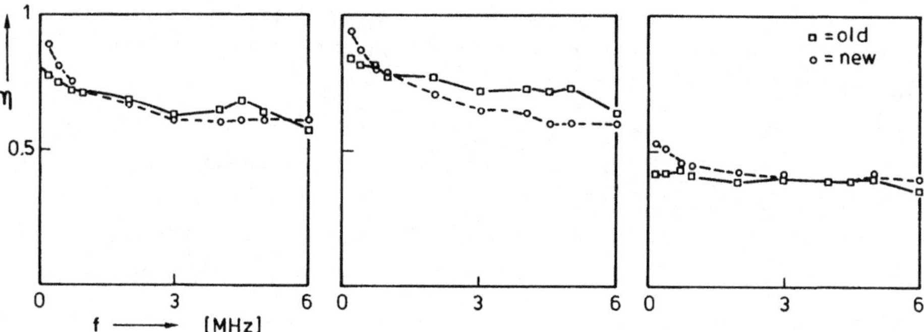

Fig. 8.51. Comparison between results obtained by the old [8.19] and new (quick) efficiency measurement on various heads. N = new, O = old.

Measuring time

With regard to measuring time the 'quick efficiency measurement' lives up to its name: to measure the efficiency at one frequency it takes about 90 seconds, whereas with the old method it would take almost a quarter of an hour to determine one efficiency value.

Another noteworthy fact is the following: To determine the intersection current I we only need three measuring points, namely the impedance at a very low current (to determine the horizontal line) and two $|\hat{Z}|$ values at two higher currents to calculate the hyperbola. So it is not absolutely necessary to measure the whole $|\hat{Z}|(|\hat{I}|)$ curve and therefore the method could be made even quicker. Determining one efficiency value would now take no longer than about 10 seconds.

We also examined a hardware implementation of the measurement, whereby the amplitude of the hf current is swept with a period of 50 ms. It then depends on the chosen number of data points to be measured, each transferred in a finite time to the computer for acquisition, when a following measurement can be started. Times much smaller than 1 second are easily obtained for one efficiency value when only three measuring points are transferred, and about 10 seconds when 200 measuring points are transferred.

8.2.6 Conclusions

For the new method of measuring the efficiency of a head, described in the foregoing, which is based on the saturation behaviour of head materials, it is necessary to know the gap length and the value of the saturation magnetization.

The whole measurement takes place via the head winding terminals by measuring $|\hat{Z}|(|\hat{I}|)$. In principle the write efficiency is measured. The method also gives information about details of the head, such as possible short circuits over the gap in e.g. metal-in-gap heads. In spite of the simplicity of the method and underlying model, the reliability of the efficiency determination seems satisfactory for most types of video and data-storage heads. Since no accurate methods of measuring efficiencies of short-gap heads exist with which we can compare our results, it is difficult to claim a specific accuracy. A number of comparisons with the results using the method described in [8.19] and with results from playback behaviour (as a function of frequency) of heads only indicate that the accuracies are usually at least better than 20%.

To get rid of sometimes very irregular $|\hat{Z}|(|\hat{I}|)$ curves that make an efficiency determination inaccurate or impossible, the biasing method described turned out to be very successful. Obviously irreproducible domain wall displacements caused the irregularities.

Possibly the most interesting application of the method, with its very short measuring time, is the fast ranking of heads in development and production.

8.3 Comparison of measurements with model calculations

In this section the previous comparisons will be extended with comparisons between calculated and measured efficiencies. In the meantime some head types are analysed and compared concerning their efficiencies and electrical impedance, which are important in the signal-to-noise ratio upon playback.

However, first an experimental check of the model calculations will be carried out by way of unwinding an originally symmetrical head, and measuring and calculating the resulting, smaller than perhaps expected, inductance change.

8.3.1 Inductance change by unwinding one core half

It is well known that the stray flux of a uniformly wound solenoid (of uniform shape) is minimal with respect to other (non-uniform) windings with the same number of turns. Analogously, the stray flux of a head (which is non-uniform, especially because of the gap) is reduced when two windings are used, one around each core half, instead of placing all turns around one core half. (The field efficiency value is not highly dependent on the stray flux and so is not noticeably increased). Hence the inductance value ($\sim N^2$) after unwinding one core half will be reduced by less than a factor of 4, i.e. normalized to the same number of turns the inductance is increased (and consequently the signal-to-noise ratio will be reduced).

This change of the inductance is measured by unwinding the 8 turns on one side of a (metal-in-gap) ferrite video head with well-known gap length and permeability (on the other side 9 windings have been placed accidentally), see results in Table 8.12. In the last column the value of L is given for the case $L \sim N^2$, which is indeed considerably less than measured for the asymmetrically-wound head core with $N = 9$. The agreement with more accurately calculated values in the third and fifth columns is very good (we did not select the head!). For the calculations in the third column , i.e. for the symmetrical head, we used the reluctance circuit in fig. 8.3b, as also used for all previous calculations, and which is equivalent to the reduced circuit in fig. 8.3a in the case of symmetrical heads.

Table 8.12. Comparison between measurements on one and the same MIG ferrite video head
a)   with about equal number of windings around both core halves ($N_1 = 9$, $N_2 = 8$).
b)   with $N_1 = 9$ and $N_2 = 0$, see third and fourth column.
   The dimensions etc. of the head are as for the ferrite head in Table 8.8, except that:
− $g = 0.2$ μm, $W = 22$ μm and $b_1 = 50$ μm,
−   at 0.75 MHz; $\mu_x' = \mu_y' = 800$ and $\mu_x'' = \mu_y'' = 400$, according to Fig. 3 in reference [8.32],
− $N = 17$ or $N_1 = 9$ and $N_2 = 8$, see text.

F [MHz]	L [μH]				
	$a: N_1 = 9, N_2 = 8$		$b: N_1 = 9, N_2 = 0$		
	meas.	calc.	meas.	calc.	simply* calc.
4.5	1.85	1.85	0.72	0.70	0.52
0.75	1.97	1.96	0.76	0.74	0.55

* Assuming simply $L \sim N^2$, i.e. $(9/17)^2 \times$ measured inductance for $N_1 = 9$ and $N_2 = 8$.

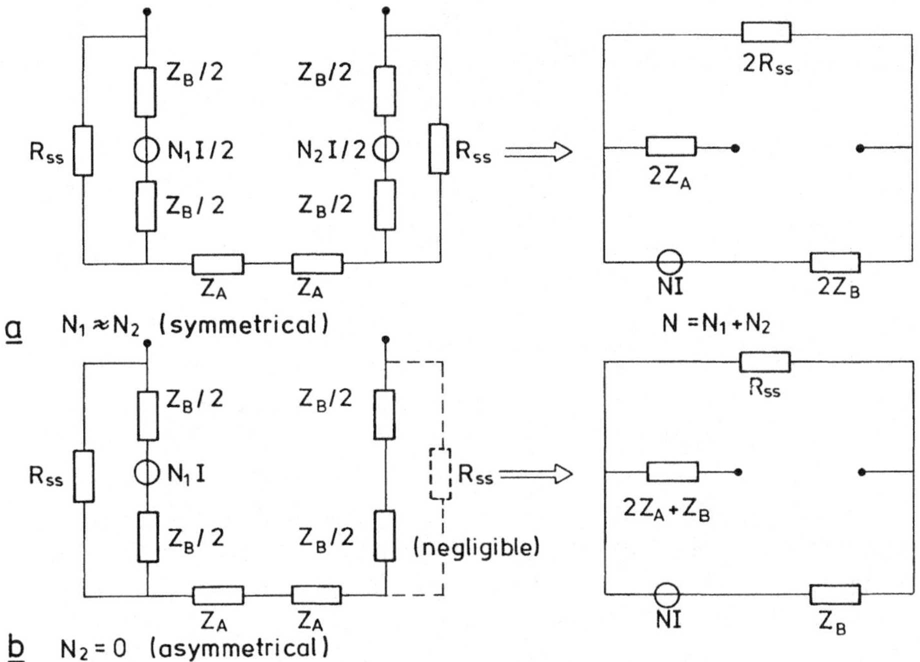

Fig. 8.52. The reduction of part of the reluctance circuit given in Fig. 8.3a in the case of
a) about equal number of turns around both core halves, see also Fig. 8.3b,
b) all turns around one core half.
The remaining part of the circuits remained unchanged. The dominating change is the reduction of the stray reluctance from $2R_{ss}$ to R_{ss}, and hence increase of the stray flux and inductance (for equal total number of turns).

In Fig. 8.52a the reduction of the lower part of this circuit is sketched again, in order to compare it with the deviating reduction when a magnetic source is placed on only one side of the head, the latter representing the asymmetrically-wound head (superscript as). So the necessary changes in the calculations (compare the reduced circuits in fig. 8.52a and b) read:

$$Z_B \rightarrow Z_B^{as} = \tfrac{1}{2}Z_B$$

$$Z_A \rightarrow Z_A^{as} = Z_A + Z_B^{as}$$

$$R_{ss} \rightarrow R_{ss}^{as} = \tfrac{1}{2}R_{ss} \tag{8.90}$$

The resulting inductance values are given in the fifth column of Table 8.12. Further the calculations showed that *the field efficiency decreased by less than a percent* (from 0.551 to 0.550), but the *flux efficiency decreased considerably* (from 0.41 to 0.31), as expected. The Q factor increased (from 4.9 to 6.3), because the stray flux through the air is almost free of any phase lag.

8.3.2 Amorphous sandwich head

A 3-fold laminated ribbon head, see SEM picture in Fig. 8.53, has been used to compare measured and calculated efficiency and inductance at different frequencies. The permeability of one lamella of 6.6 μm thickness, made of the original ribbon by polishing, was measured accurately; values are listed in Table 8.13. The magnetostriction constant, λ, of the ribbon ($Co_{70.3} Fe_{4.7} Si_{15} B_{10}$ at. percent) is rather low ($\lambda = -0.17\ 10^{-6}$) so that it is reasonable to assume that these measured permeabilities still hold for the carefully manufactured laminated head.

With these permeability values and the geometry of the head, see caption of table 8.13, the efficiency etc. values have been calculated. Together with the experimentally obtained values they are listed in table 8.13. The agreement between absolute values and of variations is satisfactory. Only the calculated Q factor is clearly too high. This is also observed for other types of heads, although to a lesser extent and is

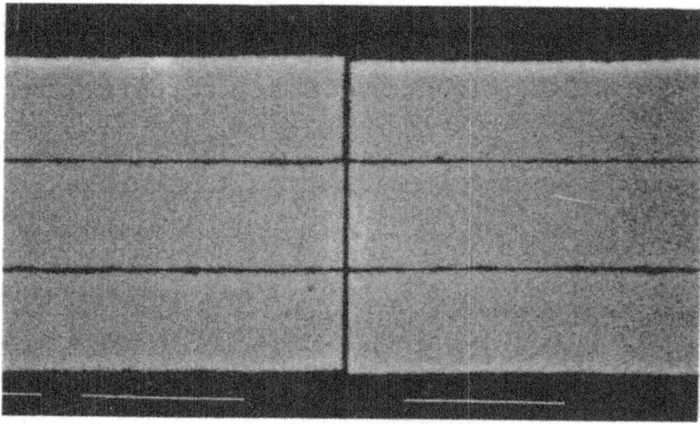

Fig. 8.53. Top view of the 3-fold laminated ribbon head. In this SEM picture, the SiO_2 gap is visible as the vertical black line. Scale: 10 μm/div.

probably caused (partly) by inaccuracies in the model; it is not excluded, however, that the μ''/μ' ratio increases after some steps in the manufacturing of the head.

8.3.3 Metal-in-gap ferrite head

For model calculations it would be practical to have some $\mu'(\omega)$ and $\mu''(\omega)$ curves in stock for the different materials. However a problem arises because of the small dimensions, especially the core (and track) width, of the head. Because of this, the shape as well as the absolute value of the permeability curves are very much influenced by the type of machining (sawing, etching, grinding, lapping or polishing) and machining conditions (diamond-particle size, pressure, speed etc.) of the side surfaces of the head. This is visible in the permeability curves in Figs. 8.54 and 8.55, which have been reproduced from figs. 7.3 and 7.7 in the thesis of Visser [8.12]. The reasons for these variations are the different characteristics of wall-pinning points at the machined side faces of the head and the differences in balancing stress in the 'bulk' of the head, the latter generated by the residual surface stress due to the machining of the surface.

It is to be expected that cold-machining processes (like sawing and grinding etc.) result in tensile balancing stresses (because of a plastically stretched surface layer under compressive stress), and that hot-machi-

Table 8.13. Measured and calculated values for the amorphous head of Fig. 8.53. The configuration etc. is as for the amorphous head in Table 8.8, except that:
– $g = 0.2$ μm and $W = 17$ μm
–   the permeabilities are as given below
– $N = 21$
Further the measured saturation magnetization, $B_s = 0.7\ T$, is used in the efficiency determination.

f [MHz]	μ'	μ''	L[μH]		ΔL[μH]		Q		$|\eta|$	
			meas.	calc.	meas.	calc.	meas.	calc.	meas.	calc.
0.2	3500	750	–	2.25	–	0.77	–	14.2	0.60	0.58
0,75	3000	1000	2.32	2.16	0.9	0.71	5	8.6	0.53	0.55
4.5	2225	1065	2.05	1.96	0.68	0.57	3.8	5.4	0.45	0.49
6.0	1900	1000	1.90	1.85	0.56	0.50	3.3	4.6	0.42	0.46

ning processes like laser cutting (and sputtering) result in compressive balancing stresses, because of the shrunk surface under tensile stress. It is clear that the effects increase with decreasing core width. During the development of heads it seems more appropriate to determine the permeability values iteratively from measured data of the completed head, than to use curves measured on e.g. rings as depicted in figs. 8.54 and 8.55, because of the influence of the subsequent technology steps in the manufacturing of a head. In an iterative proces (Gauss-Seidel) we determined the μ' and μ'' values that fit, according to the model, the measured L and Q data; see Fig. 8.56. So the first and second plots, $L(f)$ and $Q(f)$, in fig. 8.56 represent by definition both the calculated and measured data. The iteratively obtained μ' and μ'' values, given in the third plot of fig. 8.56, approach the experimental data on 150 μm thick rings lapped with diamond particles from 10-15 μm, see curves 3 in fig. 8.54, and those of rings sawn with a blade with 20 μm diamond

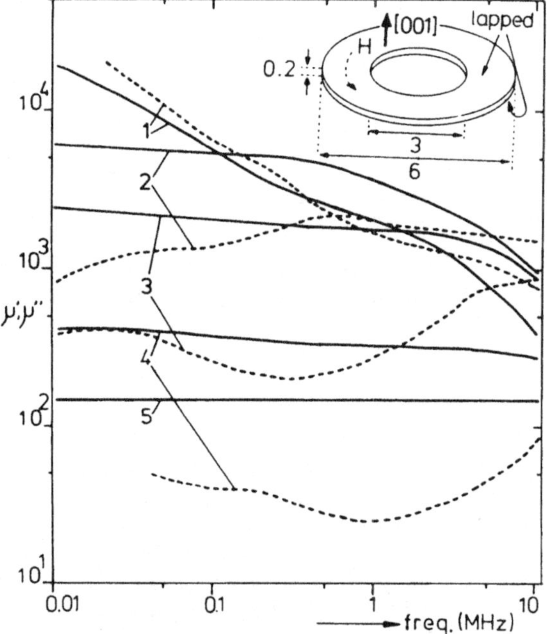

Fig. 8.54. Permeability of etched and lapped rings of $Mn_{0.59}Zn_{0.35}Fe^{II}_{0.06}Fe^{III}_2O_4$, solid curves μ', broken curves μ'' (at 50°C). Measuring field 3 A/m r.m.s. 1) As etched. 2) to 5): as lapped with diamond 4-7 μm, 10-15 μm, 20-40 μm and 16 μm alumina. The small μ'' values of (5) are omitted. Inset: dimensions of the ring (in mm). (Reproduced from Fig. 7.3 in [8.12], courtesy of E.G. Visser.)

particles, see curves 3a and 3b in fig. 8.55, while the head was actually sawn to a core width of 200 μm using a blade with 4-8 μm synthetic diamond particles, and laser-cut to a track width of about 20 μm.

The calculated efficiency values that correspond, by way of the model, to the permeability values obtained, are given in the last plot of fig. 8.56, together with the experimentally determined efficiencies. (It is noted that in the evaluation of the efficiencies from experimental data the saturation magnetization of the core and not of the thin (1.15 μm) metal (NiFe) layers 'in' the gap has been used, as explained in Sect. 8.2.) The frequency dependence of the experimentally obtained efficiencies is larger.

The μ' and μ'' values were fitted on the measured L and Q values. Another possibility would e.g. be to fit them on the measured $|\eta|$ and Q values. It is hard to say which choice is better, since L and Q are most accurately determined experimentally but in the model less accu-

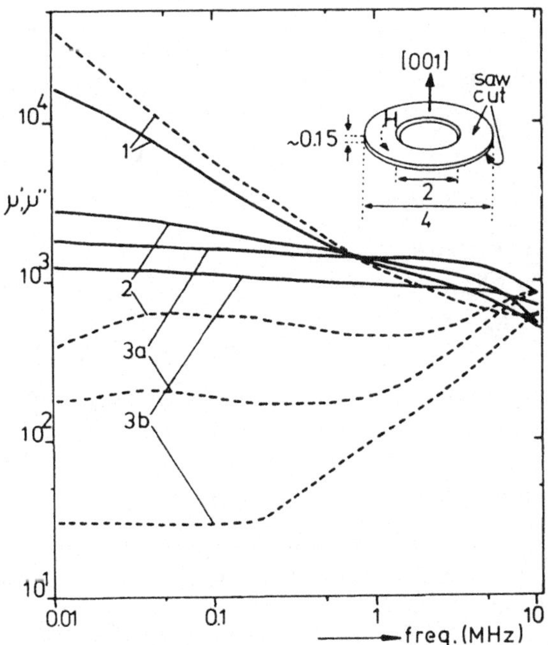

Fig. 8.55. Permeability of etched and saw-cut rings of $Mn_{0.59}Zn_{0.35}Fe^{II}_{0.06}Fe^{III}_2O_4$, solid curves μ', broken curves μ'' (at 30°C). Measuring field 3 A/m r.m.s. 1) As etched. 2) As cut with a blade having 20 μm diamond particles. 3a) and 3b): as cut with a blade having 8 μm diamond particles. (Reproduced from Fig. 7.7 in [8.12], courtesy of E.G. Visser.)

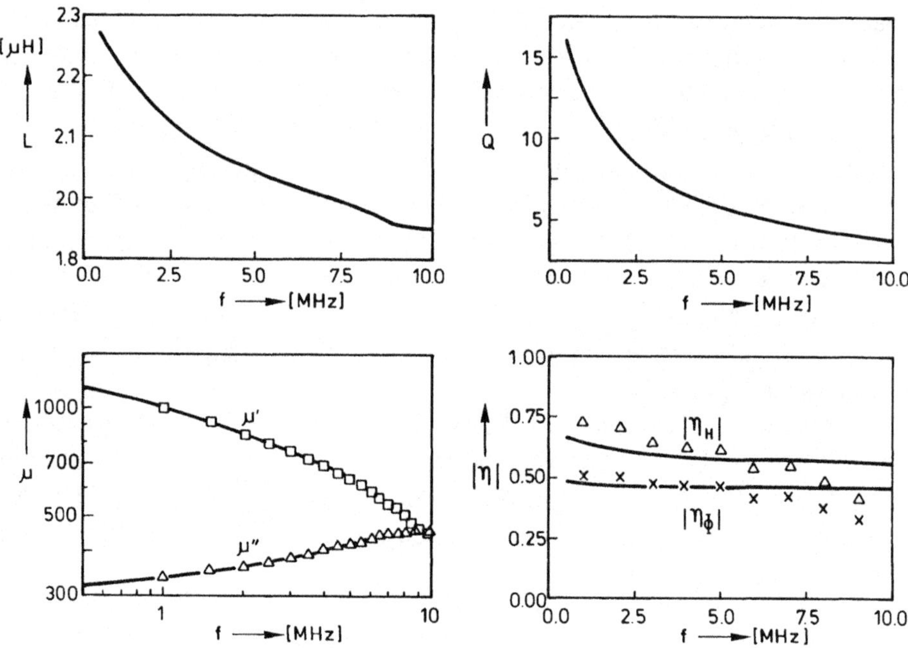

Fig. 8.56. Permeability values (3rd plot) calculated iteratively from the measured L and Q values (1st and 2nd plots). The η values as related to the permeability values are denoted by full curves in the 4th plot, while the experimentally obtained values are indicated by triangles and crosses. The input data of the MIG head used in the iterative calculation are: $\theta_1 = 90°$, $\theta_2 = 45°$, $b_1 = 26$ µm, $b_3 = 330$ µm, $b_4 = 3000$ µm, $W = 20$ µm, $W_1 = 200$ µm, $h = 45$ µm, $h_4 = 500$ µm, $h_5 = 2800$ µm, $N = 17$ and $g = 0.20$ µm.

rately described, because of the strong influence of the difficult stray flux. On the other hand $|\eta|$ in the model is described rather accurately, because the less-accurately described stray flux hardly influences $|\eta|$; the experimental determination, however, is not accurate.

The inaccuracy in the determination of the non-magnetic gap length by way of a gap-null experiment and of the optical determination of especially the gap height, limit the accuracy of both methods. This is demonstrated in Fig. 8.57 which gives the iteratively calculated values of the permeability of the core of the present MIG head if the gap height were 35 µm instead of 45 µm or if the gap length were 0.257 µm instead of 0.20 µm, both giving the same increase (28.7%) of the gap reluctance. The permeability values obtained, μ' and μ'', then change by more than 60% and 200% respectively! This is easily understood for the present high field-efficiency head, since almost all core flux is determined by the gap reluctance.

Hence small inaccuracies in the assumed gap reluctance result in large inaccuracies in the core reluctance and consequently in the μ' and μ'' values. Low efficiency heads are less sensitive to inaccuracies in the input parameters, especially those concerning the gap, but the theoretical model is less accurate for these heads, and, when the low field-efficiency head is a result of a small gap length, this small gap length can not be measured accurately enough by the gap-null method.

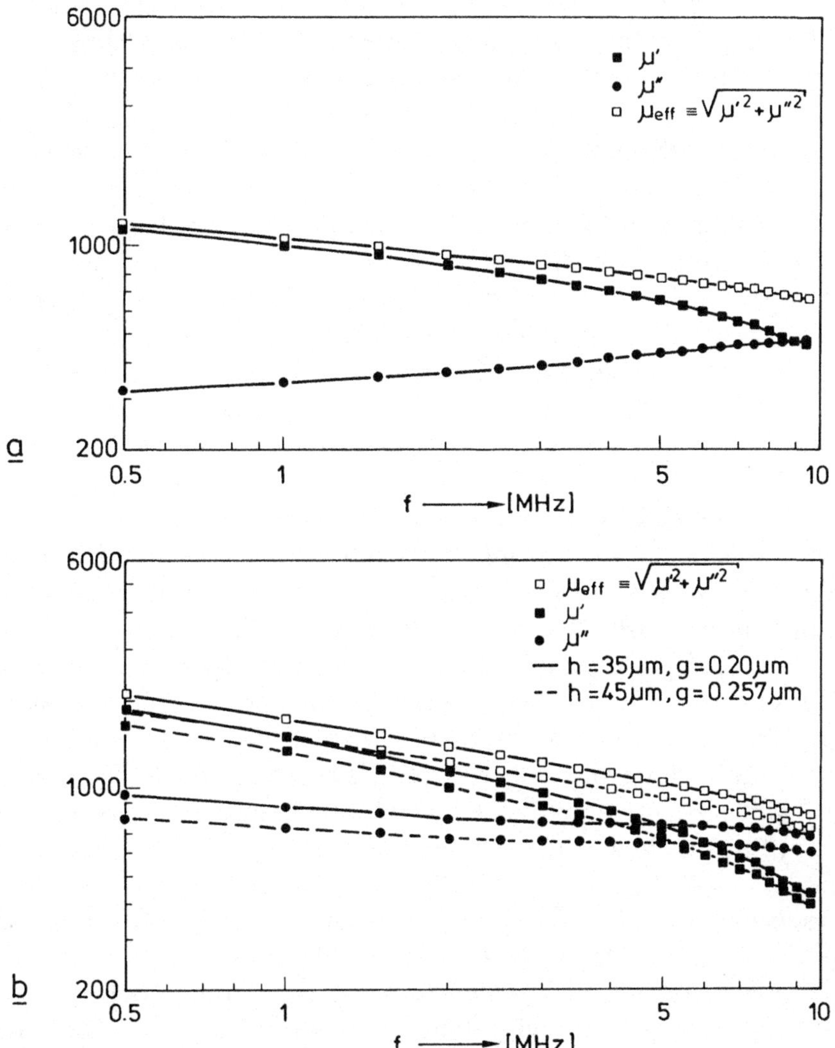

Fig. 8.57. Double-log plots of the permeability, showing the (large) influence of input data concerning the gap on the iteratively obtained permeability values.
a) The same μ', μ'' plots as in Fig. 8.56, except that the double-log scale deviates and μ_{eff} is added.
b) The permeability values when the gap height is smaller or the gap length larger.

In conclusion, a conventional video head is never suitable for a very accurate determination of the permeability values. When one needs accurate information about the permeability values after all technological steps necessary to manufacture a head, a ring measurement does not provide the right information either. A 'head' with no gap at all, or with a very small gap, preferably completely determined by well-known film thicknesses in order to include the effects of fabricating a gap, must be used for the best possible accuracy. Nevertheless, the results up till now are more than good enough to reveal whether drastic things have happened during the manufacture of the head.

8.3.4 Miniaturization

Dimensional considerations are of importance in, for instance, the miniaturization of heads.

When reducing all dimensions, all ratios remain constant. Hence the efficiencies, which depend on ratios only, stay constant. As a consequence, when one does not reduce:
- the core width, W_1, then the efficiency increases.
- the track width, W, or the gap height, h, then the efficiency decreases.
- the gap length, g, then the efficiency (strongly) increases.

Thus, the efficiency will increase notably after a strong miniaturization leaving W_1, W, h and g unchanged. Limited miniaturization under these conditions will hardly have any effect on the efficiencies.

All flux components will, in contrast, change proportionally with the linear dimensions of the head. Hence the inductance will decrease proportionally to the (linear) miniaturization factor. Also when the gap permeance is not reduced (i.e. $W \times h/g$ remained unchanged) and the core width W, remained unchanged, a considerable reduction of the (stray) flux and hence of the inductance is still possible by miniaturization.

Measurements on a mini-head, according to Fig. 8.58 (full line), show this clearly, as can be seen from the results in Table 8.14. The actual mini-head configuration is reduced to the configuration depicted by the dashed line in fig. 8.58, with dimensions given in Table 8.14, in order to make calculations possible with the same scheme as used before.

This results in calculated values, also given in Table 8.14, for both the mini-head configuration and the convenient ferrite head configura-

Fig. 8.58. Side view of the minihead configuration (full lines) and simplified model (dashed lines) used to calculate roughly the head's inductance, quality factor and efficiencies; see also Table 8.14.

Table 8.14. Calculated and measured results of the minihead, with configuration(s) depicted in Fig. 8.58.

Between the brackets the calculated results of the convenient ferrite head, with configuration defined in Table 8.8 have been given (except that $W = 10$ μm and $g = 0.24$ μm equal those of the minihead). Minihead parameters used in the calculations are: $\theta_1 = 90°$, $\theta_2 = 37.25°$, $b_1 = 10$ μm, $b_3 = 215$ μm, $b_4 = 565$ μm, $W = 10$ μm, $W_1 = 180$ μm, $h = 35$ μm, $h_4 = 250$ μm, $h_5 = 500$ μm, $N = 18$ and $g = 0.24$ μm.

μ'	μ''	L [μH]		Q		$|\eta|$		$|\eta_\Phi|$
		meas.*	calc.	meas.*	calc.	meas.*	calc.	calc.
550	450	0.90	0.79 (1.71)	7.8	6.1 (8.0)	0.68	0.73 (0.68)	0.55 (0.24)
800	400	0.93	0.82 (1.76)	4 13.9**	10.7 (13.8)	0.74	0.76 (0.71)	0.56 (0.24)

* Measured at 4.5 MHz (μ', μ'' assumed 550, 450) and at 0.75 MHz (μ', μ'' assumed 800, 400) respectively.
** The quality factor of the head with perfectly conducting windings, Q_{head}, obtained by subtracting the measured dc resistance ($R_{dc} = 0.78$ Ω) due to wire and soldering, from $R = \omega L/Q$ and subsequently calculating $Q_{\text{head}} = \omega L/(R - R_{dc})$. At 4.5 MHz the resistance of the windings (including small eddy current effects) is negligible with respect to the resistance due to μ''.

tion from Table 8.8. The measured inductance of the head is somewhat higher ($+0.2$ µH), because of the somewhat underestimated contribution of the stray flux due to the too small surface of the back part of the core in the calculation; see also Sect. 8.1.5. The calculated quality factor is a little too low for the same reason, since the stray flux is almost free of any phase lag.

As expected, the calculated (field) efficiency, η, is not significantly increased by the underestimated stray flux.

It is noted that the reduction of the inductance (i.e. stray flux) is mainly caused by locating the coil closer to the gap and hardly at all by reducing the gap surface (by way of reducing W from 20 µm to 10 µm), since the outer surfaces and hence the stray flux of the head are hardly reduced by the latter. For the (field) efficiency this gap-surface reduction is at least as important, because of the importance of the ratio of gap reluctance to core reluctance in its determination.

8.3.5 Simulation by way of physically admissible permeability versus frequency curves

For model simulations of hypothetical heads it is advisable to take account of basic physical constraints on $\hat{\mu}(\omega) = \mu'(\omega) - j\mu''(\omega)$. Firstly, there exists something like a finite permeability $\times$ bandwidth product which is proportional to the saturation magnetization in eddy-current free samples.

A second constraint is that eddy currents in electrically conducting magnetic materials reduce the bandwidth.

Thirdly, due to causality, there exists a relation between the real and the imaginary part of the complex permeability versus frequency curve. For linearly-acting materials these are the well-known Kramers-Kronig or causality relations.

It is the purpose of this section to give relations for the above effects and to derive relations for combinations of the above effects. The use of $\hat{\mu}(\omega)$ curves obtained with the aid of those expressions enables us to predict roughly the optimal behaviour of heads manufactured from material with given saturation magnetization, resistivity, lamination thickness and geometry. The applicability of the expressions and the use of $\hat{\mu}(\omega)$ curves in the analytical head model are illustrated by a few examples, using measurements on a single-crystalline ferrite ring, an amorphous ribbon and a sputtered crystalline Co alloy.

Finally, the method of incorporation of the eddy current expression into the analytical head model for both electrically-conducting 'bulk' heads and (laminated) sandwich heads is outlined.

Uniform rotation in eddy-current-free samples

A description of $\hat{\mu}(\omega)$ for uniform rotation magnetization has been given in appendix 8.1 and for the relative permeability component normal to the easy axis direction (and magnetization) this leads for zero demagnetizing fields to:

$$\hat{\mu} \equiv \mu' - j\mu'' = 1 + \chi_{r0} \frac{\left\{1 - \left(\dfrac{\omega}{\omega_{res}}\right)^2\right\} - j\alpha \dfrac{\omega}{\omega_{res}}\left\{1 + \left(\dfrac{\omega}{\omega_{res}}\right)^2\right\}}{\left\{1 - \left(\dfrac{\omega}{\omega_{res}}\right)^2\right\}^2 + \left\{2\alpha \dfrac{\omega}{\omega_{res}}\right\}^2},$$

$$(8.90)$$

where $\omega_{res} = \gamma H_{an}$ is the (ferromagnetic) resonance frequency,
$\gamma \approx 1.5 \cdot 10^5$ [m/As] is the gyromagnetic constant,
H_{an} is the anisotropy field,
$\chi_{r0} = \gamma M_s/\omega_{res} = M_s/H_{an}$ is the rotational susceptibility at zero frequency, and
α is Gilbert's damping constant ($\alpha \ll 1$).

Eddy-current expression for complex intrinsic permeability

Eddy currents give rise to an 'effective' relative permeability

$$\hat{\mu} \equiv \mu' - j\mu'' = \frac{\hat{\mu}_i}{\hat{k}} \frac{(\sinh \hat{k} + \sin \hat{k}) - j(\sinh \hat{k} - \sin \hat{k})}{\cosh \hat{k} + \cos \hat{k}}. \qquad (8.91)$$

where $\hat{\mu}_i$ = the intrinsic relative permeability of the material at the considered frequency,
$\hat{k} \equiv l/\hat{\delta}$ the ratio of the lamella thickness to the skin depth,
l = the thickness of the lamellae,
$\hat{\delta} \equiv \sqrt{2\varrho/(\omega\hat{\mu}_i\mu_o)}$ the complex 'skin depth' and
ϱ = the specific resistance of the laminations.

This expression is a generalization, for complex $\hat{\mu}_i$, of the result obtained by e.g. Olsen [8.22] for real permeabilities (and hence real δ and k).

Often the real dc permeability is substituted for $\hat{\mu}_i$, but this is only correct at frequencies where relaxation and resonance processes are not yet of influence. At video frequencies these processes are usually of importance.

Dispersion in the anisotropy and demagnetizing field

The uniform-rotation model predicts decays of $\chi'(\omega)$ and $\chi''(\omega)$ beyond the resonance frequency that are (much) steeper than experimentally observed; see e.g. references [8.23] and [8.24]. This problem can be overcome by introducing a probability density, $p(\omega_{res})$, which accounts for the effect that the anisotropy field is not uniform. Hence both ω_{res} and $\chi_{r0} = \gamma M_s/\omega_{res}$ vary from place to place. Dispersion in the direction of the anisotropy field is neglected. Averaging then leads to an approximate expression for the effective susceptibility:

$$\langle \hat{\chi}(\omega) \rangle = \int_{-\infty}^{\infty} \frac{\gamma M_s}{\omega_{res}} \frac{\left\{ 1 - \left(\dfrac{\omega}{\omega_{res}} \right)^2 \right\} - j\alpha \dfrac{\omega}{\omega_{res}} \left\{ 1 + \left(\dfrac{\omega}{\omega_{res}} \right)^2 \right\}}{\left\{ 1 - \left(\dfrac{\omega}{\omega_{res}} \right)^2 \right\}^2 + \left\{ 2\alpha \dfrac{\omega}{\omega_{res}} \right\}^2} p(\omega_{res})\, d\omega_{res} \,.$$

$$(8.92)$$

The relations between $\langle \chi'(\omega) \rangle$ and $\langle \chi''(\omega) \rangle$ in this expression are given (for small α, say ≤ 0.1) by the well-known Kramers-Kronig (or Causality) relations

$$\chi'(\omega) = \frac{2}{\pi} \int_0^{\infty} \frac{\omega_1 \chi''(\omega_1)}{\omega_1^2 - \omega^2}\, d\omega_1 \qquad (8.93a)$$

and

$$\chi''(\omega) = -\frac{2}{\pi} \int_0^{\infty} \frac{\omega \chi'(\omega_1)}{\omega_1^2 - \omega^2}\, d\omega_1, \qquad (8.93b)$$

because the χ' and χ'' and hence $\langle\chi'\rangle$ and $\langle\chi''\rangle$ functions are obtained from a physically admissible model, (such that causality is assured) and the material acts linearly.

Results of (8.92) in the case of symmetrical Gaussian probability densities, with mean width, Δ, and central angular frequency, ω_0,

$$p(\omega_{\text{res}}) = \frac{1}{\Delta}\, e^{-\frac{\pi}{\Delta^2}(\omega_{\text{res}} - \omega_0)^2}, \tag{8.94}$$

are shown in Fig. 8.59, curves $--$ and $-----$.

Also for high values of Δ and α, a sign change occurs and the decay beyond the roll-off frequency is much too steep. Usually one observes decays of roughly 14-20 dB/dec. and no sign change for $\mu'(\omega)$, excepted at extremely high frequencies, and a decay of 10-15 dB/dec. for $\mu''(\omega)$, while $\mu' \approx \mu''$ just beyond the roll-off frequency. See for example the permeability versus frequency curves of $Ni_\delta Fe_{1-\delta}Fe_2O_4$, measured by Smit and Wijn (Fig. 48.1 in [8.23]), and the spectrum of the MnZn-ferrite ring in Fig. 8.60. The reason is that in an actual sample (in zero dc field) there will be regions where the demagnetizing field approximates M_s, see [8.23], page 98. This demagnetizing field adds to the crystal-

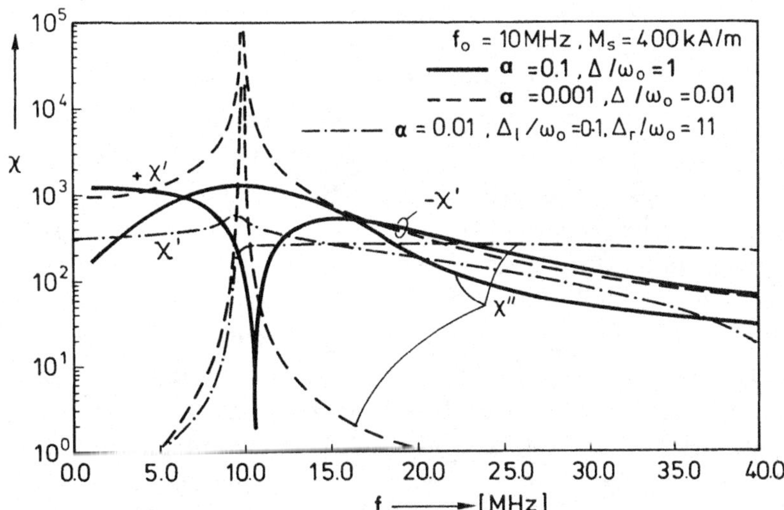

Fig.8.59. Results from Eq. (8.92) ('rotational-permeability model including dispersion in the strength of the anisotropy field').
Curves —— and ---; Symmetrical Gaussian probability densities. Curves -·-·; Asymmetrical 'Gaussian' probability density, i.e. with widths $\Delta = \Delta_l$ for $\omega_{\text{res}} < \omega_0$ in (8.94) and $\Delta = \Delta_r$ for $\omega_{\text{res}} > \omega_0$ in (8.94) and amplitude $2/(\Delta_l + \Delta_r)$ instead of $1/\Delta$.

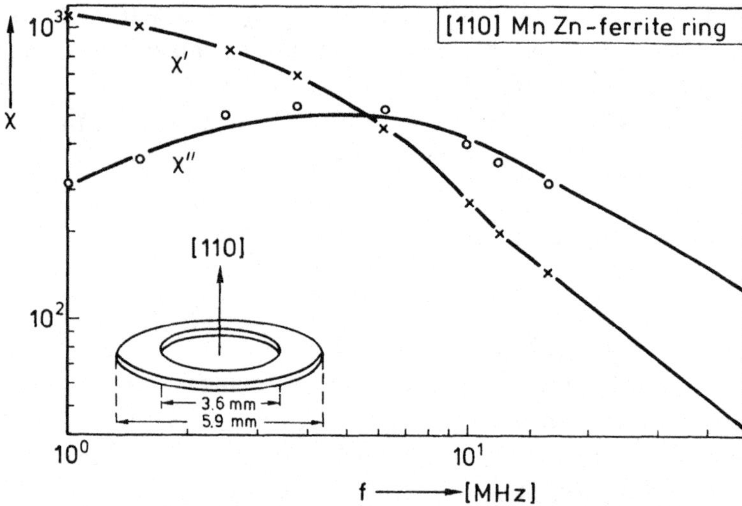

Fig. 8.60. The susceptibility of saw-cut [110]-oriented MnZn-ferrite rings. The crosses (and circles) indicate the measuring points, except those at 100 kHz ($\mu' = 1214$) and 10 kHz ($\mu' = 1417$).

anisotropy field $\boldsymbol{H}_{an}$, see expression (a8.10) in appendix 8.1. Consequently the range of possible values of ω_{res} reads:

$$\gamma H_{an} \leqslant \omega_{res} \leqslant \gamma(H_{an} + M_s) = (\chi_{r0} + 1)\gamma H_{an}. \qquad (8.95)$$

Since $\chi_{r0} >>> 1$ for the soft-magnetic materials of interest, the dispersion in ω_{res} due to the dispersion in the demagnetizing field must be expected to be much larger than the dispersion in H_{an}. A strongly-asymmetrical probability density $p(\omega_{res})$ with a very long tail on the hf side roughly describes this behaviour [8.25]. It leads to more realistic curves for $\mu'(\omega)$ which drop considerably slower beyond ω_0 and to negative values of μ' at radial frequencies far above ω_0, see Fig. 8.59 curve $-\cdot-\cdot$.

Measured permeability versus frequency curves do not only deviate from results obtained by the uniform-rotation model with symmetrical probability density, but also deviate from results from a simple relaxation model ($M + (\mathrm{d}M/\mathrm{d}t)/\omega_{rel} = \chi_0 H$), for which $\alpha = 1$ in $\hat{\chi} = \chi_0/(1 + \mathrm{j}(\omega/\omega_{rel})^\alpha) = \chi_0/(1 + (\omega/\omega_{rel})^{2\alpha}) - \mathrm{j}\,\chi_0(\omega/\omega_{rel})^\alpha/(1 + (\omega/\omega_{rel})^{2\alpha})$. Only for much smaller α ($\simeq 0.5$) is there a reasonable correspondence with experimental curves of samples free of eddy currents. Introduction of a 'mass' term ($\sim \mathrm{d}^2M/\mathrm{d}t^2$) in the relaxation model may recover the resonance effect (peak) in the permeability versus frequency curve.

A drawback of making $\alpha \neq 1$ is that the Kramers-Kronig relations are not valid anymore.

Use of Kramers-Kronig relation

Another way to obtain physically admissible $\chi'(\omega)$, $\chi''(\omega)$ combinations for model calculations is to assume a proper $\chi'(\omega)$ curve, for instance the previous $\chi'(\omega)$ with $\alpha = 0.4$, and to calculate $\chi''(\omega)$ using the Kramers-Kronig relation (8.93b). In order to avoid the singularity in the integral this expression is rewritten, by using $\int_0^\infty (\omega_1^2 - \omega^2)^{-1}d\omega_1 = 0$, to

$$\chi''(\omega) = -\frac{2}{\pi} \int_0^\infty \frac{\omega[\chi'(\omega_1) - \chi'(\omega)]}{\omega_1^2 - \omega^2}\,d\omega_1. \qquad (8.96)$$

This simplifies the numerical integration.

The trapezoidal rule is used to carry out this integration up to $\omega_1 = 10\omega$, which assures enough accuracy in the case of practical $\chi'(\omega_1)$ curves. On the hf side the data are extended by assuming that the decay is linear on a double-log scale, with a steepness given by the last (measured) input-data points. On the lf side and for intermediate data points linearity is assumed. A hundred integration steps usually suffice.

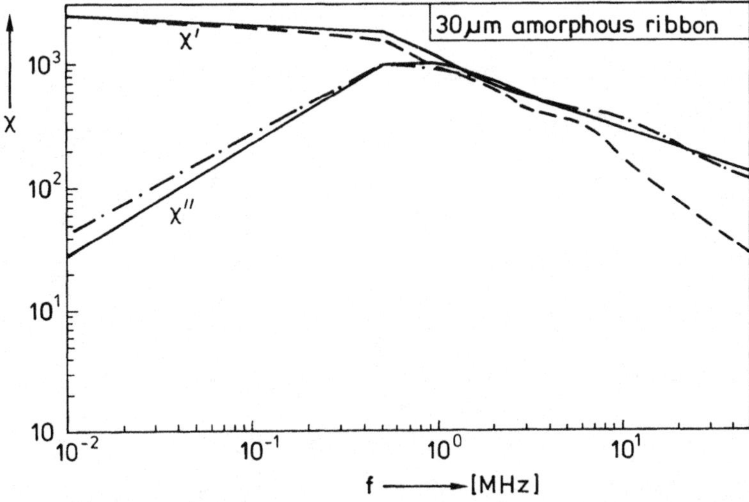

Fig. 8.61. The in-plane susceptibility of 30 μm thick amorphous ribbon as measured on strips, and as calculated with the eddy-current expression (8.91), assuming that the intrinsic permeability is constant. --- χ' as measured, -·- χ'' from Kramers-Kronig relation, —— χ' and χ'' from eddy current expression ($l = 30\,\mu$m, $\varrho = 10^{-6}\,\Omega$m, $\hat{\mu}_i(\omega) = \chi_{dc} + 1 = 2457$).

An example is given in Fig. 8.60, based on the χ'' measurement (by way of measuring L) on small [110] – oriented MnZn – ferrite rings. The orientation corresponds to the orientation in VHS video heads. The χ'' curve obtained corresponds well with the measured χ'' values.

In Fig. 8.61, the χ'' has been calculated from χ' by eq. (8.96) for an amorphous ribbon, made by melt-spinning, with composition $Co_{70}Fe_5Si_{15}B_{10}$, a thickness, l, of 30 μm and specific resistance, ϱ, of 10^{-6} Ωm (further $M_s = 600$ kA/m, i.e. $\mu_0M_s = 0.75$ T). The eddy-current calculation is based on the dc value of the measured χ' component in the plane of the film (measurements by de Wit and Jager). The calculated χ'' values correspond to the measured values (not shown). Most of the roll-off is apparently caused by eddy currents. This is because of the large thickness of the non-laminated ribbon.

Laminations and eddy currents

Fig. 8.62 shows the in-plane susceptibility, as measured by de Wit and Jager on a very thin (1.08 μm) crystalline Co alloy of Dirne and Brouha, see also [8.26], made by sputtering and of composition $Co_{86}Fe_6B_6Si_2$ ($\mu_0M_s \simeq 1.5$ T). Due to the twinned (180°) cubic [111]

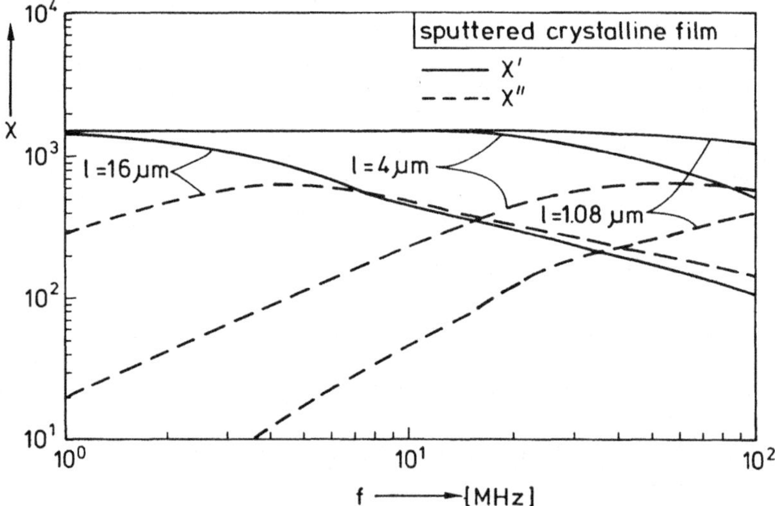

Fig. 8.62. The in-plane susceptibility of a crystalline film (easy axis perpendicular to the direction of the applied field) made by sputtering:
– as measured on a 5 mm × 34 mm strip with a film thickness of only 1.08 μm,
– as calculated (eddy currents, assuming $\varrho = 10^{-6}$ Ωm in Eq. 8.91) for 4 and 16 μm thick lamellae, by assuming that the intrinsic susceptibility equals that of the 1.08 μm thin film.

(columnar) texture perpendicular to the film plane, there are 2-times three weak 'easy axes', in the plane of the film. Because of the small thickness, the eddy currents are negligible over a wide frequency range. Also the resonance and roll-off frequency will be large compared to the usual $\omega_{res} \approx \gamma H_{an}$, due to the strong demagnetizing field for a rotation perpendicular to the (easy) plane of the film, as explained in Appendix 8.1. Therefore these values may be interpreted as intrinsic material parameters and are thus particularly suited as input for model calculations, where the number of laminations or the lamination thickness, l, is varied. An increase of the lamella thickness may decrease the roll-off frequency due to ferromagnetic-resonance effects, but this decrease is unlikely to be so drastic that it may overshadow the strongly increased eddy-current effects.

In contrast to the susceptibility versus frequency curves of the eddy-current-free samples, where ferromagnetic-resonance effects and relaxation effects usually dominate the hf behaviour and decays of $\chi''(\omega)$ and $\chi'(\omega)$ are in the ranges 10-15 dB/dec. and 14-20 dB/dec. respectively and 'hence' χ'' dominates beyond the roll-off frequency, we have in samples of highly-conductive material, where eddy-current effects dominate, a slow decrease of 10 dB/dec. since $\hat{\delta} \sim \omega^{1/2}$, in both $\chi'(\omega)$ and $\chi''(\omega)$. This means that in the limit of very high frequencies or very small lamella thickness the resonance effects and hence χ'' always dominate. Before this happens and beyond the roll-off frequency due to the eddy currents, χ' will equal χ''.

Implications of eddy currents on analytical head model

Before applying the results in the Figs. 8.60, 8.61 and 8.62 as input in the analytical head model, in order to obtain the head's electrical and magnetic parameters like L, φ, η etc., some notes have to be made.

In the case of sandwich heads ($W_1 = W$), the eddy-current expression (8.91) can directly be applied in the model, independent of the number of laminations, and hence the susceptibility values as given in the figures 8.61 and 8.62. Only in the neighbourhood of the gap, where h is not much larger than W, is the effective permeability underestimated by (8.91). It is also of importance to note that on both sides of the gap a skin must be formed parallel to this gap (except for unrealistically short gap lengths), because the magnetic scalar potential at the centre of the gap surface can only be little less than at the edges. So almost the same flux density crosses the gap at the centre as near the edges. Hence R_g must retain its original value and not be increased to a value corresponding to only the skin depth further away from the gap!

In 'bulk' heads, i.e. heads with $W_1 > W$ and no laminations, made of e.g. amorphous or crystalline metallic materials, which possess eddy currents at higher frequencies, the effective permeability depends on the local width of the core. Hence the model is not directly applicable to this case.

When, however, the skin depth is small compared to the cross-sectional dimensions of the head, a reasonable analytical approximation is possible. In that case, the h.f. approximation for the skin effect, $\hat{\mu} = \hat{\mu}_i(1 - j)/\hat{k}$ see (8.91), has to be replaced by $\hat{\mu} = \hat{\mu}_i(1 - j)(1/\hat{k}_1 + 1/\hat{k}_2)$, where $\hat{k}_1 \equiv \hat{\delta}/l_1$ and $\hat{k}_2 = \hat{\delta}/l_2$ with l_1 and l_2 the local cross-sectional dimensions and $\hat{\delta} \equiv \sqrt{2\varrho/(\omega\hat{\mu}_i\mu_0)}$ the complex 'skin depth'. Applied to region E, see also Fig. 8.7, this leads to

$$\hat{\mu} = \hat{\mu}_i \; \frac{1 - j}{2} \cdot \frac{\hat{\delta}2(\theta'r' + \theta''r'')}{\theta'r'\theta''r''} \;, \qquad (8.97)$$

i.e. $\hat{\mu}$ is proportional to the ratio of the complex 'cross-sectional area', $2\hat{\delta}(\theta'r' + \theta''r'')$, to the cross-sectional area, $\theta'r'\theta''r''$.

Hence (8.4) is replaced by

$$dZ = \frac{dr}{\mu_0\hat{\mu}_i(1 - j)\hat{\delta}(\theta'r' + \theta''r'')} \;. \qquad (8.98)$$

The integral value, Z, is easily calculated, since r' and r'' differ by only a constant from r, and leads, after substitution of the dimensions W_1, W, h_2 and h_1, and angles $\theta_1 = \theta'$ and $\theta_2 = \theta''$ of region E in the resulting expression, to:

$$Z_E \simeq \frac{1}{\mu_0\hat{\mu}_i(1 - j)\hat{\delta}(\theta_1 + \theta_2)} \ln \frac{W_1 + h_2}{W + h_1} \;. \qquad (8.99)$$

Region F can be treated in the same way and leads to

$$Z_F \simeq \frac{1}{\mu_0\hat{\mu}_i(1 - j)\hat{\delta}\theta_2} \ln \frac{W + h_1}{W + h_0} \;. \qquad (8.100)$$

The above results resemble (8.3a); see also Fig. 8.5. The minimum and maximum radii, r_{min} and r_{max}, are only replaced by the minimum and maximum circumferences, and the constant core width, l, is replaced by the complex skin depth, $\hat{\delta}$.

For the other regions, A to D, (8.91) is still applicable, since one of its cross-sectional dimensions (the core width) is small compared to the other. Taking this second dimension into account refines the calculation.

The application of the above concept is restricted to skin depths that are small compared to the track width, because of the hf approximation used in the regions E and F.

Analytical head-model calculations

Results of model calculations, based on the (intrinsic) susceptibility spectra of the MnZn ferrite of Fig. 8.60, the amorphous ribbon of Fig. 8.61 and the sputtered crystalline film of Fig. 8.62, are given in Figs. 8.63, 8.64 and 8.65 respectively.

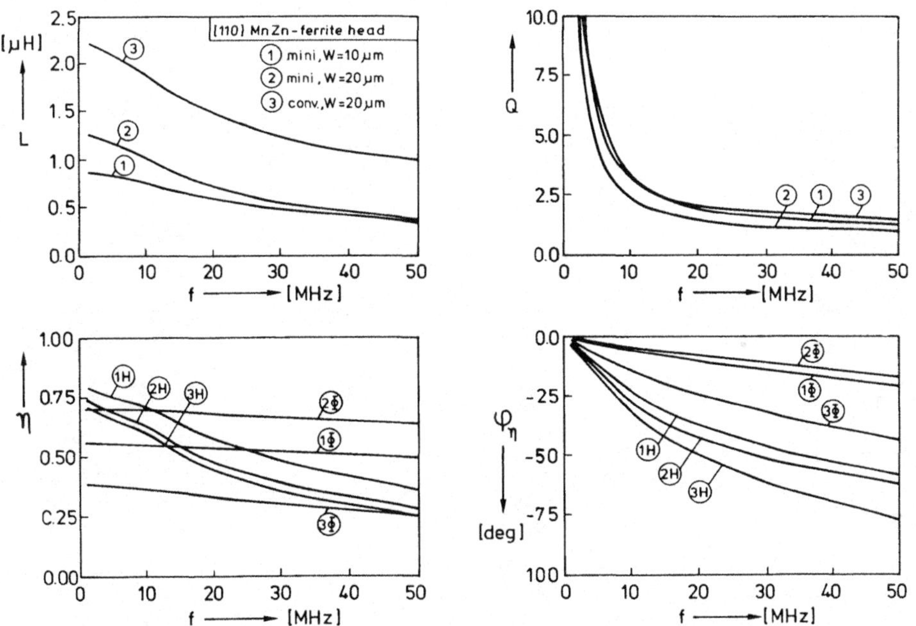

Fig. 8.63. The induction, L, quality factor, Q, and complex field (H) and flux (Φ) efficiency, $\hat{\eta}$ = $\eta\, e^{i\varphi}$, as resulting from model calculations on heads with different geometry and made of [110]-oriented MnZn ferrite with $\hat{\chi}$ values as given in Fig. 8.60.
The geometry of the minihead (curves 1 and 2) has been defined in Table 8.14, and that of the conventional head in Table 8.8. The track width is indicated in the figure.

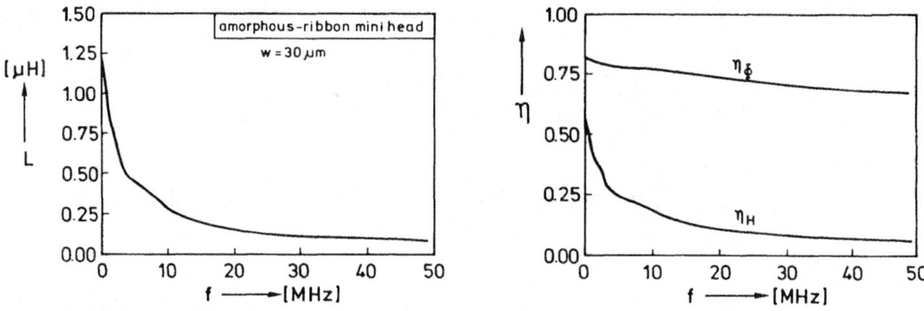

Fig. 8.64. Calculated L and η curves of a minihead based on the (measured curve of the) amorphous ribbon of Fig. 8.61.
No extrapolations have been made to heads with smaller track widths by recalculating the eddy currents, because ferromagnetic resonance may limit the permeability in these cases.

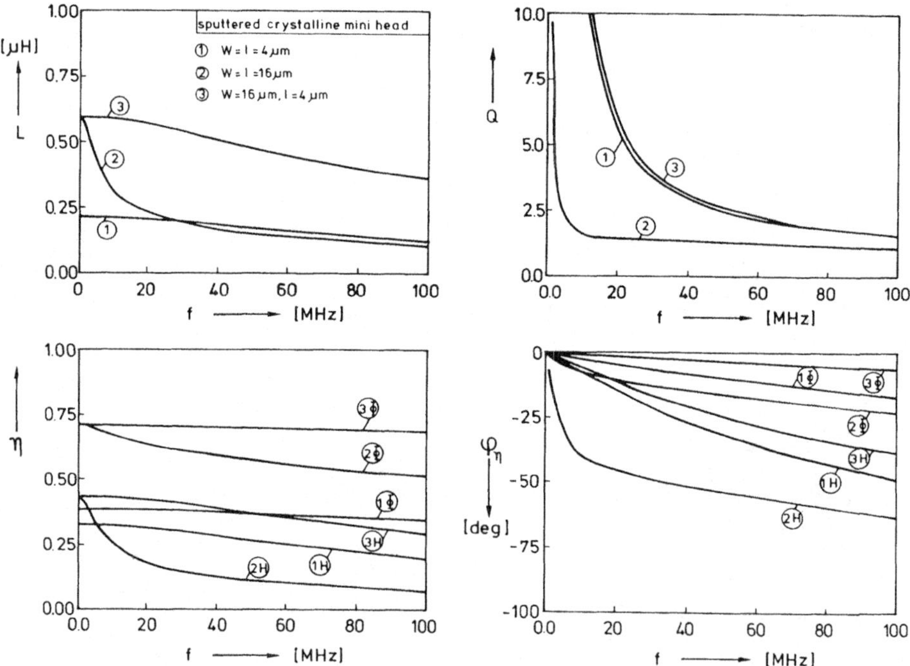

Fig. 8.65. Calculated minihead parameters.
The calculations are based on the measured permeability values of the sputtered crystalline film and calculated eddy-current effects, shown in Fig. 8.62. The extrapolations to larger lamellae thickness are allowed, since it is not likely that the ferromagnetic resonance frequency will decrease strongly as a result of this increase.

In Fig. 8.64 no extrapolations are shown to heads with smaller track widths, by recalculating the eddy current effect, because ferromagnetic resonance may limit the susceptibility in these cases, instead of the eddy currents.

Extrapolations to larger track widths and lamella thicknesses are less risky, because a large reduction of the roll-off frequency due to ferromagnetic-resonance effects with increasing lamella thickness is not likely. This extrapolation was already carried out for the susceptibility spectrum of the very thin sputtered film in Fig. 8.62 and leads, after application in the head model, to heads with parameters as shown in Fig. 8.65.

The laminated implementation, 3, of this head shows a 'reasonable' efficiency up to the highest frequency, in contrast to the unlaminated implementation, 2, and the unlaminated amorphous-ribbon head of Fig. 8.64. The head implementation with a smaller track width, 1, is worse, because of the relatively large influence of the stray flux on the potential decrease in the core, as explained before. The flux efficiency, η_ϕ, and the quality factor, Q, of the laminated head are high, and the phase lags are small, consistent with the good magnetic properties of the laminated film.

A maximum flux efficiency means a minimum percentage of stray flux and hence a minimum induction, L. The latter is of importance for minimum (thermal) noise, introduced by the 'equivalent current-noise source' of the pre-amplifier, and of importance for the highest-possible resonance frequency of the input circuit of the pre-amplifier.

The tremendous improvements in the head's hf performance when laminations are applied is best illustrated by curves 1 and 3 in Fig. 8.65. A factor of 4 decrease of the lamination thickness causes the head's parameters L, η, Q and φ at the frequency $4^2 \times f$ to be still as good as those of the unlaminated head at frequency f. This is clearly visible in Fig. 8.65, and originates from the fact that the ratio of lamination thickness to 'skin depth', $\hat{k} = l/\hat{\delta}$, is proportional to the lamination thickness and only proportional to the square root of the frequency $(\hat{\delta} = \sqrt{2\varrho/(\omega\hat{\mu}_i\mu_0)})$, while resonance and relaxation effects are negligible.

The hf behaviour of the MnZn-ferrite head is reasonable, in spite of its low susceptibility at high frequencies, because the core width is much larger than the track width.

Application of the minihead geometry especially reduced the induction, L, due to the strongly reduced outer surface of the head, and hence reduced stray flux. Consequently the flux efficiency increased considerably. The field efficiency, however, is not significantly increased because of the dominating influence of the gap and the narrow bridge and its surrounding on the magnetic-potential loss in the core.

8.4 Comparison of model with FEM calculations

The results obtained with the analytical (video) head model predicted the experimental results satisfactorily. Still a check with results obtained by the finite-element method (FEM) seems desirable to get more insight into the accuracy of the calculations when applied to different video-like heads. The available 2-dimensional (MAGGY) and 3-dimensional (PADDY) finite-element packages for magnetostatic problems do not use complex permeabilities. Hence checks on the quality factor are impossible.

Although graphic interactive programs (GEMMY) are available to generate input for those packages, the generation is still so complex for video-like heads, and hence so time-consuming, that we restricted ourselves for simplicity to only three head types with a core width equal to the track width.

The first head type (MV in Table 8.15) is more or less equivalent to the conventional ferrite head (Table 8.8), but the core width and track width are so large with respect to the lateral core dimensions that the head can be treated 2-dimensionally. The results from the analytical calculations (A) are very close to the results from the numerical FEM (N), see Table 8.16 or the bar diagrams in Fig. 8.66, even when the relative permeability is very low (200 instead of 800).

The second head type (PV) has a relative core width, W_1, and track width, W, of 20 μm and further equals the preceding 'head', see Table 8.15. Hence a 3-dimensional (PADDY) FEM package is used. The agreement between N and A is reasonable (see PV800 in Table 8.16), but obviously worse than for the preceding 2-dimensional head. By modifying and increasing the 'mesh line density' (i.e. the number of meshes) in the neighbourhood of the gap (see PVf800, where f means fine), a closer agreement is obtained. Also the calculation with the fine mesh will have at least a few percent inaccuracy. We did not investigate this further because of expanding CPU (Central Processing Unit) times;

see last column, with increasing number of mesh lines N_x, N_y and N_z.

With the same number of mesh lines, runs were carried out with a relative permeability of 4000 (PVf4000) and with a reduced gap height of 10 μm (PVfh800). As expected, the analytical results for the head with the highest flux and field efficiency, i.e. PVf4000, approaches most closely the results from FEM.

Finally we carried out 3-dimensional calculations on miniheads. The configuration used in the FEM calculation (PADDY) is depicted in Fig. 8.67. The corresponding dimensions as used in the analytical model are given in the last column of Table 8.15. It is noted that the 'length' of the back yoke in the analytical model equals the core length of 565

Table 8.15. Input parameters in the analytical head-model calculations. As input for the FEM calculations the same configurations have been used, except that the length of the back yoke of the minihead in the FEM calculations is 3000 μm instead of 565 μm (see Fig. 8.67) which is more in agreement with the actual configuration of our experimental miniheads, but hardly influences the results. Results of the 'analytical' and 'numerical' calculations are collected in Table 8.16 and Fig. 8.66.

Input parameter	Head type		
	MV..	PV..	PM..
θ_1*	90°	0.0637°	90°
θ_2	45°	45°	37.25°
b_1*	20 μm	20 μm	10 μm
b_3	300 μm	300 μm	215 μm
b_4	3000 μm	3000 μm	565 μm
W	20000 μm	20 μm	20 μm
W_1	20001 μm	20.1 μm	20.1 μm
h	35 μm	35 μm**	20 μm
h_4	500 μm	500 μm	250 μm
h_5	3000 μm	3000 μm	3000 μm
N	18	18	18
μ***	see Table 8.16	see Table 8.16	see Table 8.16
g	0.2 μm	0.2 μm	0.2 μm

* These input parameter values are of no importance, since W approaches W_1.

** Except for head PVfh800, where $h = 10$ μm.

*** The check of the analytical head model is restricted to isotropic non-complex permeabilities, because of limitations of the FEM packages.

MV Conventional ferrite head dimensions, see Table 8.8, except that the core and track width, W_1 and W, are large compared to the lateral dimensions of the core. Hence the head can be treated 2-dimensionally.

PV Conventional ferrite head dimensions ($g = 0.2$ μm however).

PM Conventional minihead dimensions, see Table 8.14 ($g = 0.2$ μm, $h = 20$ μm, $W_1 \approx W = 20$ μm and $h_5 = 3000$ μm however).

Table 8.16.

Head	η_H			$L\,[\mu H]$			η_Φ			$N_x \times N_y \times N_z$	totcpu
	N	A	D	N	A	D	N	A	D		[min]
MV800	0.33	0.32	− 3%	0.50*	0.49*	− 1%	0.96	0.94	− 2%	28×67	0.37
MV200	0.10	0.10	+ 4%	0.18*	0.17*	− 8%	0.91	0.91	0%	28×67	0.37
PV800	0.25	0.21	−16%	1.15	0.90	−22%	0.32	0.33	+ 6%	$21 \times 39 \times 27$	12
PVf800	0.23	0.21	− 8%	1.05	0.90	−14%	0.31	0.33	+ 6%	$27 \times 51 \times 26$	22
PVfh800	0.36	0.31	−13%	0.91	0.78	−14%	0.16	0.17	+ 3%	$27 \times 51 \times 26$	22
PVf4000	0.62	0.62	0%	1.98	1.89	− 4%	0.45	0.47	+ 4%	$27 \times 51 \times 26$	23
PM800	0.24	0.29	+20%	0.49	0.57	+17%	0.42	0.42	0%	$20 \times 41 \times 26$	11
PM2000	0.45	0.53	+17%	0.81	0.86	+ 6%	0.46	0.51	+11%	$20 \times 41 \times 26$	11
PM4000	0.60	0.70	+17%	1.00	1.06	+ 6%	0.49	0.54	+11%	$20 \times 41 \times 26$	11

* Per 20 μm of track width

N = Numerical FEM package, i.e. MAGGY for 2-dim. magnetostatic problems (heads MV800 and MV200) and PADDY for 3-dim. magnetostatic problems (heads PV800 and so on).

A = Analytical head model.

D = Difference between result obtained by the analytical and the FEM calculations.

N_x, N_y, N_z = Number of mesh lines in the x and y (MAGGY) or x, y and z (PADDY) direction.

totcpu = Total central-processing-unit time, necessary for the numerical FEM computation on an IBM3081. For the 'analytical' head model this time is less than a millisecond (excluding about 1 sec. I/O time), i.e. more than a million times faster!

Head = Head identification. M stands for MAGGY, P for PADDY, f for fine (more meshes), v for V2000 or conventional ferrite head, h for reduced gap height (10 μm instead of 35 μm) and the second M for minihead.

The value of the isotropic and non-complex relative permeability, μ_r, is given by the number in the head identification.

The remaining input parameters of the heads are given in Table 8.15

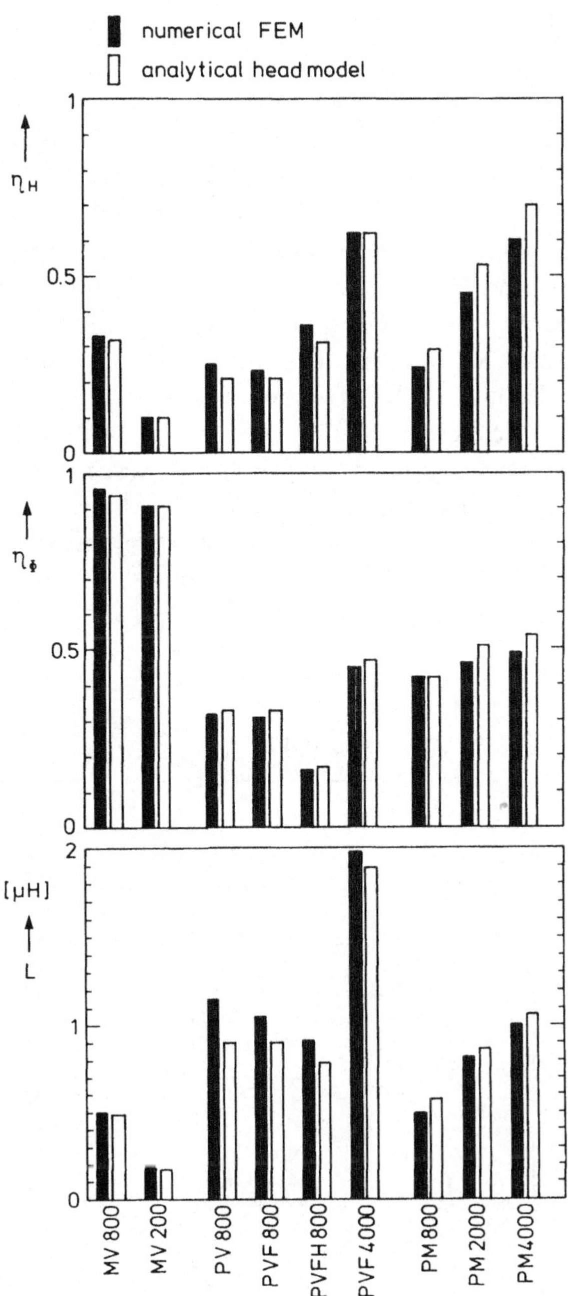

Fig. 8.66. Bar diagrams of the results collected in Table 8.16.

µm, and hence differs from the actual 3000 µm used in the FEM calculation (Fig. 8.67). This simplification was necessary because of the low
flexibility of the present 3-dimensional analytical video-head model.
The analytical results differ by at most 20% from the more accurate
results of the finite-element calculation.

The CPU time, excluding the I/O (input/output) time for the analytical calculation of as many as 200 heads, is less than 0.1 sec. on an IBM
main-frame computer, i.e. less than a millisecond for each head. This
is more than a million times shorter than for the corresponding 3-dim.
FEM calculation on the same main frame; see Table 8.16.

The time necessary to generate input for the analytical model is very
short; a minute or so, whereas this costs hours for a simple video head
($W = W_1$) and some days for a video head with a core width, W_1,
unequal to the track width, W.

The accuracy and flexibility of the analytical program are less than
those of the FEM, but satisfactory for the development and evaluation
of video-like heads. In conclusion, the analytical model with respect to
the FEM

– cannot take non-linear effects into account,

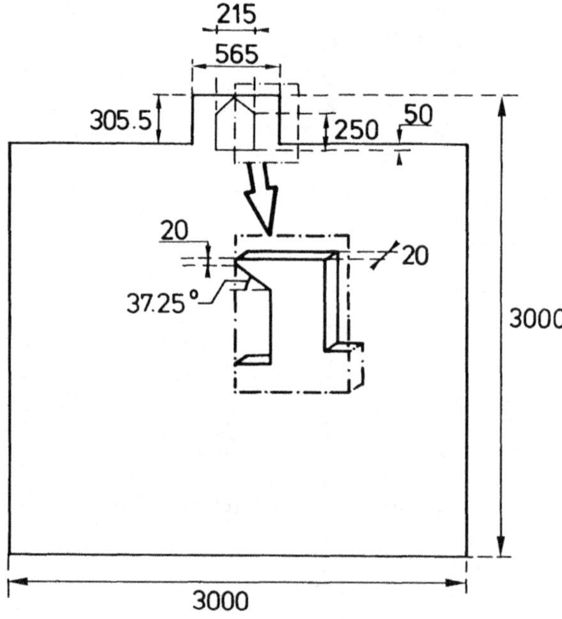

Fig. 8.67. Side view of the miniheads, PM800, PM2000 and PM4000, as used in the 3-dimensional
finite-element (PADDY) calculations.

- is restricted to video-like heads,
- is less accurate (accuracy better than 20 percent),
- is more than a million times faster (CPU time),
- has dedicated input generation (10^2 to 10^3 times faster)
- uses complex permeabilities, i.e. also calculates the (core)losses and the important quality factor, Q, of the head. The complex behaviour is essential for accurate calculations at video and higher frequencies.

Appendix 8.1 Tensor character of stress and frequency dependence of rotational permeability

Tensor character of stress

In a MnZn monocrystal [8.12] two magnetostriction constants, λ_{100} and λ_{111}, suffice, because of cubic symmetry. With these two, the relative elongation for any direction of the spontaneous magnetization can be described (if the crystal is free to move in these directions). When, under the influence of stress, the crystal further elongates (or contracts), another change in electron energy takes place (which usually does not possess cubic symmetry). When the strain $\varepsilon = \sigma/E$ (where E is Young's modulus of elasticity) induced by this (externally applied) stress exceeds the strains (elongation λ_{100} as well as elongation λ_{111}) induced by the spontaneous magnetization, a linear relation between stress and strain can be expected. Then the following phenomenological expression can be derived, see [8.27] or [8.12], for this extra energy density E_{MS} dependent on the direction of magnetization:

$$E_{\mathrm{MS}} = -\tfrac{3}{2}\lambda_{100} \sum_{i=1}^{3} \sigma_{ii}\alpha_i^2 - 3\lambda_{111} \sum_{i>j=1}^{3} \sigma_{ij}\alpha_i\alpha_j \qquad (\mathrm{a}8.1)$$

where the α_i's are the direction cosines (M_x/M_s, M_y/M_s and M_z/M_s) of $\vec{M}$, so $\sum_{i=1}^{3} \alpha_i^2 = 1$, and the σ_{ij}'s are the elements of the stress tensor. E_{MS} is usually called the magnetostrictive energy density. The total energy density, $E_{\mathrm{tot\,an}}$, follows by adding the (first order) crystal anisotropy-energy density for a uniaxial material,

$$E_{\mathrm{K}} = K_1 \sum_{i>j=1}^{3} \alpha_i^2 \alpha_j^2, \qquad (a8.2)$$

to E_{MS}. K_1 in our ferrite equals about -50 J/m^3 at room temperature.

Biaxial stress in (001) plane

The stress tensor that represents equal stress in all directions of the (001) plane is given by

$$\boldsymbol{\sigma} = \sigma \begin{pmatrix} 100 \\ 010 \\ 000 \end{pmatrix}, \qquad (a8.3)$$

where $\sigma > 0$ represents tensile and $\sigma < 0$ represents compressive stress. This follows by inspection of the forces on any element with surface dA, σndA, with $\boldsymbol{n}$ varying in the (001) plane. Hence, the total anisotropy energy in the case of this biaxial stress becomes:

$$E_{\mathrm{tot\,an}} = K_1 \sum_{i>j=1}^{3} \alpha_i^2 \alpha_j^2 - \tfrac{3}{2} \lambda_{100} \sigma (\alpha_1^2 + \alpha_2^2). \qquad (a8.4)$$

When the magnetostrictive energy term is large compared to the total anisotropy energy term, then, using the constraint $\alpha_1^2 + \alpha_2^2 + \alpha_3^2 = 1$:

$$E_{\mathrm{tot\,an}} = C + \tfrac{3}{2} \lambda_{100} \sigma \alpha_3^2. \qquad (a8.5)$$

Obviously, minimum energy is obtained for
1) $\boldsymbol{M}$ lying in the stressed (001) plane ($\alpha_3 = 0$) when $\lambda_{100}\,\sigma > 0$,
2) $\boldsymbol{M}$ in the [001] direction ($\alpha_3 = 1$) when $\lambda_{100}\,\sigma < 0$.
In 8X1 ferrite (with $\lambda_{100} \simeq -1.\,10^{-5}$) situation 2 occurs when $\sigma > 0$, i.e. in the case of biaxial tensile stress in the (001) plane, and situation 1 occurs in the case of biaxial compressive stress.
In both cases the (static) rotational susceptibility χ_{r0} (for fields, $H_\perp$, perpendicular to the easy plane or easy axis) follows by minimizing the

total free energy (including the magnetostatic free energy density $E_M = -\mu_0\, \boldsymbol{H}\cdot\boldsymbol{M}$), which for all above cases can be rewritten to:

$$E_f = C - \mu_0 H_\perp M_\perp + \tfrac{3}{2}\,|\lambda_{100}\,\sigma|(M_\perp/M_s)^2. \tag{a8.6}$$

Hence

$$\frac{\partial E_f}{\partial M_x} = -\mu_0 H_\perp + \tfrac{3}{2}\,|\lambda_{100}\,\sigma|\,\frac{2M_\perp}{M_s^2} = 0, \tag{a8.7}$$

and

$$\boxed{\;\chi_{r0} \equiv \frac{M_\perp}{H_\perp} = \frac{\mu_0 M_s^2}{3|\lambda_{100}\,\sigma|}\;}. \tag{a8.8}$$

Frequency dependence of rotational susceptibility

In the empirical equation of motion of Landau and Lifshitz [8.28] (see also expressions (23.10) and (23.11) in [8.23])

$$\dot{\boldsymbol{M}} = \gamma \boldsymbol{M} \times (\boldsymbol{H} + \boldsymbol{H}_{an}) - \frac{\alpha\gamma}{M_s}\,\boldsymbol{M} \times (\boldsymbol{M} \times (\boldsymbol{H} + \boldsymbol{H}_{an})) \tag{a8.9a}$$

$$= \gamma(1 + \alpha^2)\boldsymbol{M} \times (\boldsymbol{H} + \boldsymbol{H}_{an}) - \frac{\alpha}{M_s}\,(\boldsymbol{M} \times \dot{\boldsymbol{M}}), \tag{a8.9b}$$

valid for small damping, the spontaneous magnetization is treated as equivalent to the motion of a spinning top.

The magnetic field $\boldsymbol{H}$ (including demagnetizing fields) plus the anisotropy (crystal and stress- induced) field $\boldsymbol{H}_{an}$ is the total or equivalent 'field', $\boldsymbol{H}_{equiv}$, experienced by the spontaneous magnetization, M_s. Further γ is the gyromagnetic ratio ($\approx 1.5 \cdot 10^5$ m/As) and α is Gilbert's damping constant (typically $\alpha \approx \mathcal{O}(10^{-2})$).

For large values of the damping constant, the above Landau Lifshitz equation leads to absurdities, as pointed out by Gilbert [8.33, 8.34] and Kikuchi [8.35], see also a recent paper of Mallinson [8.36] who draws attention to Kikuchi's paper. These absurdities will be dealt with shortly. In addition it will be shown that Gilbert's form of the equation, see later on, leads to a step response that corresponds to the step response of damped systems like e.g. the LCR circuit and pendulum.

Consider for this purpose the case of a time-independent equivalent field in the z direction with amplitude H_z. The solution of (a8.9), assuming $M_x = M_0$ and $M_y = 0$ at $t = 0$, than reads:

$$M = \left(M_0 e^{-t/\tau}\cos \omega_{\text{res}}t, \; -M_0 e^{-t/\tau}\sin \omega_{\text{res}}t, \; M_s\sqrt{1 - \frac{M_0^2}{M_s^2}\, e^{-2t/\tau}} \right),$$

$$(a8.10)$$

where $\tau = T/2\pi\alpha = 1/\alpha\gamma H_z \sim t_s$, $T = 2\pi/\gamma H_z$ is the period of one oscillation and t_s is a switching time (related to the time constant τ). Eq. (a8.10) follows most easily by inspection from the original form of the Landau-Lifshitz equation (a8.9a), since the right-hand members in this equation represent the orthogonal components of $\dot{M}$ in azimuthal (undamped precession) and polar (due to damping) direction.

In (a8.10) the period remains unaltered and the switching time decreases as the damping increases (assuming γ independent of α). In all known oscillating circuits, however, where damping is introduced by adding a dissipative term to the second order differential equation describing the undamped oscillation, an increase in the period is calculated. This is in agreement with the observations: the period of a pendulum increases as you damp the movement, and the period of the oscillation in an $(L + R)/\!/C$ circuit increases as you increase the dissipative element R. Moreover, at a certain critical damping, i.e. as $\tau = CR = 2/\omega_{\text{res}}$, the switching time (stepresponse) is optimal: lower damping leads to oscillations that are less damped before a new equilibrium is established, while a larger damping leads to a slower movement to the new equilibrium.

The slight change we have to make in (a8.9) to arrive at analogous results is to replace γ by $\gamma/(1 + \alpha^2)$ so that $\tau = T/2\pi\alpha = (1 + \alpha^2)/\alpha\gamma H_z \sim t_s$. Now τ and t_s are optimal (minimal) when $\alpha = 1$, i.e. $\tau = 2/\gamma H_z = 2/\omega_{\text{res}}$! The Landau-Lifshitz equation is now replaced by

$$\boxed{\; \dot{M} = \frac{\gamma}{1 + \alpha^2}\, M \times (H + H_{\text{an}}) - \frac{\alpha\gamma}{(1 + \alpha^2)M_s}\, M \times (M \times (H + H_{\text{an}})) \;}$$

$$(a8.11a)$$

$$= \gamma M \times (H + H_{\text{an}}) - \frac{\alpha}{M_e}\, (M \times \dot{M}). \qquad\qquad (a8.11b)$$

The second equation is written in the original form of Gilbert, the first in the physically-preferable 'orthogonal' form.

Assume $\boldsymbol{H}_{an} = (0, 0, H_{an})$, and assume a sinusoidal magnetic field $\boldsymbol{H}$ that is small and perpendicular to this anisotropy field, e.g. $\boldsymbol{H}(t) = (H_x(t), 0, 0)$ with $H_x \ll H_{an}$. Hence $M_z(t) \simeq M_s$; changes in M_z are negligible compared to those in M_x. After introducing complex quantities, $H_x(t) = Re\{\hat{H}_x e^{j\omega t}\}$, $M_x(t) = Re\{\hat{M}_x e^{j\omega t}\}$ and $M_y(t) = Re\{\hat{M}_y e^{j\omega t}\}$, expression (a8.11b) is rewritten to:

$$j\omega(\hat{M}_x, \hat{M}_y, 0) = \gamma(\hat{M}_y H_{an}, M_s \hat{H}_x - \hat{M}_x H_{an}, 0) - \frac{\alpha}{M_s} j\omega(-M_s \hat{M}_y, M_s \hat{M}_x, 0).$$

$$(a8.12)$$

So

$$j\omega\hat{M}_x = \gamma\hat{M}_y H_{an} + j\omega\alpha\hat{M}_y \qquad (a8.13a)$$

and

$$j\omega\hat{M}_y = \gamma(M_s\hat{H}_x - \hat{M}_x H_{an}) - j\omega\alpha\hat{M}_x. \qquad (a8.13b)$$

Elimination of $\hat{M}_y$ and introduction of $\omega_{res} \equiv \gamma H_{an}$ yields:

$$\hat{\chi}_x \equiv \frac{\hat{M}_x}{\hat{H}_x} = \frac{\gamma M_s}{\omega_{res}} \frac{(1 + j\alpha\omega/\omega_{res})}{(1 + j\alpha\omega/\omega_{res})^2 - (\omega/\omega_{res})^2},$$

$$\simeq \frac{\gamma M_s}{\omega_{res}} \frac{1 + j\alpha(\omega/\omega_{res})}{1 - (\omega/\omega_{res})^2 + j2\alpha\omega/\omega_{res}} \qquad (a8.14)$$

when $\alpha \ll 1$.

Hence, using again $\alpha \ll 1$:

$$\boxed{\hat{\chi}_x = \frac{\gamma M_s}{\omega_{res}} \frac{\{1 - (\omega/\omega_{res})^2\} - j\alpha\omega/\omega_{res}\{1 + (\omega/\omega_{res})^2\}}{\{1 - (\omega/\omega_{res})^2\}^2 + \{2\alpha\omega/\omega_{res}\}^2}}. \qquad (a8.15)$$

From (a8.13a) it follows that

$$\hat{\chi}_{yx} \equiv \frac{\hat{M}_y}{\hat{H}_x} = \frac{\hat{M}_y}{\hat{M}_x}\hat{\chi}_x = \frac{j\omega}{\omega_{res} + j\omega\alpha}\hat{\chi}_x. \qquad (a8.16)$$

Hence a precession movement is carried out by the magnetization vector at non-vanishing frequencies.

A magnetization-independent anisotropy field (in the direction of the easy axis) was introduced to describe the action of the easy axis. This was only possible because the stiffness for movements of the magnetization vector away from this easy axis was independent of the direction of this movement. In general cases the stiffness depends on the instantaneous direction of the rotation of the magnetization vector.

Effect of demagnetizing field; ferroxplana

For example, when the dimensions of a sample are different in the x and y directions, a demagnetizing field $H_{dem} = (-N_xM_x, -N_yM_y, -N_zM_z)$ exists that depends on the instantaneous direction of the magnetization and a corresponding torque is exerted on the 'magnetization vectors'. The (torque) equation of motion (without damping, $dJ/dt = \mu_0 M \times H_{equiv}$ see also (a8.11), where J is the angular momentum per unit volume, which is equal to $\mu_0 M/\gamma$ when γ is the magneto-mechanical ratio, and H_{equiv} is the total or equivalent 'field' as experienced by the magnetization vectors, i.e. caused by both the demagnetizing fields of magnetic origin, and the anisotropy field, H_{an}, in the easy axis direction, z, of mechanical origin) then leads to [8.29] [8.30]:

$$\omega_{res} = \gamma\sqrt{\{H_{an} + (N_x - N_z)M_z\}\{H_{an} + (N_y - N_z)M_z\}}$$

$$\equiv \gamma\sqrt{H_{zx}H_{zy}}. \tag{a8.17}$$

Note that H_{zx} and H_{zy} are not the components of the equivalent field H_{equiv}; H_{zx} is the (value of the) 'effective field' *in the z direction* for a rotation in the zx plane and H_{zy} is the (value of the) 'effective field' *in the z direction* for a rotation in the zy plane.

If the xz plane is an easy plane with a (small) anisotropy in the plane itself, described by an 'effective field' $H_{\parallel}$ for a (small) movement of the magnetization vector in this plane away from the z axis (easy axis), and described by an 'effective field' $H_{\perp}$ for a (small) movement away from the z axis perpendicular to the xz plane, then (a8.17) reads:

$$\boxed{\omega_{res} = \gamma\sqrt{H_{\parallel}H_{\perp}}}. \tag{a8.17}$$

The dc susceptibility is unaffected and still reads $\chi_{r0\perp} = M_s/H_\perp$ (smaller than $\gamma M_s/\omega_{res}$ now) for a small rotation in the direction perpendicular to the easy plane and $\chi_{r0\parallel} = M_s/H_\parallel$ (larger than $\gamma M_s/\omega_{res}$ now) for a small rotation in the plane, away from the easy axis. The precession movement of the rotating magnetization vectors is strongly elongated in the easy plane direction. This explains the relatively large and small resonance frequencies (relatively small and long cycle of a rotation) when magnetization changes are considered in the plane and perpendicular to the plane, respectively, compared with $\omega_{res} = \gamma H_{an}$ in the case of a uniform (in two dimensions) anisotropy.

In ferroxplana [8.31] for instance, $\chi_{r0\parallel} = M_s/H_\parallel$ is high, because of the low crystal (hexagonal) anisotropy in the plane of the flat ferroxplana particles, while at the same time ω_{res} is high, because of the strong demagnetizing field perpendicular to the plane of the flat particles, such that $H_\perp$ approaches M_s.

References

[8.1] Eric R. Katz, *Numerical analysis of ferrite recording heads with complex permeability*, IEEE Trans. Magn., Mag-16, 1404 (1980).

[8.2] Finn Jorgensen, *The Complete Handbook of Magnetic Recording*, TAB Books Inc., Blue Ridge Summit, (1980).

[8.3] Stig Berglund, *The permeance in magnetic circuits*, Journal of Magnetism and Magnetic Materials 19, 323 (1980).

[8.4] Sam Enger, *Magnetberäkningar Permeauser I Luftgap*, Dissertation, KTH, Stockholm, (1977).

[8.5] J. Corcoran and N. Pope, *Transmission line model for magnetic heads including complex permeability*, Journal of Magnetism and Magnetic Materials 54-57, 1591 (1986).

[8.6] D.E. Heim, *Flux propagation in thin-film magnetic structures*, J. Appl. Phys. 59, 864 (1986).

[8.7] S.P. Timoshenko and J.N. Goodier, *Theory of Elasticity*, McGraw-Hill Kogakusha, Ltd, Tokyo etc., 3rd edition, p. 39.

[8.8] Leon I. Maissel and Reinhard Glang, *Handbook of Thin Film Technology*, McGraw-Hill Book Company, 1970. See chapter 12 equation (25).

[8.9] Francis I. Baratta, *When is a beam a plate?*, Communications of the American Ceramic Society, C-86, May (1981).

[8.10] K.R. Bijkerk, A. van Veen, G.J. van der Kolk and T. Minemura, *Crystallization of sputtered $Cu_{60}Zr_{40}$ studied by thermal desorption*, submitted for publication to Vacuum, (1988).

[8.11] Kazuhiro Saito and Taiich Mori, *A magnetic film, a method of producing the magnetic film, and a thin-film magnetic head*, European patent application 0220012.

[8.12] E.G. Visser, *Magnetization processes in stressed monocrystalline manganese zinc ferrite*, Thesis, University of Technology, Eindhoven, the Netherlands (1983).

[8.13] E.P. Valstyn, *Optimization of ferrite heads for thin media*, IEEE Trans. Magn., Vol. Mag-22, 847 (1986).

[8.14] Philips Data Handbook C5 (1985), Components and Materials, Ferroxcube for Power, Audio/Video and Accelerators.

[8.15] R.F. Hoyt, D.E. Heim, J.S. Best, C.T. Horng, and D.E. Horne, *Direct measurement of recording head fields using a high resolution inductive loop*, J. Appl. Phys., Vol. 55, 2241-2244 (1984).

[8.16] R.S. Indeck, et al., *Direct measurement of external magnetic fields from a single-pole type recording head using a microloop*, Digest of the Intermag Conf. April 29-May 2, 1985, St Paul, MN, USA.

[8.17] J.H.J. Fluitman, *Recording head field measurement with a magnetoresistive transducer*, IEEE Trans. Magn., Vol. Mag-14, 433-435 (1978).

[8.18] C.D. Lustig, A.W. Baird, W.F. Chaurette, H. Minden and W.T. Maloney, *High resolution magnetic field measurement system for recording heads and disks*, Rev. Sci. Instrum. 50(3), 321-325 (1979).

[8.19] E.G. Visser, L.R.M. van Rijn, and H.J.F. Maas, *An improved measurement of the absolute efficiency of magnetic heads by saturating the gap field*, IEEE Trans. Magn., Mag-21, 1283-1288 (1985).

[8.20] J.P. Morel, private communication, Philips Videq Head Laboratory, Eindhoven, the Netherlands.

[8.21] J.J.M. Ruigrok, U.E. Enz and C.W.M.P. Sillen, *Magnetic transducing head having clad core faces*, European patent application 0246706.

[8.22] E. Olsen, *Applied Magnetism*, Philips Technical Library (1966).

[8.23] J. Smit and H.P.J. Wijn, *Ferrites*, Philips' Techn. Libr. (1959).

[8.24] R. Wood, D.A. Lindholm and R.M. Haag, *On the bandwidth of magnetic record/reproduce heads*, IEEE Trans. Magn., Mag-21, 1566 (1985).

[8.25] A. Broese van Groenou, private communication, Philips Research Laboratories, Eindhoven, the Netherlands.

[8.26] F.W.A. Dirne and M. Brouha, *Soft-magnetic properties of microcrystalline Co-Fe-Si-B alloys prepared by sputtering*, IEEE Trans., Mag-24, 1862 (1988).

[8.27] S. Chikazumi, *Physics of Magnetism*, R.E. Krieger, New York (1978).

[8.28] L. Landau and E. Lifshitz, Phys. Z. Soviet Un., *On the theory of the dispersion of magnetic permeability in ferromagnetic bodies*, Vol. 8, 153-169 (1935).

[8.29] Ch. Kittel, *On the theory of ferromagnetic resonance absorption*, Phys. Rev. 73, 155 (1948).

[8.30] Landolt-Börnstein, *Numerical data and functional relationships in science and technology*, Volume 12, Springer-Verlag, Berlin, Heidelberg, New York (1982).

[8.31] J. Verweel, *Magnetic properties of some ferroxplana single crystals*, Thesis, University of Amsterdam, the Netherlands (1966).

[8.32] D. Stoppels, *Frequency dependence of the complex permeability of monocrystalline MnZn ferrous ferrites*, J. Appl. Phys., Vol. 52, 2433-2435 (1981).

[8.33] T.L. Gilbert, *A Lagrangian formulation of the gyromagnetic equation of the magnetization field*, Phys. Rev., Vol. 100, 1243 (1955).

[8.34] T.L. Gilbert and J.M. Kelly, Proceedings of the Pittsburgh Conference on Magnetism and Magnetic Materials, Am. Inst. Electr. Engrs., 253 (1955).

[8.35] R. Kikuchi, *On the minimum of magnetization reversal time*, J. Appl. Phys., Vol. 27, 1352-1357 (1956).

[8.36] J.C. Mallinson, *On damped gyromagnetic precession*, IEEE Trans. Magn., Mag-23, 2003-2004 (1987).

Chapter 9

Probe-type heads: models for efficiency and auxiliary-pole effect

9.1 Introduction

In this chapter useful definitions for the efficiency of main and auxiliary poles of single-sided probe heads are given, which are based on the general ideas on efficiency outlined in Sect. 5.3. Approximate expressions are derived for these efficiencies and for auxiliary-pole effects for different types of symmetric single-sided PHs. It is noted that the backlayer of the tape forms an essential part of the PH. The four types of PH are depicted in Figs. 9.1-9.4. The first, denoted PH, has no return yoke, while the permeability of the backlayer, μ_2, and that of the probe, μ, are assumed to be infinitely high. The second, denoted

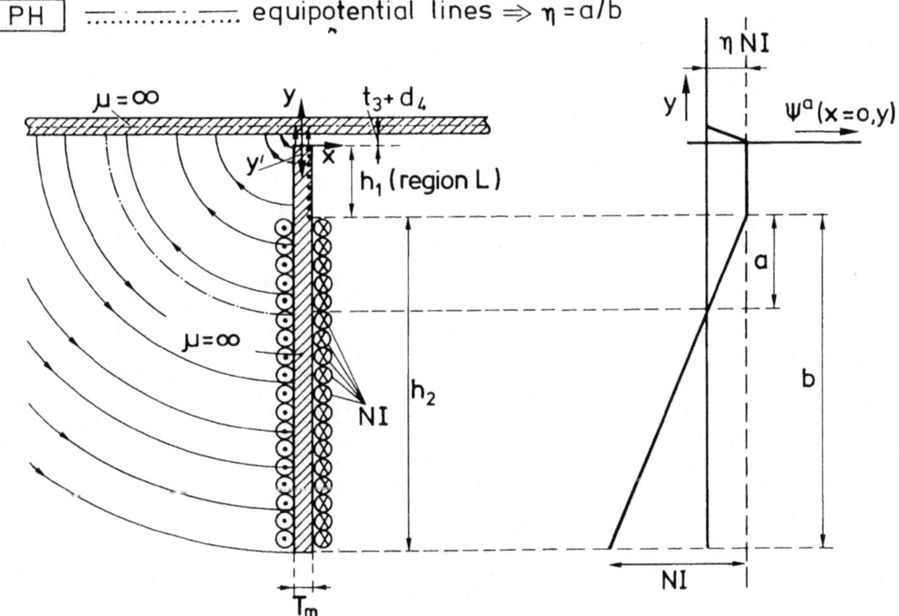

Fig. 9.1. Single-sided probe-type head (PH). Denoted are circular flux paths, as assumed in the analytic calculations. As a consequence of the assumed infinite permeability of the probe the potential Ψ^a as a function of y is described by straight lines.

PHRY, contains a return yoke which is represented by the infinitely-permeable material, and has a probe consisting of three parts, I, II and III, which are allowed to have different permeabilities and 'thicknesses'. The permeability and thickness of the tape (usually the backlayer) can also be chosen freely in this case. The analytic calculations are

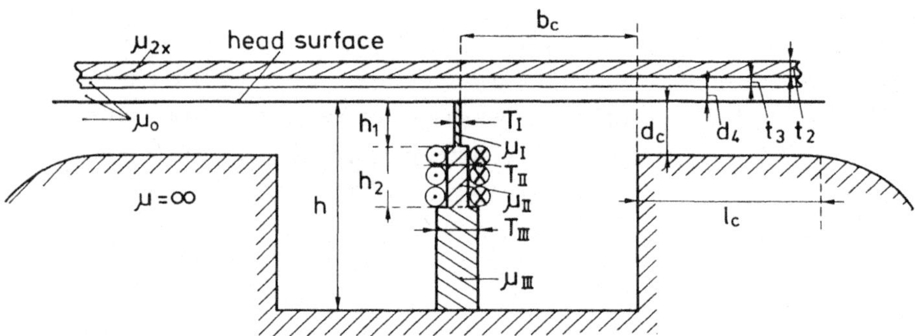

Fig. 9.2. Single-sided probe-type head with return yoke (PHRY), case 1: $b_c \geq h_1 + h_2 + t_3 + d_4$. Only the basic dimensions and quantities are denoted in the figure. μ_I, μ_{II} and μ_{III} denote the y-components of the permeabilities in the three regions of the probe.

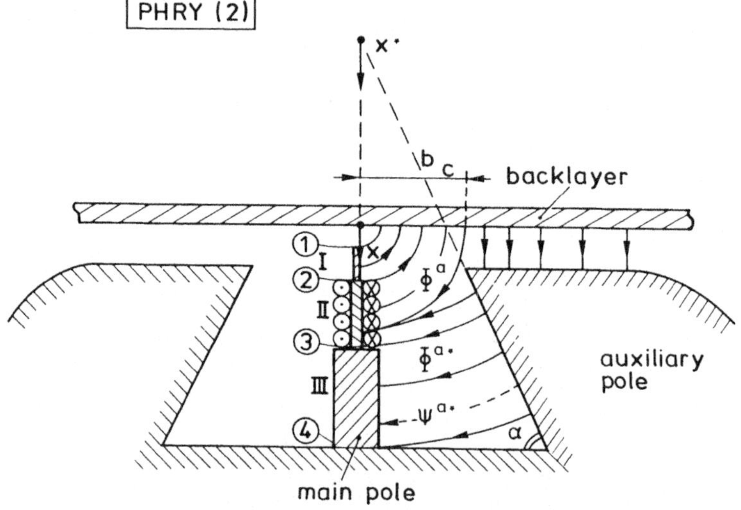

Fig. 9.3. Single-sided probe-type head with return yoke (PHRY), case 2: $h_1 + t_3 + d_4 \leq b_c \leq h_1 + h_2 + t_3 + d_4$. For this 'case 2', the angle α may influence the results noticeably when b_c is small, in contrast to 'case 1' (large b_c).

based upon the same assumptions, i.e. circular and straight flux paths, as used for the RH-efficiency calculations. An essential difference compared with the RH calculations is that the heads are not built up from a discrete number of lumped reluctances, but are treated as a continuum by solving the differential (or integral) equations valid in the different regions. This is desirable in order to obtain an acceptable accuracy in the case of an open structure like the PH or a closed structure containing thin films (probe and backlayer), like the PHRY. Due to the extra mathematical difficulty of the PHRY in particular (with its finite permeability), most calculations are carried out for 2-dimensional

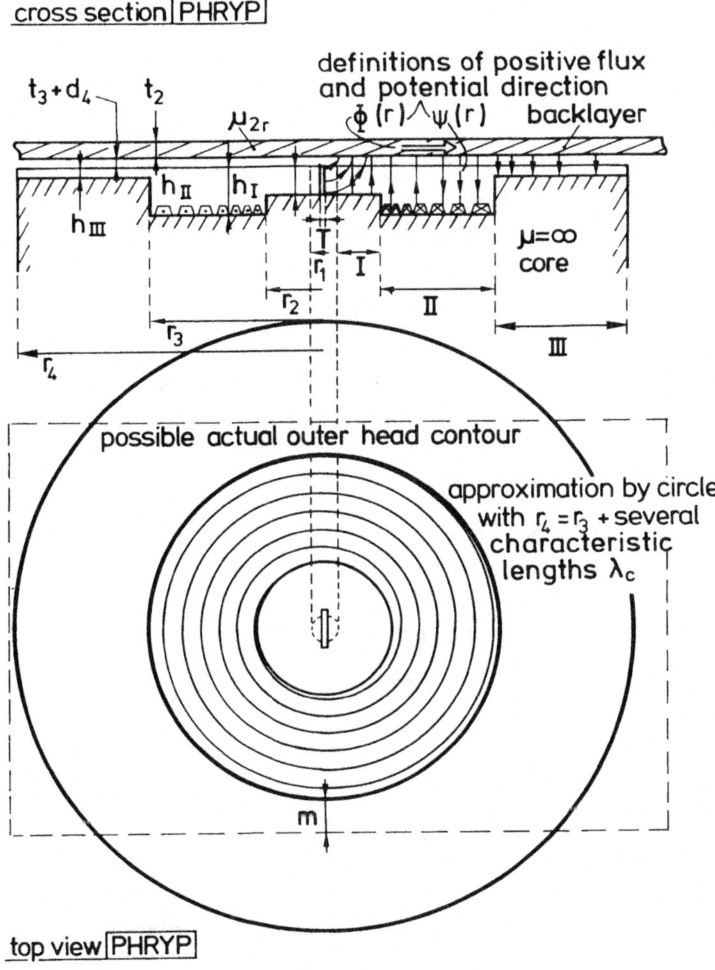

Fig. 9.4. An almost circular-symmetric planar (P) probe-head design with the possibility of a return yoke ($r_4 > r_3$), abbreviated PHRYP.

configurations, such that no side effects influence the results. The present calculations on the PHRY are in fact transmission-line calculations, but are carried out on a non-uniform transmission line formed by the backlayer and probe and the diverging distance in between. Therefore the solutions are not e-powers but turn out to be Bessel functions. For the more bulky RHs, this transmission-line description is not necessary, but can improve the accuracy somewhat. This has been carried out, though not exactly, by Corcoran and Pope [9.1], for a 3-dimensional RH of a simpler structure than our RH in chapter 8.

Because of their 'straight-line' approximation, introduced after a change of variable, they arrived at the usual transmission-line equations and results (combinations of e-powers or hyperbolic functions) of uniform transmission lines.

The third configuration, depicted in Fig. 9.3, is an extension of the configuration in Fig. 9.2. Only the length of the coil chamber, b_c, is smaller than $h_1 + h_2$, which requires other boundary conditions. This complicates the calculations.

The last configuration is given in Fig. 9.4 and has circular symmetry. For this reason the calculations of this 3-dimensional configuration could be carried out '1-dimensionally'. The applicability of results of these calculations to actual heads of such design is therefore more accurate than application of the other results to heads with cross-sections as given in Figs. 9.1 to 9.3 but with track width W, which will in practice not be much wider than all other dimensions. This inaccuracy probably leads to a larger underestimation of the efficiencies according as W becomes smaller. The calculations are especially meant for a qualitative discussion of different probe-head designs.

We applied some of the concepts and ideas developed in this chapter to the design of high-performance probe heads for perpendicular recording on double-layer media, see Zieren et al. [9.2], [9.3] and [9.4]. Especially the proposed efficiency definition in this chapter (which is in fact the usual efficiency definition in longitudinal recording, see Section 9.2), turned out to be very convenient in the understanding of weaknesses in earlier probe-head designs.

For the practical side of the design I therefore refer to the above-mentioned articles. The elaborate diverging-transmission-line description is omitted in the calculations in those papers at the cost of some accuracy.

9.2 *GLF*s, *SF*s, ηs and the auxiliary-pole effect in PHs

9.2.1 Potential and flux

In Fig. 9.5 a probe head and different approximations for the potentials and field on the pole levels have been sketched.

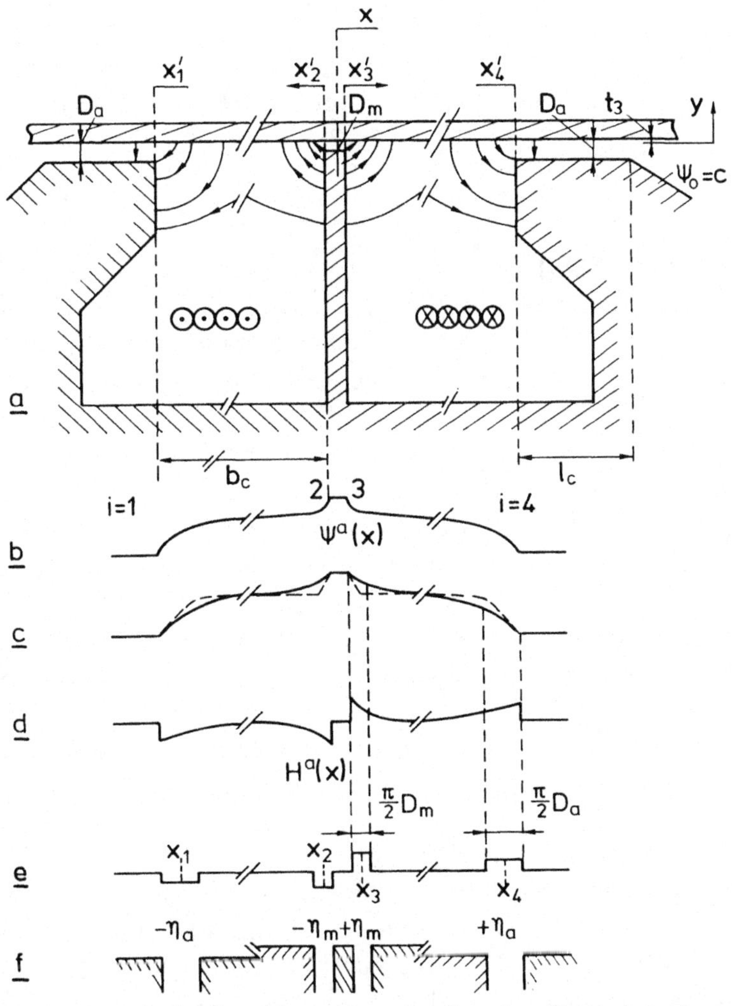

Fig. 9.5. Sketch of circular flux lines (a) near the main pole and auxiliary poles and corresponding approximations (c) and (d) of the scalar potential $\Psi^a(x)$ and field $H^a(x)$ at main-pole surface level ($y = -D_m$) and auxiliary-pole surface level, respectively.
The potential (c) consists of four 'half-arctan' functions with steepnesses proportional to $1/D$. The actual potential (b) is infinitely steep near the edges and so H_x is infinite at the edges. Nevertheless the field is further approximated (e) by four rectangles, corresponding to the straight-line approximation depicted by the dashed curve in c, and corresponding to the gaps configuration given in f.

Flux lines at the outer edges of the auxiliary pole are usually completely negligible, since the characteristic length of the transmission line formed by the backlayer and auxiliary pole, $\lambda_c = \sqrt{\mu_{2x} t_2 D_a} \simeq 20 \sqrt{D_a}$, is usually much smaller than the length of the auxiliary pole, l_c. Consistent with the approximation for the flux path being circular, as sketched in Fig. 9.5, while the potentials in backlayer and poles near the edges are assumed not to vary much, we obtain for the potentials at the pole levels:

$$\Psi_i(x_i')/NI = C_i \pm \frac{2}{\pi} \eta_i \arctan{(x_i'/D_i)} \qquad x_i' \geq 0$$

$$= C_i \qquad\qquad\qquad\qquad x_i' \leq 0 \qquad (9.1)$$

where $\eta_i \equiv |\Delta\Psi_i|/NI$ (see also (5.17)) equals the potential difference over the considered region i ($= 1, 2, 3$ or 4) divided by the number of ampère turns (IN), and x_i' is the distance from the considered pole edge and points away from the edge (see Fig. 9.5a). $C_2 = C_3$ (and $C_1 = C_4$) and the sign in regions 1, 2, 3 and 4 is $+, -, -, +$ times sign(I) respectively (with the positive current direction as depicted by the cross and the dot in Fig. 9.5a).

Close to the edges the flux lines are not circular. For an accurate description of this part of the curves, results from conformal mapping can be used. Westmijze's conformal transformations for an infinitely 'thin' and 'semi-infinite' gap [9.5] directly apply to the PHs (by a 90° rotation of his RHs) for the cases $T \ll D_m$ and $T \gg D_m$ respectively. The latter case always applies to the auxiliary pole(s). Explicit expression for the SFs of the poles can then be derived along the same lines as pointed out by Westmijze, but they will contain rather complicated integrals.

For intermediate T values, expressions for the potential can be derived by using the Schwarz-Christoffel transform as worked out by Steinback et al. [9.6]. Zieren et al. [9.2] obtained close agreement between SFs obtained by FEM calculations and those obtained by using Steinback's analytical method.

As depicted in Fig. 9.5b, the steepness of these accurate potentials near the edges is infinite; see also Fig. 6.2.1. Far away from the edges, all the above approximations are invalid, because:

− far away from edges the flux lines are not circular (see Fig. 9.5) and

– the potential in the backlayer increases monotonically, because of finite backlayer permeability, from auxiliary poles to main pole, while in addition the potential at pole surface level ($y = t_3 - D_i$) approaches the potential in the backlayer far away from the edges.

9.2.2 Definitions of ηs

The influence of the finite backlayer permeability on the potential in the backlayer can usually be neglected over small distances, i.e. a few times D_i, such that the potential difference $- \int H_y dy$ between backlayer and pole surface equals NI times the previously defined $\Delta\Psi_i$. Hence the efficiencies $\eta_i \equiv \Delta\Psi_i / NI$ equal

$$
\eta_i = \frac{1}{NI} \int_{\text{pole}}^{\text{backlayer}} H_y(x_i = 0) dy \quad , \tag{9.2}
$$

which will be used as the definition for η_i in all actual calculations on η_i in this chapter.

Our efficiency definition differs from those used by other authors. Dugas, Bonin and Judy [9.7] for instance define the efficiency as the fractional amount of the single bit flux which links the coil. This efficiency, however, depends on the chosen bit shape and length. Nakamura et al. [9.10] defined the efficiency as the output of the head for a given permeability of the core divided by the output for infinite permeability. This definition, however, has the drawback that a head with a poor output due to a not well elaborated geometry, i.e. not due to a low permeability, still gets an efficiency of almost 1. None of the above drawbacks appear in our definition. In our approach the bit shape influences the output via the gap loss function and hardly at all via the efficiency. *Our definition corresponds to the conventional definition of the efficiency of a ring head in the case of a single-layer (perpendicular or longitudinal) medium.* Our efficiency is a function only of the frequency, because of the frequency-dependent permeability of the core and the backlayer, and is not a function of the wavelength. *Heads with a bad geometry as well as heads with an insufficient permeability lead to a low efficiency and an ideal head leads to an efficiency of 1.*

9.2.3 Definitions of *SF*s and *GLF*s

When the potential distribution is divided into four parts, which is desirable for non-symmetric PHRYs, then analogously to (5.21):

$$SF = \sum_{i=1}^{4} \eta_i\, GLF_i\, e^{-jkx_i} \equiv \sum_{i=1}^{4} SF_i \qquad (9.3)$$

where x_i, is the origin, chosen in the evaluation of GLF_i, and η_i equals the asymptotic value of the potential difference divided by NI, $\Delta\Psi_i^a/NI$, over the corresponding section of the head in the 'fictitious write situation' denoted by the superscript a, as introduced in chapter 3; see Fig. 9.5. From now on we shall often omit the superscript a.

Each individual GLF_i has a maximum absolute value of 1, obtained at infinite wavelength, because each is derived from a potential function that is either monotonically increasing or decreasing and scaled to $\Delta\Psi_i/NI$. For a 'sense' function to be a loss function this is desirable.

For the considered probe heads with one coil twisted around the probe:

$$\eta_1 + \eta_2 \text{ and } \eta_3 + \eta_4 \leq 1,\ \text{i.e.} \sum_{i=1}^{4} \eta_i \leq 2. \qquad (9.4)$$

This means that the SF_m ($= SF_2 + SF_3$) of the main pole may approach 2.

If two GLFs are combined to 1, for instance to study interference effects between signals 'sensed by the left and right sides of the main pole', then the GLF can best be defined as

$$GLF \equiv \frac{\sum \eta_i\, GLF_i\, e^{-jkx_i}}{\sum \eta_i} = \frac{SF}{\sum \eta_i} \qquad (9.5)$$

and consequently

$$SF = GLF \cdot \eta \quad \text{with} \quad \eta = \sum \eta_i = \sum |\Delta\Psi_i|/NI, \qquad (9.6)$$

such that again $|GLF| \leq 1$, as required. At infinite wavelengths (9.5) reads

$$GLF(k = 0) = \frac{\sum \eta_i \, \text{sign} \, (-d\Psi_i/dx)}{\sum \eta_i}. \qquad (9.7)$$

The sign changes according to $--++$ for $i = 1, 2, 3, 4$, which means that the *GLF* of the main pole as well as the *GLF* of the complete probe head vanish at large wavelengths because of symmetry ($\eta_1 = \eta_4$ and $\eta_2 = \eta_3$) and have a maximum at some intermediate wavelength, as was sketched in Fig. 6.2.II.

9.2.4 Straight-lines approximation

The above results can easily be made more quantitative if one uses the straight-lines approximation for the potential, depicted by the dashed lines in Fig. 9.5c as suggested in Sect. 5.3, which corresponds to the fields and gaps configuration shown in Fig. 9.5e and f. This leads at pole-surface levels (see also (5.21)) to:

$$GLF_1 = -\text{sinc}\left(\frac{kD_a}{4}\right) e^{jk\left(\frac{T}{2} + b_c - \frac{\pi D_a}{4}\right)}$$

$$GLF_2 = -\text{sinc}\left(\frac{kD_m}{4}\right) e^{jk\left(\frac{T}{2} + \frac{\pi D_m}{4}\right)}$$

$$GLF_3 = \text{sinc}\left(\frac{kD_m}{4}\right) e^{-jk\left(\frac{T}{2} + \frac{\pi D_m}{4}\right)}$$

$$GLF_4 = \text{sinc}\left(\frac{kD_a}{4}\right) e^{-jk\left(\frac{T}{2} + b_c - \frac{\pi D_a}{4}\right)}, \qquad (9.8)$$

where $\text{sinc} \, x \equiv \sin \pi x / \pi x$.

Transformation to main-pole surface level

To transform GLF_1 and GLF_4 to main-pole surface level, statement (4.14) can be used, which transforms the Fourier transform of the head's potential or of any of its derivatives, $(jk)^n \Psi(k)$ to another y-level, when the decay of $\mathcal{B}_y^b(k)$ between the two y-levels is known. This decay follows from (2.46) when '(2.49)' is substituted or follows from (6.25) when t_3 is replaced by D_m and $t_3 + d_4$ by D_a. Hence (subscript a stands for auxiliary pole(s)):

$$\frac{GLF_a(k, y = t_3 - D_m)}{GLF_a(k, y = t_3 - D_a)} = \frac{(1 - e^{-2kD_m})e^{-2k(D_a - D_m)}}{1 - e^{-2kD_a}}. \tag{9.9}$$

The overall $SF = SF_m + SF_a$ is thus, according to (9.3), approximated at main-pole surface level ($y = t_3 - D_m \equiv - d_4$) by:

$$SF(k, y = -d_4) = -2j\eta_m \text{sinc}\left(\frac{kD_m}{4}\right) \sin\left\{k\left(\frac{T}{2} + \frac{\pi D_m}{4}\right)\right\} -$$

$$2j\eta_a \frac{(1 - e^{-2kD_m})e^{-k(D_a - D_m)}}{1 - e^{-2kD_a}} \text{sinc}\left(\frac{kD_a}{4}\right) \sin\left[k\left(\frac{T}{2} + b_c - \frac{\pi D_a}{4}\right)\right].$$

$$\tag{9.10}$$

9.2.5 Half-arctan approximation

When the more accurate half-arctan functions (9.1) are substituted in (5.18) and the results in (9.3), while using (9.9), then after some manipulations:

$$SF = -2j\eta_m\left[\sin\left(k\frac{T}{2}\right)\frac{2}{\pi}\int_0^\infty \frac{D_m \cos kx}{x^2 + D_m^2}\,dx + \right.$$

$$\left. \cos\left(k\frac{T}{2}\right)\frac{2}{\pi}\int_0^\infty \frac{D_m \sin kx}{x^2 + D_m^2}\,dx\right] -$$

$$2j\eta_a \frac{(1 - e^{-2kD_m})e^{-k(D_a - D_m)}}{1 - e^{-2kD_a}} \times$$

$$\left[\sin\left\{k\left(\frac{T}{2} + b_c\right)\right\}\frac{2}{\pi}\int_0^\infty \frac{D_a \cos kx}{x^2 + D_a^2}\,dx - \right.$$

$$\left. \cos\left\{k\left(\frac{T}{2} + b_c\right)\right\}\frac{2}{\pi}\int_0^\infty \frac{D_a \sin kx}{x^2 + D_a^2}\,dx\right]. \tag{9.11}$$

The Fourier-cosine integrals are solved by using the theorem of residues and result in the well-known 'distance-loss' factor e^{-kD} related to a complete arctan (see e.g. equations (5.64) + (5.65) + (5.72)). The sine transform is more complicated but has been found in [9.12], and leads to:

$$\frac{2}{\pi} \int_0^\infty \frac{D \sin kx}{x^2 + D^2}\, \mathrm{d}x = \frac{1}{\pi} \left[e^{-kD} Ei(kD) - e^{kD} Ei(-kD) \right]$$

$$(9.12)$$

where $Ei(x) \equiv - \int_{-x}^\infty \frac{e^{-t}}{t}\, \mathrm{d}t$, for which Cauchy's principal value must be calculated when $x > 0$.

For this function, Ei, a numerical routine is available in the NAG (fortran) library, namely S13AAF, for $x < 0$ only. For $x > 0$ the principal value of the integral between $-x$ and $+x$ equals $\int_0^x \frac{2 \sinh t}{t}\, \mathrm{d}t$ (of which the integrand is regular at $t = 0$) and is calculated 'stepwise' using Romberg's method, while for the remaining interval routine S13AAF is used.

9.2.6 Rules of thumb for auxiliary-pole effects

Although (9.10) is not accurate, it is more suitable for deriving rules of thumb and for discussing qualitatively the auxiliary-pole read effects than any of the more accurate expressions or methods.

The right-hand term in (9.10) oscillates relatively quickly in the k domain, because of the large coil-chamber length, b_c, relative to T and D_m (and D_a). This is comparable to the effects described in Sect. 5.3 and demonstrated in Fig. 5.2. Thus a mean curve (read by the main pole) is disturbed by 'oscillations' (due to the auxiliary pole reading). For the present cases this is shown in Fig. 9.6. The lowest curve in Fig. 9.6a is calculated by using the more realistic arctan potentials, i.e. by using (9.11), instead of the linear potentials (9.10). Obviously the absolute value of the output is overestimated and the gap-null frequencies are underestimated by the simplest (linear-potential) model, whilst the

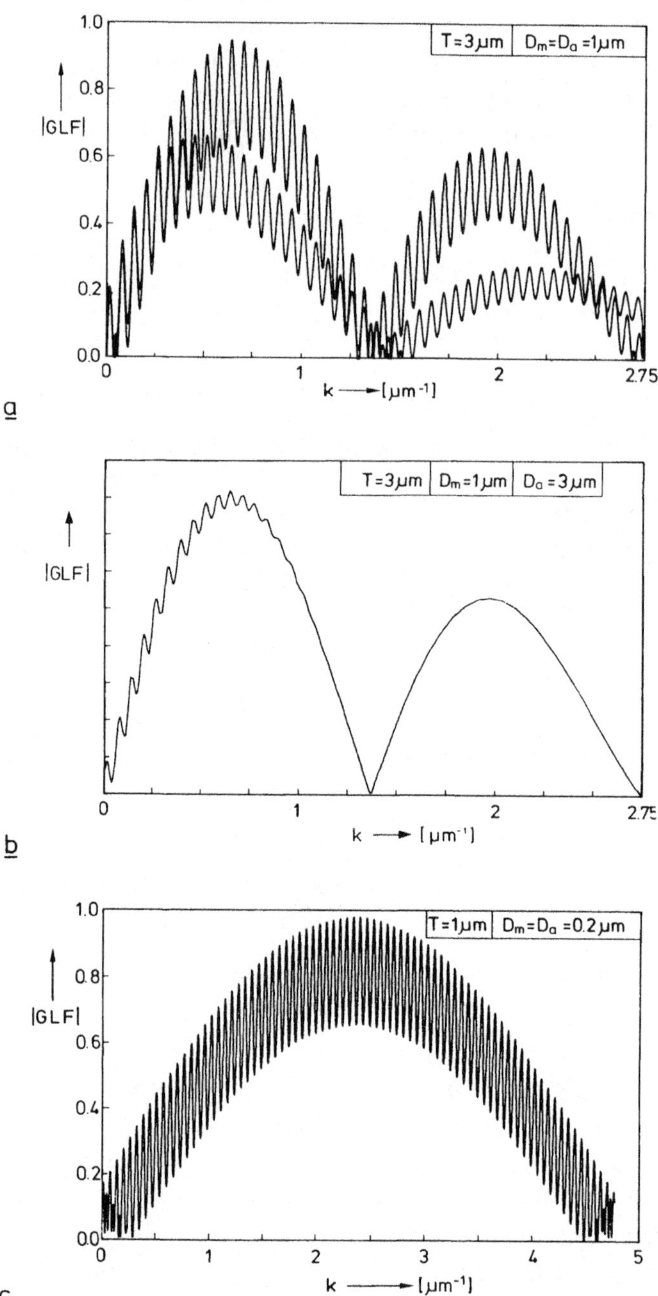

Fig. 9.6. The absolute value of the gap loss function, $|GLF|$ according to (9.5) when the rough approximation (9.10) for the sensitivity function is used, for $\eta_a/\eta_m = 0.2$ and $b_c = 100$ μm. The n-th zero appears at $k_n = 2\pi n/(T + \pi D_m/2)$ in this approximation. The 'oscillations' are due to the auxiliary pole.

The lower curve in Fig. 9.6 is based on the more-accurate half-arctan potential distribution.

periodicity of the oscillations is almost unchanged, as is to be expected because of the differences between the potential distributions in Fig. 9.5c. The amplitude of the oscillations quickly decreases at smaller wavelength; see Fig. 9.6b. Fig. 9.6c shows that for a more realistic, i.e. larger, coil-chamber length to pole-tip length (and head-backlayer distance) ratio, b_c/T, the number of periods within the frequency band is larger. For the present purpose, the ratio of 'oscillation amplitude' to 'mean amplitude', defined as $a(k)$, is of importance and this ratio is much better predicted by the simplest model. In this context it must be noted that the arctan approximation, see Fig. 9.7a, in turn corresponds rather well with the exact results from conformal mapping, shown in Fig. 9.7b (after Baker et al. [9.8]). In this verification no comparison is possible between the absolute values of the output voltage, because of the lack of information about Baker's probe-head configuration and therefore unknown efficiency and unknown efficiency change with D_m/T. The gap-null wavelength in the arctan approach is larger, especially at large relative head-tape distances, D_m/T. This is also apparent in Fig. 9.8. For the exact (conformal mapping) solution in this figure, the Schwarz-Christoffel transform was calculated by Luitjens [9.11], using the results of Steinback et al. [9.6]. Luitjens [9.11] also verified that the results are consistent with results of Minuhin (fig. 14 in [9.9]). The results in Fig. 9.8 moreover show that the correspondence between the

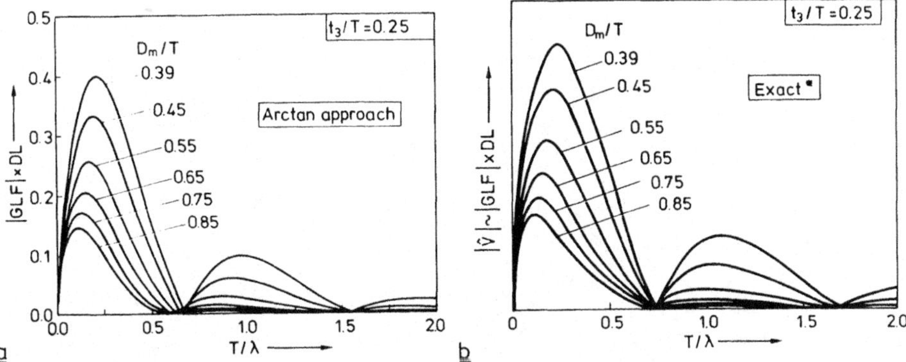

Fig. 9.7. The output voltage for fixed t_3/D_m (proportional to $GLF \times DL$ according to Eq. (5.6) etc., with DL given in Eq. (4.76b)) as a function of T/λ at various values of D_m/T. Further, $\eta_a = 0$ (no auxiliary pole).
a) Half-arctan model, i.e. (9.11) is used in (9.5).
b) Exact result from conformal mapping, after Baker et al. [9.8].

* We do not know the scale factor(s) for the y axis in b because of lack of information about Baker's probe-head configurations and hence about the efficiency (efficiencies).

absolute values is reasonable for a wide range of D_m/T values, except near the gap-null frequency.

Hence the approximation (9.10) can safely be used for the present purpose and so we obtain for the ratio, $a(k)$, between the extrema of the oscillations (i.e. $\left|\sin\{k(T/2 + b_c - \pi D_a/4\}\right| = 1$) and the signal read by the main pole:

$$a(k) = \frac{\eta_a}{\eta_m}\frac{D_m}{D_a}\frac{\left|\sin(\pi k D_a/4)\right|}{\left|\sin(\pi k D_m/4)\right|}\frac{(1 - e^{-2kD_m})e^{-k(D_a - D_m)}}{(1 - e^{-2kD_a})\left|\sin\left\{k\left(\dfrac{T}{2} + \dfrac{\pi D_m}{4}\right)\right\}\right|}.$$

$$(9.13)$$

For $D_a \geqslant D_m$ it is easily checked from (9.12) and (9.10) that the auxiliary-pole effects are most pronounced at the long-wavelength side of the main-pole's sensitivity function, SF_m; see also Fig. 9.6. (This means a larger a value is found for the same values of SF_m or SF_m times a distance factor.)

Criteria for probe heads not to suffer from auxiliary-pole read effects thus read:

$$a(\lambda_{\max}) \leqslant d \, , \qquad\qquad (9.14)$$

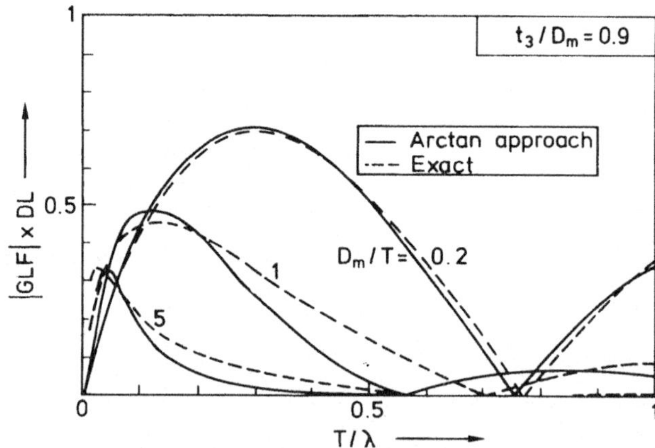

Fig. 9.8. $GLF \times DL$ (proportional to the output voltage, see caption of Fig. 9.7), for fixed t_3/D_m, using the half-arctan approach, see caption of Fig. 9.7. The exact result is calculated by Luitjens [9.11] using the results of Steinback et al. [9.6].

where *d* is the maximum allowable disturbance (oscillation) relative to the mean signal sensed by the main pole.

When the auxiliary pole is wider than the main pole, extra 'spurious' signals may be picked up from adjacent tracks. The contribution of these spurious signals (which can be included in $a(k)$) depends on phase relations between signals in adjacent tracks and hence on azimuthal angle differences etc..

Criterion for $D_a = D_m$

At the long-wavelength side of SF_m

$$\sin\left\{k\left(\frac{T}{2} + \frac{\pi D_m}{4}\right)\right\} \simeq \frac{\pi(T + \pi D_m/2)}{\lambda} . \tag{9.15}$$

Hence at the maximum wavelength, λ_{max}, that has to be reproduced, a good criterion is:

$$a(\lambda_{max}) \simeq \frac{\eta_a}{\eta_m} \frac{\lambda_{max}}{(\pi/2)\lambda_p} < d, \tag{9.16a}$$

i.e.

$$\boxed{\eta_a/\eta_m < (\pi d/2)(\lambda_p/\lambda_{max})} , \tag{9.16b}$$

where $\lambda_p \equiv 2(T + \pi D_m/2)$ equals about the wavelength for which, according to (9.10), SF_m has its main maximum.

If for example $\lambda_{max} / \lambda_p = 5$ and $d = 0.1$ are required, then η_a/η_m must be smaller than 0.03! This seems a very difficult requirement, and therefore $D_a \gg D_m$ might be necessary.

Criterion for $D_a \gg D_m$

At the long-wavelength side is $\sin(\pi k D_m/4)$ now in addition approximated by $\pi k D_m/4$. Further $\sin(\pi k D_a/4)$ oscillates quickly too, because of the large values of D_a relative to D_m, and thus must be taken 1 in the evaluation of $a(k)$. Finally the distance-loss factor in (9.13) decreases to $2kD_m e^{-kD_a}$ and $8/\pi^2$ is approximated by 1. Hence, a good criterion is

$$a(\lambda_{max}) \simeq \frac{\eta_a}{\eta_m} \frac{D_m}{D_a} e^{-\pi\left(\frac{\lambda_p}{\lambda_{max}}\right)\left(\frac{D_a}{T+\pi D_m/2}\right)} \frac{2\lambda_{max}}{\lambda_p} < d, \qquad (9.17a)$$

i.e.

$$\boxed{\frac{\eta_a}{\eta_m} < \frac{d}{2}\left(\frac{\lambda_p}{\lambda_{max}}\right)\left(\frac{D_a}{D_m}\right) e^{\pi\left(\frac{\lambda_p}{\lambda_{max}}\right)\left(\frac{D_a}{T+\pi D_m/2}\right)}} \qquad (9.17b)$$

For example, when $\lambda_{max}/\lambda_p = 5$ and $d = 0.1$ are desired and $D_a : D_m : T$ is only 10:1:2, then η_a/η_m might already be as large as 0.6 and for 20:1:2 even as large as 6!

So, for $D_a \gg D_m$, ordinary requirements seem easily fulfilled, even when one takes account of the fact that η_a will be larger and η_m will be smaller according as D_a is larger.

For $\lambda_{max} > 4\pi D_a$, the factor $1 - e^{-2kD_a}$ in (9.13) is not allowed to be taken 1 and has to be added at the right-hand side of (9.17b), which might make the requirement very hard to fulfill. However $4\pi D_a$ will always be very large with respect to T and D_m, since choosing $D_a > D_m$ in a head design (meant for in-contact recording) automatically means $D_a \gg D_m$, otherwise wear would soon make D_a equal to D_m.

This also implies that criteria for only $D_a = D_m$ and $D_a \gg D_m$ will always suffice for practical head designs.

Of course the accuracy, especially at very short and very long wavelengths, is not expected to be very good. However for the present purpose these regions are of less importance and in addition the accuracy required for 'rules of thumb' is low. A reliable criterion for the head not to write or erase with the auxiliary pole-edges is much easier to give since a field at the auxiliary pole edges $H_y < \frac{1}{2}H_c$ is hardly able to write any signal into the tape or erase any signal from the tape. So requiring a factor of 4 difference in H_y fields at main and auxiliary pole certainly suffices. *Therefore the criterion not to write or erase with the auxiliary pole* is

$$\boxed{\frac{\eta_a D_m}{\eta_m D_a} \leqslant \frac{1}{4}} \qquad (9.18)$$

In the following sections the auxiliary pole's efficiency, η_a, will often be calculated in addition to the main pole's efficiency, η_m, so that the auxiliary pole write and read effects can be easily discussed for various head designs.

9.3 Probe head without return yoke (PH)

The configuration is given in Fig. 9.1.

For simplicity, $\mu = \infty$ is assumed in the backlayer and pole, such that the flow of flux through the layer and pole does not contribute to the potential. Consequently the potential over region L, with length h, is a constant and increases linearly over region h_2; see Fig. 9.1. Hence, the potential equals

$$\Psi^a = \frac{y' - h_1}{h_2} NI - \Phi_L^a R_L, \tag{9.19}$$

where Φ_L^a is the flux that emanates from region L and R_L is the reluctance between region L and the backlayer. The assumption of circular flux paths leads to

$$P_L \equiv \frac{1}{R_L} \simeq \mu_0 W \left(\frac{T_m}{t_3 + d_4} + \frac{4}{\pi} \ln \left(\frac{h_1 + t_3 + d_4}{t_3 + d_4} \right) \right) \tag{9.20}$$

and to

$$\Phi_L^a = \int_{y' = h_1}^{h_1 + h_2} \Psi^a dP \tag{9.21}$$

where Ψ^a is given by (a9.19) and dP by

$$dP = \frac{2\mu_0 W}{(y' + t_3 + d_4)\pi/2} dy. \tag{9.22}$$

Introduction of $\eta_m \equiv \int_0^{t_3 + d_4} H_y^a/NI \, dy = \Phi_L^a \, R_L/NI$ in this equation leads to the equation:

$$\eta_{\mathrm{m}} = \frac{4\mu_0 W R_{\mathrm{L}}}{\pi h_2} \int_{h_1}^{h_1 + h_2} \frac{y' - h_1}{y' + t_3 + d_4} \, \mathrm{d}y' - \eta_{\mathrm{m}} \frac{4\mu_0 W R_{\mathrm{L}}}{\pi} \int_{h_1}^{h_1 + h_2} \frac{\mathrm{d}y'}{y' + t_3 + d_4} \, .$$

$$(9.23)$$

After solving the integrals, rearranging terms and substituting for R_{L} (9.20), the following explicit expression for η_{m} results:

$$\eta_{\mathrm{m}} = \frac{1 - \dfrac{h_1 + t_3 + d_4}{h_2} \ln \left(\dfrac{h_1 + h_2 + t_3 + d_4}{h_1 + t_3 + d_4} \right)}{\dfrac{\pi T_{\mathrm{m}}}{4(t_3 + d_4)} + \ln \left(\dfrac{h_1 + h_2 + t_3 + d_4}{t_3 + d_4} \right)} \, . \qquad (9.24)$$

From this expression it follows that h_1 must be made as small as possible. The greatest inaccuracy when applying this expression to actual heads is caused by the assumption that the problem is 2-dimensional, i.e. $W \gg h_1 + h_2$, a problem that also arises when the problem is solved numerically by a 2-dimensional finite-element method (FEM) as applied in for instance the Maggy package, see e.g. Zieren et al. [9.2]. By introducing the phenomenological term, described in Sect. 8.1.5, in the analogous 3-dimensional calculation this problem can be solved for $W < h_1 + h_2$ too. The results obtained with this concept (see following section) are however such that this 'effort' does not seem worthwhile.

9.4 Results for PH

Efficiency results according to (9.24) are shown in Fig. 9.9. The efficiency decreases with the height of the unwound part, h_1, as well as the height of the wound part, h_2, of the probe. Large variations in the thickness, T_{m}, of the probe have only a slight influence on the efficiency. A large T_{m} reduces the reluctance between upper probe surface and backlayer and consequently the 'gap' reluctance (between upper probe surface and backlayer). This is the same effect that an increase of the gap height, h_1, has on the efficiency of a ring head, with this difference that h is large and so the effect of T_{m} is much smaller. In all cases shown, the PH efficiency is very low, in spite of the assumption that the permeabilities in probe and backlayer are infinite. Neglecting the influence of T_m at very small head-backlayer distances, $t_3 + d_4$, only the geometric ratios $h_1/(t_3 + d_4)$ and $h_2/(t_3 + d_4)$ are responsible for this: these ratios determine the finite parallel reluctances between left and right probe surface and backlayer.

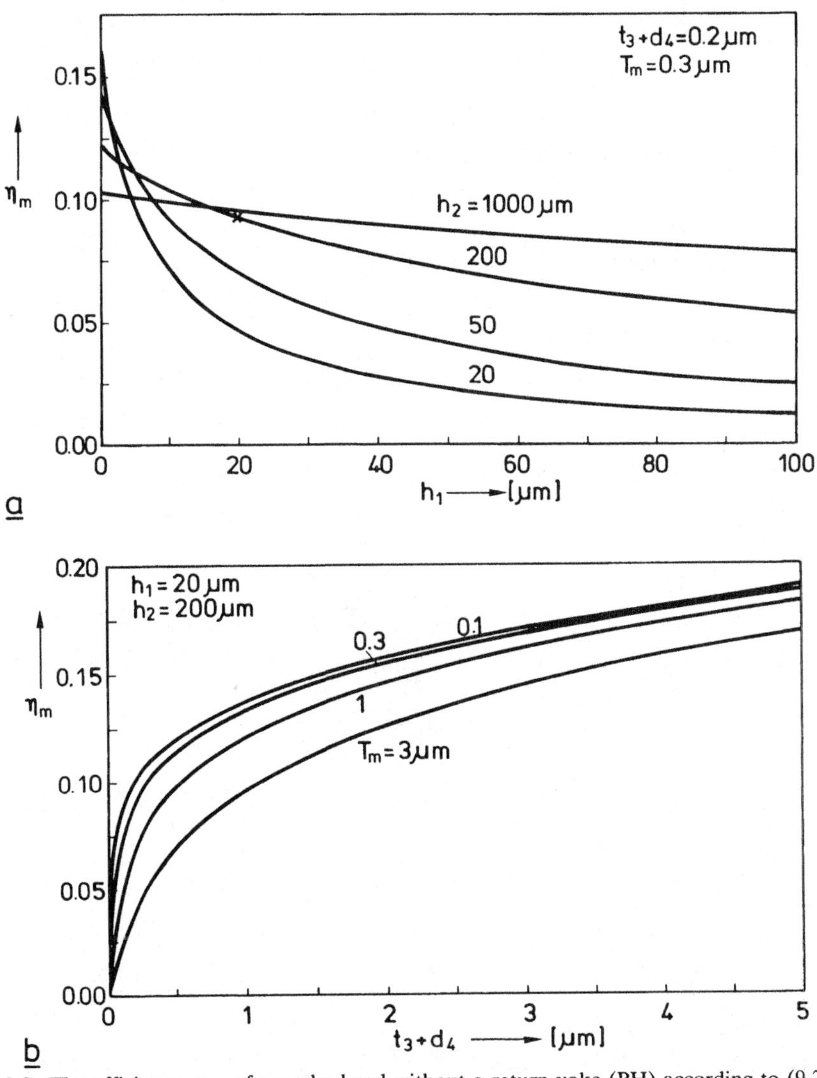

Fig. 9.9. The efficiency, η_m, of a probe head without a return yoke (PH) according to (9.24)
a) as a function of the height of the free (unwound) front part of the probe, h_1, at various heights of the wound part, h_2, of the probe,
b) as a function of the probe-backlayer distance, $t_3 + d_4$, at various thicknesses, T_m, of the probe.

Practically (and theoretically!) *it is impossible to obtain a good efficiency* with this head concept.

9.5 Probe head with return yoke (PHRY)

Because it would take up too much space to give the complete derivations that lead to expressions for the efficiencies, only some main steps are shown.

9.5.1 Case 1: long coil-chamber length

A long coil-chamber length is here defined by $b_c \geqslant h_1 + h_2 + t_3 + d_4$. The steps leading to the relevant expression for this case are outlined in this section.

Fig. 9.10 gives a number of quantities between brackets that are not given in Fig. 9.2. These quantities are derived from the basic quantities and are of help in the derivations and in the presentation of the final η-calculation scheme.

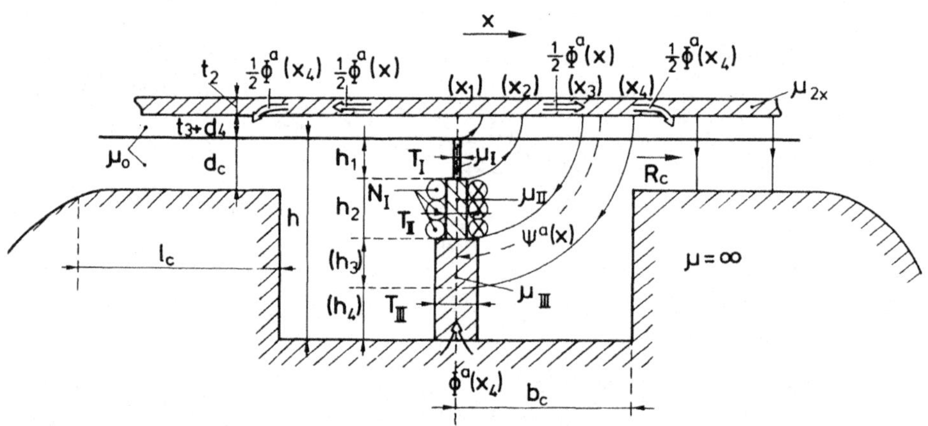

Fig. 9.10. Basic (without brackets) and derived dimensions and quantities (between brackets) used in the derivations of the expressions for the efficiencies η_m and η_a of the 2-dimensional PHRY in case $b_c \geqslant h_1 + h_2 + t_3 + d_4$.
$\Rightarrow$ defines positive flux direction, $\Phi > 0$.
$\leftarrow--$ defines positive magnetic-scalar-potential-difference direction, $\Psi > 0$, i.e. at the arrow point the potential is higher than on the other side of the arrow.
This $\Psi(x)$ is not the potential at a fixed point relative to the (zero) potential at a fixed point, but is defined as the potential somewhere in the probe relative to the potential at the corresponding place, i.e. following the quarter circles, in the backlayer!

With reference to Fig. 9.10, the main pole and backlayer form together a transmission line. The distance between both parts, main pole and backlayer, of this transmission line is however not constant but diverges. In contrast, the transmission line formed by backlayer and auxiliary pole is uniform. The decay of the potential and flux in the latter is then described by the well-known e-powers. The solution to non-uniform transmission lines must be described by other functions. We will show that a transmission line with linearly increasing distance in between, and with the simplification of flux lines being circular, leads to differential equations with Bessel functions as solutions.

Region I.

$$\Psi = \Psi(x_1) + \int_{x_1}^{x} \Phi(x')\, d(\tfrac{1}{2}R_2 + R_{\mathrm{I}}) \qquad (9.25)$$

where R_2 and R_{I} are the reluctances of backlayer and main pole (region I).

$$\Phi = \Phi(x_1) + \int_{x_1}^{x} \Psi(x')\, d(2P) \qquad (9.26)$$

where P is the permeance between backlayer and main pole.

Region II

$$\Psi = \Psi(x_2) + \int_{x_2}^{x} \Phi(x')\, d(\tfrac{1}{2}R_2 + R_{\mathrm{II}}) - \frac{x - x_1}{h_2}\, NI. \qquad (9.27)$$

In this region Ψ will change its sign. Φ is as before.

Region III

$$\Psi = \Psi(x_3) + \int_{x_3}^{x} \Phi(x')\, d(\tfrac{1}{2}R_2 + R_{\mathrm{III}}) - NI. \qquad (9.28)$$

Ψ has the opposite sign of that in region I. For $x < x_4$ the flux, Φ, is as before. Beyond x_4 the flux is constant. After differentiation of the expressions for the potential and flux in each region, simultaneous differential equations are obtained. Another differentiation of the first and substitution in the latter leads to three second-order differential equations for the potentials:

$$x \frac{d^2 \Psi_i}{dx^2} - C_{ai} C_b \Psi_i = 0 \tag{9.29}$$

where i = I, II and III,

$$C_{ai} = \frac{1}{\mu_0 W} \left(\frac{1}{2\mu_{2x} t_2} + \frac{1}{\mu_i T_i} \right)$$

and $C_b = 4\mu_0 W/\pi.$

After a change of variable, $x' \equiv 2j\sqrt{C_{ai} C_b x}$, the above equation reduces to a standard differential equation, given by the equations (8.401) and (8.494.1) in reference [9.14]; or see [9.15], with solution

$$\Psi_i = z_i \{ C_{i1} K_1(z_i) + C_{i2} I_1(z_i) \} \tag{9.30}$$

where $z_i \equiv 2\sqrt{C_{ai} C_b x}$.

The flux is proportional to the first derivative of the potential according to (9.25) and (9.28), whereby for the flux in region II an extra constant due to the last term in (9.27) must be added. Carrying out the differentiations on (9.30), while using the 'recursion formulas' (8.486.11 + 2 + 17 + 8) in [9.14], then leads to:

$$\Phi_i = 2C_b \{ -C_{i1} K_0(z_i) + C_{i2} I_0(z_i) \} \qquad\qquad i = I,\ III \tag{9.31a}$$

$$\Phi_i = 2C_b \{ -C_{i1} K_0(z_i) + C_{i2} I_0(z_i) \} + \frac{C_d}{C_{aII}} \qquad\qquad i = II \tag{9.31b}$$

where $C_d \equiv NI/h_2$.

The six constants are solved by the six boundary conditions: one at the beginning (x_1) and one at the 'end' (x_4) of the diverging transmission line, and two at each boundary, x_2 and x_3, in between. At the beginning the 'gap' reluctance, R_g, and at the end the sum of the reluctance of the uniform transmission line formed by the backlayer and the auxiliary pole, R_c, plus the reluctance of the main pole beyond x_4, define relations between Ψ and Φ.

At the two boundaries in between, continuity of flux and potential leads to relations between fluxes as well as potentials in neighbouring regions.

For the 'gap' reluctance an expression analogous to (8.12) is used

$$R_g = \frac{1}{\mu_0 W} \frac{1}{\dfrac{T_I}{t_3 + d_4} + \dfrac{1}{\pi}} \tag{9.32}$$

where $1/\pi$ accounts for the spread of flux within x_i. The reluctance of the uniform transmission line equals the characteristic impedance

$$R_c = \frac{1}{\mu_0 W} \sqrt{\frac{t_3 + d_4 + d_c}{\mu_{2x} t_2}} \tag{9.33}$$

if the characteristic length

$$\lambda_c = \sqrt{\mu_{2x} t_2 (t_3 + d_4 + d_c)} \tag{9.34}$$

is smaller than the auxiliary pole length, l_c.

This will be assumed in all calculations and checked afterwards.

Since the efficiency of the main pole follows from the potential at the beginning of the diverging transmission line, i.e.

$$\boxed{\eta_m \simeq \Psi(x_1)/NI = z_{I1}\{C_{I1}K_1(z_{I1}) + C_{I2}I_1(z_{I1})\}/NI} \tag{9.35}$$

with $z_{I1} \equiv 2\sqrt{C_{aI}C_b x_1}$,
only the constants C_{I1} and C_{I2} need to be derived.

This derivation from the boundary conditions is however so very lengthy that we will not reproduce it here.

Only the results from these calculations as used in a Fortran program are reproduced in App. 9.1. There NI and W are consequently omitted (or taken 1) because of linearity and 2-dimensionality of the problem, respectively.

9.5.2 Case 2: short coil-chamber length

A short coil-chamber length is here defined by $h_1 + t_3 + d_4 < b_c < h_1 + h_2 + t_3 + d_4$.

Only essential differences with respect to the derivations for case 1 will be discussed.

The configuration was given in Fig. 9.3. Because of the smaller coil chamber length, b_c, (compare with case 1) circular flux lines emanating from the coil region now reach the auxiliary pole and will be taken into account. As a consequence the angle α may influence the results. In case 1, b_c was so large that it sufficed to describe this part of the head with a series reluctance only. Now this part is described by a (non-uniform) transmission line with angle α.

Beyond $x = b_c$, the differential equations differ from those of case 1, in that:

1) a new coordinate $x^* = x + b_c \tan \alpha - (t_3 + d_4 + d_c)$ instead of x (see Fig. 9.3) defines the distance to the point $x^* \equiv 0$ where the non-uniform transmission line with angle α diverges virtually.

2) The angle $\pi/2 - \alpha$ replaces the angle of $\pi/2$ in the expression for the parallel permeance P, hence $C_b \rightarrow C_b^* \equiv 2\mu_0 W/(\pi/2 - \alpha)$.

3) The potential at the new 'boundary' $x = b_c$ jumps by an amount $\frac{1}{2}\Phi (b_c) R_c$, where the potential beyond $x = b_c$ is defined as the potential of the main pole with respect to the auxiliary pole and not with respect to the backlayer; this leads to two extra boundary conditions and one extra second-order differential equation for the second part of region II.

4) Beyond $x = b_c$, consequently the term $\frac{1}{2}R_2$ vanishes in the differential equations for the potential, see (9.25), (9.27) and (9.28), hence beyond $x = b_c$,

$$C_{ai} \rightarrow C_{ai}^* = \frac{1}{\mu_0 W} \frac{1}{\mu_i T_i} \qquad i = \text{II, III,} \qquad (9.36)$$

together with 1) and 2) this leads to a replacement of the arguments z_i into $z_i^* = 2\sqrt{C_{ai}^* C_b^* x^*}$ in the Bessel functions.

5) The boundary condition at the end of the main pole reduces to $\Psi(x_4^*) = 0$.

Since after writing out all the boundary conditions two more linear equations are obtained with respect to the six in case 1, no attempt is made to solve these equations, without making mistakes, by hand. The matrix representation of these 8 coupled linear equations reads:

$$
\begin{pmatrix} a_{11} & & a_{18} \\ & & \\ & & \\ & & \\ & & \\ & & \\ & & \\ a_{81} & & a_{88} \end{pmatrix}
\begin{pmatrix} C_{\mathrm{I}1} \\ C_{\mathrm{I}2} \\ C_{\mathrm{II}1} \\ C_{\mathrm{II}2} \\ C_{\mathrm{II}1}^{*} \\ C_{\mathrm{II}2}^{*} \\ C_{\mathrm{III}1}^{*} \\ C_{\mathrm{III}2}^{*} \end{pmatrix}
=
\begin{pmatrix} b_1 \\ \\ \\ \\ \\ \\ \\ b_8 \end{pmatrix}.
\tag{9.37}
$$

The constants $C_{\mathrm{I}1}$, etc. are solved by the numerical Fortran routine FO4ATF in the NAG library. The coefficients a_{ij} and b_{j} are read from the boundary equations and are given in App. 9.2. Most of them are zero. The efficiencies read:

$$
\boxed{\eta_{\mathrm{m}} \simeq \Psi(x_1)/NI = z_{\mathrm{I}1}\{C_{\mathrm{I}1}K_1(z_{\mathrm{I}1}) + C_{\mathrm{I}2}\, I_1\, (z_{\mathrm{I}1})\}/NI}
\tag{9.38}
$$

and

$$
\boxed{\eta_{\mathrm{a}} = \frac{\tfrac{1}{2}\Phi(b_{\mathrm{c}})R_{\mathrm{c}}}{NI} = \frac{[-C_{\mathrm{b}}C_{\mathrm{III}1}K_0(z_{\mathrm{IIc}}) + C_{\mathrm{b}}C_{\mathrm{III}2}I_0(z_{\mathrm{IIc}}) + C_{\mathrm{d}}/2C_{\mathrm{aII}}]R_{\mathrm{c}}}{NI}}
$$

$$
\tag{9.39}
$$

where $z_{\mathrm{I}1} \equiv 2\sqrt{C_{\mathrm{aI}}C_{\mathrm{b}}x_1}$ and $z_{\mathrm{IIc}} \equiv 2\sqrt{C_{\mathrm{aII}}C_{\mathrm{b}}b_{\mathrm{c}}}$.

One practical point is that W can be omitted because of the 2-dimensional character of the problem and NI can be omitted because of the linearity of the problem, if both are consequently left out in all expressions, which is done in the presentation of the complete results in App. 9.2. The results given in App. 9.1 and 9.2 represent almost exactly the body of the Fortran program we used for the calculations on PHRYs, discussed in the following section.

9.6 Results for PHRY

9.6.1 Main-pole efficiency

In order to show the influence of variations of the PHRY parameters on the efficiency, we start with the 'rather arbitrary' parameter values given in Table 9.1. Each parameter is varied subsequently, while taking for all the other parameters the standard values given in this Table, which leads to a number of plots that will be discussed. The calculation of 100 points (one plot) takes only a fraction of a second on a larger computer.

To simulate a probe head without a return yoke, i.e. a PH comparable to the previous 'idealized' PH but further as general as the present PHRY, it was only necessary to leave the return yoke in the PHRY calculations out of account. (This is easily done by taking the reluctance of the 'transmission line' between auxiliary-pole surface and backlayer very large, e.g. $R_c = 10^{20}$, and b_c large enough to guarantee that all flux lines, assumed to be circular, leaving the probe also reach the backlayer.) It has been checked that, for a high permeability in backlayer and probe, the results of the previous idealized PH in Sects. 9.3 and 9.4, are exactly reproduced.

First the efficiency decrease due to increasing h_1 and h_2 is shown in Fig. 9.11. Even for small h_1 and h_2 the efficiency is small, as was also observed for the idealized PH in Fig. 9.9a. The efficiency of the present PH in Fig. 9.11 is about half as large as that of the PHRY in Fig. 9.11.

Table 9.1. Standard values as used in all calculations on PHRYs and PHs unless otherwise noted in the figures.

$t_3 + d_4$	0.2 μm	μ_{II}	400
t_2	0.5 μm	T_{III}	20 μm
μ_{2x}	500	h	200 μm
T_I	0.3 μm	μ_{III}	300
h_1	20 μm	b_c	200 μm*
μ_I	500	d_c	15 μm*
T_{II}	5 μm	ℓ_c	500 μm*
h_2	150 μm	α	85°†

*   For the PH the values of b_c, d_c and ℓ_c have no meaning.

†   The coil chamber angle α is only meaningful for PHRYs with $b_c < h_1 + h_2 + t_3 + d_4$. This is the case for the dotted part of the curve in Fig. 9.12 and all curves in Figs. 9.13 and 9.14.

In Fig. 9.12 the influence of the head-backlayer distance, $t_3 + d_4$, is made visible for two rather extreme values of the pole-tip thickness, T_I. Qualitatively the results are as for the idealized PH, which has an efficiency which is only a little higher than that of the present more realistic PH. *The PHRY has a much higher efficiency than the PH, but at usual $t_3 + d_4$ values the efficiency is still very low.*

The intersection point between the curves for $T_I = 0.3$ μm and $T_I = 3$ μm (visible for the PHRY) show that for a finite permeability of the pole tip ($\mu_I = 500$), a larger pole thickness may be more profitable for the efficiency (of course not for the gap-loss function), in contrast to the idealized PH; see also Sect. 9.4.

The optimum thickness of the pole tip, T_I, follows in more detail from Fig. 9.13. The optimum shifts to lower T_I values according as the permeability increases, as must be expected:
- left of the maxima the increasing reluctance of the pole reduces the efficiency,
- right of the maximum the decreasing 'gap' (between upper pole-tip surface and backlayer) reluctance reduces the efficiency.

For the present PHRYs (with $t_3 + d_4 = 0.2$ μm) the optima are reached at usual values of T_I, i.e. $0.2 \sim 0.5$ μm.

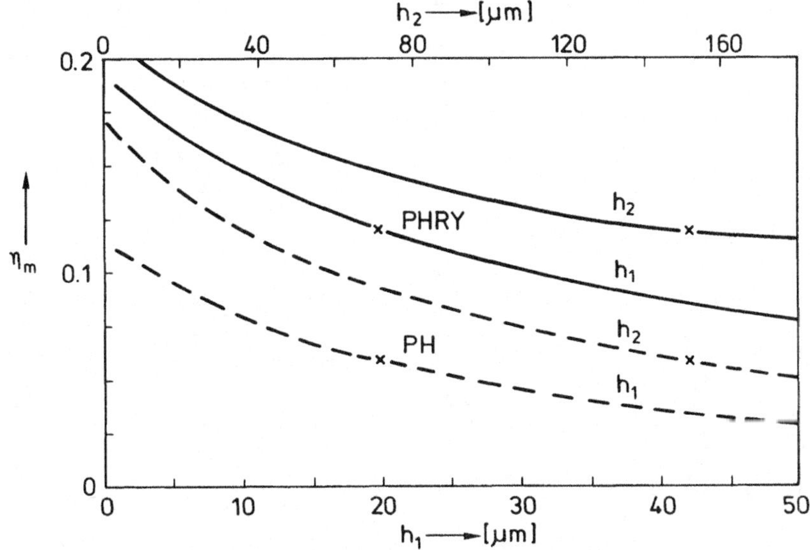

Fig. 9.11. The efficiency of the PHRY and PH as a function of the height, h_1, of the unwound main-pole tip and the height, h_2, of the wound part of the main pole. Other values as given in Table 9.1.

Fig. 9.14 shows the increase of the efficiency due to increasing the permeabilities in the main pole (μ_I, μ_{II} and μ_{III}) and in the backlayer (μ_{2x}). The influence of μ_{2x} is most severe. The probe permeabilities μ_{II} and μ_{III}, if larger than 200, do not influence the results noticeably (mainly because of the assumed large T_{II} and T_{III} values), while the

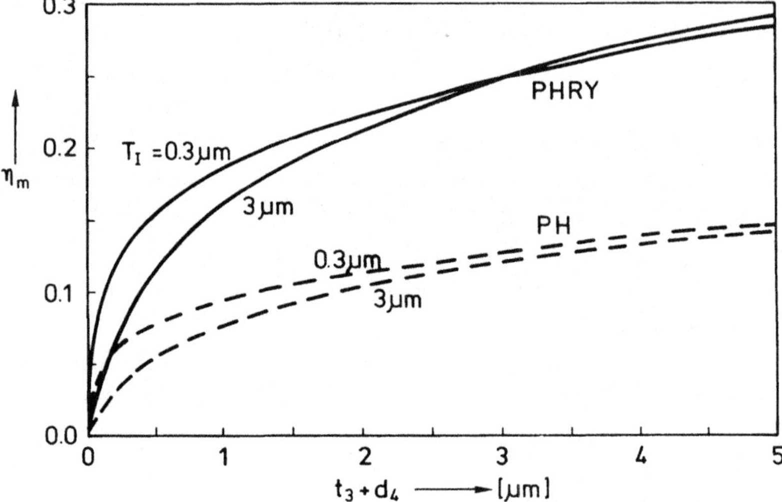

Fig. 9.12. The efficiency of the PHRY and PH as a function of the pole-tip surface – backlayer distance, $t_3 + d_4$, at two 'extreme' values of the pole-tip thickness, T_I. Other values as given in Table 9.1.

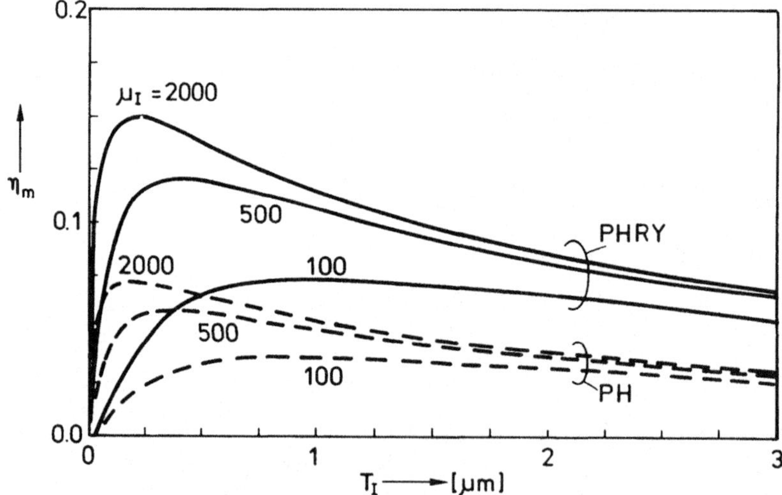

Fig. 9.13. The efficiency of the PHRY and PH as a function of the pole-tip thickness, T_I, at different pole-tip permeabilities, μ_I. Other values as given in Table 9.1.

main-pole tip permeability μ_I has to be larger than, say, 500 before this can be said.

Since it is hard to increase μ_{2x} and/or t_2 drastically (because of magnetic and mechanical problems respectively), the coil-chamber length b_c must be reduced to obtain almost the same dramatic increase in the efficiency as is obtained by increasing $\mu_{2x}t_2$; see Fig. 9.15.

A reduction of the distance d_c ($+ t_3 + d_4$) between the auxiliary-pole surface and the backlayer reduces the characteristic reluctance R_c as well as the characteristic length λ_c of the corresponding section of the head almost in proportion to $\sqrt{d_c}$. It depends on the point of view whether the (small) decrease of η in Fig. 9.15 is to be attributed to the increase of R_c or to the increase of the effective coil-chamber length b_c due to λ_c. The latter point of view, however, makes it clear that the region over which the influence of the finite permeance of the backlayer extends, i.e. ca. $b_c + \lambda_c$, can only be successfully reduced, by way of reducing b_c, as long as $\lambda_c \lesssim b_c$! If in addition $b_c \lesssim (t_2 \, \mu_{2x}/T_I\mu_I) \times h_1 \simeq$ 33 μm, then a simultaneous reduction of the pole-tip reluctance also becomes important. For the standard parameter values ($d_c = 15$ μm etc.) it follows that $\lambda_c = \sqrt{\mu_{2x}t_2(d_c + t_3 + d_4)} = 61$ μm and so the increase of the efficiency by reducing b_c is limited for both reasons (see Fig. 9.15).

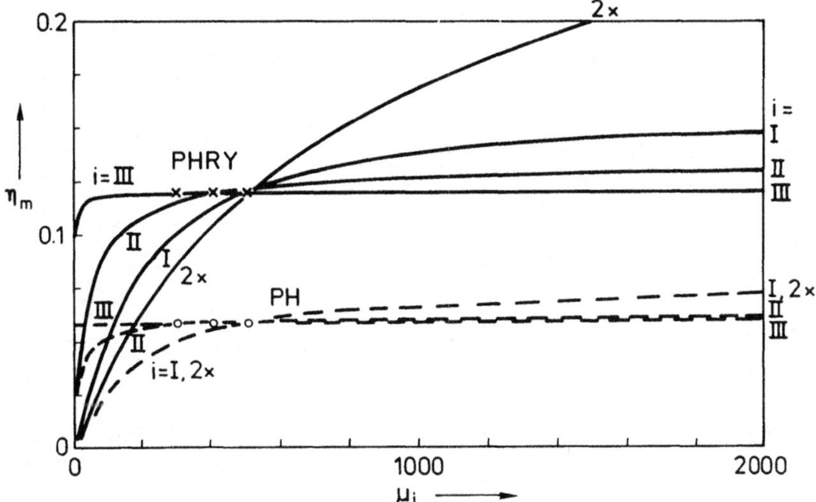

Fig. 9.14. The efficiency of the PHRY and PH as a function of the main-pole permeabilities, μ_I, μ_{II} and μ_{III} and the backlayer permeability, μ_{2x}. Other values as given in Table 9.1.
The (almost complete) coincidence of the curves for μ_I and μ_{2x} is no more than a 'pure coincidence'.

Fig. 9.16a shows that, for small values of b_c, a considerable increase of the efficiency is obtained only when the reduction of d_c (from 15 via 3 to 0 μm, in order to reduce λ_c) is accompanied by an increase of μ_I. Although $T_{II} \gg T_I$, the increase of μ_{II} was also of importance, because $h_2 \gg h_1$ too.

It is also read from the figure that the increase of the main-pole permeabilities hardly influences the ratio η_a/η_m, i.e. η_a is increased considerably as well. It is clear that, in contrast, a change of d_c mainly influences η_a.

When the permeance of the main pole is small over a large height and b_c is small, then the head acts like a thin-film head, the efficiency of which usually decreases with decreasing 'gap length' (b_c) between thin film (main pole) and substrate (auxiliary pole); see e.g. chapter 11. This effect is just visible in Fig. 9.17 below $b_c \simeq 40$ μm. In order to show this effect, T_{II} had to be reduced to 1 μm, which also explains the very low efficiency, which is comparable with that of the magnetoresistive thin-film heads in chapter 11.

From the previous results, especially those in Fig. 9.16a, it is clear that a high efficiency for the present head concept, PHRY, is still hard to obtain. Only when b_c is made rather small, when d_c is small and

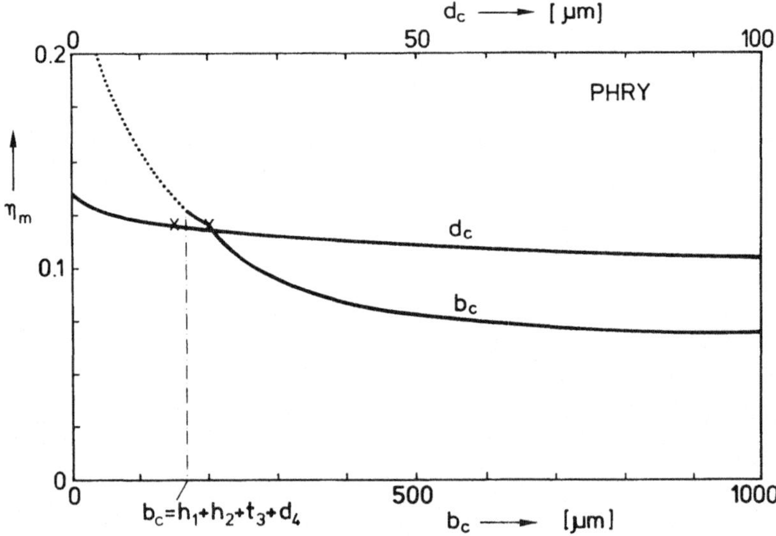

Fig. 9.15. The efficiency of the PHRY as a function of the coil chamber length, b_c, and as a function of the distance between the auxiliary-pole surface level and the main-pole surface level, d_c. Parameter values and a remark concerning the dotted part of the curve $\eta_m(b_c)$ are given in Table 9.1.

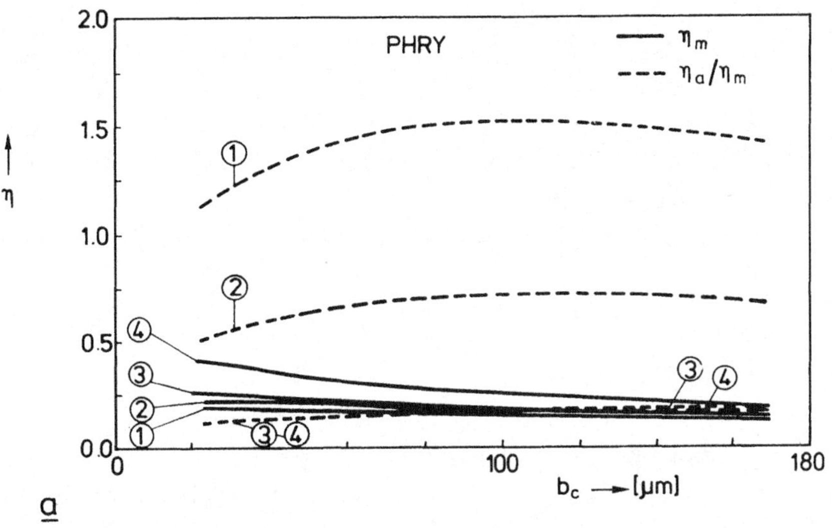

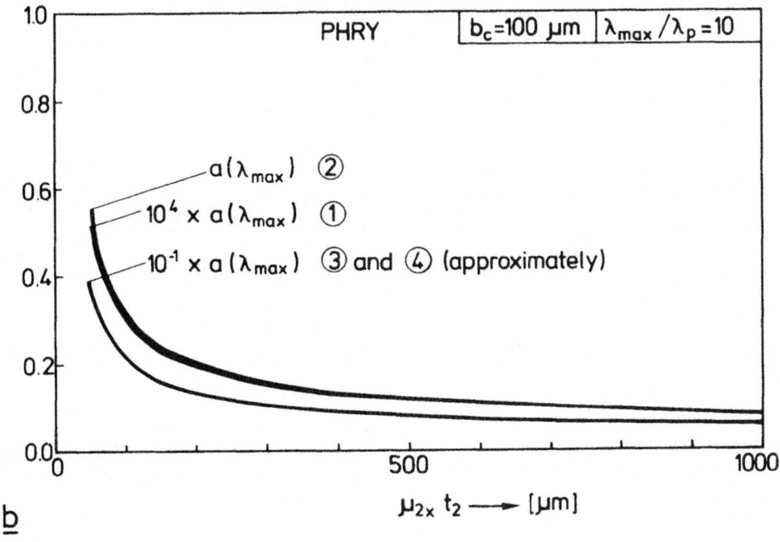

Fig. 9.16.

a) The efficiency of the PHRY as a function of the coil chamber length b_c and the importance of a simultaneous increase of μ_I (and μ_{II}) with the reduction of d_c.

b) Auxiliary-pole read effects represented by the perturbation quantity $a(\lambda_{max})$ at $b_c = 100$ μm and $\lambda_{max}/\lambda_p = 10$ as a function of $\mu_{2x}t_2$.

1) $d_c = 15$ μm $\mu_I = 500$ $\mu_{II} = 400$ ($\equiv$ standard)
2) $d_c = 3$ μm $\mu_I = 500$ $\mu_{II} = 400$
3) $d_c = 0$ μm $\mu_I = 500$ $\mu_{II} = 400$
4) $d_c = 0$ μm $\mu_I = \mu_{II} = 1000$

Other values as given in Table 9.1 unless indicated.

permeabilities are as high as possible in practice, may an acceptable efficiency be reached. When the angle α (see Fig. 9.3) is reduced, then b_c can be further reduced without involving problems with the previously mentioned 'thin-film head' effects. However d_c must then become so small (for $b_c + \lambda_c$ to be effectively reduced) that auxiliary write/read effects may become severe.

This will be investigated with the aid of the ratio η_a/η_m plotted in Fig. 9.16a and Fig. 9.18, using the criteria derived in Sect. 9.2.

9.6.2 Auxiliary-pole effects

The auxiliary-pole effects can easily be predicted using the very simple expressions derived in Sect. 9.2 once η_a/η_m has been calculated. For the PHRY configurations ① to ④ in Fig. 9.16a with $d_c = 15$ and 3 μm and the two configurations with $d_c = 0$ μm, we read η_a/η_m values of approximately 1.5, 0.7 and 0.16 respectively. Application of (9.17a) to the first two configurations and of (9.16a) to the last two configurations, where $D_m = D_a$, gives for $\lambda_{max}/\lambda_p = 10$:

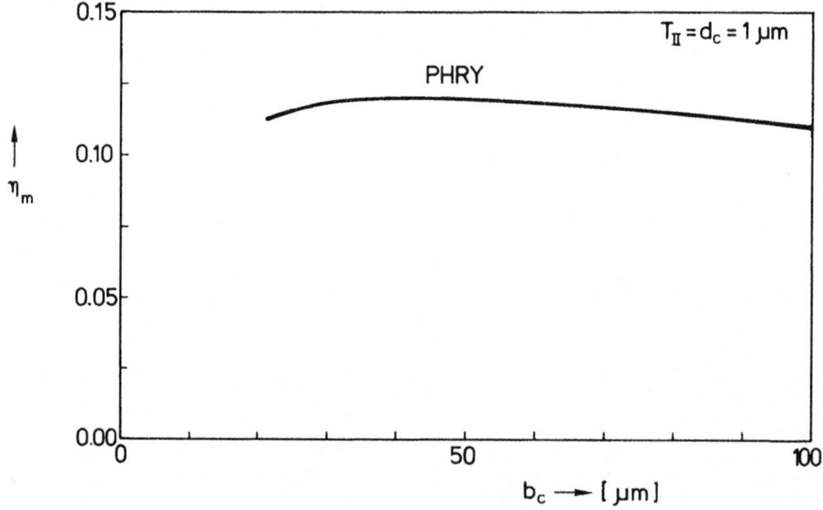

Fig. 9.17. The reduction of the efficiency at lower b_c ('thin-film head effect') made visible in a PHRY by choosing a thin main pole ($T_{II} = 1$ μm). The optimum coil chamber length, $b_{c\ opt}$, depends on all permeabilities and dimensions. Not indicated values as given in Table 9.1.

① $d_c = 15\,\mu$m: $\qquad a(\lambda_{\max}) \approx 1.66\,10^{-4} \ll 0.2 \;\Rightarrow$ no auxiliary-pole read effects

$$\frac{\eta_a}{\eta_m}\frac{D_m}{D_a} = 0.02 \ll \tfrac{1}{4} \qquad\Rightarrow$$ no auxiliary-pole write effects

② $d_c = 3\,\mu$m: $\qquad a(\lambda_{\max}) \approx 0.17 < 0.2 \qquad\Rightarrow$ doubtful read effects

$$\frac{\eta_a}{\eta_m}\frac{D_m}{D_a} = 0.044 \ll \tfrac{1}{4} \qquad\Rightarrow$$ no write effects

③ and ④ $d_c = 0\,\mu$m: $\quad a(\lambda_{\max}) \simeq 1 \gg 0.2 \qquad\Rightarrow$ unacceptable read effects

$$\frac{\eta_a}{\eta_m} = 0.16 < \tfrac{1}{4} \qquad\Rightarrow$$ acceptable write effects.

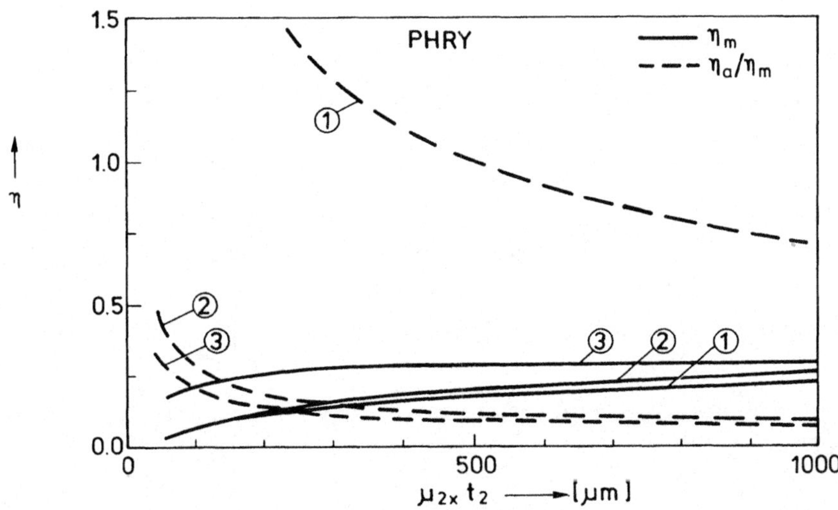

Fig. 9.18. The ratio η_a/η_m in the PHRY, important in the determination of auxiliary-pole effects; and the efficiency of the main pole, η_m, as a function of the permeability and thickness of the backlayer at different coil chamber lengths, b_c, and auxiliary pole levels, d_c.
1) $d_c = 15\,\mu$m $b_c = 170\,\mu$m
2) $d_c = 0\,\mu$m $b_c = 170\,\mu$m
3) $d_c = 0\,\mu$m $b_c = 21\,\mu$m
Other values as given in Table 9.1.

See also the plots of $a(\lambda_{\max})$ as a function of $\mu_{2x}t_2$ in Fig. 9.16b.

So, only the configuration with the worst efficiency η_m, see curve ① in Fig. 9.16a, is completely free of any auxiliary pole effects. The only configuration with an acceptable main-pole efficiency, i.e. configuration ④, will show unacceptably strong auxiliary-pole read effects.

In Fig. 9.18 the ratio η_a/η_m is given as a function of the backlayer permeability and thickness, where b_c is 170 μm or 21 μm and d_c is 15 or 0 μm. Only at impractically low $\mu_{2x}t_2$ values (where in addition the calculations become more inaccurate), is a considerable rise of the ratio η_a/η_m observed. At the largest $\mu_{2x}t_2$ shown (1000 μm), the effects have been reduced by a factor of 2 with respect to the standard situation for which $\mu_{2x}t_2 = 250$ μm.

Thus, even in the most profitable situation for η_a/η_m (and also for η_m), i.e. $\mu_{2x}t_2$ large and b_c small, *the technologically most attractive configuration, i.e. $d_c = 0$ μm, will fail because of auxiliary-pole effects.* The auxiliary-pole effects are strong for $d_c = 0$ μm at wavelengths of interest, since $GLF_a(k)$ is of the order of $GLF_m(k)$ when $d_c = 0$ μm, in contrast to the case that $d_c \gg 0$ μm. For $d_c = 0$ μm this auxiliary-pole effect is often observed experimentally; see for instance Fig. 4 in reference [9.13].

9.7 Planary probe head with return yoke (PHRYP)

Also for this configuration the calculations will not be shown in detail. Only main steps and essential differences with the previous calculations will be outlined. When approximations have to be made this will of course be told, such that in principle every result can be checked by the reader, as for the previous calculations.

Approximation for the reluctance, R_T, between the main pole and backlayer

The present planar head, given in Fig. 9.4, is meant to be a head with a relatively short main pole, say $h_1 \approx 20$ μm. The losses due to the finite main pole permeability are then small and its permeability can be taken infinite. For the same reason the losses over a distance of about h_1 in the backlayer can be neglected. Thus the total reluctance from the beginning of the main pole up to a few tens of µms away from the main pole in the backlayer, R_T, can be approximated by using

$$R_T^{-1} = P_T = \frac{4\mu_0(W + T)}{\pi} \ln\left(\frac{h_I + t_3 + d_4}{t_3 + d_4}\right) + 2\mu_0 h_I + \mu_0 W\left(\frac{T}{t_3 + d_4} + \frac{1}{\pi}\right)$$

$$(9.40)$$

where
– the first term (on the right-hand side) represents the permeance of the volumetric region that extends from the four side surfaces of the main pole to the backlayer, according to (8.2); the assumed quarter-circular flux lines in this region are sketched in Fig. 9.4;
– the second term is the phenomenological term that accounts for the spread of flux at the edges, see Sect. 8.1.5;
– the third term is the 'gap' between upper main pole surface and backlayer permeance, see also (9.32).

Symmetry considerations

Region I starts at $r = r_1$, where r_1 is chosen such that the circumference $2\pi r_1$ of a virtual cylinder (dotted in Fig. 9.4) equals the circumference of the main pole $2(W + T)$, i.e.:

$$r_1 \equiv \frac{W + T}{\pi} .$$

$$(9.41)$$

In order to reduce the head configuration to a pure circular-symmetric configuration, the potential at the circumference of this cylinder in the backlayer (circle) is approximated by a constant value. This is easily done by assuming that the reluctance from this circle to the beginning of the main pole equals R_T, which is a reasonable approximation as outlined before. Also the outer edge of the head must be a circle before the symmetry of the head is completely circular. This is denoted by the radius r_4. For a practical head configuration, a circular outer circumference is usually technologically more complicated than a rectangular one.

It is important to note that strong deviations from the circle, e.g. a rectangular outer circumference, have hardly any influence on the potentials and fluxes in the regions I and II, when the minimum distance (denoted m in Fig. 9.4) from actual outer edge to r_3 is more than a few characteristic distances, $\lambda_c = \sqrt{\mu_{2r}t_2(h_{III} + t_3 + d_4)}$, of the 'transmission-line' formed by backlayer and auxiliary pole surface. (It is alternatively checked afterwards that small variations in r_1 and extremely large variations of r_4 indeed have no effect on the results.) Furthermore, the flux lines in region I (excepted those already accounted for in R_T) and in the regions II and III are approximated by straight vertical lines, since the head is planar.

So for almost any 3-dimensional planar head configuration, consisting of a cylindrical coil chamber, carrying a spiral coil, and having its main pole in the centre of this chamber, only one coordinate, r, suffices to describe the head.

Regions I and III

Instead of (9.25) we now have in region I (and analogously in region III)

$$\Psi = \Psi(r_1) + \int_{r_1}^{r} \Phi(r')\, \mathrm{d}R_2, \qquad (9.42)$$

where $\mathrm{d}R_2 = \mathrm{d}r'/(\mu_0\mu_{2r}t_2 2\pi r')$ is the reluctance of an infinitesimal part of the backlayer.

In contrast to $\mathrm{d}R_2/\mathrm{d}x$ in the 2-dimensional PHRY, this $\mathrm{d}R_2/\mathrm{d}r$ is dependent on (inversely proportional to) the radius.

Instead of (9.26) we have:

$$\Phi = \Phi(r_1) + \int_{r_1}^{r} \Psi(r')\, \mathrm{d}P, \qquad (9.43)$$

where $\mathrm{d}P = \mu_0 2\pi r'\,dr'/(h_T + t_3 + d_4)$ is the permeance of a circular strip with infinitesimal thickness dr' between backlayer and head.

Now r' appears in the numerator, in contrast to the P in the PHRY, where x' appeared in the denominator.

These differences will lead to a different second-order differential equation with essentially different solutions from those for the PHRY, namely:

$$\frac{d^2\Psi_i}{dr^2} + \frac{1}{r}\frac{d\Psi_i}{dr} - C_a C_{bI}\Psi_i = 0 \qquad i = I, III \qquad (9.44)$$

with solutions

$$\Psi_i = C_{i1}K_0(z_i) + C_{i2}I_0(z_i) \qquad (9.45)$$

$$\Phi_i = \frac{z_i}{C_a}\left[-C_{i1}K_1(z_i) + C_{i2}I_1(z_i)\right] \qquad (9.46)$$

where $z_i \equiv r\sqrt{Cz_a C_{bi}}$,

$$C_a \equiv \frac{1}{\mu_0\mu_{2r}t_2 2\pi} \qquad (= r'dR_2/dr') \ ,$$

and $\quad C_{bi} \equiv \frac{\mu_0 2\pi}{h_i + t_3 + d_4} \qquad \left(= \frac{1}{r'}\,dP/dr'\right).$

Equation (9.44) with solution (9.45) followed from the standard differential equation (8.401), see also (8.494.1), in reference [9.14] after substitution of $z \equiv r\sqrt{C_a C_{bi}}$ in (9.44). In the derivation of the flux expression (9.46) use is made of the recursion formulas (8.486.11) and (8.486.2) in [9.14].

Region II with minimum impedance winding

In this region

$$\frac{d\Psi}{dr} \simeq C_a\frac{\Phi}{r} - I\frac{dN(r)}{dr} \qquad (9.47a)$$

and

$$\frac{1}{r}\frac{d\Phi}{dr} = C_{bII}\Psi \qquad (9.47b)$$

where $dN(r)/dr$ is the winding density and I the current flowing in each turn.

We assume a very special winding density, namely

$$\frac{dN(r)}{dr} = \frac{C_d}{Ir} \tag{9.48}$$

with $C_d = NI/\ln(r_3/r_2)$, where N equals the total number of turns $\int_{r_2}^{r_3} dN(r)$ in region II, i.e. *the distance between the turns is proportional to the radius* (see also Fig. 9.4).

Further we assume that a fixed part, α, of this distance is filled by the conductor.

For this choice it can be proved that
1) the resistance of each turn is equal,
2) *the resistance* of the complete N-turn winding *is minimal* with respect to any other N-turn winding,
3) *the efficiency is higher* than for a head with a constant winding density.

Point 2 follows after minimizing the resistance of two neighbouring turns, the first located between r_1 and r_2, and the second located between r_2 van r_3, with the constraint that $r_3 - r_1 \equiv C$ is constant and r_2 is free. Hence, with $P \sim r_{max}/r_{min}$ being the permeance of a turn according to (8.2), the extrema in the total resistance of the two turns follow from

$$\frac{d}{dr_2}\left\{ \frac{1}{\ln\left(\alpha\, \frac{r_2}{r_1}\right)} + \frac{1}{\ln\left(\alpha\, \frac{r_1 + c}{r_1}\right)} \right\} = 0 \tag{9.49}$$

and after a while the minimum from

$$\frac{r_3}{r_2} = \frac{r_2}{r_1}, \tag{9.50}$$

i.e. the conductor width must be proportional (or the winding density inversely proportional etc.) to the radius.

Point 3 is a direct consequence of the fact that the efficiency of turns located nearer to the main pole is higher.

The difference between the equations in (9.47) and the corresponding equations in the regions I and II (which have not been shown, but follow directly from (9.42) and (9.43)), is the addition of the source (I) term in the right-hand side of (9.47a). The second-order differential equation follows after differentiation of Φ in (9.47a) and substitution in (9.47b). An extra advantage of the special choice of the winding density according to (9.47) is that the source term only adds a constant to Φ, and thus vanishes after the differentiation. Hence (9.44) and (9.45) also apply to region II, i.e.:

$$\psi_{II} = C_{II1}K_0(z_{II}) + C_{II2}I_0(z_{II}).$$

(9.51)

The flux follows after differentiation etc., from (9.47b):

$$\Phi_{II} = \frac{z_{II}}{C_a}\left[-C_{II1}K_1(z_{II}) + C_{II2}I_1(z_{II})\right] + \frac{C_d}{C_a}$$

(9.52)

and differs from the structure of (9.46) in the last 'source' term (z_{II} is defined below (9.46)).

Boundary conditions

The boundary conditions are:

$$\Psi_I(r_1) = R_T\Phi_I(r_1)$$

(9.53)

$$\Psi_I(r_2) = \Psi_{II}(r_2) \text{ and } \Phi_I(r_2) = \Phi_{II}(r_2)$$

(9.54)

$$\Psi_{II}(r_3) = \Psi_{III}(r_3) \text{ and } \Phi_{II}(r_3) = \Phi_{III}(r_3)$$

(9.55)

$$\Phi_{III}(r_4) = 0,$$

(9.56)

i.e. six equations from which, after writing them out and putting the coefficients (given in App. 9.3) in a [6 × 6] matrix, the six unknown constants have been solved.

The efficiencies read:

$$\boxed{\eta_m = \frac{\Psi_I(r_1)}{NI} = \frac{C_{I1}K_0(z_{I1}) + C_{I2}I_0(z_{I1})}{NI}}$$

(9.57)

and

$$\boxed{\eta_a = \frac{-\Psi_{II}(r_3)}{NI} = \frac{-\left[C_{II1}K_0(z_{II3}) + C_{II2}I_0(z_{II3})\right]}{NI}}\,. \qquad (9.58)$$

Because of the linearity of the problem, NI can be neglected, if this is consistently done in all expressions. This is done in the presentation of the complete results in App. 9.3. This appendix represents almost exactly the body of the Fortran program used for the calculations on PHRYPs in the following section. An essential difference compared with the previous (2-dim.) configurations is that W now influences the results because of the 3-dimensional character of the PHRYP.

9.8 Results for PHRYP

9.8.1 Main-pole efficiency

The efficiency of PHRYPs is much higher than of all previous 2-dimensional probe-type heads; see Fig. 9.19a. This is true for the standard PHRYP configuration, whose parameters are given in the caption of Fig. 9.19a, and for all configurations in which the inner and outer radii of the coil, r_2 and r_3, remain unchanged or have been reduced. This can be concluded from the other curves in Fig. 9.19a where all parameters, except r_2 and r_3, have been changed relatively to the standard values. It is also shown in this figure that an increase of r_3 enhances the efficiency, but only significantly in the low backlayer-permeability region; obviously the backlayer is not the limiting factor when r_3 is as small as 200 µm.

On the other hand, a reduction of the outer radius of the head, r_4, to r_3 (i.e. the auxiliary pole is left out), as well as an increase of the auxiliary pole to backlayer distance ($h_{III} = 10$ µm), reduce the efficiency at especially larger backlayer permeabilities and thicknesses, because at low $\mu_{2r}t_2$ values the backlayer permeance dominates.

An increase of the main-pole thickness reduces the efficiency, because of the lower 'gap' reluctance, as already explained for the PHRY.

Fig. 9.20 shows the considerable reduction of the efficiency when r_2 and/or r_3 are increased. Poor results for $r_2 \geqslant 100$ µm and $r_3 \geqslant 500$ µm

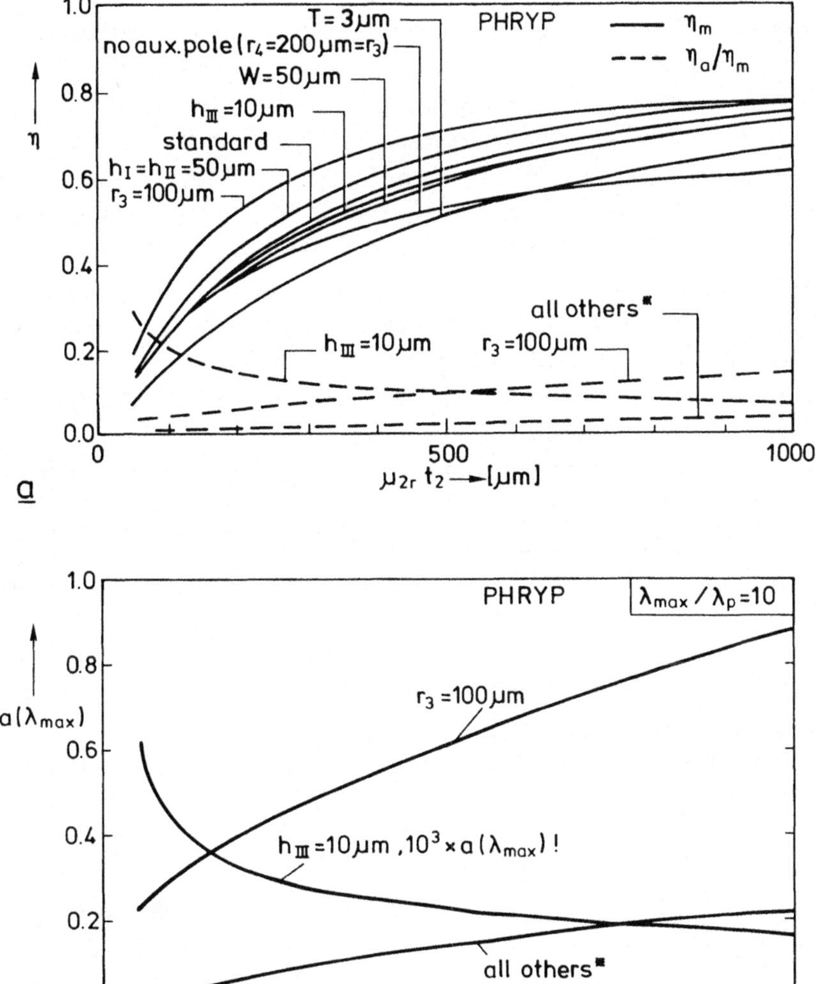

Fig. 9.19.
a) The ratio η_a/η_m in PHRYPs, important in the prediction of auxiliary-pole effects, and the efficiency, η_m, as a function of the permeability and thickness of the backlayer.
b) Auxiliary-pole read effects represented by the ratio $a(\lambda_{max})$.

Only the values that deviate from the following standard values are given at each curve.
Standard values:

h_I	$= 20\,\mu m$	T	$= 0.3\,\mu m$	r_2	$= 30\,\mu m$
h_{II}	$= 25\,\mu m$	$t_3 + d_4$	$= 0.2\,\mu m$	r_3	$= 200\,\mu m$
h_{III}	$= 0\,\mu m$	W	$= 20\,\mu m$	r_4	$= 500\,\mu m$

* Except for the case $r_4 = r_3$, since this means that no auxiliary pole is present and so η_a and $a(\lambda_{max})$ are meaningless.

are obtained (even for a large backlayer thickness and permeability). This result is important if one considers using the thinnest manageable commercially-available wire (diameter including insulation about 30 μm) with smallest possible inner turn radius of, say, $r_2 = 200$ μm (only 7 times the diameter of the wire) and $r_3 = 500$ μm (i.e. 10 turns). So a thin-film or thick-film technique is necessary to obtain the small dimensions and good results given in Fig. 9.19a. With these techniques very high efficiency heads can be constructed that can compete with even the best ring heads!

When no thin-film or thick-film coil is used, a high efficiency is still possible when the part of the core-winding hole that actually contains the coil is located further away from the head's tape-facing surface, such that the outer radius, r_3, of the upper part of the coil-winding hole can be reduced to the necessary value. The head is then no longer planar.

The reason for the promising efficiencies of PHRYPs is the *optimal utilization of the backlayer*. The circular structure and the relatively small track width, W, compared to the circumference of the auxiliary pole, $2\pi r_3$, assure the almost lowest backlayer reluctance (an ellipsoid or the like with smallest dimension in the trackwidth direction will be even slightly more efficient).

Much of the advantage of the PHRYP may already be obtained when

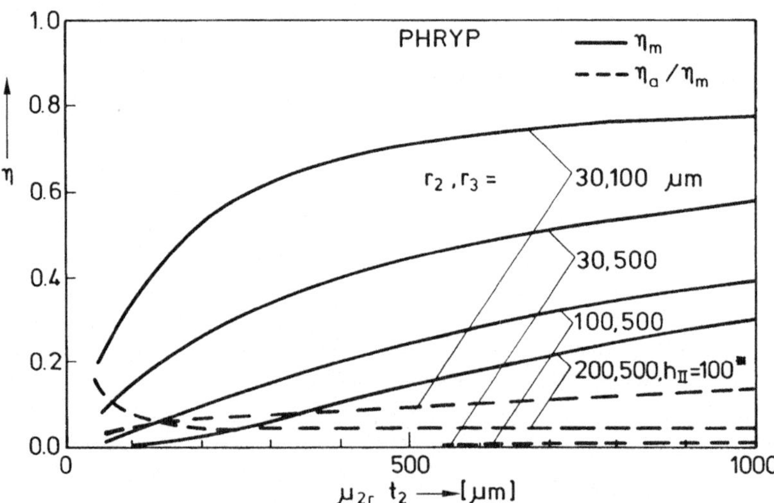

Fig. 9.20. See caption of Fig. 9.19. In contrast to the chosen parameter variations in Fig. 9.19, now just those parameters have been changed that influence the main-pole efficiency strongly.

only part of the circle is used, or when the auxiliary pole edges form straight lines with widths much larger than the track width.

The necessity of using the '3-dimensional' configurations discussed in this section (and the advantage of choosing the best of these) is greater according as $\mu_{2x}t_2$ is smaller.

9.8.2 Auxiliary-pole effects

Another advantage of the large $2\pi r_3/W$ ratio is the small reluctance between backlayer and auxiliary pole relative to the 'gap' reluctance (between main pole and backlayer). Because of this, the ratio of the potentials (flux times reluctance) over the auxiliary-pole backlayer 'gap' and over the main-pole backlayer gap, i.e. η_a and η_m, was expected to be much smaller than for the 2-dimensional PHRYs. This is indeed proved by the dashed curves in Fig. 9.19a, which show η_a/η_m ratios that are almost an order of magnitude smaller than those of the PHRYs given in Figs. 9.16a and 9.18. As a result, the read effect, denoted by the perturbation quantity $a(\lambda_{max})$, *is an order of magnitude smaller than for the PHRYs;* compare for instance the curve indicated by 'all others' in Fig. 9.19b with curves 3 and 4 in Fig. 9.16b.

At a smaller r_3, the ratio $2\pi r_3/W$ is smaller and η_a/η_m becomes larger and soon results in a too large perturbation $a(\lambda_{max})$; see Fig. 9.19b. When the distance between backlayer and auxiliary pole, D_m, is small, as is the case for the dashed curves indicated by 'all others', then a large $\mu_{2r}t_2$ increases both η_m and η_a considerably, but η_a of course more strongly than η_m, such that the auxiliary-pole effects increase. This is in contrast to the observation for large auxiliary-pole gaps (see curve for $h_{III} = 10$ μm, for which η_a is already maximum at low backlayer permeabilities and thicknesses).

It must be noted that the PHRYP reads neighbouring tracks by its auxiliary pole, but with increasing azimuth according as the tracks are further away. Assuming no phase relations between the different tracks, the power of the signals read from the individual tracks can be superposed analogously to noise. So the *'effective' $a(\lambda_{max})$ is larger than calculated* (reading of neighbouring tracks was not taken into account in Sect. 9.2), but the correction factor is smaller or even much smaller than $\sqrt{2r_3/W}$.

9.8.3 Concluding remarks

–   The PHRYP is favourable in all respects. Both a much higher efficiency as well as much smaller auxiliary-pole effects have been predicted.
–   A thin-film conductor winding is desirable for the highest efficiency, because of the necessary small inner and outer radii of the coil and the desirable increase of the conductor diameter with the radius.
–   The ratio r_3/W must not be made too small, because of auxiliary pole read effects.
–   No problems with auxiliary-pole write effects are expected for the usually not too small r_3/W ratios, because of the low η_a/η_m values (< 0.2).

Appendix 9.1 Body of the PHRY program (long coil-chamber length; case 1)

Long coil-chamber lengths, b_c, are defined by:

$$b_c \geq h_1 + h_2 + t_3 + d_4$$

The following layout and text are based on the sequence of the formulas and the text as we used in a Fortran program, but must be rewritten a little for use in an actual (Fortran) program.

Input:

$\mu_0, \pi, h_1, h, b_c, d_c, T_I, T_{II}, T_{III}, t_2, t_3, d_4, l_c, \mu_{2x}, \mu_I, \mu_{II}, \mu_{III}$

Derived coordinates and measures:

$x_1 = t_3 + d_4$

$x_2 = x_1 + h_1$

$x_3 = x_2 + h_2$

$h_4 = h + t_3 + d_4 - b_c$

$h_3 = h - (h_1 + h_2 + h_4) \quad \text{if } h_4 \geq 0.$

$$h_3 = h - (h_1 + h_2) \qquad \text{if } h_4 < 0.$$

If $h_3 < 0$ then '$b_c < h_1 + h_2 + t_3 + d_4$ not allowed' is printed.

$$x_4 = x_3 + h_3$$

Impedances and derived quantities:

$$R_c = \frac{1}{\mu_0} \sqrt{\frac{t_3 + d_4 + d_c}{\mu_{2x}t_2}}$$

$$\lambda_c = \sqrt{\mu_{2x}t_2(t_3 + d_4 + d_c)}$$

If $\lambda_c > l_c$ then 'R_c is not approximated well by the used characteristic impedance of the uniform transmission line' is printed.

$$R_g = \frac{1}{\mu_0} \frac{1}{T_I/(t_3 + d_4) + 1/\pi}$$

$$R_4 = \frac{1}{\mu_0} \frac{h_4}{\mu_{T_{III}}T_{III}} \quad \text{if } h_4 \geq 0$$

$$R_4 = \frac{1}{\mu_0} \cdot \frac{-h_4}{\mu_{2x}t_2} \quad \text{if } h_4 < 0$$

$$C_{a_i} = \frac{1}{\mu_0} \left(\frac{1}{2\mu_{2x}t_2} + \frac{1}{\mu_i T_i} \right) \quad i = \text{I, II, III}$$

$$C_b = \frac{4\mu_0}{\pi}$$

The track width has been omitted in the above expression because the PHRY is 2-dimensional.

Normalized coordinates:

$$z_{ij} = 2\sqrt{C_{a_i}C_b x_j} \quad ij = \text{I1, I2, II2, II3, III3, III4}$$

Some constants and factors:

$$N_I = I_1(z_{II2}) K_0(z_{II2}) + I_0(z_{II2}) K_1(z_{II2})$$

$$C_I = \sqrt{C_{a_I}/C_{a_{II}}}K_1(z_{I2}) I_0(z_{II2}) + K_0(z_{I2}) I_1(z_{II2})$$

$$D_\mathrm{I} = \sqrt{C_{a_\mathrm{I}}/C_{a_\mathrm{II}}}\, I_1(z_{\mathrm{I}2})\,I_0(z_{\mathrm{II}2}) - \ I_0(z_{\mathrm{I}2})\,I_1(z_{\mathrm{II}2})$$

$$F_\mathrm{I} = \sqrt{C_{a_\mathrm{I}}/C_{a_\mathrm{II}}}\, K_1(z_{\mathrm{I}2})\,K_0(z_{\mathrm{II}2}) - \ K_0(z_{\mathrm{I}2})\,K_1(z_{\mathrm{II}2})$$

$$G_\mathrm{I} = \sqrt{C_{a_\mathrm{I}}/C_{a_\mathrm{II}}}\, I_1(z_{\mathrm{I}2})\,K_0(z_{\mathrm{II}2}) + \ I_0(z_{\mathrm{I}2})\,K_1(z_{\mathrm{II}2})$$

$$C_\mathrm{II} = \sqrt{C_{a_\mathrm{II}}/C_{a_\mathrm{III}}}\, K_1(z_{\mathrm{III}3})\,I_0(z_{\mathrm{III}3}) + \ K_0(z_{\mathrm{III}3})\,I_1(z_{\mathrm{III}3})$$

$$D_\mathrm{II} = \sqrt{C_{a_\mathrm{II}}/C_{a_\mathrm{III}}}\, I_1(z_{\mathrm{III}3})\,I_0(z_{\mathrm{III}3}) - \ I_0(z_{\mathrm{III}3})\,I_0(z_{\mathrm{III}3})$$

$$F_\mathrm{II} = \sqrt{C_{a_\mathrm{II}}/C_{a_\mathrm{III}}}\, K_1(z_{\mathrm{III}3})\,K_0(z_{\mathrm{III}3}) - \ K_0(z_{\mathrm{III}3})\,K_1(z_{\mathrm{III}3})$$

$$G_\mathrm{II} = \sqrt{C_{a_\mathrm{II}}/C_{a_\mathrm{III}}}\, I_1(z_{\mathrm{III}3})\,K_0(z_{\mathrm{III}3}) + \ I_0(z_{\mathrm{III}3})\,K_1(z_{\mathrm{III}3})$$

$$C_{12} = \frac{2R_\mathrm{g}C_\mathrm{b}I_0(z_{\mathrm{I}1}) - z_{\mathrm{I}1}I_1(z_{\mathrm{I}1})}{2R_\mathrm{g}C_\mathrm{b}K_0(z_{\mathrm{I}1}) + z_{\mathrm{I}1}K_1(z_{\mathrm{I}1})}$$

$$A = (2R_4 + R_\mathrm{c})\,C_\mathrm{b}I_0(z_{\mathrm{III}4}) + z_{\mathrm{III}4}I_1(z_{\mathrm{III}4})$$

$$B = (2R_4 + R_\mathrm{c})\,C_\mathrm{b}K_0(z_{\mathrm{III}4}) - z_{\mathrm{III}4}K_1(z_{\mathrm{III}4})$$

$$E_\mathrm{II} = I_1(z_{\mathrm{III}3})/(2C_{a_\mathrm{II}}C_\mathrm{b}h_2)$$

$$H_\mathrm{II} = K_1(z_{\mathrm{III}3})/(2C_{a_\mathrm{II}}C_\mathrm{b}h_2)$$

$$E_\mathrm{I} = I_1(z_{\mathrm{II}2})/(2C_{a_\mathrm{II}}C_\mathrm{b}h_2)$$

$$H_\mathrm{I} = K_1(z_{\mathrm{II}2})/(2C_{a_\mathrm{II}}C_\mathrm{b}h_2)$$

$$I = (C_{12}F_\mathrm{I} + G_\mathrm{I})/N_\mathrm{I}$$

$$K = (C_{12}C_\mathrm{I} + D_\mathrm{I})/N_\mathrm{I}$$

$$C_{12} = \frac{A\{(E_\mathrm{I}F_\mathrm{II} - H_\mathrm{I}G_\mathrm{II})/N_\mathrm{I} + H_\mathrm{II}\} - B\{(E_\mathrm{I}C_\mathrm{II} - H_\mathrm{I}D_\mathrm{II})/N_\mathrm{I} - E_\mathrm{II}\}}{B(KC_\mathrm{II} + ID_\mathrm{II}) - A(KF_\mathrm{II} + IG_\mathrm{II})}$$

$$C_{\mathrm{I}1} = C_{12}C_{12}$$

$$\eta_\mathrm{m} = z_{\mathrm{I}1}\{C_{\mathrm{I}1}K_1(z_{\mathrm{I}1}) + C_{\mathrm{I}2}I_1(z_{\mathrm{I}1})\}$$

NI has been omitted, i.e. has been taken 1, in the above expressions, because of linearity.

Appendix 9.2 Body of the PHRY program (short coil-chamber length; case 2)

Short coil-chamber lengths, b_c, are defined by:

$$h_1 + t_3 + d_4 < b_c < h_1 + h_2 + t_3 + d_4.$$

Input:

$\mu_0, \pi, h_1, h_2, h, b_c, d_c, T_{\mathrm{I}}, T_{\mathrm{II}}, T_{\mathrm{III}}, t_2, t_3, d_4, l_c, \alpha, \mu_{2x}, \mu_{\mathrm{I}}, \mu_{\mathrm{II}}, \mu_{\mathrm{III}}$

Derived coordinates and measures:

$$x_1 = t_3 + d_4$$

$$x_2 = x_1 + h_1$$

$$x_3 = x_2 + h_2$$

$$b_c^* = b_c + b_c \tan \alpha - (t_3 + d_4 + d_c)$$

$$x_2^* = x_2 + b_c \tan \alpha - (t_3 + d_4 + d_c)$$

$$x_3^* = x_3 + b_c \tan \alpha - (t_3 + d_4 + d_c)$$

$$x_4^* = h - d_c + b_c \tan \alpha$$

Checks:

If $b_c > h_1 + h_2 + t_3 + d_4$ then go to case 1.

If $b_c < h_1 + t_3 + d_4$ then '$b_c < h_1 + t_3 + d_4$ is not allowed' is printed.

If $0 > b_c \tan \alpha > d_c - h$ then 'angle is larger than geometrically possible' is printed.

Impedances and derived quantities:

$$R_c = \frac{1}{\mu_0} \sqrt{\frac{t_3 + d_4 + d_c}{\mu_{2x} t_2}}$$

$$\lambda_c = \sqrt{\mu_{2x} t_2 (t_3 + d_4 + d_c)}$$

If $\lambda_c > l_c$ then 'R_c is not approximated well by the used characteristic impedance of the uniform transmission line' is printed.

$$R_g = \frac{1}{\mu_0} \frac{1}{T_I/(t_3 + d_4) + 1/\pi}$$

$$C_{a_I} = \frac{1}{\mu_0}\left(\frac{1}{\mu_1 T_I} + \frac{1}{2\mu_{2x}t_2}\right)$$

$$C_{a_{II}} = \frac{1}{\mu_0}\left(\frac{1}{\mu_{II} T_{II}} + \frac{1}{2\mu_{2x}t_2}\right)$$

$$C^*_{a_{II}} = \frac{1}{\mu_0} \frac{1}{\mu_{II} T_{II}}$$

$$C^*_{a_{III}} = \frac{1}{\mu_0} \frac{1}{\mu_{III} T_{III}}$$

$$C_b = \frac{4\mu_0}{\pi}$$

$$C^*_b = \frac{4\mu_0}{\pi - 2\alpha}$$

The track width has been omitted in the above expression because the PHRY is 2-dimensional.

Normalized coordinates:

$$z_{I1} = 2\sqrt{C_{a_I} C_b x_1}$$

$$z_{I2} = 2\sqrt{C_{a_I} C_b x_2}$$

$$z_{II2} = 2\sqrt{C_{a_{II}} C_b x_2}$$

$$z_{IIC} = 2\sqrt{C_{a_{II}} C_b b_c}$$

$$z^*_{IIC} = 2\sqrt{C^*_{a_{II}} C^*_b b^*_c}$$

$$z^*_{II3} = 2\sqrt{C^*_{a_{II}} C^*_b x^*_3}$$

$$z^*_{III3} = 2\sqrt{C^*_{a_{III}} C^*_b x^*_3}$$

$$z^*_{III4} = 2\sqrt{C^*_{a_{III}} C^*_b x^*_4}$$

Matrix and vector coefficients unequal to zero:

$$a_{11} = 2R_g C_b K_0(z_{I1}) + z_{I1} K_1(z_{I1})$$

$$a_{12} = -2R_g C_b I_0(z_{I1}) + z_{I1} I_1(z_{I1})$$

$$a_{21} = z_{I2}K_1(z_{I2})$$

$$a_{22} = z_{I2}I_1(z_{I2})$$

$$a_{23} = -z_{II2}K_1(z_{II2})$$

$$a_{24} = -z_{II2}I_1(z_{II2})$$

$$a_{31} = -K_0(z_{I2})$$

$$a_{32} = I_0(z_{I2})$$

$$a_{33} = K_0(z_{II2})$$

$$a_{34} = -I_0(z_{II2})$$

$$a_{43} = z_{IIC}K_1(z_{IIC}) - C_bR_cK_0(z_{IIC})$$

$$a_{44} = z_{IIC}I_1(z_{IIC}) + C_bR_cI_0(z_{IIC})$$

$$a_{45} = -z^*_{IIC}K_1(z^*_{IIC})$$

$$a_{46} = -z^*_{IIC}I_1(z^*_{IIC})$$

$$a_{53} = -2C_bK_0(z_{IIC})$$

$$a_{54} = 2C_bI_0(z_{IIC})$$

$$a_{55} = 2C^*_bK_0(z^*_{IIC})$$

$$a_{56} = -2C^*_bI_0(z^*_{IIC})$$

$$a_{65} = z^*_{II3}K_1(z^*_{II3})$$

$$a_{66} = z^*_{II3}I_1(z^*_{II3})$$

$$a_{67} = -z^*_{III3}K_1(z^*_{III3})$$

$$a_{68} = -z^*_{III3}I_1(z^*_{III3})$$

$$a_{75} = -K_0(z^*_{II3})$$

$$a_{76} = I_0(z^*_{II3})$$

$$a_{77} = K_0(z^*_{III3})$$

$$a_{78} = -I_0(z^*_{III3})$$

$$a_{87} = K_1(z^*_{III4})$$

$$a_{88} = I_1(z^*_{III4})$$

$$b_3 = 1/(2C_{a_{II}}C_b h_2)$$

$$b_4 = -R_c/(2C_{a_{II}} h_2)$$

$$b_5 = (1/C^*_{a_{II}} - 1/C_{a_{II}})/h_2$$

$$b_7 = -1/(2C^*_{a_{II}} C^*_b h_2)$$

NI has been omitted, i.e. has been taken 1, in the above expression, because of linearity.

Efficiency determination:

The matrix is solved (for our Fortran program by the NAG library routine F04ATF) for the unknown constants; see (9.37). The four necessary constants are put in the efficiency expressions for η_m and η_a, (9.38) and (9.39), while omitting *NI* and plotted or printed.

Appendix 9.3: Body of the PHRYP program.

Input:

$$\mu_0, \pi, h_I, h_{II}, h_{III}, T, t_2, t_3, d_4, r_2, r_3, r_4, W, \mu_{2r}$$

Derived coordinates and check:

$$r_1 = (W + T)/\pi$$

If $r_1 > r_2$ then '$r_1 > r_2$ is not allowed' is printed.

Impedances and derived quantities:

$$R_T = \cfrac{1}{\cfrac{4\mu_0(W + T)}{\pi} \ln\left\{\cfrac{h_I + t_3 + d_4}{t_3 + d_4}\right\} + 2\mu_0 h_I + \mu_0 W\left(\cfrac{T}{t_3 + d_4} + \cfrac{1}{\pi}\right)}$$

$$C_a = \cfrac{1}{\mu_0\mu_{2r}t_2 2\pi}$$

$$C_{bi} = \cfrac{2\pi\mu_0}{h_i + t_3 + d_4} \qquad i = \text{I, II, III}$$

Normalized coordinates:

$$z_{I1} = \sqrt{C_a C_{b_I}}\, r_1$$

$$z_{I2} = \sqrt{C_a C_{b_I}}\, r_2$$

$$z_{II2} = \sqrt{C_a C_{b_{II}}}\, r_2$$

$$z_{II3} = \sqrt{C_a C_{b_{II}}}\, r_3$$

$$z_{III3} = \sqrt{C_a C_{b_{III}}}\, r_3$$

$$z_{III4} = \sqrt{C_a C_{b_{III}}}\, r_4$$

Matrix and vector coefficients unequal to zero:

$$a_{11} = K_0(z_{I1}) + R_T\ \frac{z_{I1}}{C_a}\ K_1(z_{I1})$$

$$a_{12} = I_0(z_{I1}) - R_T\ \frac{z_{I1}}{C_a}\ I_1(z_{I1})$$

$$a_{21} = +K_0(z_{I2})$$

$$a_{22} = +I_0(z_{I2})$$

$$a_{23} = -K_0(z_{II2})$$

$$a_{24} = -I_0(z_{II2})$$

$$a_{31} = -z_{I2}K_1(z_{I2})$$

$$a_{32} = +z_{I2}I_1(z_{I2})$$

$$a_{33} = +z_{II2}K_1(z_{II2})$$

$$a_{34} = -z_{II2}I_1(z_{II2})$$

$$a_{43} = +K_0(z_{II3}) \qquad or^{\otimes} = +R_c z_{II3} K_1(z_{II3})/C_a$$

$$a_{44} = +I_0(z_{II3}) \qquad or = -R_c z_{II3} I_1(z_{II3})/C_a$$

$$a_{45} = -K_0(z_{III3}) \qquad or = 0$$

$$a_{46} = -I_0(z_{III3}) \qquad or = 0$$

$$a_{53} = -z_{II3}K_1(z_{II3}) \quad or = 0$$

$$a_{54} = +z_{II3}I_1(z_{II3}) \quad or = 0$$

$$a_{55} = +z_{\text{III3}}K_1(z_{\text{III3}}) \; or = 0$$

$$a_{56} = -z_{\text{III3}}I_1(z_{\text{III3}}) \;\; or = 0$$

$$a_{65} = -K_1(z_{\text{III4}}) \qquad or = 0$$

$$a_{66} = +I_1(z_{\text{III4}}) \qquad or = 0$$

$$a_3 \; = -1/\ln(r_3/r_2)$$

$$a_4 \; = 0 \qquad\qquad\quad or \; R_\text{c}/(C_\text{a}\ln(r_3/r_2))$$

$$a_5 \; = +1/\ln(r_3/r_2) \quad or \; 0$$

NI has been omitted, i.e. has been taken 1, in the above expressions, because of linearity.

Efficiencies:

η_m and η_a are calculated by way of (9.57) and (9.58), while omitting *NI*, after determination of the four necessary constants by way of the NAG library Fortran routine F04ATF from the matrix equation (like (9.37)) with above coefficients.

$^{\otimes}$ If $z_{\text{III4}} \gg 30$ then the determinant of the matrix and/or the values of I_0, I_1, K_0, $K_1 \simeq \text{e}^{\pm z_{\text{III4}}}$ are too extreme, such that errors are detected by the numerical routines. However the reluctance of section III is well approximated by the characteristic reluctance of a uniform transmission line with width $2\pi r_3$ and distance $t_3 + d_4 + h_{\text{III}}$,

$$R_\text{c} = \frac{1}{2\pi r_3\mu_0} \sqrt{\frac{t_3 + d_4 + h_{\text{III}}}{\mu_{2\text{r}}t_2}},$$

see (8.40), if the widening of section III over one characteristic distance,

$$\lambda_\text{c} = \sqrt{\mu_{2\text{r}}t_2(t_3 + d_4 + h_{\text{III}})},$$

is small. This is the case when, say,

$$r_3 > 4\lambda_\text{c},$$

assuming that $r_4 > r_3 + \lambda_\text{c}$.

So, if $z_{\text{III4}} > 30$ and the latter conditions are fulfilled, then the [6 × 6] matrix is reduced to a [4 × 4] matrix, with matrix elements as given above after '*or*'.

References

[9.1] J. Corcoran and N. Pope, *Transmission line model for magnetic heads including complex permeability*, Journal of Magn. and Magnetic Materials 54-57, 1591 (1986).

[9.2] V. Zieren, S.B. Luitjens, C.P.G. Schrauwen, J.P.C. Bernards, R.W. de Bie and M. Piena, *Properties of one-sided probe heads on double-layer perpendicular recording media*, IEEE Trans. Magn., Mag-22, 370-372 (1986).

[9.3] V. Zieren, J.J.M. Ruigrok, M.J. Piena, S.B. Luitjens, C.W.M.P. Sillen and J.P.M. Verbunt, *Efficiency improvement of one-sided probe heads for perpendicular recording on double-layer media*, IEEE Trans. Magn., Mag-23, 2479-2481 (1987).

[9.4] V. Zieren, S.B. Luitjens, M.J. Piena, R.W. de Bie, C.P.G. Schrauwen and J.P.C. Bernards, *High performance heads for perpendicular recording*, accepted for publication in IEEE Trans. Magn., Mag-25, Jan (1989).

[9.5] W.K. Westmijze, *Field configuration around the gap and the gap-length formula*, Philips Res. Rep. 8, see p. 168 and p. 164, 161-183 (1953).

[9.6] M. Steinback, J.A. Gerber and Th.J. Szczech, *Exact solution for the field of a perpendicular head*, IEEE Trans. Magn., Mag-17, 3117 (1981).

[9.7] Matthew Dugas, Wayne Bonin, and Jack Judy, *A finite element analysis of the MSP single pole perpendicular recording head*, IEEE Trans. Magn., Mag-21, 1554 (1985).

[9.8] Bill Baker et al., *Analytic model for single pole heads*, J. Appl. Phys. 57, 3985 (1985).

[9.9] Vadim B. Minuhin, *Theory of playback process with soft magnetic underlayer*, IEEE Trans. Magn., Mag-21, 28 (1985).

[9.10] K. Nakamura, N. Echigo, H. Yohda, S. Mitani and N. Kaminaka, *Single-turn perpendicular thin film heads*, IEEE Trans. Magn., Mag-23, 2482 (1987).

[9.11] S.B. Luitjens, *Private communication* (1987).

[9.12] Harry Bateman, *Tables of integral transforms*, volume 1, p. 8 eq. 11 and p. 65 eq. 14, McGraw-Hill Book Company, Inc., New York Toronto London (1954).

[9.13] T. Ozeki, T. Sakata, J. Toriu and K. Momiyama, *Fabrication of vertical recording heads*, IEEE Trans. Magn., Mag-21, 1557 (1985).

[9.14] I.S. Gradshteyn and I.M. Ryzhik, *Table of integrals series and products*, Academic Press, New York etc. (1965).

[9.15] E. Kamke, *Differentialgleichungen, Lösungen und Lösungsmethoden*, p. 440 expression (2.162.1a), Akademische Verlagsgesellschaft Becker & Erler Kom.-Ges., Leipzig, 1942.

Chapter 10

Cross measurements in magnetic recording: a new way of determining head performance

This chapter also appeared, in slightly reduced form, as the following paper:
- Jaap J.M. Ruigrok, *Cross measurements in magnetic recording: a new way of determining head performance*, IEEE Trans. on Magn., Mag-20, 875-877 (1984).

Section 10.6 is added to show, once more, the general applicability of the method and the transparency of the results.

10.1 Abstract

A description is given of how to separate the recording and playback performance of an inductive head relative to that of a 'reference' head. The resulting relative recording and relative playback figures are completely insensitive to azimuthal differences between recording and playback head and are independent of errors in track adjustment. Second order effects can now be measured. In addition, when the measuring system carries out the simple calculations necessary for obtaining these figures, a considerable amount of time and effort for interpretation is saved.

The method is applied to a metal-in-gap head, a high-saturation metal (ribbon) head and to ferrite heads with various gap lengths.

10.2 Introduction

Heads are often measured as combined recording-playback heads. The disadvantage of this method is that it is unclear how strongly the recording and how strongly the playback behaviour of the investigated head influenced the results.

A following step is using a fixed playback head when the recording performance of one or more heads on a particular tape has to be investigated. Conversely a fixed recording head is used when one or more playback heads have to be examined. This method has important advantages over the first, although its accuracy at especially high frequencies is poor, because of hardly avoidable differences in azimuthal angle (alignment loss) and gap irregularities [10.1].

A more accurate and quick determination of recording as well as playback performance of heads being developed is very important, no matter if the head will be developed as a combined recording-playback head or as a separate recording or playback head. This is of both practical importance (the development of heads) and scientific interest (the understanding of head performance).

The following simple method gives figures that are insensitive to differences in azimuthal angle, gap irregularities and track following. We propose this method as a new way of separating recording and playback performance.

The method is applied to a metal-in-gap head, a high-saturation metal (ribbon) head and to ferrite heads with various gap lengths.

10.3 Recording (R) and playback (P) figures

The playback signal is called U_{ij} and is a result of the recording performance R_i of head i with track width W_i on a particular medium and the playback performance P_j of head j with track width W_j. All four 'cross measurements' that are possible with two heads, U_{11}, U_{12}, U_{22} and U_{21} have to be carried out.

The easiest way to understand the expressions representing relative recording and relative playback performance is to state them first and to discuss them afterwards. The recording and/or playback performance of head 1 relative to head 2, R_{12} and/or P_{12}, can be deduced from the four preceding measurements by ($\equiv$ means per definition)

$$\boxed{R_{12} \equiv \frac{R_1}{R_2} = \sqrt{\frac{U_{12}U_{11}W_2}{U_{21}U_{22}W_1}}} \qquad (10.1)$$

$$P_{12} \equiv \frac{P_1}{P_2} = \sqrt{\frac{U_{11}U_{21}W_2}{U_{22}U_{12}W_1}} \,. \qquad (10.2)$$

These expressions are easily verified by splitting all U_{ij}'s in (10.1) and (10.2) in their essential factors. For example:

$U_{12} =$ the product of

R_1 = recording performance of head 1 per meter track width on the actual medium (including recording distance loss)
L_a = loss due to differences in azimuthal angle between head 1 and 2
L_i = loss due to geometrical irregularities of the gaps
L_t = loss due to errors in the track adjustment
P_2 = playback sensitivity of head 2 per meter track width on the actual medium (including playback distance loss)
W_s = the smallest of the two track widths, i.e. W_1 or W_2.

If $W_1 < W_2$ the ratio W_1/W_2 has to be added as an extra factor before U_{22} and vice versa, since U_{22} is the only favoured measurement, see expressions (10.1) and (10.2). The loss factors L_a, L_i and L_t all cancel in (10.1) and (10.2). The only figures that do not cancel are R_1^2 in the numerator and R_2^2 in the denominator of the square-root of (10.1). In (10.2) under the square-root sign only P_1^2/P_2^2 do not cancel. From now on R_{ij} and P_{ij} will often be abbreviated as R and P.

10.4 Cross-accuracy figure (C)

Although the above arguments indicate independence of R and P figures from the alignment, gap irregularities and track-following, we will show this experimentally. For that purpose, the cross-accuracy figure $C \equiv L_a \cdot L_i \cdot L_t$ has to be determined, in order to use it as a parameter in the experiments. In an analogous way as before, the four measurements (U_{ij}) can be combined so that only the square of the product of the 3 loss factors $L_a^2 \cdot L_i^2 \cdot L_t^2$ i.e. C^2 remains. The proper combination gives

$$C \equiv L_a \cdot L_i \cdot L_t \quad \begin{cases} = \sqrt{\dfrac{U_{12}U_{21}W_1}{U_{11}U_{22}W_2}} \leqslant 1 \text{ if } W_1 \geqslant W_2 \\[2em] = \sqrt{\dfrac{U_{12}U_{21}W_2}{U_{11}U_{22}W_1}} \leqslant 1 \text{ if } W_1 \leqslant W_2 \end{cases} \qquad (10.3)$$

When track following is perfect and gap irregularities are negligible, $C(f)$ determines alignment loss as a function of frequency.

10.5 Examples

All following measurements are only examples of the application of the method and not descriptions of different types of heads. The remarks concerning the different tape-head combinations only serve the illustrations. In the examples we chose heads with interesting R and P figures to compare them with ordinary MnZn-ferrite heads.

Table 10.1 represents the typical print-out of an automatic measurement based on the preceding principles. Head 1 is the head to be investigated and head 2 serves as the reference head. The track widths W_1 and W_2 were 18 and 29 μm respectively. The medium was a high-coercive video tape. The head-tape velocity was $v = 3.14$ m/s in all experiments. The write current was optimized in all direct-recording experiments.

Table 10.1. Typical print out of an automatic measurement of performance figures

Freq.	U_{11}	U_{12}	U_{22}	U_{21}	C	R_{12}	P_{12}	$R_{12} \cdot P_{12}$
8.0	1.78	1.26	4.87	1.37	0.57	0.74	0.80	0.59
7.0	3.63	3.48	7.69	2.50	0.71	1.03	0.74	0.76
6.0	6.00	4.83	11.95	4.82	0.72	0.90	0.90	0.81
5.0	11.32	10.86	19.68	9.55	0.87	1.03	0.90	0.93
4.0	18.18	16.75	28.31	15.94	0.91	1.04	0.99	1.03
3.0	36.87	31.85	41.60	24.90	0.90	1.34	1.04	1.39
2.0	72.40	57.15	49.12	36.91	0.98	1.92	1.24	2.37
1.0	88.02	69.48	44.44	41.78	1.09	2.30	1.39	3.19
.5	79.97	61.55	37.34	36.58	1.10	2.41	1.43	3.45

R results, also plotted in Fig. 10.1, indicate superior record performance of the high-saturation head 1 relative to the ferrite head 2, especially at low frequencies where the recording depth was large.

Tables 10.2a and b show the results of cross-measurement analysis on a metal-in-gap head with a 21 μm track width (W_1) and a reference ferrite head with a 40 μm track width (W_2). The metal-in-gap head is described in chapter 12 and in the corresponding paper [10.2] and also in paper [10.3]. The optimal currents of the two heads I_1 and I_2 are automatically measured and listed in the table. The R figures here are therefore representative of 'optimal' direct recording. In table 10.2a, corresponding to the circles in the Figs. 10.2a-c, the alignment between head 1 and head 2 is almost perfect. Only gap irregularities have a small effect in C at the highest frequencies. In table 10.2b however, the azimuthal difference is about one degree, resulting in a very low value for C. From figure 10.2b it is obvious that R_{12} does not suffer from a large error in alignment. The small erratic differences in R_{12}, P_{12} (and $R_{12} \cdot P_{12}$) are due to the irreproducibility of recording measurements in general. The systematic difference in P_{12} at decreasing frequencies is a result of gradual improvement of the playback performance of the fer-

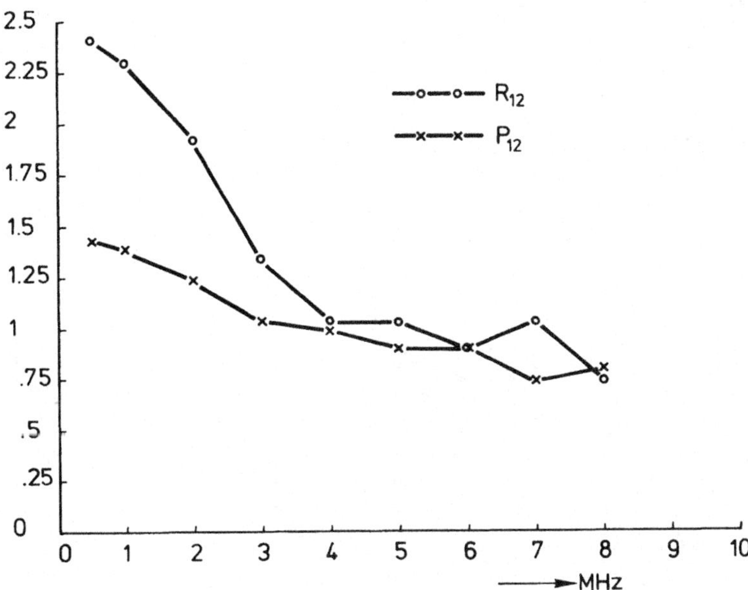

Fig. 10.1. Frequency dependence of recording (R) and playback (P) performance of a high-saturation head relative to a ferrite head tape with H_c = 90 kA/m (1140 Oe).

rite head (head 2) and slight deterioration in the playback performance of head 1, which took place during the measurements. This follows from the increased values of U_{22} and slightly decreased values of U_{11}, see table 10.2a and b. The bumps in the relative playback performance R_{12} are due to interference between primary gap and residual gaps between the metal layers and the core: a detailed explanation can be found in chapter 12 or the corresponding paper [10.2].

Table 10.2a.   Recording figures for small errors in alignment ($C \to 1$)

Freq.	U_{11}	I_1	U_{12}	U_{22}	I_2	U_{21}	C	R_{12}	P_{12}	$R_{12} \cdot P_{12}$
7.0	1.36	28.04	3.04	5.10	28.20	0.88	0.86	1.32	0.38	0.51
6.5	0.72	38.38	3.84	6.12	18.53	0.37	0.78	1.53	0.15	0.22
5.5	1.60	17.82	6.42	8.75	22.16	1.05	0.96	1.46	0.24	0.35
5.0	8.61	20.48	7.62	9.30	25.17	4.55	0.91	1.72	1.03	1.76
4.5	13.15	22.80	8.65	11.55	24.84	7.60	0.91	1.57	1.38	2.17
4.0	11.47	23.76	10.54	13.06	36.23	6.59	0.94	1.64	1.02	1.67
3.5	12.42	23.41	12.31	15.83	90.04	7.08	0.92	1.61	0.93	1.49
3.0	27.18	29.22	16.11	17.09	104.20	12.86	0.92	1.95	1.56	3.03
2.5	41.58	24.41	18.00	17.07	64.31	17.79	0.93	2.17	2.14	4.64
2.0	50.75	30.73	25.01	22.44	136.30	23.39	0.99	2.15	2.01	4.31
1.5	36.13	43.76	20.94	19.40	181.91	16.21	0.96	2.14	1.66	3.55
1.0	31.94	58.92	20.54	17.35	176.78	14.97	1.03	2.19	1.60	3.51
.5	23.66	67.79	11.98	10.19	266.86	9.18	0.93	2.40	1.84	4.42

Table 10.2b.   Recording figures for large errors in alignment (C small)

Freq.	U_{11}	I_1	U_{12}	U_{22}	I_2	U_{21}	C	R_{12}	P_{12}	$R_{12} \cdot P_{12}$
7.0	1.85	17.79	1.68	7.36	17.04	0.42	0.31	1.38	0.35	0.48
6.5	1.06	17.18	1.76	7.99	17.13	0.20	0.28	1.49	0.17	0.25
5.5	0.99	15.72	3.90	11.31	18.87	0.28	0.43	1.52	0.11	0.17
5.0	9.81	20.56	3.16	12.69	22.52	2.06	0.32	1.50	0.98	1.47
4.5	12.96	20.44	4.71	14.61	23.62	3.37	0.40	1.54	1.1	1.69
4.0	10.84	21.69	8.79	17.96	30.42	2.94	0.50	1.85	0.62	1.15
3.5	12.65	29.83	9.98	19.48	87.32	3.32	0.51	1.93	0.64	1.24
3.0	29.71	30.62	15.74	22.75	52.54	11.08	0.70	1.88	1.32	2.49
2.5	45.35	24.65	22.83	24.02	110.66	18.59	0.86	2.10	1.71	3.60
2.0	46.75	32.40	27.13	25.54	135.23	21.08	0.96	2.12	1.65	3.49
1.5	37.07	39.91	28.02	24.15	182.37	16.55	0.99	2.22	1.31	2.92
1.0	37.14	46.55	24.76	21.80	174.11	18.71	1.04	2.07	1.57	3.25
.5	35.32	67.59	22.68	15.36	214.01	17.08	1.17	2.41	1.82	4.38

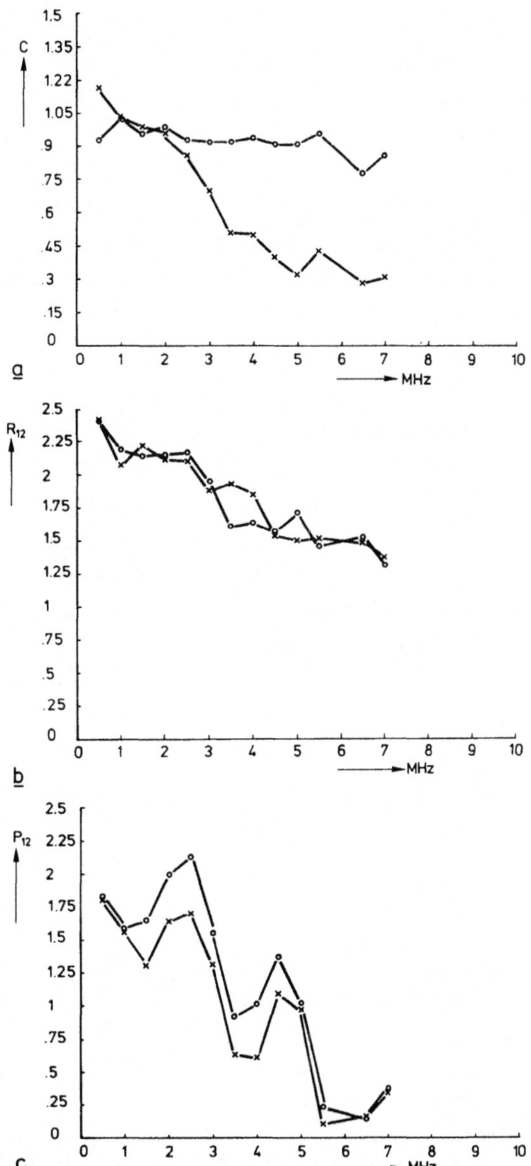

Fig. 10.2. Plots of C, R_{12} and P_{12} in Table 10.2a and b.

a) Cross accuracy C when the heads are well (o) and badly (x) aligned.

b) Independence of relative recording performance (R_{12}) on cross-accuracy (C) for an arbitrary pair of heads.
   o: C = high, x: C = small.

c) Independence of relative playback performance (P_{12}) on crossaccuracy (C) for the same arbitrary pair of heads.
   o: C = high, x: C = small.

10.6 Application to ferrite head with varying gap length

It is generally known that a long gap length improves the recording of long-wavelength information and deteriorates the playback of small-wavelength information. No method seems to be more suitable for separating these read and write effects quantitatively and in an accurate way than the proposed cross measurement method. The precise read and write effects of heads with various gap length have been investigated by Dr. W. Tuma, using ferrite video heads with dimensions about equal to those of the ferrite head given in table 8.8 and a gap length of 0.18, 0.29, 0.40, 0.50, 0.70, 0.91 and 1.56 µm respectively. The reference head is a tilted-sputtered-sendust (TSS) head [12.8], which is a kind of metal-in-gap (MIG) head [10.2] with a very thick metal layer. Metal-powder (MP) tape with a high-coercive field ($H_c = 115$ kA/m) and a remanence, M_r, of 164 kA/m is used. The results, in Fig. 10.3, speak for themselves. They tell us for instance that,
–   according as the wavelength increases, both the optimal recording gap length and the optimal playback gap length increase;
–   a gap length of 0.5 µm is about optimal for recording on MP tape of signals in the wide video band (ranging from about 0.5 to 8 MHz) but still 20% worse than that of the TSS head;

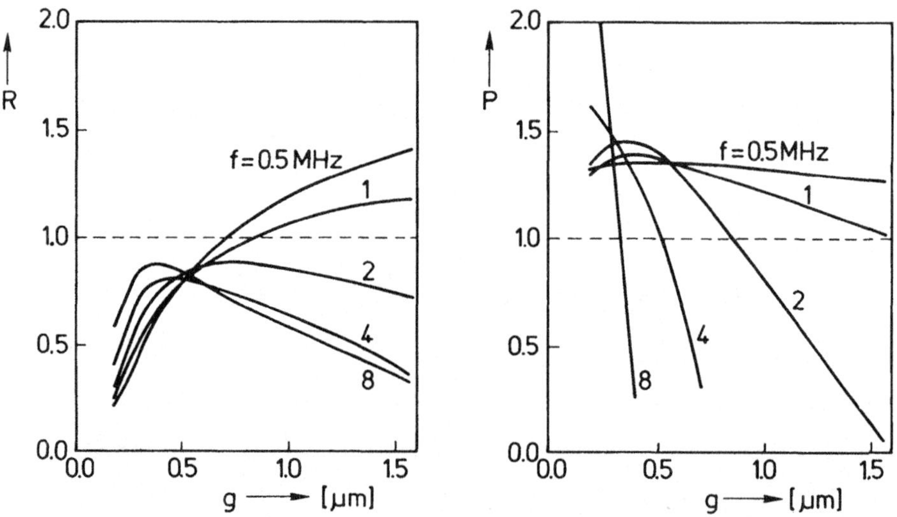

Fig. 10.3 Gap length and frequency dependence of the recording (R) and playback (P) performance of a ferrite head with varying gaplength relative to that of a TSS head, measured on high-coercive MP tape at a head-to-tape velocity of 3.14 m/s.

— a gap length of 0.3 µm is about optimal for playback of the whole video spectrum and the playback performance of the ferrite head is then about 40% better than that of the TSS head.

The most important conclusion is that application of a larger gap length in a head with a lower saturation never results in a complete compensation of the deteriorated recording performance.

10.7 Discussion

In this section we discuss the significance of the R, P and C figures.

R and P figures not only characterize heads, but also compare the performance of head 1 on tape x with the performance of head 2 on tape x. R and P figures include effects such as the effective head-tape distance. This means that a difference in damaged layers at the surfaces of the heads, due to a different interaction (chemical as well as abrasive) between head 1 and head 2 and the tape is included. The formation of damaged layers is expected to depend on the type of tape used. This may influence P (and R to a lesser extent). When this effect is present it can be accurately measured by comparing $P(x_1)$ on tape x_1 with $P(x_2)$ on tape x_2, since different magnetic characteristics of both tapes do not influence P, supposing the heads behave linearly in the read mode. In chapter 12 and in reference [10.2] it is experimentally shown for one case that P did not depend on the magnetic characteristics of the tape or on recording conditions such as direct or bias recording. The R figure is of course strongly affected by the magnetic characteristics of the tape, especially by the coercivity when heads with high and low saturation are used; see for instance reference [10.2] or chapter 12.

Side fringes influence the R, P and C values when $W_1 + \lambda \equiv < W_2$, and vice versa. The head with the smallest track width reads a signal, recorded with the other head, as if its track width equals [10.4] $W + 0.24\lambda$. Only for large wavelength/track-width ratios must this effect be taken into acount. When $W_1 > W_2 + \lambda$, side fringing response only favours U_{12}. We ignore side writing, which usually is a smaller effect, and can be measured as reported by Hoyt and Sussner [10.5]. As a result, R_{12} and C are about $(1 + 0.12\lambda/W_2)$ times too large and P_{12} the same factor too small. When $W_2 > W_1 + \lambda$, then P_{12} and C are about $(1 + 0.12\lambda/W_1)$ times too large and R_{12} the same factor too small.

$R_{12} \cdot P_{12}$ is not affected. In all our experiments the effect is visible in C only at 1 and 0.5 MHz. If necessary a refinement can be introduced in the expressions for R, P and C.

The cross-accuracy figure C gives information about side fringing response (low frequencies), gap irregularities and alignment loss (high frequencies), and track adjustment (frequency independent). The different frequency dependences make a partial separation of these loss factors possible. The gap-irregularity loss can only be measured accurately when the alignment is carefully optimized. Accurate track adjustment is simple, especially when $W_1 \neq W_2$.

R_{is} and P_{is} figures of heads to be investigated (subscript i) can also be given relative to a (hypothetical or) existing 'standard' head (subscript s). Then it is necessary to make one measurement of the R_{rs} and P_{rs} figures of the reference head(s) used (subscript r) relative to the standard head, on the tape on which the measurements will be carried out. Multiplication of $R_{ir} \cdot R_{rs}$ gives R_{is}. In this case it is even more important that neither degradation nor improvement of the reference head(s) takes place during the measurements.

All calculations necessary for obtaining R, P and C figures are simple, once the four cross measurements (U_{ij}) have been carried out. The method can be easily implemented on an (automatic) testing system. Most of the interpretation is then carried out by computer.

10.8 Conclusions

Measurements with two heads at the same time, a head to be investigated and a reference head, make it possible to determine recording and playback performance separately.

The relative recording and relative playback figures R and P are independent of adjustments such as the alignment of azimuthal angles and track following, and independent of gap irregularities. This makes the measurements more precise, with the result that smaller effects can be measured.

The cross-accuracy factor C gives information about gap irregularities and alignment, track adjustment and side fringing effects.

The ease and speed of measurements are of equal practical importance. Let the testing system, as in our case, do the simple calculations necessary for obtaining R and P, and the bulk of the interpretation is already done.

The examples showed that especially the long-wavelength recording performance improves when either a metal-in-gap head, a high-saturation (ribbon) head or a large-gap length ferrite head is used instead of an ordinary ferrite head. However, the small-wavelength recording performance of the large-gap length head is degraded. Moreover, at a smaller gap length where the recording of both long- and short-wavelength information is satisfactory, this performance is still worse than that of the high-saturation head types.

It will be clear that this new analysing method is not restricted to recording/playback measurements. It is applicable to all reciprocal transducers, i.e. transducers that can operate like actuators as well as (linear) sensors.

References

[10.1] J.C. Mallinson, *Gap Irregularity Effects in Tape Recording*, IEEE Trans. on Magn., Mag-5, 71 (1969).

[10.2] J.J.M. Ruigrok, *Analysis of metal-in-gap heads*, IEEE Trans. on Magn., Mag-20, 872-874 (1984).

[10.3] F.J. Jeffers, R.J. McClure, W.W. French, N.J. Griffith, *Metal-in-gap record head*, IEEE Trans. on Magn., Mag-18, 1146-1148 (1982).

[10.4] A. van Herk, *Analytical Expressions for Side Fringing Response and Crosstalk with Finite Head and Track Widths*, IEEE Trans. on Magn., Mag-13, 1764-1766 (1977).

[10.5] R.F. Hoyt and H. Sussner, *Precise side writing measurements using a single recording head*, IEEE Trans. on Magn., Mag-20, 909-911 (1984).

Chapter 11

Analytical description of thin-film yoke magnetoresistive heads

The performance of thin-film read heads using magnetoresistive elements is explained by means of a one-dimensional transmission line model. With this method it turns out to be possible to obtain analytic expressions for the response. Two common configurations we consider are the two-legged magnetoresistive heads with and without barberpole stripes. A general expression for the harmonic distortions, neglecting domain walls, is derived. The influence of bias and measuring currents as well as saturation of the magnetoresistive element is taken into account. As a verification of theory, transfer characteristics of actual head configurations are calculated with the aid of the derived equations and compared with experiments. With the use of the equations it is possible to design heads with the required performance as long as domain walls can be neglected.

This chapter appeared in Philips Journal of Research [11.16].

11.1 Introduction

R.P. Hunt [11.1] introduced the magnetoresistive read-out transducer (MRE) in 1970. Anderson, Bajorek and Thompson [11.2] described the response of the vertical MRE by means of a numerical relaxation technique. They included the non-uniform demagnetizing field of a rectangular strip instead of the uniform demagnetizing field of an ellipsoid. Initially, exchange was taken into account, but could be neglected later on. Casselman and Hanka [11.3] tried to find the functional dependence on MRE dimensions by numerically solving the model equation of Anderson et al. Until now, the MRE incorporated in a magnetic yoke, with [11.4] or without barberpole stripes, hereafter called barberpole yoke magnetoresistive head (BYMRH) or yoke magneto resistive head (YMRH), has not been described extensively.

A simplified exploded view of the BYMRH is shown in Fig. 11.1. We will not consider side fringes, e.g. side writing and reading, and thus reduce the problem to a two-dimensional one with cross-section as in Fig. 11.2a. The thin-film head (TFH) is processed on a magnetic (ferrite) or non-magnetic (silicon) wafer. In front of the gap, a contra-coil (I_c) is sputtered to carry out wafer tests. In some experiments a bias coil (I_b) is placed between the MRE and the lower flux-guide in order to allow freedom in choosing the sense current I_m, which also biases the element [11.5], because of the asymmetrical location in the yoke.

11.2 Magnetic material equations

To describe the behaviour of thin-film magnetic recording heads (TFHs) it is important to know the material equations. We restrict ourselves to describing such low frequencies that we can disregard eddy current losses and skin effect in the conductors and the magnetic materials. The flux guides have dimensions comparable with those of the domains. The thickness p of the flux guides is in addition too big to

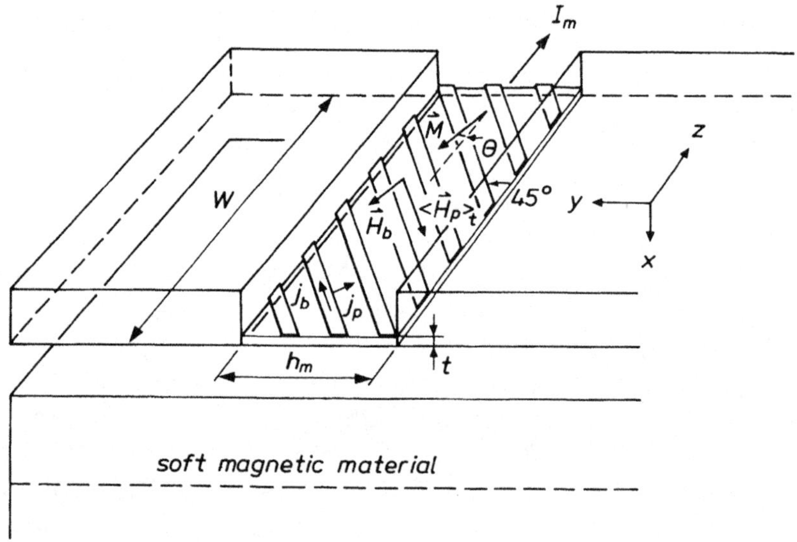

Fig. 11.1. Exploded view of a barberpole magnetoresistive head (BYMRH). Overlap regions are not drawn.

suppress domain wall formation. For these reasons it is hard to introduce a constant permeability μ in the material. However, experiments on the wafer with electroplated permalloy shields of comparable dimensions show that the flux guide can be described with a μ of about 1000, subject to the restriction that μ increases with decreasing trackwidth [11.6]. Our sputtered flux guides show permeabilities up to 2000, corresponding with a lower observed anisotropy field H_{an}.

The dimensions of the MRE ($\approx 20 \times 0.05 \times 200$ to $600\ \mu m^3$) are such that, for this part of the read head, the single-domain configuration is most probable. In either case no closure domains are observed. The

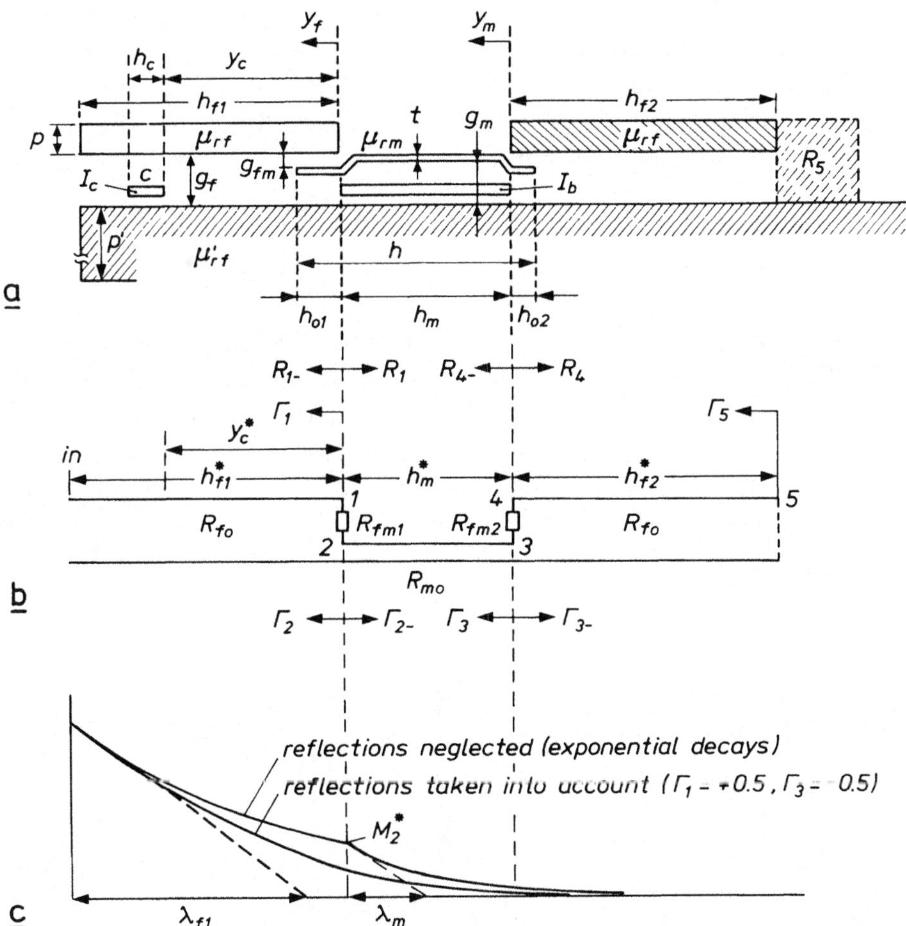

Fig. 11.2. *a*) Cross section of (B)YMRHs with bias (I_b) and contra coil (C). *b*) Transmission line model of (B)YMRH. *c*) Exact and approximated flux decay in (B)YMRH.

y-component of the current through the permalloy MRE of the BYMRH in fig. 11.1 gives rise to extra 'repulsive' forces. These forces reduce the permeability component in the y-direction $\mu_{rm} = B_s/\mu_0 H_{an}$ according to

$$\mu_{rm} = \frac{B_s}{\mu_0 H_{an} + \mu_0 H_z/\cos\theta} \,. \qquad (11.1)$$

This expression is derived from the torque equation of the single domain MRE in the case of the easy axis being in the z-direction, leaving out of account the very small exchange forces. The reduction of sensitivity due to the repulsive force has been shown experimentally in ref. [11.7]. In the calculations we will neglect hysteresis, as shown in Fig. 11.3 for hard axis magnetization; e.g. the dotted curve or a reduced but constant value will be used for μ_{rm}. These simplifications enable us to describe the head linearly until saturation is reached.

11.3 One dimensional approach of (B)YMRH

The first one-dimensional description of a TFH was given by Paton [11.8]. The same concept is used by many others; see for instance ref. [11.9]. More suited for taking account of reflections at boundaries between sections of the head is the transmission line description [11.10, 11.5] which makes use of distributed elements, as illustrated in Fig.

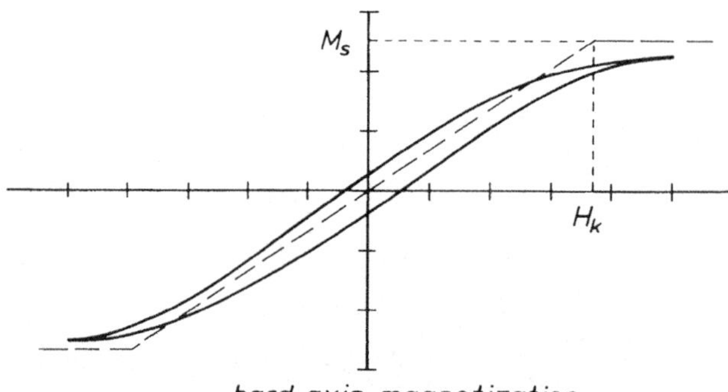

Fig. 11.3. Example of hard axis magnetization with approximation (dashed curve) used in the calculations.

11.2b. This concept is very useful because of the extreme height-thickness ratios of all head sections.

With the aid of some fundamental concepts of transmission line theory analytical expressions will be derived for the decay of flux at the different homogeneous head sections. The overlap regions h_{o1} and h_{o2} are approximated by discrete instead of distributed elements (see Fig. 11.2b), which can easily be taken into account in a one-dimensional transmission line description. As in all one-dimensional models, side fringes, e.g. side writing and reading, are neglected. We are especially interested in the flux distribution at the magnetoresistive element in order to calculate the response of the head.

Although the derivations will be complete, it may be desirable for those who are not familiar with the transmission line concept to continue at this distribution (11.9). It is noted that 'heights' signed with a dot are given relative to (divided by) their characteristic lengths $\lambda = \sqrt{\mu_r p g}$ and magnetizations signed with a dot are given relative to their saturation magnetization. Integer subscripts correspond to the considered boundary (1-5), letter subscripts to the sections between, see fig. 11.2b. When a minus sign is added to a subscript a (fictitious) reversed operation is considered, otherwise flux from pole tip towards MRE is considered.

11.3.1 Transmission line calculations

By analogy with resistances in electric transmission lines magnetic resistances (reluctances) will be defined as $R \equiv \Delta U/\Phi$, where U is the magnetic potential (A) and Φ the magnetic flux (Wb). In all following calculations, Φ's and R's will be expressed for a trackwidth of 1 m and the R's in addition for a height of 1 m.

Consider the decay of flux in for instance the flux-guide section of fig. 11.2b with series reluctance $R_f = 1/(\mu_0\mu_{rf}p) + 1/(\mu_0\mu'_{rf}p')$ A/Wb and shunt reluctance $R_{gf} = g_f/\mu_0$ Am^2/Wb:

$$\frac{\partial U}{\partial y_f} = \Phi R_f \qquad \frac{\partial \Phi}{\partial y_f} = \frac{U}{R_{gf}}. \qquad (11.2 + 3)$$

The solution of this simultaneous differential equation is

$$U = R_{\text{fo}}\big[\Phi_+\exp(y_f/\lambda_f) + \Phi_-\exp(-y_f/\lambda_f)\big], \qquad (11.4a)$$

$$\Phi = \Phi_+\exp(y_f/\lambda_f) - \Phi_-\exp(-y_f/\lambda_f), \qquad (11.4b)$$

where

$$\boxed{\lambda_f \equiv \sqrt{\frac{R_{\text{gf}}}{R_f}} = \sqrt{\frac{\mu'_{\text{rf}}p'\mu_{\text{rf}}pg_f}{\mu'_{\text{rf}}p' + \mu_{\text{rf}}p}}} \; . \qquad (11.4c)$$

is called the characteristic length (loss 8.7 db/λ) and

$$\boxed{R_{\text{fo}} = \sqrt{R_{\text{gf}}R_f} = \left(\frac{1}{\mu_0}\right)\sqrt{\left(\frac{1}{\mu'_{\text{rf}}p'} + \frac{1}{\mu_{\text{rf}}p}\right)g_f}} \qquad (11.4d)$$

is called the characteristic reluctance of flux-guide section, λ's and R_o's of the different head sections in the case of a TFH on a ferrite wafer ($\mu'_{\text{rf}}p' \gg \mu_{\text{rf}}p$) are given in table 11.1.

Φ_-/Φ_+, the reflection coefficient at the 'end' ($y = 0$) of a section follows from the continuity of flux and potential. At intersections ($y = 0$) this continuity leads to

$$\frac{\Phi_+R_o + \Phi_-R_o}{\Phi_+ - \Phi_-} = R_L, \qquad (11.5)$$

where R_L is the 'load' reluctance due to all following sections. Hence

$$\boxed{\Gamma \equiv \frac{\Phi_-}{\Phi_+} = \frac{R_L - R_o}{R_L + R_o}} \; . \qquad (11.6)$$

Applied in (11.4b) for a section with length h this gives the transfer rate

$$\boxed{\frac{\Phi(y)}{\Phi(h)} = \frac{\exp(y^*) - \Gamma\exp(-y^*)}{\exp(h^*) - \Gamma\exp(-h^*)}} , \qquad (11.7)$$

where * means relative to (divided by) the characteristic length of the section. The input reluctance of this section serves in turn as a load impedance for the foregoing section and follows from eqs (11.4a), (11.4b) and (11.6)

$$R = R_\mathrm{o} \frac{1 + \Gamma \exp(-2h^*)}{1 - \Gamma \exp(-2h^*)} . \tag{11.8}$$

Substituting $R_5 = \infty$ or 0 (see fig. 11.2a) in eq. (11.6) and applying the result to (11.8) and adding the reluctance of the second overlap region yields the load reluctance of the MRE section. All necessary quantities

Table 11.1 Recording transmission line quantities

$\lambda_\mathrm{f} = \sqrt{\mu_\mathrm{rf} p\, g_\mathrm{f}}$ [m]

$\lambda_\mathrm{m} = \sqrt{\mu_\mathrm{rm} t\, g_\mathrm{m}}$ [m]

$R_\mathrm{fo} = \dfrac{1}{\mu_0} \sqrt{\dfrac{g_\mathrm{f}}{\mu_\mathrm{rf} p}}$ [Am/Wb]

$R_\mathrm{mo} = \dfrac{1}{\mu_0} \sqrt{\dfrac{g_\mathrm{m}}{\mu_\mathrm{rm} t}}$ [Am/Wb]

$R_\mathrm{fm1} = \dfrac{g_\mathrm{fm}}{\mu_0 h_\mathrm{o1}}$ [Am/Wb]

$R_\mathrm{fm2} = \dfrac{g_\mathrm{fm}}{\mu_0 h_\mathrm{o2}}$ [Am/Wb]

$\Gamma_5 = +1$ if open and -1 if closed second flux-guide

$R_3 = R_\mathrm{fm2} + R_\mathrm{fo}\, \dfrac{1 + \Gamma_5 \exp(-2h^*_\mathrm{f2})}{1 - \Gamma_5 \exp(-2h^*_\mathrm{f2})}$ [Am/Wb]

$\Gamma_3 = \dfrac{R_3 - R_\mathrm{mo}}{R_3 + R_\mathrm{mo}}$

α_m1 from Table a11.1

$R_1 = R_\mathrm{fm1} + R_\mathrm{mo}\, \dfrac{1 + \Gamma_3 \exp(-2h^*_\mathrm{m})}{1 - \Gamma_3 \exp(-2h^*_\mathrm{m})}$ [Am/Wb]

$\Gamma_1 = \dfrac{R_1 - R_\mathrm{fo}}{R_1 + R_\mathrm{fo}}$

$\eta_1 = \dfrac{1 - \Gamma_1}{\exp(h^*_\mathrm{f1}) - \Gamma_1 \exp(-h^*_\mathrm{f1})}$

$\eta_\mathrm{w} = \alpha_\mathrm{m1} \eta_1$ write efficiency of gap, appendix 11.2.

---

are found by applying this value again to (11.6), (11.8) and (11.7) etc. with appropriate λ_m and so working back to the front of the head; see table 11.1. Once the reflection Γ_3 at the end of the MRE section and the characteristic length of this section are known, the decay of flux or magnetization in the MRE section is known:

$$\boxed{\frac{M^*(y_m^*)}{M_2^*} = \frac{\exp(y_m^*) - \Gamma_3 \exp(-y_m^*)}{\exp(h_m^*) - \Gamma_3 \exp(-h_m^*)} \equiv f_m(y_m^*)} \quad , \qquad (11.9)$$

with

$$M_2^* \equiv \frac{M_2}{M_s} = \frac{\eta_1 \Phi_{in}}{\mu_0 t M_s}$$

and

$$\eta_1 \equiv \frac{\Phi_2}{\Phi_{in}}$$

given in Table 11.1.

11.4 Application on (B)YMRHs

In order to calculate resulting resistance changes in the YMRH and BYMRH it is necessary to integrate the microscopic equation of resistivity changes $\Delta\varrho$ in permalloy [11.11]

$$\frac{\Delta\varrho}{\Delta\varrho_{max}} = \cos^2\theta_{Mj} \qquad (11.10)$$

with θ_{Mj} the angle between magnetization M and current density j_p.

For a uniform current distribution in the z-direction (YMRH), using $\Delta R \ll R$, we obtain

$$\frac{\Delta R}{\Delta R_{max}^*} = -\langle \sin^2\theta \rangle_m = -\langle M^{*2} \rangle_m \qquad (11.11)$$

where θ is the angle between $\boldsymbol{M}$ and the z-axis, ΔR is measured in relation to $R(\theta = 0)$, $\langle \quad \rangle_{\mathrm{m}}$ means averaged over h_{m}, M is the y-component of $\boldsymbol{M}$, $\Delta R_{\mathrm{max}}^* \equiv (h_{\mathrm{m}}/h)\Delta R_{\mathrm{max}}$ with $\Delta R_{\mathrm{max}}/R \approx \Delta\varrho_{\mathrm{max}}/\varrho \lesssim 2\%$ in 80:20 NiFe and $h \equiv h_{\mathrm{o1}} + h_{\mathrm{m}} + h_{\mathrm{o2}}$.

For a uniform current distribution making an angle of 45° with the z-axis (BYMRH) integration of (11.10) yields

$$\frac{\Delta R}{\Delta R_{\mathrm{max}}^*} = \pm \langle \sin\theta \cos\theta \rangle_{\mathrm{m}} = \pm \langle M^* \sqrt{1 - M^{*2}} \rangle_{\mathrm{m}} \qquad (11.12)$$

The sign depends on the direction of the barberpole stripes ($\pm$ 45°), the side of the MRE where barberpole stripes are located and/or the side of the MRE where soft magnetic material is located. Note that currents flowing above the MRE are less effective in biasing the element. For the configuration and current direction of Fig. 11.1 the $+$sign is valid if $\boldsymbol{M}$ corresponds to the absolute minimum free energy, i.e. the z-component of $\boldsymbol{M}$ points towards the negative z-direction, and θ is taken relative to this direction and measured clockwise.

11.4.1 YMRH response

The YMRH response integral follows by substituting (11.9) in (11.11)

$$\frac{\Delta R}{\Delta R_{\mathrm{max}}^*} = - \langle f_{\mathrm{m}}^2 M_2^{*2} \rangle_{\mathrm{m}}. \qquad (11.13)$$

Evaluation of the integral yields

$$\boxed{\frac{\Delta R}{\Delta R_{\mathrm{max}}^*} = -\alpha_{\mathrm{m2}} M_2^{*2}} \qquad (11.14)$$

with $\alpha_{\mathrm{m2}} \equiv \langle f_{\mathrm{m}}^2 \rangle_{\mathrm{m}}$ given in table a11.1.

Magnetic saturation is reached suddenly when $M_2^* = 1$ in our model, assuming M_2^* saturates before the tip of the flux-guide i.e. $\eta_1 p > t$. The efficiency η_1 of the flux-guide is given in table I. A further increase of Φ_{in} will not result in a noticeable change of magnetization at $y_{\mathrm{m}} < h_{\mathrm{m}}$ since $\mu_{\mathrm{rm}} \gg 1$ before saturation: ΔR remains constant.

In a YMRH a bias is necessary to obtain a linear response. A nearly homogeneous bias in the MRE results from I_b (see fig. 11.2a) even in the extreme cases, when $h_m^* \gg 1$ because of the homogeneous current distribution, and when $h_m^* \ll 1$ because of negligible flux leakage across the MRE.

Parts of the current I_m flowing in the overlap regions h_{o1} and h_{o2} result in magnetizations with opposite exponential decays. Their sum will be approximated by a homogeneous magnetization in the h_m region too, with value equal to the magnetization at $y_m = h_m$. The bias M_c^* at $y_m = h_m$ due to I_c (see fig. 11.2a) decays like M_2^* with $f_m(y_m^*)$. Thus, M_c^* can be superimposed on M_2^* in (11.14). Expressions for the total homogeneous bias M_b^* and the inhomogeneous bias M_c^* are derived with the aid of a reciprocity theorem in appendix 11.2. Inclusion of both homogeneous and inhomogeneous bias yields

$$
\boxed{
\begin{aligned}
\frac{\Delta R}{\Delta R_{\max}^*} &= -\left\langle \left[M_b^* + f_m(M_2^* + M_c^*) \right]^2 \right\rangle_m \\
&= -(2\alpha_{m1}M_b^* + 2\alpha_{m2}M_c^*)M_2^* - \alpha_{m2}M_2^{*2} \\
&\quad -(M_b^* + 2\alpha_{m1}M_b^{*2}M_c^* + \alpha_{m2}M_c^{*2})
\end{aligned}
}
\tag{11.15}
$$

with $\alpha_{m1} \equiv \langle f_m \rangle_m$ given in table a11.1.

The small-signal response, proportional to $2\alpha_{m1}M_b^* + 2\alpha_{m2}M_c^*$ is maximal for $M_b^* = 1$ with $M_c^* = 0$, since $\alpha_{m1} > \alpha_{m2}$ (see appendix 11.1). A homogeneous bias is apparently favourable for maximum sensitivity. For reproducing large signals a bias $M_b^* + M_c^* = \frac{1}{2}$, or $M_b^* = \frac{1}{2}$ and $M_c^* = 0$ for more sensitivity, is favourable. Saturation is then obtained when $M_2^* = \pm\frac{1}{2}$. No further change of ΔR results from a further increase of Φ_{in} (or I_c).

11.4.2 Exact BYMRH response

An exact solution of the BYMRH response integral can be obtained by expanding the integrand in a power series. Because of anti-symmetry in the case of optimal biasing an odd power series in M_2^* will be obtained after some mathematics

$$\frac{\Delta R}{\Delta R^*_{\text{max}}} = \pm \sum_{n=1,3,5}^{\infty} c_n M_2^{*n} \alpha_{\text{m}n} \qquad (11.16)$$

with $c_1 = 1$, $c_n = (n-4)c_{n-2}/(n-1)$ for $n = 3, 5, 7$ etc. and $\alpha_{\text{m}n} \equiv \langle f_{\text{m}}^n \rangle_{\text{m}}$ the over h_{m} averaged n^{th} power of $f_{\text{m}}(y_{\text{m}}^*)$. A relation is derived for $\alpha_{\text{m}n}$ in appendix I. Expressions for $\alpha_{\text{m}n}$, $n = 1, 2, 3, ..., 9$ are given in Table a11.1. $\alpha_{\text{m},\,n+1} \leqslant \alpha_{\text{m},\,n} \leqslant 1$ holds, see appendix 11.1, and $c_n < 0$ if $n > 1$. When there is *no flux leakage across the MRE*, $\alpha_{\text{m}n} = 1$, i.e. $h_{\text{m}}^* = 0$, and (11.16) relaxes to the *typical barberpole response*

$$\frac{\Delta R}{\Delta R^*_{\text{max}}} = \pm \sum c_n M_2^{*n} = \pm M_2^* \sqrt{1 - M_2^{*2}} = \pm \tfrac{1}{2} \sin (2 \arcsin M_2^*),$$

$$(11.17)$$

see the theoretical curves for $h_{\text{m}}^* = 0$ in Fig. 11.4. $\alpha_{\text{m}n}$ and thus the response are functions of the reflection coefficient at the MRE section Γ_3 and the length of the MRE related to the characteristic length of that section h_{m}^*. The curves in fig. 11.4 show the normalized response (11.16) as a function of Γ_3 and h_{m}^*. A reduction of Γ_3 i.e. an easier flow of flux in the back flux-guide and so less magnetic potential difference across the gaps reduces the flux leakage across the gap g_{m}. Thus the homogeneity of flux across the MRE decreases according to h_{m}^* and or Γ_3 increase and consequently these curves do not show the typical barberpole response anymore.

By substitution of $M_2^* \equiv \hat{M}_2^* \cos \omega t$ and $\Delta R = \sum_{n=1}^{\infty} \Delta \hat{R}_n \cos n\omega t$ in

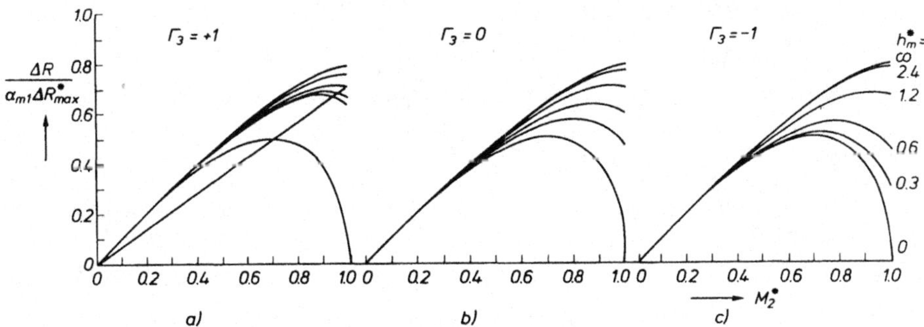

Fig. 11.4. Normalized response as a function of its variables Γ_3 and h_{m}^*.

(11.16) a relation for the harmonic distortions d_n for small signals is found

$$d_n \equiv \frac{\Delta \hat{R}_n}{\Delta \hat{R}_1} = - \frac{\alpha_{mn}|c_n|\hat{M}_2^{*(n-1)}}{\alpha_{m1}2^{n-1}}. \qquad (11.18)$$

The accuracy is better than $\hat{M}_2^{*2}/(1 - \hat{M}_2^{*2}) \times 100$ percent. In Table 11.2 the absolute values of harmonic distortions, expressed in dB, $20 \log |d_n|$, are presented for different values of Γ_3 and h_m^*, corresponding with fig. 11.4, curves a, b and c. These values are calculated for $\hat{M}_2^* = 1$ and so have to be reduced by $(n - 1)20 |\log \hat{M}_2^*|$ to arrive at usual values of $\hat{M}_2^*$ ($\lesssim 0.5$). According as homogeneity decreases, distortion decreases for fixed values of $\hat{M}_2^*$, since α_{mn}/α_{m1} decreases. However the sensitivity and maximum attainable resistance change ($\approx \alpha_{m1}\Delta R_{max}^*$ according to (11.21) in sec. 11.4.3) decreases too.

11.4.3 Approximation of BYMRH response

Substitution of (11.9) in (11.12) delivers the BYMRH response integral

$$\frac{\Delta R}{\Delta R_{max}^*} = \pm \langle f_m M_2^* \sqrt{1 - f_m^2 M_2^{*2}} \rangle_m, \qquad (11.19)$$

which can be easily rewritten to

Table 11.2   BYMRH distortions in case of perfect magnetic materials in dB

h_m^*	$\Gamma_3 = +1$				$\Gamma_3 = 0$				$\Gamma_3 = -1$			
	d_3	d_5	d_7	d_9	d_3	d_5	d_7	d_9	d_3	d_5	d_7	d_9
0	-18	-42	-60	-76	-18	-42	-60	-76	-18	-42	-60	-76
0.3	-24	-52	-72	-90	-20	-47	-67	-84	-19	-43	-62	-78
0.6	-24	-52	-73	-91	-22	-50	-70	-89	-20	-46	-65	-83
1.2	-25	-53	-74	-92	-25	-53	-74	-92	-23	-52	-73	-92
2.4	-26	-55	-76	-94	-27	-55	-76	-95	-27	-56	-77	-95
∞	-28	-56	-77	-95	-28	-56	-77	-95	-28	-56	-77	-95

$$\frac{\Delta R}{\Delta R_{\mathrm{max}}^{*}} = \pm \left\langle \tfrac{1}{2}\sin\left(2\arcsin\left(f_{\mathrm{m}}M_{2}^{*}\right)\right)\right\rangle_{\mathrm{m}}. \qquad (11.20)$$

This complicated integral can be simplified by assuming that only the first part $\alpha_{\mathrm{m1}}h_{\mathrm{m}}$ of the MRE is effective for flux from Φ_{in} or I_{c}. The whole MRE section is effective in the case of the homogeneous M_{b}^{*}. Thus, both M_{b}^{*} and $M_{2}^{*} + M_{\mathrm{c}}^{*}$ are fully effective in the region $\alpha_{\mathrm{m1}}h_{\mathrm{m}}$ and only M_{b}^{*} in the remaining region $(1 - \alpha_{\mathrm{m1}})h_{\mathrm{m}}$. Hence

$$\frac{\Delta R}{\Delta R_{\mathrm{max}}^{*}} \approx \pm\, \alpha_{\mathrm{m1}}\,\tfrac{1}{2}\sin\left(2\arcsin\left(M_{2}^{*} + M_{\mathrm{c}}^{*} + M_{\mathrm{b}}^{*}\right)\right) \pm$$

$$(1 - \alpha_{\mathrm{m1}})\,\tfrac{1}{2}\sin\left(2\arcsin M_{\mathrm{b}}^{*}\right). \qquad (11.21)$$

This approximation is qualitatively satisfactory in the region of operation $|M_{2}^{*} + M_{\mathrm{c}}^{*} + M_{\mathrm{b}}^{*}| < \tfrac{1}{2}\sqrt{2}$. Eq. (11.21) is exact for small signals. When the magnetization reaches saturation and α_{m1} is small ($\lambda_{\mathrm{m}} < h_{\mathrm{m}}$) the response saturates at a high level instead of decreasing to its initial value, as followed from the exact solution of the integral.

The BYMRH response at small signals $\alpha_{\mathrm{m1}}M_{2}^{*}$ is equal to the small-signal response of the YMRH with the most favourable bias for big signals ($M_{\mathrm{b}}^{*} = \tfrac{1}{2}$), and half the YMRH's maximal small-signal response ($M_{\mathrm{b}}^{*} = 1$).

11.5 Comparison of theory with experiments

11.5.1 Shape of response

With a simple computer program we have calculated the head perfor-mance of many practical configurations using the tabulated expressions.

A barberpole configuration with a big gap g_{m} of 5 µm and closed second flux-guide resulted in theoretical values of $\Gamma_{3} = -0.16$ and $h_{\mathrm{m}}^{*} = 0.35$, which begins to show the typical barberpole response given theoretically as well as experimentally (driven by I_{c}) in Fig. 11.5, curves b. The typical barberpole response is approached even better when the source current is applied to the bias coil in the gap of the MRE of a BYMRH with a much smaller gap $g_{\mathrm{m}} = 1.3$ µm and open second flux-guide, since this kind of drive is fairly homogeneous (sec. 11.4.1). This is shown in fig. 11.5, curve c. Fig. 11.5a shows the result of a strongly decaying magnetization in the same head, when driven from the front coil c. In the theoretical calculations $\mu_{rf} = 2000$ and 1200 are used. This change has little influence on Γ_3. As shown in fig. 11.5, this does not influence the normalized curves (shape) visually. The sensitivity of course strongly depends on μ_{rf}, especially when y_c^* in a wafertest, or h_{f1}^* in recording are big, e.g. p, μ_{rf} and g_f are small. Hysteresis is not drawn in fig. 11.5, but was present. Fig. 11.6 shows hysteresis in a BYMRH with $\Gamma_3 = +0.26$ and $h_m^* = 0.56$, driven by a small coil positioned near the front of the head instead of a magnetized tape. The curves in figs. 11.5 and 11.6 are seen to coincide with the theoretical curves almost completely as long as M_2^* does not exceed $\frac{1}{2}$.

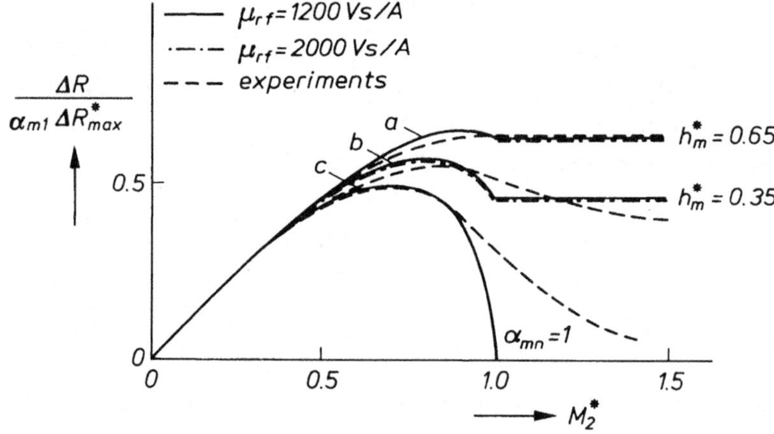

Fig. 11.5. Wafer experiments on two head configurations ($a + c$, small gap; b, big gap) for two different kinds of drive ($a + b$, contra-coil; c, bias coil). $a + c$: $\Gamma_3 = 0.28$ ($\mu_{rf} = 1200$), $\Gamma_3 = 0.24$ ($\mu_{rf} = 2000$), respectively. b: $\Gamma_3 = -0.16$ ($\mu_{rf} = 1200$), $\Gamma_3 = -0.20$ ($\mu_{rf} = 2000$), respectively.

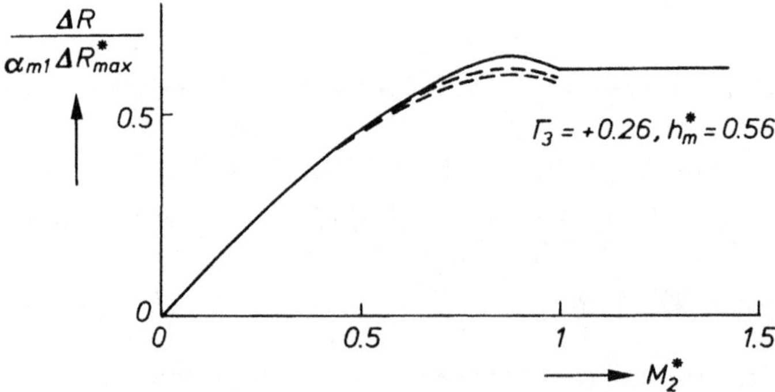

Fig. 11.6. Hysteresis in the response of a BYMRH when M_2^* is driven up to saturation ($\hat{M}_2^* = 1$).

Because of coercivity, the response is not only affected by hysteresis, but can also be affected by domain wall jumps. In our heads jumps are invisible on the oscilloscope trace. To show the typical relation between periodic domain wall jumps in the response and the frequency spectrum we therefore consider a poor head which is much more affected by jumps of the domain walls in the flux guide. In the Figs. 11.7 and 11.8 the response and spectrum of this poor head are given for $\hat{M}_2^* = \frac{1}{2}$ and only the spectrum for $\hat{M}_2^* = \frac{1}{4}$. Most of the jumps took place obviously at fixed values of the 1 kHz periodic signal, since the oscilloscope picture was the result of a long exposure time. Two differences between experimental and calculated distortions of this head ($\Gamma_3 = +0.28$, $h_m^* = 1.13$) are obvious: first the only slow decrease of amplitude of harmonic distortions with harmonic number and consequently the presence of

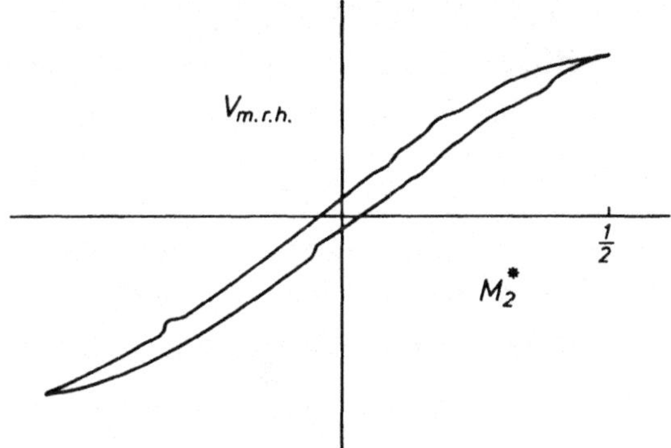

Fig. 11.7. Measured response of a poor MRH.

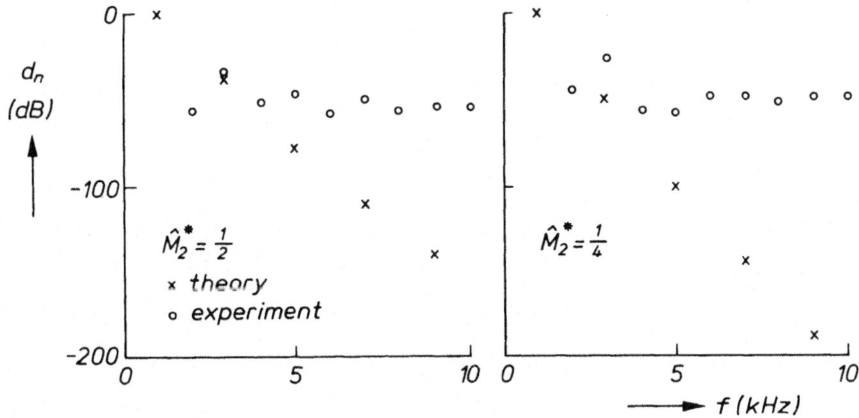

Fig. 11.8. The strong deviations of observed higher harmonic distortions from theoretical values due to the jumps in fig. 11.7.

even harmonics which are both general properties of jumps, and second the, at most, slow decrease of harmonic distortions with amplitude $\hat{M}_2^*$.

Hysteresis only contributes to lower odd harmonics. Theoretical values of distortions are therefore absolute minima, which are useful in efforts to achieve a more perfect material, e.g. if the magnetic materials in the head are required to be free of (moving) domain walls, in which case only rotation of magnetization would occur until suddenly saturation is reached.

11.5.2. Efficiency and recording experiments

So far no attention has been paid to the absolute values of sensitivity. Only the shapes of transfer curves have been considered. When using the contracoil c as the input we observed mostly good correspondence ($\leqslant 30\%$) between observed and theoretically calculated sensitivities while using $\mu_{\mathrm{rf}} \approx 2000$ for sputtered and $\mu_{\mathrm{rf}} \approx 1000$ for electroplated permalloy flux guides, and $\mu_{\mathrm{rm}} \approx B_s/(\mu_0 H_k + \mu_0 H_z) \approx 2000$ to 4000 for (bias) sputtered MREs. When the normalized flux-guide height h_{f}^* is relatively large, however, small tolerances in g_{f}, p and/or μ_{rf} have strong influences on efficiency. This is demonstrated in the results of recording experiments [11.7] in Fig. 11.9. If the characteristic length of this head,

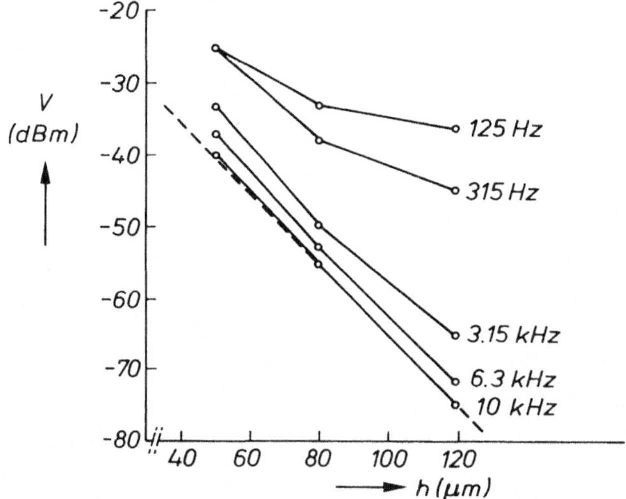

Fig. 11.9. Influence of flux-guide height on BYMRH sensitivity. The reference tape was recorded in accordance with DIN 45513.

lapped to $h_f = 100$ μm, were mismatched by only 20%, the output would be mismatched by about 8.7 dB $\times$ 20 μm/$\lambda_f = 10$ dB or by a factor of 3. The parameters in this recording experiment are: $\mu_{rf} = 900$ (electroplated flux-guide), $\mu_{rm} = 2000$; $g_f = 0.9$, $g_{fm} = 0.3$, $g_m = 0.9$, $p = 0.38$, h_{f1} = variable, $h_{f2} = 165$, $h_{o1} = 15$, $h_{o2} = 15$, $h_m = 10$, $t = 0.05$ μm; $\Gamma_5 = +1$; $\mu_0 M_s = 1$ Wb/m^2; $R = 70$ Ω; $\Delta R_{max}/R = 1.35\%$ (deduced from measurements on equivalent but unshielded MREs); $I_m = 20$ mA; short circuit flux from tape at 10 kHz, $\Phi_{sc} = 3.3$ nW/m $+2$ dB resp. -3 dB tolerance; gain of amplifier $A = 2950$; velocity of tape $v = 4.75$ cm/s.

For this head a write efficiency $\eta_w = 0.023$ is calculated for $h_{f1} = 50$ μm. We will take into account gap loss and distance loss by using the Karlqvist approximation [11.12, 11.13] $\exp(-ka)\sin(kg/2)/(kg/2) = 0.63$ for $\lambda = 4.75$ μm and $a \approx 0.3$ μm, so by neglecting the influence of the potential distribution at the outer edge of the pole tip p. Unlike a high-efficiency ferrite head ($p \to \infty$, $\mu \to \infty$), a TFH does not short circuit or load the tape noticeably. The magnetization in the tape M does not reverse to the remanence magnetization M_r at $H = 0$ but remains at the magnetization at $H \approx -H_c$ for $\lambda < 10$ times the tape thickness. When the relative permeability of the tape $\mu_{rt} = 1.7$ and $M_r \approx 3 H_c$ a magnetization of only $M = M_r - (\mu_{rt} - 1)H_c = 0.77 M_r$ results. For the actual magnetization ($\ll M_r$) the same reduction is assumed. The average flux through the MRE $\langle \Phi \rangle_m$ divided by the short circuit flux of the tape Φ_{sc} is then calculated to be approximately $0.77 \times 0.63 \times 0.023 = 0.0112$. From fig. 11.6 and given values for $R_{max} \times I_m$, h_m/h, t and $\mu_0 M_s$ a ratio $\langle \Phi \rangle_m/\Phi_{sc} = 0.0086$ is destillated. Most heads showed a good correspondence between the measured and calculated values. However the above head was very suitable for testing the theory and model, because the calculated characteristic length of 17.4 μm (electroplated flux-guide, $\mu_{rf} = 900$; $g_f = 0.9$, $p = 0.38$ μm) used in the calculations could be compared with the actual value by lapping the head from $h_{f1} = 120$ μm to $h_{f1} = 50$ μm and turned out to be almost exact (measured 17.5 μm). It may therefore be concluded that the theory can be applied too in short-wavelength recording.

The influence of all head parameters on head characteristics can easily be calculated by changing the desired parameter. The influence of overlap height h_{o1} on bias sensitivity is shown in Fig. 11.10.

Head parameters were: $\mu_{rf} = 1000$, $\mu_{rm} = 2000$; $g_f = 1$, $g_{fm} = 0.2$, $g_m = 0.8$, $p = 0.8$, $t = 0.05$, $h_{f1} = 30$, $h_{o2} = 5$, $h_m = 10$, $h_{f2} = 50$, $h_c \to 0$,

$y_c = 15$ μm, $\Gamma_5 = -1$; $I_c = 1$, $I_m = 5$ mA; $\mu_0 M_s = 1$ Wb/m^2; $\Delta R_{max} = 1.8$ Ω.

For this configuration an overlap $h_{o1} \approx 5$ μm seems favourable, at least for bias sensitivity. Results for very high and very low values of h_{o1} must not be taken seriously.

Fig. 11.11 shows the influence of a decreasing t on sensitivity, assuming that $I_m^2 R$, $\Delta R_{max}/R$ and the flux source Φ_{in} remain constant. The magnetic potential difference across the MRE and hence $\alpha_{m1} M_2^*$ approaches its maximum (obtained when $R_m = \infty$ or $t \to 0$) when the magnetic series reluctance of the MRE $R_m = 1/(\mu_0 \mu_{rm} t)$ reaches the

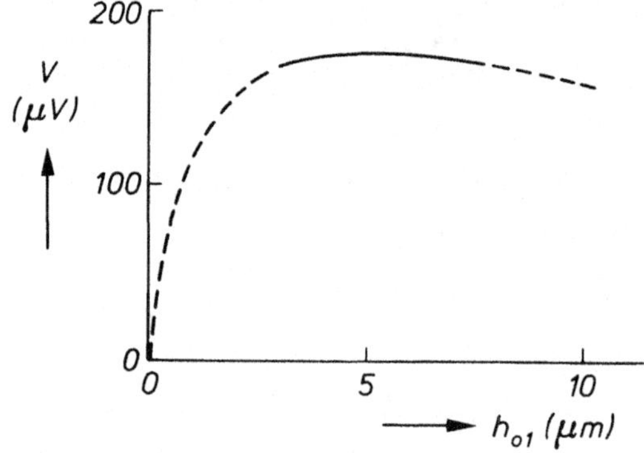

Fig. 11.10. The influence of overlap height h_{o1} on the bias sensitivity, represented by the voltage change over the MRE, V, due to a change of the bias current, I_b, of 1 mA.

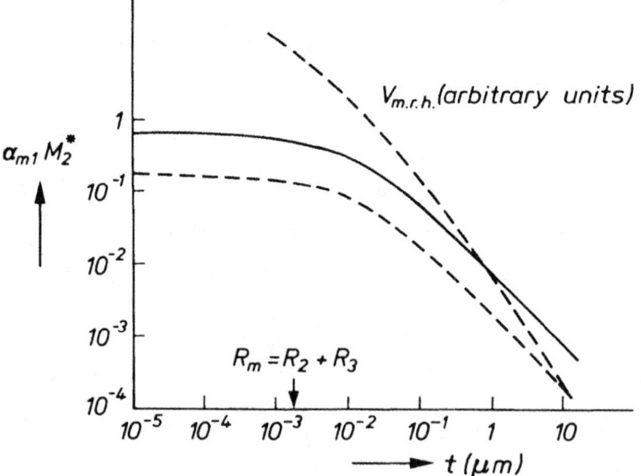

Fig. 11.11. Influence of t on sensitivity when power consumption in the MRE is kept constant and $\Delta\varrho_{max}/\varrho$ is assumed constant.

'internal reluctance' of the source Φ_{in} measured across the MRE, being $R_{2-} + R_3$.

Head parameters were: $\mu_{rf} = 2000$, $\mu_{rm} = 4000$; $g_f = 1.1$, $g_{fm} = 0.5$, $g_m = 0.8$, $p = 1$, $h_{f1} = 50$, $h_{o1} = 5$, $h_{o2} = 5$, $h_m = 10$, $h_{f2} = 50$, $h \rightarrow 0$, $y_c = 35$ μm; $\Gamma_5 = -1$; $\mu_0 M_s = 1$ Wb/m^2; $\Phi_{in} = 100$ nW/m.

At very small values of t, $I_m^2 R$ must be decreased because j_p approaches the maximum current density and $\Delta\varrho_{max}/\varrho$ will decrease [11.14] in practice.

11.6 Conclusions

The transmission line description turned out to be very useful for finding analytical solutions to linear 'one-dimensional' configurations. This also applied when integrations of fluxes etc. had to be carried out. Analytical solutions are preferable to numerical solutions because of the explicit and transparent dependence of head performance on head parameters. The sensitivity of (B)YMRHs is described very well with the theory. The factors that determine the accuracy in the case of low efficiency heads are technological tolerances and the lack of exact knowledge of permeabilities.

The theoretical shape of response curves of BYMRHs is only a function of normalized height h_m^* of the MRE and of the reflection coefficient Γ_3 at the end of this head section. The neglection of (moving and pinning) domain walls in the calculations led to smooth response curves, which seem to coincide quantitatively with observed curves if $M_2^* \lesssim \frac{1}{2}$ and qualitatively for higher M_2^*. Closer observation, however, may show jumps, usually at a number of fixed values of magnetization, which are responsible for higher harmonic distortions decaying only slowly with harmonic number. Differences between theoretical values of distortions and the observed spectrum are therefore an indication of the quality of the magnetic materials in the head.

We have restricted the description to wafer tests and short-wavelength recording. A following step in the calculations on (B)YMRHs would be to describe long-wavelength recording (and even an approximation of side reading) by an analogous calculation of the flux distribution in the head – and thus the potential distribution on the head – in a fictitious write situation, including the fringe at the MRE, and transformation of this potential to pole surface level.

Appendix 11.1 Homogeneity coefficients

The homogeneity coefficients α_{mn} are defined by

$$\alpha_{mn} \equiv \langle f_m^n \rangle_m \equiv \frac{1}{h_m^*} \int_0^{h_m^*} \left(\frac{\exp(y_m^*) - \Gamma_3 \exp(-y_m^*)}{\exp(h_m^*) - \Gamma_3 \exp(-h_m^*)} \right)^n dy_m^* \qquad (a11.1)$$

With the aid of the binomial theorem followed by integration the following relation is found:

$$\alpha_{mn} = \frac{1}{h_m^*[\exp(h_m^*) - \Gamma_3 \exp(-h_m^*)]^n} \times$$

$$\left\{ \frac{1}{n}[\exp(nh_m^*) - 1)] - \frac{n}{n-2} \Gamma_3[\exp((n-2)h_m^*) - 1] \right.$$

$$+ \frac{n(n-1)}{(n-4)2!} \Gamma_3^2[\exp((n-4)h_m^*) - 1]$$

$$\left. - \,...(-1)^n \frac{1}{n} \Gamma_3^n[1 - \exp(-nh_m^*)] \right\} \qquad (a11.2)$$

When n is an even number, the central term in (a11.2) is undefined. Separate calculation of this term or a determination of the limit $n' \to n$ of this term is then necessary. For the BYMRH only odd n values occur. For all possible values of $h_m^*(\geq 0)$ and $\Gamma_3(|\Gamma_3| \leq 1)$ it can be proved that $\alpha_{m,\,n+1} \leq \alpha_{mn} \leq 1$. Limits of α_{mn} are

$$\alpha_{mn} = \frac{1}{nh_m^*}(1 - \exp(-nh_m^*)) \text{ if } |\Gamma_3| << \exp(h_m^*), \qquad (a11.3)$$

$$\alpha_{mn} = \frac{1}{nh_m^*} \text{ if } nh_m^* >> 1. \qquad (a11.4)$$

α_{mn} values for $n = 1, 3, 5, 7$ and 9 (BYMRH) as well as α_{m2} (YMRH) are given in Table a11.1. For $n > 9$ (a11.4) is used in the computer calculations, except when $h_m^* \to 0$.

Table a11.1 Homogeneity coefficients α_{mn}.

$$\alpha_{mn} = \frac{1}{h_m^*(\exp(h_m^*) - \Gamma_3\exp(-h_m^*))^n} \times$$

$n = 1$: $\left(\exp(h_m^*) - 1\right) - \Gamma_3(1 - \exp(-h_m^*))$,

$n = 2$: $\tfrac{1}{2}(\exp(2h_m^*) - 1) - 2\Gamma_3 h_m^* + \tfrac{1}{2}\Gamma_3^2(1 - \exp(-2h_m^*))$,

$n = 3$: $\tfrac{1}{3}(\exp(3h_m^*) - 1) - 3\Gamma_3(\exp(h_m^*) - 1)$
$\qquad\qquad + 3\Gamma_3^2(1 - \exp(-h_m^*)) - \tfrac{1}{3}\Gamma_3^3(1 - \exp(-3h_m^*))$,

$n = 4$: $\tfrac{1}{4}(\exp(4h_m^*) - 1) - \tfrac{4}{2}\Gamma_3(\exp(2h_m^*) - 1)$
$\qquad\qquad + 6\Gamma_3^2 h_m^* - \tfrac{4}{2}\Gamma_3^3(1 - \exp(-2h_m^*)) + \tfrac{1}{4}\Gamma_3^4(1 - \exp(-4h_m^*))$,

$n = 5$: $\tfrac{1}{5}(\exp(5h_m^*) - 1) - \tfrac{5}{3}\Gamma_3(\exp(3h_m^*) - 1) + 10\Gamma_3^2(\exp(h_m^*) - 1)$
$\qquad\qquad - 10\Gamma_3^3(1 - \exp(-h_m^*)) + \tfrac{5}{3}\Gamma_3^4(1 - \exp(-3h_m^*))$
$\qquad\qquad - \tfrac{1}{5}\Gamma_3^5(1 - \exp(-5h_m^*))$,

$n = 6$: $\tfrac{1}{6}(\exp(6h_m^*) - 1) - \tfrac{6}{4}\Gamma_3(\exp(4h_m^*) - 1) + \tfrac{15}{2}\Gamma_3^2(\exp(2h_m^*) - 1)$
$\qquad\qquad - 20\Gamma_3^3 h_m^{*4} + \tfrac{15}{2}\Gamma_3^4(1 - \exp(-2h_m^*)) - \tfrac{6}{4}\Gamma_3^5(1 - \exp(-4h_m^*))$
$\qquad\qquad + \tfrac{1}{6}\Gamma_3^6(1 - \exp(-6h_m^*))$,

$n = 7$: $\tfrac{1}{7}(\exp(7h_m^*) - 1) - \tfrac{7}{5}\Gamma_3(\exp(5h_m^*) - 1) + \tfrac{21}{3}\Gamma_3^2(\exp(3h_m^*) - 1)$
$\qquad\qquad - 35\Gamma_3^3(\exp(h_m^*) - 1) + 35\Gamma_3^4(1 - \exp(-h_m^*))$
$\qquad\qquad - \tfrac{21}{3}\Gamma_3^5(1 - \exp(-3h_m^*)) + \tfrac{7}{5}\Gamma_3^6(1 - \exp(-5h_m^*))$
$\qquad\qquad - \tfrac{1}{7}\Gamma_3^7(1 - \exp(-7h_m^*))$,

$n = 8$: $\tfrac{1}{8}(\exp(8h_m^*) - 1) - \tfrac{8}{6}\Gamma_3(\exp(6h_m^*) - 1) + \tfrac{28}{4}\Gamma_3^2(\exp(4h_m^*) - 1)$
$\qquad\qquad - \tfrac{56}{2}\Gamma_3^3(\exp(2h_m^*) - 1) + 70\Gamma_3^4 h_m^* - \tfrac{56}{2}\Gamma_3^5(1 - \exp(-2h_m^*))$
$\qquad\qquad + \tfrac{28}{4}\Gamma_3^6(1 - \exp(-4h_m^*)) - \tfrac{8}{6}\Gamma_3^7(1 - \exp(-6h_m^*))$
$\qquad\qquad + \tfrac{1}{8}\Gamma_3^8(1 - \exp(-8h_m^*))$,

$n = 9$: $\tfrac{1}{9}(\exp(9h_m^*) - 1) - \tfrac{9}{7}\Gamma_3(\exp(7h_m^*) - 1) + \tfrac{36}{5}\Gamma_3^2(\exp(5h_m^*) - 1)$
$\qquad\qquad - \tfrac{84}{3}\Gamma_3^3(\exp(3h_m^*) - 1) + 126\,\Gamma_3^4(\exp(h_m^*) - 1)$
$\qquad\qquad - 126\Gamma_3^5(1 - \exp(-h_m^*)) + \tfrac{84}{3}\Gamma_3^6(1 - \exp(-3h_m^*))$
$\qquad\qquad - \tfrac{36}{5}\Gamma_3^7(1 - \exp(-5h_m^*)) + \tfrac{9}{7}\Gamma_3^8(1 - \exp(-7h_m^*))$
$\qquad\qquad - \tfrac{1}{9}\Gamma_3^9(1 - \exp(-9h_m^*))$.

Appendix 11.2 Derivation of bias expressions and read efficiency

First a useful reciprocity relation will be derived from the reciprocity theorem in network analysis [11.15]. It states that the forward transfer impedance in a linear single source network with arbitrary input and output terminals equals the reverse transfer impedance (output and input interchanged).

In the YMRH with flux guide divided into an infinite number of elements dy_f, with magnetic series resistance $R_f dy_f$, the electric current (magnetic potential source) $dI_c \equiv \sigma_c dy_f$ can be transformed with Norton's theorem into a flux source

$$\Phi(y_f) = \frac{\sigma_c}{R_f}, \tag{a11.5}$$

across $R_f dy_f$ (input terminals). The output potential difference, dU_1, is measured across the gap at $y_f = 0$.

In a fictitious reversed situation $dU_- = \Phi_-(y_f)r_f dy_f$, measured across the input terminals, is due to a flux source Φ_{1+-} across the output terminals. According to reciprocity

$$\frac{dU_1}{\Phi(y_f)} = \frac{dU_-}{\Phi_{1+-}}. \tag{a11.6}$$

Integration, while using the superposition theorem, gives

$$U_1 = \frac{1}{\Phi_{1+-}} \int \sigma_c \Phi_-(y_f) dy_f. \tag{a11.7}$$

Evaluating the integral for a homogeneous surface current density $\sigma_c = I_c/h_c$ gives

$$\frac{U_1}{I_c} = \frac{\langle \Phi_-(y_f) \rangle_c}{\Phi_{1+-}} = \frac{\eta_{c-}\alpha_{c1-}R_1}{R_1 + R_{1-}}. \tag{a11.8}$$

The right-hand term has been expressed as a function of the reverse efficiency η_{c-} from $y_f = 0$ to y_c and the homogeneity coefficient α_{c1-}, both given in Table a11.2. Note that only the fraction $\Phi_{1-} = \Phi_{1+-}R_1/$

Table a11.2 Bias transmission line quantities

$$R_{c-} = R_{fo} \frac{1 + \exp(-2(h_{f1}^* - y_c^*))}{1 - \exp(-2(h_{f1}^* - y_c^*))} \qquad \text{[Am/Wb]}$$

$$\Gamma_{c-} = \frac{R_{c-} - R_{fo}}{R_{c-} + R_{fo}}$$

$$\eta_{c-} = \frac{1 - \Gamma_{c-}}{\exp(y_c^*) - \Gamma_{c-}\exp(-y_c^*)}$$

$$R_{cc-} = R_{fo} \frac{1 + \exp(-2(h_{f1}^* - y_c^* - h_c^*))}{1 - \exp(-2(h_{f1}^* - y_c^* - h_c^*))} \qquad \text{[Am/Wb]}$$

$$\Gamma_{cc-} = \frac{R_{cc-} - R_{fo}}{R_{cc-} + R_{fo}}$$

$$\alpha_{c1-} = \frac{\exp(h_c^*) - 1 - \Gamma_{cc-}(1 - \exp(-h_c^*))}{h_c^*[\exp(h_c^*) - \Gamma_{cc-}\exp(-h_c^*)]}$$

$$h_{o1}^* = \frac{h_{o1}}{\lambda_f}$$

$$R_{o1-} = R_{fo} \frac{1 + \exp(-2(h_{f1}^* - h_{o1}^*))}{1 - \exp(-2(h_{f1}^* - h_{o1}^*))} \qquad \text{[Am/Wb]}$$

$$\Gamma_{o1-} = \frac{R_{o1-} - R_{fo}}{R_{o1-} + R_{fo}}$$

$$\alpha_{o1-} = \frac{\exp(h_{o1}^*) - 1 - \Gamma_{o1-}(1 - \exp(-h_{o1}^*))}{h_{o1}^*[\exp(h_{o1}^*) - \Gamma_{o1-}\exp(-h_{o1}^*)]}$$

$$h_{o2}^* = \frac{h_{o2}}{\lambda_f}$$

$$R_{o2} = R_{fo} \frac{1 + \Gamma_5\exp(-2(h_{f2}^* - h_{o2}^*))}{1 - \Gamma_5\exp(-2(h_{f2}^* - h_{o2}^*))} \qquad \text{[Am/Wb]}$$

$$\Gamma_{o2} = \frac{R_{o2} - R_{fo}}{R_{o2} + R_{fo}}$$

$$\alpha_{o2} = \frac{\exp(h_{o2}^*) - 1 - \Gamma_{o2-}(1 - \exp(-h_{o2}^*))}{h_{o2}^*[\exp(h_{o2}^*) - \Gamma_{o2}\exp(-h_{o2}^*)]}$$

$$R_{1-} = R_{fo} \frac{1 + \exp(-2h_{f1}^*)}{1 - \exp(-2h_{f1}^*)} \qquad \text{[Am/Wb]}$$

$$R_2 = R_{fm1} + R_{1-} \qquad \text{[Am/Wb]}$$

$$\Gamma_{2-} = \frac{R_{2-} - R_{mo}}{R_{2-} + R_{mo}}$$

$$\eta_{m-} = \frac{1 - \Gamma_{2-}}{\exp(h_m^*) - \Gamma_{2-}\exp(-h_m^*)}$$

$$R_{4-} = R_{fm2} + R_{mo} \frac{1 + \Gamma_{2-}\exp(-2h_m^*)}{1 - \Gamma_{2-}\exp(-2h_m^*)} \qquad \text{[Am/Wb]}$$

$(R_1 + R_{1-})$ flowed into the flux guide at $y_f = 0$. From (11.3) + (11.7) applied to flux travel from left to right and simplification with the aid of (11.4c), (11.4d) and (11.6) it follows that

$$U_1 = R_{gf} \left(\frac{\partial \Phi}{\partial y_f} \right)_{y_f = 0} = \frac{1 + \Gamma_1}{1 - \Gamma_1} \cdot \frac{R_{gf} \Phi_c(0)}{\lambda_f} = R_1 \Phi_c(0), \tag{a11.9}$$

as is to be expected.

Substitution of (a11.8) in (a11.9) and solving for $M_c^* = \Phi_c(0)/(\mu_0 t M_s)$ yields,

$$M_c^* = \frac{\eta_{c-} \alpha_{c1-}}{\mu_0 t M_s} \cdot \frac{I_c}{R_1 + R_{1-}}. \tag{a11.10}$$

This bias decays like M_2^* in the MRE. The bias due to I_b and I_m, however, is approximately homogeneous, as explained in sec. 11.4.1. This bias M_b^* can be derived in a manner analogous to that given in (a11.10). The part of the current I_m that flows through the permalloy in the region h_m can be divided in two equal parts, one flowing above the MRE and the other flowing in the gap of the MRE. The part of the current flowing above the MRE is not effective in biasing the element and the same applies to the current through the barberpole stripes if located above the MRE.

Hence only one quarter of the current in this region is effective in case of the BYMRH and about a half in case of the YMRH. The effective part can be added to I_b. Note that currents flowing in the overlap regions, however, will contribute fully to the magnetization in the region h_m as a consequence of the model proposed in fig. 11.2b: flux generated or flowing in the overlap region flows mainly through the flux guide. After some mathematics we found for the approximate bias (at $y_f = 0$)

$$M_b^* \approx \frac{I_b + (h_m/h)\frac{1}{4}I_m}{R_2 + R_{2-}} \cdot \frac{\alpha_{m1}}{\mu_0 t M_s}$$
$$+ \frac{(h_{o1}/h)I_m}{R_1 + R_{1-}} \cdot \frac{\alpha_{o1-}}{\mu_0 t M_s}$$
$$+ \eta_{m-} \frac{(h_{o2}/h)I_m}{R_4 + R_{4-}} \cdot \frac{\alpha_{o2}}{\mu_0 t M_s}, \tag{a11.11}$$

with newly introduced quantities given in Table a11.2.

The frequency response to tape signals follows from the reciprocity theorem of recording theory. The potential distribution at pole surface level is then scaled with the efficiency in a write situation $\eta_\mathrm{w} = Hg_\mathrm{f}/I_\mathrm{w}$.

In our simplified model, the fictitious current I_w flows tightly around the h_m region. The potential across the gap now follows in analogy with (a11.8) from

$$Hg_\mathrm{f} = \langle \Phi \rangle_\mathrm{m} \, \frac{I_\mathrm{w}}{\Phi_\mathrm{in}} = \eta_1 \alpha_\mathrm{m1} I_\mathrm{w} \qquad \text{(a11.12)}$$

showing that $\langle \Phi \rangle_\mathrm{m}/\Phi_\mathrm{in} = \eta_1 \alpha_\mathrm{m1}$ *can be used for determining the write efficiency* $\eta_\mathrm{w} \equiv Hg_\mathrm{f}/I_\mathrm{w}$ of the gap.

References

[11.1] R.P. Hunt, *A magnetoresistive readout transducer*, IEEE Trans. Magn. Mag-7, 150, (1970).

[11.2] R.L. Anderson, C.H. Bajorek and D.A. Thompson, *Numerical analysis of a magnetoresistive transducer for magnetic recording applications*, A.I.P. Conf. Proc. 10, 1445, (1972).

[11.3] T.N. Casselman and S.A. Hanka, *Calculation of the performance of a magnetoresistive permalloy magnetic field sensor*, IEEE Trans. Magn. Mag-16, 461, (1980).

[11.4] K.E. Kuijk, W.J. van Gestel and F.W. Gorter, *The barberpole, a linear magnetoresistive head*, IEEE Trans. Magn. Mag-11, 1215, (1975).

[11.5] D.A. Thompson, *Magnetoresistive transducers in high-density magnetic recording*, A.I.P. Conf. Proc. 24, 528, (1974).

[11.6] W.F. Druyvesteyn, E.L.M. Raemaekers, R.D.J. Verhaar, J. de Wilde, J.H.J. Fluitman and J.P.J. Groenland, *Magnetic behaviour of narrow track thin-film heads*, J. Appl. Phys. 52 (3), 2462, (1981).

[11.7] W.F. Druyvesteyn, J.A.C. van Ooijen, L. Postma, E.L.M. Raemaekers, J.J.M. Ruigrok and J. de Wilde, *Magnetoresistive heads*, IEEE Trans. Magn., Mag-17, 2884-2889, (1981).

[11.8] A. Paton, *Analysis of the efficiency of thin-film magnetic recording heads*, J. Appl, Phys. 42, 5868, (1971).

[11.9] R.E. Jones, *Analysis of the efficiency and inductance of multiturn thin-film magnetic recording heads*, IEEE Trans. Magn. Mag-14, 509, (1978).

[11.10] For instance: T.S. Saad, *Microwave Engineers' Handbook 1*, Artech House Inc., or H. Mooijweer, *Microwave Techniques*, MacMillan, London, (1971).

[11.11] J. Smit, *Magnetoresistance of ferromagnetic metals and alloys at low temperatures*, Physica 17, 612, (1951).

[11.12] O. Karlqvist, *The magnetic calculation of the magnetic field in the ferromagnetic layer of a magnetic drum*, Trans. Roy. Inst. Techn. Stockholm 86, 3, (1954).

[11.13] W.K. Westmijze, *Studies on magnetic recording*, Philips Res. Rep. 8, 148, (1953).

[11.14] E.N. Mitchell, *Compositional and thickness dependence of the ferromagnetic anisotropy in resistance of iron-nickel films*, J. Appl. Phys. 35, 2604, (1964).

[11.15] J.A. Edminister, *Electric circuits*, McGraw-Hill Book Company NY, 158, (1965).

[11.16] J.J.M. Ruigrok, *Analytical description of thin-film yoke magnetoresistive heads*, Philips J. Res. 36, 289, (1981).

Chapter 12

Metal-in-gap heads

In this chapter two published papers are incorporated:

- Jaap J.M. Ruigrok, *Analysis of metal-in-gap heads*, IEEE Trans. Magn., Mag-20, 872 (1984).
- C.W.M.P. Sillen, J.J.M. Ruigrok, A.\Broese van Groenou, U. Enz, *Permalloy/sendust metal-in-gap head*, IEEE Trans. Magn., Mag-24, 1802 (1988).

With respect to the original papers, which were limited to three journal pages, more attention is paid to model calculations and comparison with experiments. For this purpose the sections 12.1.3 and 12.1.7 are added. Also added, to section 12.2.3, are the results on the gradual-gap calculations in chapters 5 and 7 in order to explain some early experimental results which are added to the discussion in section 12.1.7.

12.1 Analysis of metal-in-gap heads

12.1.1 Abstract

In this section some anomalous effects are described in the recording and playback performance of heads with thin metal, especially sendust, layers on one or both gap surfaces. The most apparent effect, bumps (undulations) in the output spectrum while reading with this metal-in-gap head, is discussed in detail. Under normal recording conditions no bumps are observed when another head is used for playback. The magnetic behaviour of the interface between metal layer and head-core material can be characterized with the expressions derived in the paper. Analogously the magnetic gap length of very small gaps (even below 200 Å) can be determined accurately with the aid of the derived expressions when a larger gap is incorporated in the same head for that purpose.

12.1.2 Introduction

Pole tip saturation is a well known problem in magnetic recording on high coercive tape [12.1, 12.2]. A metal-in-gap (MIG) head [12.3] may combine the excellent playback performance of MnZn-ferrite heads with the good recording performance of a high-saturation metal. The thin metal film, e.g. sendust with thickness p_1 or p_2, is deposited on one or both gap surfaces by sputtering or electroplating, as depicted in Fig. 12.1a.

The boundary between metal film and core material (e.g. ferrite) may exhibit poor magnetic performance. The corresponding gaps g_1 and g_2, depicted in figure 12.1b, cause interference bumps (undulations), see figure 12.3, in the output spectra of such a head. The aim of this article is to understand this phenomenon quantitatively in order to determine the location and magnetic length of this extra gap from the measured playback spectrum. This is necessary for good feedback to the technology. On the other hand it delivers an accurate method for investigating very small gaps, necessary for going to higher bit densities.

The above-mentioned interference effects are unwanted. Two methods for minimizing the bumps in the output spectrum will be discussed on the basis of an analytic expression that describes the bumps almost exactly.

12.1.3 Theory of playback performance

Gap-loss function

The (multi-) gap configuration used for the model calculations is shown in figure 12.1b. The distances between the centres of the residual

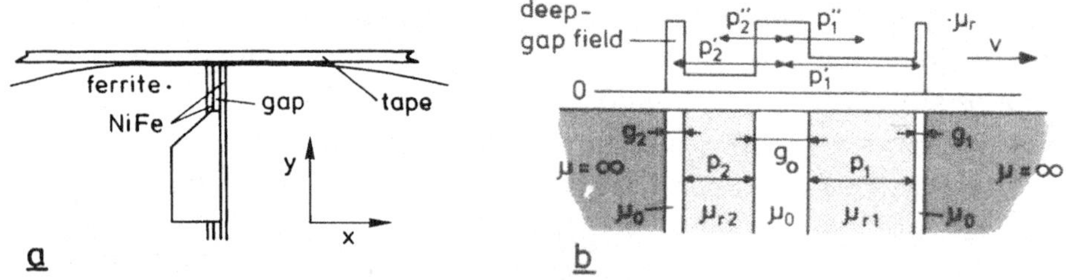

Fig. 12.1 a) Front part of a metal-in-gap head (MIG).
  b) Enlargement of the gap region of a MIG head. p_1 and p_2 denote the metal layers, e.g. sendust.

gaps g_1 and g_2 and the centre of the primary gap g_0, p_1' and p_2', differ slightly from the metal-layer thicknesses p_1 and p_2. p_1'' and p_2'' denote the distances to the centres of the metal layers. The flux, Φ, through the cross-sections is hardly dependent on x. The deep-gap field will thus be proportional to the local permeability, which is assumed to be a function of x only and is depicted in figure 12.1b for discrete steps in the permeability. The generalized gap-loss or sensitivity function follows by reciprocity, see chapter 3 or reference [12.4], and is determined by the Fourier transform of the x component of the field at the surface ($y = 0$) of the unsaturated head. According to the reciprocity theorem

$$\Phi(t) = \mu_0 \int M(r, t) \cdot H(r)\mathrm{d}r \qquad (12.1)$$

$$\Phi(\omega) = \int \Phi(t)\mathrm{e}^{-\mathrm{j}\omega t}\mathrm{d}t \qquad (12.2)$$

A number of terms, all depending on the wavenumber $k = 2\pi/\lambda = \omega/v$, with λ the wavelength and v the head-tape velocity, occur after integration. We shall only focus on the gap-loss function. This complex function, abbreviated as GLF and defined by

$$GLF(k) = \int H_x(x, y = 0)\mathrm{e}^{-\mathrm{j}kx}\mathrm{d}x \qquad (12.3)$$

can easily be calculated for the field distribution depicted in figure 12.1b (Karlqvist i.e. $\mu_\mathrm{r} = 0$ or deep-gap field approximation). With the normalization condition

$$\int H_x(x, y = 0)\,\mathrm{d}x = 1 \qquad (12.4)$$

we obtain

$$GLF = X_0 + X_1\mathrm{e}^{-\mathrm{j}kp_1'} + X_2\mathrm{e}^{\mathrm{j}kp_2'} + X_{\mathrm{p}_1}\mathrm{e}^{-\mathrm{j}kp_1''} + X_{\mathrm{p}_2}\mathrm{e}^{\mathrm{j}kp_2''} \qquad (12.5)$$

where $X_i \equiv (g_i/\Sigma g_j)\mathrm{sinc}(g_i/\lambda)$, $X_{\mathrm{p}_i} = \left[p_i/(\mu_{\mathrm{r}i}\Sigma g_j)\right]\mathrm{sinc}(p_i/\lambda)$

$$\mathrm{sinc}\, x \equiv (\sin \pi x)/(\pi x)$$

and

$$\Sigma g_j \equiv g_0 + g_1 + g_2 + p_1/\mu_{\mathrm{r}1} + p_2/\mu_{\mathrm{r}2}.$$

The absolute value $|GLF|$ and phase φ follow after splitting GLF into real and imaginary parts:

$$|GLF| = \sqrt{\mathrm{Im}^2[GLF] + \mathrm{Re}^2[GLF]} \qquad (12.6)$$

$$= \left(\left[X_1\sin(kp_1') + X_{p_1}\sin(kp_1'') - X_2\sin(kp_2') - X_{p_2}\sin(kp_2'')\right]^2 + \right.$$
$$\left.\left[X_1\cos(kp_1') + X_{p_1}\cos(kp_1'') + X_2\cos(kp_2') + X_{p_2}\cos(kp_2'') + X_0\right]^2\right)^{\frac{1}{2}}$$

$$\varphi = \arctan\left(\mathrm{Im}[GLF]/\mathrm{Re}[GLF]\right). \qquad (12.7)$$

After inverse Fourier transformation of all terms the real $\Phi(t)$ would be obtained. $|GLF|$ occurs as the gap-loss factor and φ as the extra phase shift due to the location and shape of the gaps. For asymmetrical multi-gap configurations this phase shift will be non-proportional to the frequency, i.e. phase distortion will occur in $\Phi(t)$.

The result (12.5) can be interpreted as the superposition (due to linearity of the Fourier transform) of five heads with gaps g_0, g_1, g_2, p_1 and p_2 respectively. Every fictitious head, including those with gap length p_1 and p_2, has an individual gap-null wavelength equal to its gap length (Karlqvist approximation). Every fictitious head reads with an efficiency η_i proportional to its gap reluctance R_i or 'magnetic gap length' g_i defined as

$$g_i \equiv \int (1/\mu_{ri})\mathrm{d}x , \qquad (12.8)$$

where the integration is carried out over one of the five regions. So the efficiency of the tiny gaps is small, as might be expected.

The result (12.8) neither depends on the permeability distribution nor on the model (Karlqvist, Westmijze, etc.) used for the field distribution outside the head. Only the integral values (12.8) are of importance for the efficiencies of the fictitious heads.

The field distribution and so the precise permeability distribution only influences the sinc functions and so only the high frequency playback performance. The individual gap-null wavelengths $\lambda_0 = g_0$, $\lambda_1 = g_1$ and $\lambda_2 = g_2$ in the arguments of the sinc functions can be replaced (if $\lambda \geqslant \lambda_i$, $\lambda_i \ll p_i$ and μ_{r1}, $\mu_{r2} \gg 1$) by the Westmijze value ($\lambda_i = 1.13\, g_i$) for $\mu_r = 1$ [12.4], or by the numerically obtained values of

Bertram and Lindholm [12.5] for tapes with $\mu_r > 1$ at finite distances from the head. This improves the description of the high frequency performance, although the basic idea of all these models, i.e. discrete steps in the permeability, is disputable for very small gap lengths.

The phase difference between adjacent gaps is $2\pi p'/\lambda$. So, destructive interference, see minima in figure 12.3, occurs at wavelengths λ_{pdn} or frequencies f_{pdn}, where this phase difference equals $(n - \frac{1}{2})2\pi$, hence

$$\lambda_{pdn} = p'/(n - \tfrac{1}{2}) \text{ or } f_{pdn} = (n - \tfrac{1}{2})v/p' , \qquad (12.9)$$

where v is the relative head-tape velocity.

Constructive interference occurs for

$$\lambda_{pcn} = p'/n \text{ or } f_{pcn} = nv/p'. \qquad (12.10)$$

The exact location of minima and maxima in the $|GLF|$ function and in the overall frequency response usually differs only slightly from (12.9) and (12.10), since usually the frequency dependence of all the other terms in $d\Phi/dt$ is much smaller.

In most experiments p_1/μ_{r1} and p_2/μ_{r2} are negligible, $p_1 = p_2$ and $g_1 = g_2$, hence

$$GLF(k) = X_0 + 2X_1 \cos(kp'_1). \qquad (12.11)$$

GLF is purely real now because of the symmetrical multi-gap configuration: no phase distortion occurs.

An easy rule of thumb is obtained from (12.11) for the amplitude of the bumps, ΔV, relative to the average value, $\langle V \rangle$, in GLF when λ_1, $\lambda_2 \ll \lambda$, λ_0;

$$\boxed{\frac{\Delta V}{\langle V \rangle} = \frac{2X_1}{X_0} = \frac{2g_1}{g_0 \, \text{sinc}(\lambda_0/\lambda)}} . \qquad (12.12)$$

It is important to realise that the low frequency ($\lambda \gg \lambda_0$) bumps are almost completely determined by the ratio of the magnetic gap-lengths (g_1/g_0) as defined by (12.8).

Reduction of average output due to residual gaps

Due to the reluctance of the residual gaps, the potential over the main gap, g_0, is reduced with respect to the potential in the case where no residual gaps are present. Physically one expects this potential over the main gap to be responsible for the average output $\langle V \rangle$. Consequently one must expect that the efficiency η_0, defined as

$$\eta_0 \equiv \int_{g_0} \frac{H_x(x, y = 0)\,dx}{NI} , \qquad (12.13)$$

where NI is the number of ampere-turns and g_0 means integrated over only the main gap, will determine the *average* output and will decrease with increasing residual gap-length.

Mathematically this can be understood as follows. The above efficiency definition means that in the normalization condition (12.4) too the integration must be carried out over the main gap, as explained in section 5.3, and not over the main gap and the residual gaps. As a consequence the GLF in (12.5) increases by a factor $\Sigma g_j/g_0$ and changes into a function, GLF_0, with 'average' value $X_0 = \mathrm{sinc}(g_0/\lambda)$ instead of $X_0 = f(\{g_j\})\,\mathrm{sinc}(g_0/\lambda)$. All changes in the average sensitivity of the head $\langle SF \rangle \approx \eta_0 \langle GLF_0 \rangle \approx \eta_0 X_0$ according to (5.19), are now incorporated in η_0 instead of being incorporated partly in GLF and partly in η. It is this property of η_0 that makes η_0 very suitable for investigating the influence of the residual gap-lengths on the average output. In the following, an expression for η_0 as a function of g_1 and g_2 is easily obtained with the aid of expressions derived in chapter 8.

The reluctance of the core near the gap relative to the reluctance of the main gap, $\Delta Z_{\text{int}}/R_{g0}$, is increased by an amount $((\Sigma g_j) - g_0)/g_0$ Substituting this into (8.42b) one obtains

$$\frac{\eta_0'}{\eta_0} = \frac{1}{1 + \dfrac{\eta_0}{\eta_{\text{int}}} \dfrac{(\Sigma g_j) - g_0}{g_0}} . \qquad (12.14)$$

For the conditions leading to (12.11) and (12.12) this result simplifies to

$$\frac{\eta_0'}{\eta_0} = \cfrac{1}{1 + \cfrac{\eta_0}{\eta_{\text{int}}} \cfrac{g_1 + g_2}{g_0}} \ . \qquad (12.15)$$

When $\eta_0 \to 1$ (and hence the internal efficiency $\eta_{\text{int}} \to 1$),

$$\frac{\eta_0'}{\eta_0} \approx \frac{g_0}{g_0 + g_1 + g_2} \approx \frac{\langle V \rangle}{\langle V \rangle + \langle \Delta V \rangle} \qquad (\text{for } \lambda_1, \lambda_2 \ll \lambda, \lambda_0), \qquad (12.16)$$

and when $\eta_0 \to 0$ and $\eta_{\text{int}} \neq 0$, then $\eta_0' \to \eta_0$, see Fig. 12.2.

The reduction factor for the high-efficiency head approaches the ratio between the peaks and the average value at large wavelength, while the average output of the low-efficiency head is hardly affected by the presence of residual gaps. Both extremes are easily understood when it is realized that the equivalent circuit of any high-efficiency head approximates to an ideal 'voltage source' and the equivalent circuit of the low-efficiency head (assuming $\eta_0 \ll \eta_{\text{int}}$, as is usually the case then) approximates to an ideal 'current source'.

The important conclusion from the above results is that the average

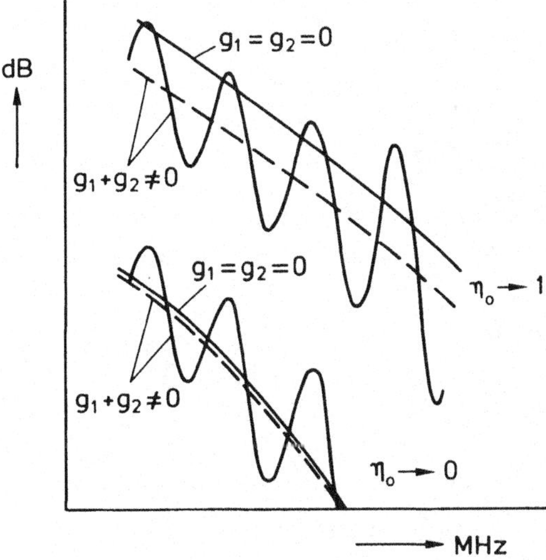

Fig. 12.2 Sketch of the reduction of the average output due to the potential loss over the residual gaps for heads with a large main-gap efficiency, $\eta_0 \to 1$, and for heads with a small main-gap efficiency, $\eta_0 \to 0$.

output of a head with a large main-gap efficiency, η_0, suffers much more from the potential loss over the residual gaps than the output of a head with a low main-gap efficiency usually does.

Thick metal layers

We have so far assumed that the metal layers are thin, so that the divergence of the flux over the layer thickness was negligible and consequently the flux density in the side gaps approximated the flux density in the metal layers and the main gap.

In the case of thick and high permeability metal layers, the divergence of the flux in the metal layer is not negligible if the metal layers
- widen [12.15] or
- cover the side faces of the track-defining part of the head (Fig. 12.11) or
- continue in the coil chamber (see also Fig. 12.11).

In the calculation of the field at $y = 0$, $H_x(x, y = 0) = B_x(x, y \uparrow 0)/\mu_0\mu_{rx}(x)$, the reduction in the flux density must now be taken into account. Consequently the peaks and dips in the output versus frequency curve due to the side gaps *decrease by a reduction factor equal to the ratio between B_x at the side gaps and B_x at the main gap.* (The same is true for the peaks and dips due to a low metal-layer permeability, considering in this case the flux density in the metal layer instead of that at the side gaps.)

12.1.4 Playback experiments

We tested the expressions by using a ferrite head with sendust (Fe, Si, Al) layers and non-magnetic spacers (SiO_2) between sendust and ferrite.

Figure 12.3a shows the output spectra for a head-tape velocity of 3.14 m/s in an experimental situation on 1140 Oe metal-powder (MP) tape. The output spectrum of the MIG head shows its second, third and fourth dips at 2.3, 3.8 and 5.3 MHz respectively, thus corresponding well according to (12.9) to $p_1' = p_2' = 2.1$ µm as measured with a scanning-electron microscope (SEM). The relative amplitude of the bumps increases with increasing frequency, corresponding to (12.12) for $g_0 = 0.40 \pm 0.04$ µm (SEM: 0.38 ± 0.01 µm) and $g_1 = 0.035 \pm 0.002$ µm (SEM: 0.02 ± 0.01 µm) for all three bumps.

Figure 12.3b shows the playback spectrum on CrO_2 tape of a MIG head with large residual gaps. The gap dimensions as determined from SEM pictures are $g_0 \approx 0.18$, $p_1 \approx 2.03$, $p_2 \approx 2.16$ and $g_1 \approx g_2 \approx 0.07$ μm. Introducing a total (recording and playback) distance loss $\exp(-2ka)$ with $a = 0.09$ μm we obtained a good fit for $g_0 = 0.2$, $p_1 = 2.03$, $p_2 = 2.08$ and $g_1 = g_2 = 0.085$ μm with the aid of (12.6). Again a difference of approximately 150 Å is measured between the magnetic gap-length and the layer thickness of the tiny gaps. The fit would have been almost perfect if $p_1 = 2.11$ and $p_2 = 2.16$ μm had been chosen, which is still within the accuracy of the SEM pictures. The slight dispersion, i.e. $p_1 \neq p_2$, is the main reason for the gradual decrease of the relative bump-amplitude with increasing frequency. Dispersion mainly affects the high frequencies, since small differences in gap locations cause large differences in phase only at the highest frequencies. A small part of this effect is caused by the individual gap-losses of g_1 and g_2 i.e. $\mathrm{sinc}(\lambda_1/\lambda)$ which appears in the numerator of (12.12) for a finite residual gap-length. Since $g_1 + g_2 \approx g_0$ for this head, almost complete destructive interference occurs in the dips, see Fig. 12.3b.

It will be clear that the above way of evaluation of the magnetic gap-lengths of the tiny gaps through the output spectrum is (much) more precise than through the SEM measurements. The accuracy approaches the accuracy of the determination of the magnetic gap-length

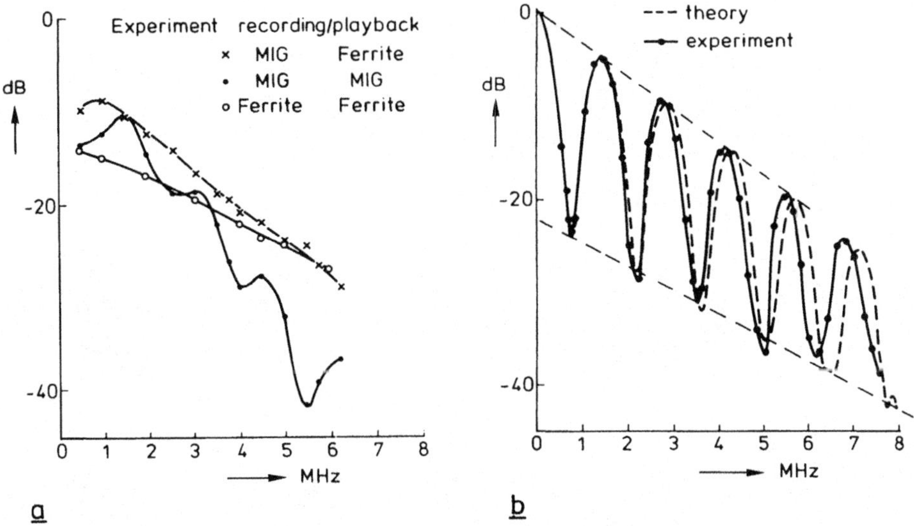

Fig. 12.3 a) Cross measurements, $g_1 + g_2 \ll g_0$. (90 kA/m (1140 Oe) tape).
b) Strong playback dips when $g_1 + g_2 \approx g_0$. (CrO_2 tape).

of the primary gap, which is about ten per cent. The magnetic gap-lengths of the tiny gaps clearly are larger than the thickness of the oxide spacer between sendust and ferrite. With only one (large) gap it would never be possible to determine that accurately the difference between layer thickness and magnetic gap-length as defined in (12.8). This method of characterizing a small gap by way of a large gap is of importance in going to higher densities where smaller gaps will be used.

By subtracting the crossed curve from the dotted curve in figure 12.3a a rough idea of the poor playback behaviour of this MIG head relative to the ferrite head is obtained. However, azimuth differences (alignment loss) and track-following and adjustment usually influence the cross measurement (crossed curve). In either case, the primary gap (g_0) of this MIG head is clearly too large, as follows from the steep decrease of the output with the frequency.

In Fig. 12.4a the precise playback behaviour (P) of a MIG head, with

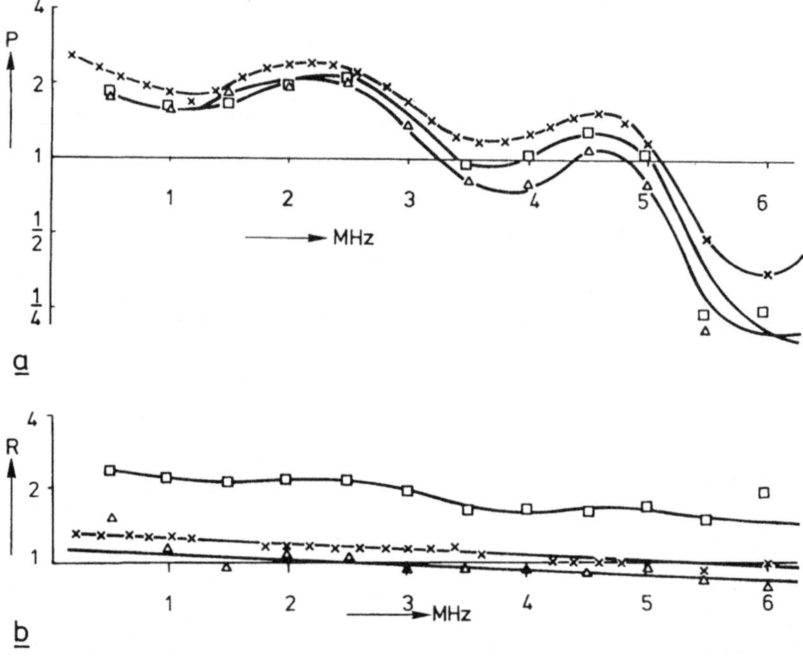

Fig. 12.4 a) Playback figures (P) of a MIG head relative to a ferrite head.
        b) Recording figures (R) of a MIG head relative to a ferrite head.
          × Bias recording on CrO_2.
          △ Direct recording on CrO_2.
          □ Direct recording on 117 kA/m (1480 Oe) MP tape.

a smaller gap than the one in figure 12.3a, is given relative to the playback behaviour of a reference ferrite head. The way we determine P (relative playback behaviour) is described in chapter 10. These figures are completely free of influences of alignment loss and track adjustment.

The aim of the measurement on different tapes and under different bias conditions, see the caption of figure 12.4, was to verify if the playback behaviour would remain unaltered. Only at higher frequencies is some difference observed between the biased and unbiased measurement, which is probably due to the irreproducibility of the measurement. No other physical reason can be found for the difference, supposing the heads act linearly.

12.1.5 Optimum recording performance

The recording process is strongly non-linear. Only fields of the order of H_c and higher are able to write information into the tape. When the field from the primary gap is optimized for recording, then the fields from the tiny gaps at the tape coating are still well below the threshold field H_c. So the recording does not suffer from the residual gaps. Reading with a ferrite head shows that indeed no interference effects occur, see crossed curve in figure 12.3a. This is confirmed again in figure 12.4b. The recording figure R, independent of alignment loss and track adjustment, is used for determining the recording performance of the MIG head relative to the ferrite head. No significant dips or peaks are observed, either in optimum direct recording on metal-powder (MP) and CrO_2 tape or in optimum bias recording on CrO_2 (bias current 20 mA turns at 10 MHz: optimum for recording of 1 MHz signals). The optimum bias signal linearizes only the writing characteristic of the primary gap.

The recording behaviour of a MIG head is clearly more favourable than that of a ferrite head on 1480 Oe MP tape. This was already the case on 1140 Oe MP tape, as shown in figure 12.3a.

12.1.6 Recording at higher currents

For recording with (bias) currents far above the optimum values, writing with the tiny gaps becomes possible if saturation around these gaps does not start too soon. We therefore selected a head which had residual gaps of about 0.07 µm (SEM), and examined the recording behaviour on a low coercive tape (CrO_2). $p_1' \approx p_2' \approx 2.25$ µm (SEM), giving playback dips at 0.7 MHz, 2.1 MHz, etc. A ferrite head was again used as the playback head. The result for increasing recording currents is given in figure 12.5a. Since the difference between primary and residual gaps is fairly small, optimum recording is already influenced by interference, see curve RD_1 for $I_{opt} = 30$ mA.

The frequency of this first recording dip equals that of the first playback dip (PD), both located at 0.7 MHz. At 137 mA a very sharp (almost completely destructive interference) dip occurs at 0.9 MHz (RD_2) instead of 0.7 MHz. This must be due to a larger shift of the recording zone of the large (primary) gap in relation to the shift of the recording zone of the small (residual) gap at increasing recording currents, see figure 12.5b. The decrease of the effective distance between the recording zones $\Delta_{rz} \equiv p_1' - p_{1r}'$ can be easily determined from the measurements by rewriting (12.9) and (12.10), to

$$\Delta_{rz} = \frac{(n - 1/2)v(f_{rdn} - f_{pdn})}{f_{rdn}f_{pdn}} = \frac{nv(f_{rcn} - f_{pcn})}{f_{rcn}f_{pcn}} \qquad (12.17)$$

where $f_{rdn}(I)$ and $f_{rcn}(I)$ are the measured frequencies at the n-th recording dip and peak and I is the recording current. At 113 mA this Δ_{rz} is calculated to be 0.50 µm. Experiments like these look interesting from a theoretical point of view, since shifts of recording zones as a function of recording current, tape and wavelength can be accurately measured in this way to verify theoretical predictions.

12.1.7 MIG-head with 3 µm sendust layer

In this section the frequency spectra obtained from cross measurements with a MIG head with a thicker sendust layer than in the previous experiments, on MP tape, are considered for the following three purposes:

- the visibility of anomalous recording effects at optimal recording,
- the demonstration of the accuracy of the playback model,
- the demonstration of the power of the cross measurement in separating recording and playback effects.

Even when the adhesion between sendust and ferrite is perfect, the residual gaps may be so large that some writing with the residual gaps already appears at optimum write currents, as in the experiment of Fig. 12.5. This might happen especially when the head-to-tape distance, a, is small. In Fig. 12.4b this is possibly the case for the MP tape, because some interference effects seem to be visible. Another reason might be that the M_r/H_c ratio of MP tapes is a little smaller and hence the optimum writing depth a little larger than in CrO_2 tape. Consequently the field from the residual gap at the head side of the coating, relative to the coercive field of the tape is a little larger (at the optimum writing

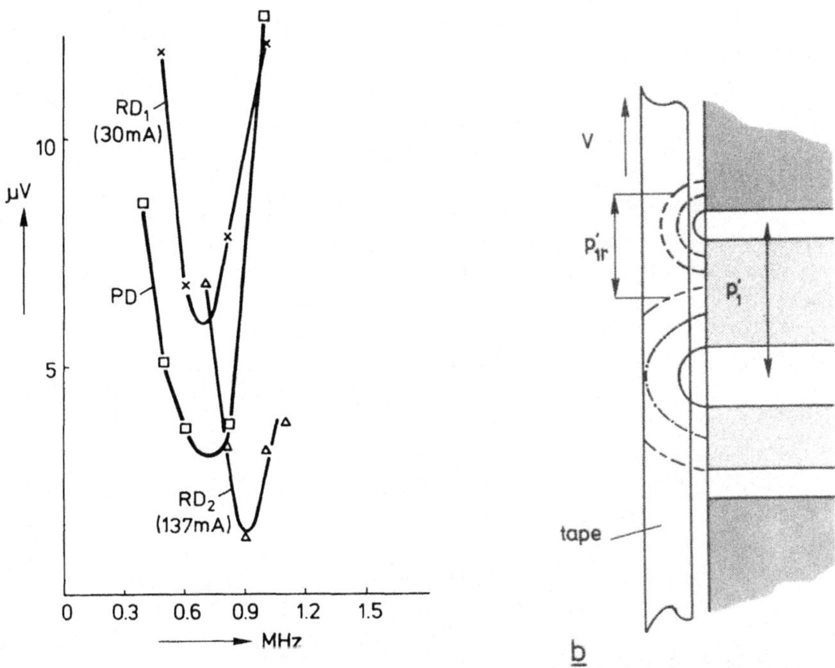

Fig. 12.5 a) Shift of the recording dip from RD_1 to RD_2 due to an increased recording current. PD is the playback dip.

b) Sketch of the decrease $\Delta_{rz} \equiv p'_1 - p'_{1r}$ of the distance between recording zones of large (primary) and small (last) gap.

—— part of the $H_x = H_c$ curve for $I = I_1$.

—·— part of the $H_x = H_c$ curve for $I = 2 \cdot I_1$.

——— part of the $H_x = H_c$ curve for $I = 3 \cdot I_1$.

current) than in CrO_2. Small changes in the field strength just below the 'threshold field' H_c may have a considerable influence on the output.

In order to examine the interference effects in the write process more closely, we used a MIG head with sendust layers of $p \simeq 3$ μm instead of $p \simeq 2$ μm. The periods in the frequency domain are therefore reduced, see e.g. Fig. 12.6, and so easier to observe. Additionally the number of measurements per MHz was increased from 2 to 10. Optimum direct recording is applied on the MP tape. This means that before each measurement the write current is increased until the maximum read signal is found. The write current at this maximum, the optimal current, is subsequently used in cross measurements. Fig. 12.6 shows the results of the four cross measurements, U_{11}, U_{12}, U_{21} and U_{22}, that are possible with the MIG head under test (head 1) and a ferrite head (head 2) which serves as the reference head. The first subscript indicates the write head and the second subscript indicates the read head. As an example, U_{12} is the e.m.f. of the ferrite head after writing with the MIG head.

The playback and recording figures, P and R, resulting from the data in Fig. 12.6, are given in Fig. 12.7 as a function of the frequency. The recording figure shows not only the superior writing performance of the sendust MIG head relative to the ferrite head on MP tape, but also small interference effects in the write process. The bumps are shifted to the right with respect to the bumps in the playback-figure curve.

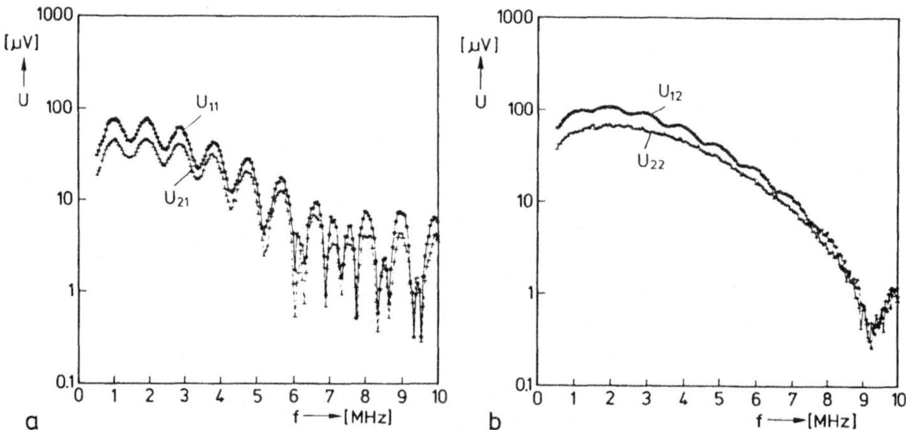

Fig. 12.6 Cross measurements of a sendust MIG head and a ferrite head against MP tape. The head-to-tape velocity equals 3.14 m/s.

Note that the gap-length of the ferrite head easily follows from the frequency (9 MHz) where the playback figure tends to infinity, corresponding to the gap-null frequency in Fig. 12.6b. In the case of the MIG head, see Fig. 12.6a and 12.7, a clear gap-null frequency or corresponding frequency where the playback figure tends to zero is not visible. To determine this and other parameters of the MIG head, simple model calculations are necessary. In Fig. 12.8 the curve that results from simple model calculations (dashed line) is added to the experimental results (continuous line). The input parameters of the model are given in the caption of Fig. 12.8 and have been found by trial and error. Note the 6% difference between p_1' and p_2', necessary to obtain the good fit. The head-to-tape distance $a = 0.06$ µm is low compared to the head-to-tape distance $a = 0.09$ µm in the experiment of Fig. 12.3, where no interference effects in the write process are visible. The sensitivity for parameter changes is very large. Hence the only small inaccuracies are from the model and originate from the assumed exponential write-spacing loss and the simplified main-gap loss function.

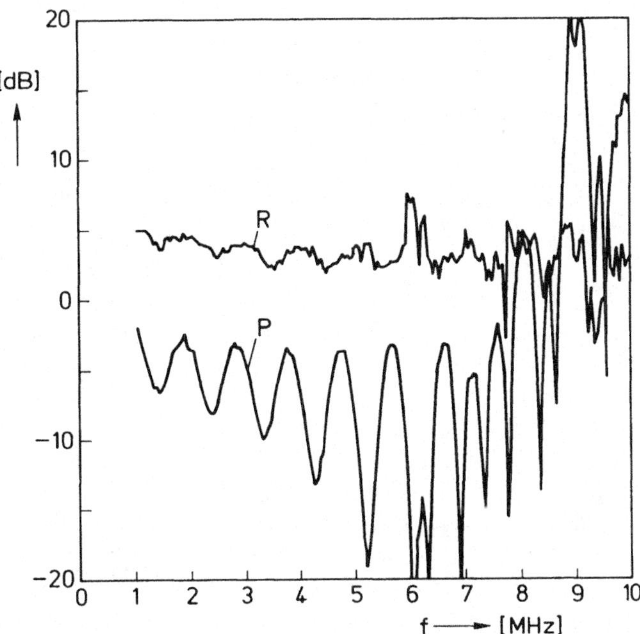

Fig. 12.7 Playback and recording behaviour of the sendust MIG head relative to the ferrite head, as calculated from the experimental data in Fig. 12.6.

12.1.8 Minimization of bumps

The best way to minimize bumps in the output spectrum is of course to minimize the gaps at the interfaces through technological solutions. Apart from these efforts it is also possible to reduce the influence of the tiny gaps by using only one sendust layer, preferably on the side where the tape leaves the gap [12.6]. This reduces the bumps by exactly 50%, see theoretical curve in Fig. 12.9. Choosing $p_1 \neq p_2$ is another alternative. However, only part of the spectrum can be equalized in this way, see curve for $p_2/p_1 = 1.25$. For these calculations all other terms resulting from (12.2) have been taken into account. It is beyond the scope of this chapter to go into details of these calculations. The simple model for the write process given in chapter 7 (direct longitudinal recording) was included. The parameters in the calculations were: $g_0 = 0.3$ μm, $g_1 = g_2 = 0.04$ μm; head-tape distance $a = 0.06$ μm; coercivity of the tape $H_c = 115$ kA/m (1480 Oe), relative isotropic permeability of the tape $\mu_{rt} = 1.75$, remanent induction of the tape B_r

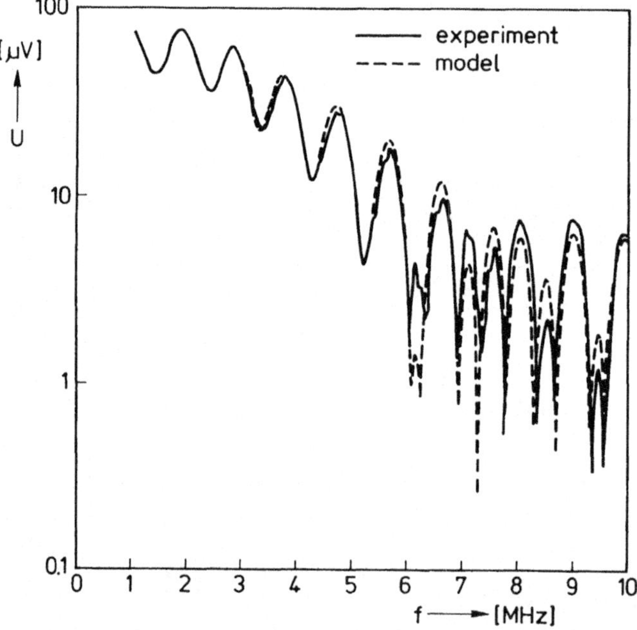

Fig. 12.8 Comparison between the measurement (playback and recording with the sendust MIG head) and model (interference effects in the write process neglected, write + read spacing loss e^{-2ka} and *GLF* given by (12.6) with $g_0 = 0.4$ μm, $g_1 = g_2 = 0.06$ μm, $p_1' = 3.05$ μm, $p_2' = 3.25$ μm and head-to-tape distance $a = 0.06$ μm).

= 0.256 Tesla (2560 Gauss) characterizing usual metal-powder tape; p_1 and p_2 are given in the legend of figure 12.9 in μm.

12.1.9 Conclusions

Metal-in-gap heads can write much stronger signals in 1140 and 1480 Oe tape than ferrite heads.

In the playback spectrum of these heads bumps often appear, due to residual gaps between metal and core. The main purpose of this article was to explain this phenomenon in detail. The excellent correspondence of the model with a large number of measurements made it possible to develop MIG heads with $p_1 \neq p_2$ giving a partly equalized spectrum. A technological solution to the residual-gap problem remains the best answer, since phase distortion is associated with asymmetrical multi-gap configurations and a non-flat amplitude spectrum with symmetrical gaps configurations. Non-sinusoidal signals are distorted by a non-linear phase shift (phase-distortion) as well as by a non-flat amplitude spectrum (without phase distortion). In the meantime residual gaps have been almost completely suppressed by adequate technological means.

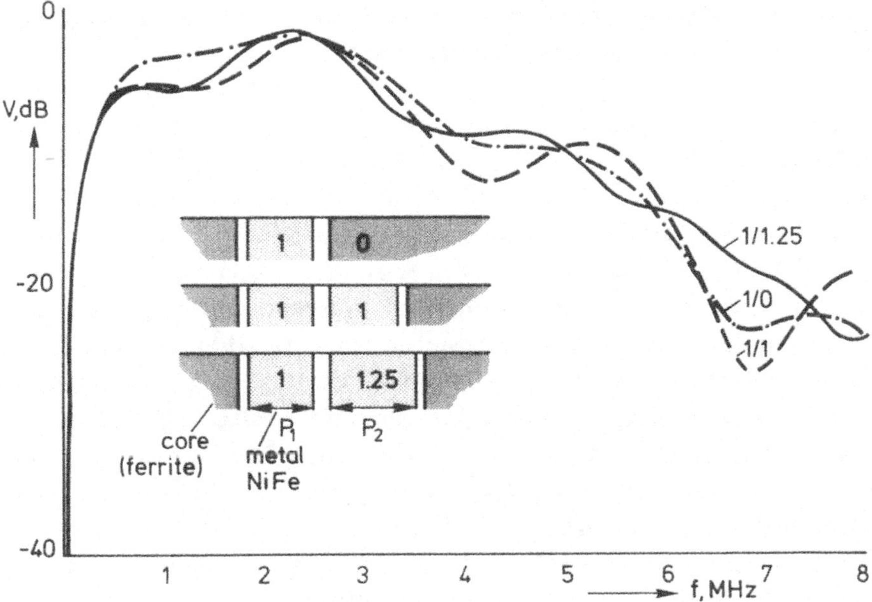

Fig. 12.9 Influences of $p_1 \neq p_2$ on the bumps in the spectrum.

The magnetic gap-length of very small gaps (even below 200 Å) can be accurately determined with the aid of the derived expressions when a larger gap is incorporated in the same head.

12.2 Permalloy/sendust metal-in-gap head

12.2.1 Abstract

A Metal-In-Gap (MIG) head with sendust gap cladding has pseudo gaps at the ferrite/sendust interfaces. The pseudo (residual) gaps are absent if permalloy is used for a gap cladding, but, in contrast to sendust, considerable differential wear of the permalloy occurs, which causes exponential playback and nearly uniform recording losses.

By first depositing a thin permalloy layer on the ferrite and subsequently a thicker sendust layer on the permalloy, no pseudo gaps occur. At the gap the relative wear is slight, due to the presence of the sendust. The gap cladding has minimal residual stress in order to avoid a decrease of head efficiency due to induced stress in the ferrite. The $SiO_2/Mo/Au$ gap is prepared by low-temperature thermodiffusion. This results in a magnetically sharp gap with a reproducible length.

With this permalloy/sendust MIG head a high-density recording head for in-contact recording has been realized.

12.2.2 Introduction

Conventional video recording is performed with tapes with a low coercive field, H_c, such as Co-γFe$_2$O$_3$ ($H_c = 55$ kA/m) and CrO$_2$ ($H_c = 52$ kA/m), in combination with conventional video heads of monocrystalline ferrite. The storage density on the tape can be enhanced by using smaller wavelengths. For this purpose, tapes with higher coercivity, such as Metal Powder (MP) tapes ($H_c = 115$ kA/m), have to be applied. With the conventional video head the signals cannot be recorded (i.e. written) on these high-coercivity tapes because of the low saturation, B_s, of the ferrite (0.5 T).

With a Metal-In-Gap (MIG) head [12.3] or a Sandwich head (e.g. [12.7]), where soft magnetic materials with a high saturation (0.8-1 T, e.g. permalloy, sendust and Co-based alloys) are applied in the gap area, a sufficiently high writing field for recording on these tapes can be provided. The preparation of the MIG head shows many similarities with that of the conventional ferrite head and is more simple than that of the Sandwich head.

In the most simple design of the MIG head the interface of the ferrite and the gap cladding material is parallel to the gap plane. Permalloy or sendust seem to be suitable for application in this type of MIG head, but the two materials each have their own disadvantage. If permalloy is used, a non-negligible level difference occurs between the permalloy and the surrounding ferrite, due to differential wear after in-contact recording with MP-tape, cf. Fig. 12.10a. A sendust MIG head shows very slight differential wear, but a thin layer with poor magnetic properties is present at the transition of the ferrite to the sendust. This results in residual (pseudo) gaps in the head, cf. Fig. 12.10b.

In the literature various geometrical solutions for avoiding the negative effect of the pseudo gaps in sendust MIG heads are proposed [12.8-12.11], see also section 12.1.8. All these solutions need more preparation stages and thicker gap claddings than the MIG head with a smooth ferrite/gap-cladding interface parallel to the gap plane. Moreover they only avoid the effect of the pseudo gaps, but these are still present and reduce the head efficiency slightly.

In a technologically more simple and magnetically essentially better solution, see also the european patent application 0246706 [12.15], the

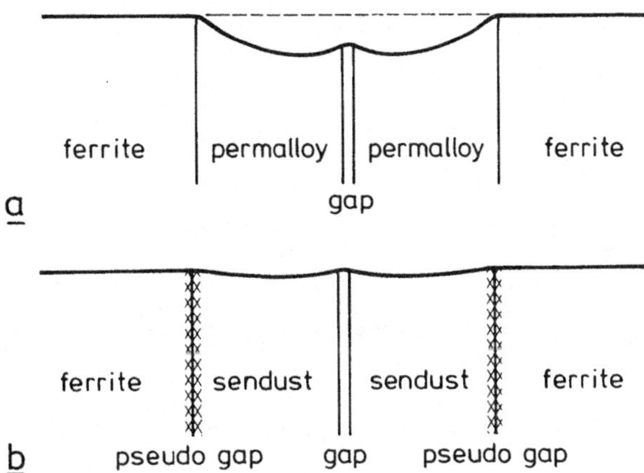

Fig. 12.10 Metal-In-Gap Head
a) Wash out of the gap cladding, e.g. Permalloy MIG head.
b) Pseudo gaps at the ferrite/gap-cladding interface, e.g. Sendust MIG head.

advantages of permalloy and sendust are combined and the disadvantages of both are eliminated, cf. Fig. 12.11. By first depositing a thin permalloy layer on the ferrite and subsequently a thicker sendust layer on the permalloy, no pseudo gaps occur at either the ferrite/permalloy or the permalloy/sendust interfaces. The relative wear of the permalloy may still be considerable, but the relative wear at the gap is slight, due to the presence of the sendust.

12.2.3 Requirements regarding the gap cladding properties

The gap cladding layers have to satisfy a number of requirements for a good head performance. Because the major part of the gap cladding will consist of sendust, it is primarily the sendust that will be tested in respect of these requirements.

Stress

Residual stresses in the sendust occur, due to the sputtering process and the difference in thermal expansion coefficients of the sendust and the ferrite. Without special care, the average stress, $\langle \sigma_m \rangle$, in the sen-

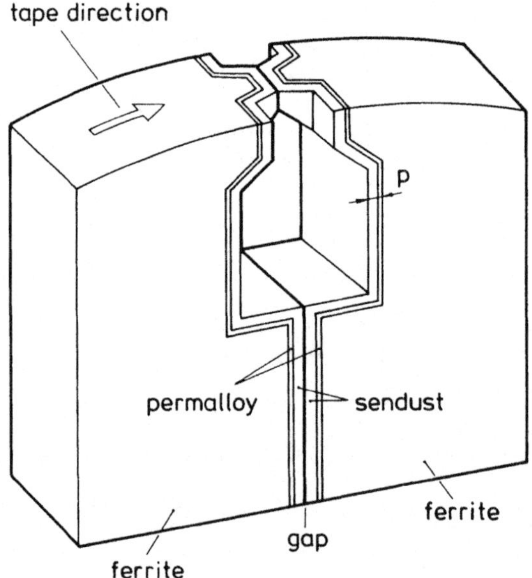

Fig. 12.11 Permalloy/Sendust Metal-In-Gap Head (PS-MIG Head)

dust may be as large as some GPa, i.e. the integral stress may be several GPa μm in thin layers. The influence of stress on the permeability and hence on the head performance depends on the magnetostriction. The reduction of the permeability in the sendust is such that it can be disregarded in the calculation of the gap-loss function, *GLF*; see the section on permeability. This reduction can also be disregarded in the calculation of the head efficiency. The balancing stress in the ferrite is spread out over a large area on the left and right of the gap, with a total length of a few times the trackwidth. Thus the change of the *GLF* is negligible and the only effect is a decrease in the efficiency, see section 8.1.6.3. The integral of the balancing stress is the negative of the integral of the residual stress $\langle \sigma_m \rangle$ $(p_1 + p_2)$ in the metal layers, but has much more influence because of the large magnetostriction constant of the MnZn ferrite ($\lambda_{100} = -10^{-5}$) compared with that of the sendust (which is of the order of -10^{-6}). The efficiency, η_0', of the head is reduced with respect to the efficiency of the head, η_0, in the absence of stresses and follows from the equations (8.42b) and (8.46) in section 8.1.6.3:

$$\frac{\eta_0'}{\eta_0} \geq \frac{1}{1 + \dfrac{3|\lambda_{100}|\,|\langle \sigma_m \rangle|}{\mu_0 M_s^2 g_0}\,(p_1 + p_2)\,\dfrac{\eta_0}{\eta_{\mathrm{int}}}} . \qquad (12.18)$$

where g_0 is the main gap length,
p_1 and p_2 are the sendust film thicknesses,
M_s the saturation magnetization of the ferrite,
η_{int} the internal efficiency of the head: this is approximately the limited efficiency of the head when g is large, see also section 8.1.6, which is about 0.9 for our MIG (ferrite) heads.

The equals sign is valid if the magnetization changes take place by rotation of the magnetization vectors out of the (001) plane and not by wall displacements. In this 'worst case' situation $\eta_0'/\eta_0 \sim 0.7$ if $\langle \sigma_m \rangle = 0.2\,\mathrm{GPa}$, $p_1 + p_2 = 6\,\mu\mathrm{m}$, $g_0 = 0.3\,\mu\mathrm{m}$, $\mu_0 M_s = 0.5\,\mathrm{T}$ and $\eta/\eta_{\mathrm{int}} = 2/3$.

Thus stresses in the metal film with integral values larger than 1 GPa μm may have a noticeable influence on the efficiency of MIG ferrite heads.

Permeability

The permeability of the gap cladding does not have to be high because the thickness of this layer is small in the present heads. The finite permeability of this layer mainly influences the *GLF* as described by eq. (12.6). When the (simple) model for the writing process (for the write current being optimized for each wavelength) is incorporated and the total efficiency of the head, η ($\equiv H(g_0 + g_1 + g_2)/NI$, is taken to be constant, then the output of the head is calculated to follow the curves in Fig. 12.12 for a 1 µm thick layer. For this thin layer a relative permeability of only 50 suffices, while for thicker layers proportionally higher permeabilities are recommended. In comparison with the occurrence of residual gaps, (dashed curve), all peaks and dips, starting with the first dip, are shifted a quarter of the period to higher frequencies. Another difference is that the amplitude of the peaks and dips decreases with frequency, whereas residual gaps result in an increase in this amplitude for equal film thicknesses.

Differential wear

Differential wear induces an increase in the head-to-tape distance, a, and hence influences the recording as well as the playback process. The

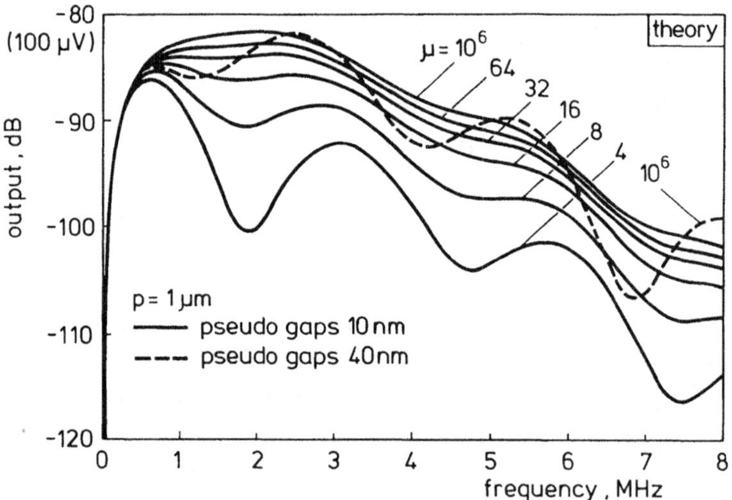

Fig. 12.12 The calculated differing pattern of peaks and dips due mainly to a low film permeability (full curves) and due only to large residual gaps (dashed curve).

Tape parameters (MP): H_c = 115 kA/m (1450 Oe), μ_r = 1.75 (isotropic), B_r = 0.256 T. Head parameters: g_0 = 0.3 µm, B_s = 0.8 T, $p_1 = p_2$ = 1 µm, η = 1, trackwidth: 20 µm, number of turns: 18; Head-to-tape distance: 0.06 µm. Relative head-to-tape velocity: 3.14 m/s.

effects are calculated, using the simple model for the assumed longitudinal recording process, described in chapter 7. The recording loss is due to an exponential loss, $e^{-0.7ka}$, which is most dramatic at the higher frequencies, and a loss due to the finite writing depth of the head, which is most dramatic at lower frequencies. Together, for the present tape and head parameters, see Fig. 12.13, they result coincidentally in an almost uniform recording loss at all frequencies of interest. At greater gap lengths and/or saturation of the head or lower coercive field of the tape, only the exponential decay is visible. The playback distance loss is almost completely determined by the common factor e^{-ka}.

The experimental results in Fig. 12.14 show a remarkable agreement with the theoretically predicted losses, in spite of the simple recording model used in the calculations (an explanation of the cross-measurement analysis method which is used to find the exact degradation in the playback (P) and recording (R) processes, is given in reference [12.12] which is incorporated in chapter 10 of this book).

Irregular and gradual gap

When the permeability of the sendust near the gap changes gradually from a small value ($\mu_r \sim 1$) at the gap-facing surface of the sendust to

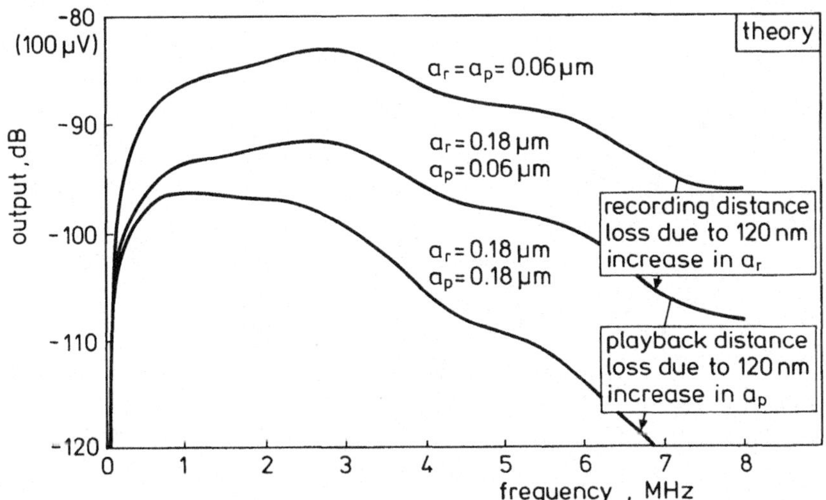

Fig. 12.13 Calculated effects of an increase in the head-to-tape distance during recording and during playback of a video-like MIG head on MP tape. (a_r and a_p: recording and playback head-to-tape distances respectively).

Tape parameters: see caption of Fig. 12.12. Head parameters: $g_0 = 0.2$ µm, residual gaps, $g_1 = g_2 = 10$ nm, $B_s = 0.8$ T. Relative head-to-tape velocity: 3.14 m/s.

a value $\gg 1$ a distance Δg away from the gap, then the resulting bad gradient degrades the playback as well as the recording of the information on the tape. This is described in sections 5.7.9 and 7.3.1 by introducing an extra effective head-to-tape distance Δa_{eff}:

$$\boxed{\Delta a_{\text{eff}} = \Delta g/\pi} \ . \tag{12.19}$$

During playback this causes a Gradual Gap Loss

$$GGL_p = e^{-k\Delta a_{\text{eff}}} \tag{12.20}$$

and during recording a loss of about

$$GGL_r \sim e^{-0.7k\Delta a_{\text{eff}}}, \tag{12.21}$$

i.e. analogous to an increase of the actual distances during playback and recording, a_p and a_r, with an amount Δa_{eff}.

The losses due to gap irregularities, GIL, are described by Mallinson [12.13], see also section 5.9.2.

$$GIL_p = e^{-(k\sigma)^2/2}, \tag{12.22}$$

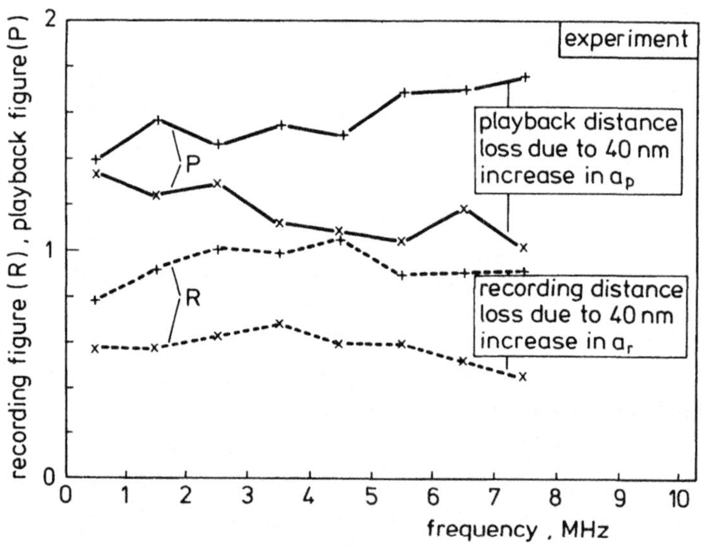

Fig. 12.14 Experimentally determined degradation of the playback figure, P, and recording figure, R, for a video-like MIG head, $g_0 = 0.2\ \mu$m, on a 115 kA/m (1450 Oe) MP tape with $B_r = 0.256$ T.

where σ is the spread of the recording- or playback- gap-centre position relative to a straight line. The assumption of a constant gap length is essential in the calculation, i.e. the gap side positions are completely correlated. When the gap is formed by bonding two blocks with irregular surfaces with spread σ, then both the gap-centre position and the gap length are irregular. This complicates the calculations of the playback effects significantly. However, for almost all wavelengths of interest ($\lambda > 1.6g$) the result is simple, as shown in section 5.9.3:

$$\boxed{GIL \sim e^{-(k\sigma)^2/4}} \, . \tag{12.23}$$

The loss (in dBs) is half as much as in the case of the constant gap length!

12.2.4 Material

The gap cladding layers are deposited on the ferrite substrate by sputtering. Because 80-85% of the gap cladding consists of sendust, the sputtering of the sendust has been optimized, with the most important parameter being the substrate temperature, T_s, (varying from 150 to 500°C). Since layer properties depend on the material, texture and roughness of the substrate, it is necessary to measure the properties of the layers on the ferrite substrate. The integral magnetic properties have been determined at approx. 300°C, which is beyond the Curie temperature of the ferrite (180°C). Hence, the ferromagnetic contribution of only the soft magnetic layer(s) on the ferrite is measured (the permeability of the permalloy/sendust layers on glass, measured at 300°C, was approx. 20-25% higher than when measured at room temperature).

Fig. 12.15 shows the relationship between the average permeability of the sendust layers, measured at 4.5 MHz and 300°C, and the residual stress. It is observed that the compressive stress decreases as T_s increases. At about 450°C the stress has fallen to zero: above this temperature the stress becomes tensile. For the layers with minimum (zero) stress, the permeability reaches its maximum value of about 800.

Stress and permeability of a double layer of 0.5 µm permalloy and 3.0 µm sendust on the ferrite substrate are comparable with those of only a sendust layer on the ferrite substrate.

The permalloy/sendust layers hardly have any magnetic anisotropy in the film plane. The saturation is of the order of 1 T and the coercivity is approx. 120 A/m. These results correspond fairly closely with the ones, given in [12.14]. It can be concluded that the properties of the double layers of permalloy/sendust, sputtered at the optimal temperature, satisfy the aforementioned requirements for application in MIG heads.

12.2.5 Head technology

The substrate material is a polished monocrystalline MnZn ferrite. Flat substrates are used when the trackwidth and coil chamber are cut by lasering after the head core preparation; otherwise the head preparation starts on a preprofiled substrate.

The permalloy and sendust layers (0.5 and 3.0 μm respectively) are sputtered with an RF diode sputtering apparatus. The composition of the permalloy target is 80/20 wt% NiFe, that of the sendust target 85.2/9.1/5.7 wt% FeSiAl. The permalloy and sendust are sputtered at 450°C at a rate of 1 and 0.5 μm/hr respectively.

For the gap preparation a multilayer of successively SiO_2, Mo and Au is sputtered on the sendust. The actual gap adhesion is ensured by

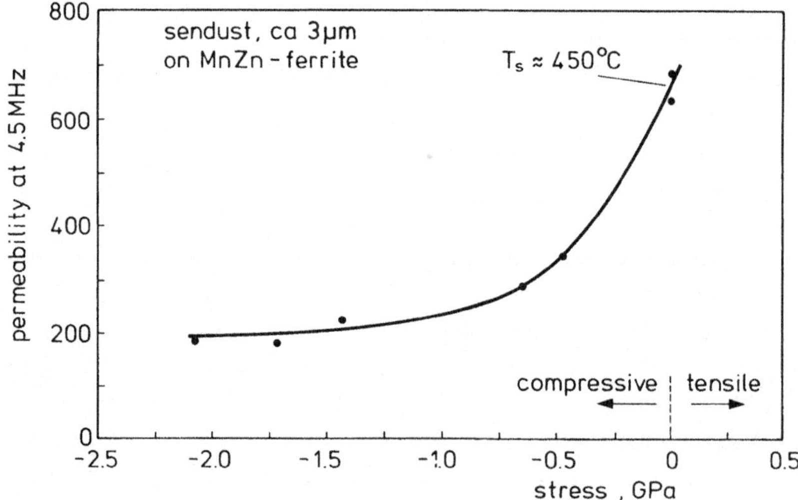

Fig. 12.15 The influence of stress on the permeability of sendust layers on ferrite (T_s: substrate temperature during sputtering).

thermodiffusion of the Au layers. The Mo layer serves as an adhesive layer for the Au. SiO_2 is used as a diffusion and electrochemical barrier between the metal layers of the gap and those of the gap cladding. The thermodiffusion process can be performed successfully at temperatures in the range of 150-400°C and at pressures of 10-70 MPa. The great advantage of this type of gap is its reproducible length and, in contrast to a glass gap, its low preparation temperature. Its magnetic sharpness is of the utmost importance for the playback and recording performance of the head, as outlined in the chapters 5 and 7 respectively. A typical gap with a length of 0.27 μm consists of 2 × (75 nm SiO_2 + 30 nm Mo + 30 nm Au) and is composed at 200°C and at 30 MPa.

In Fig. 12.16 SEM pictures are shown of permalloy/sendust MIG heads, prepared with a preprofiled substrate (a) or with lasering (b). In Fig. 12.16c, where the gap area is shown in more detail, the two gap cladding layers and the various gap layers can be distinguished.

12.2.6 Experimental results and discussion

The frequency characteristics of three different MIG heads with either permalloy, sendust or permalloy/sendust gap cladding are shown in Fig. 12.17. The curves have been obtained with fixed heads in contact with rotating MP tape (H_c = 115 kA/m) at 3.14 m/s.

The indicated gap-lengths have been determined by measuring the gap-null wavelength from a recording experiment and dividing this value by 1.15 (this factor approaches the Westmijze factor). In all three cases this is the total thickness of layers in the gap (SEM).

In the beginning of the research on MIG heads with MoAu gaps it was hard to measure the gap-null wavelength accurately because of its low value, roughly half the intended gap-length, and because of a larger head-to-tape distance than usual. Apparently the Mo layers were (poor) soft magnets and hence decreased the gap length and also made the magnetic gap more gradual. These effects disappeared after the application of the SiO_2 barrier between the Mo and the permalloy or sendust gap cladding.

The permalloy MIG head has a very high output at low frequencies, but at 4.5 MHz it has degraded significantly. This degradation is partly due to the rather large gap-length (0.35 μm), but the major cause is the occurrence of the considerable relative wear of the permalloy. In many

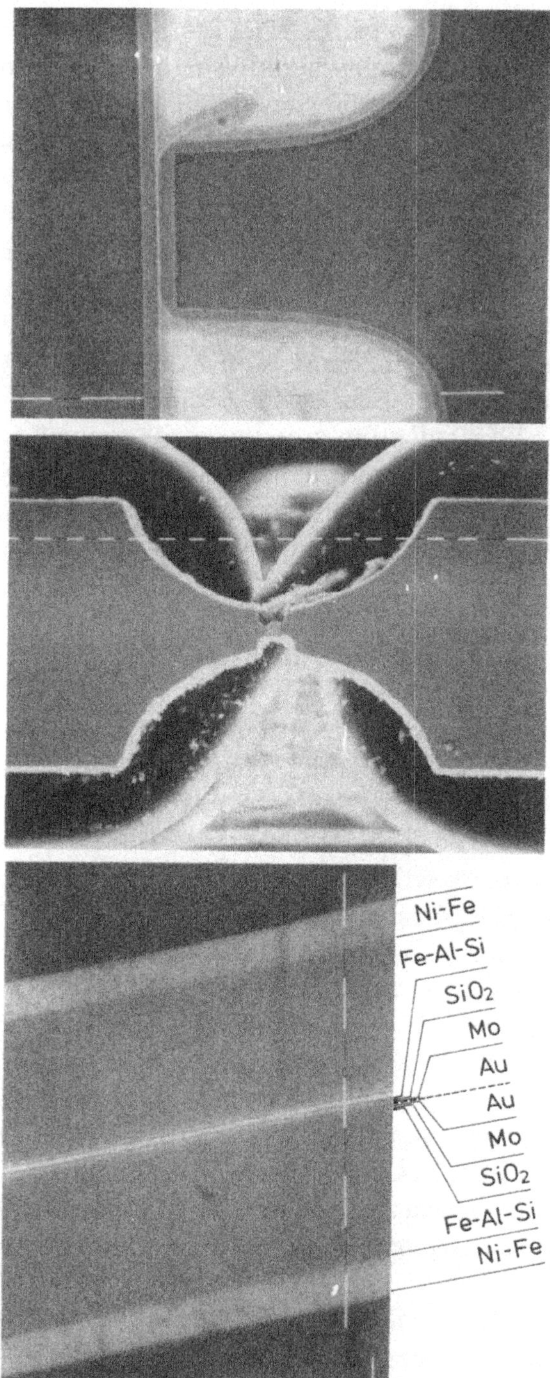

Fig. 12.16 SEM pictures of PS-MIG heads:
a) Preprofiled (10 μm/div.).
b) Lasered (10 μm/div.).
c) Gap area (1 μm/div.)

cases a differential wear in a permalloy MIG head of 60-90 nm has been observed, resulting in an additional output loss, specifically at the higher frequencies (see above).

The frequency characteristic of the sendust MIG head has bumps, but no additional decrease at high frequencies. The bumps are caused by interference of the pseudo gaps with the main gap. According to the periodicity of the bumps, which is related to the location of the pseudo gaps (see section 12.1), the pseudo gaps are found at the ferrite/sendust interface. The good high-frequency performance of the sendust head is a consequence of the slight relative wear of the sendust (20-30 nm) and the small gap-length (however, the small gap-length disturbs the low frequency performance).

In the frequency characteristic of the permalloy/sendust MIG head the absence of the bumps indicates the absence of pseudo gaps, whereas the relatively small slope of the curve points to a small head-to-tape distance and, hence, a slight differential wear near the gap. A typical example of the differential wear in the gap area is shown in Fig. 12.18.

The efficiency of the head (measured separately without using tape,

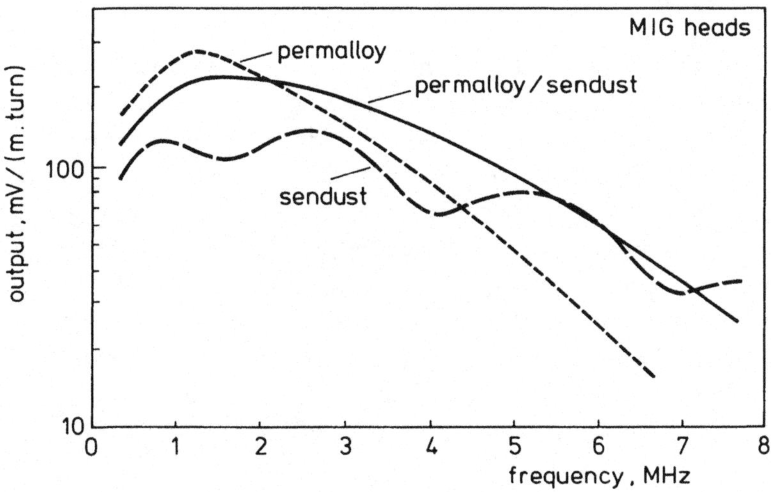

Fig. 12.17 Frequency characteristics of MIG heads (trackwidth: 20 μm)

gap cladding		gap length	trackwidth
permalloy,	3 μm	0.35 μm	preprofiled
sendust,	1 μm	0.20 μm	lasered
perm./send.,	0.5/2.5 μm	0.30 μm	lasered

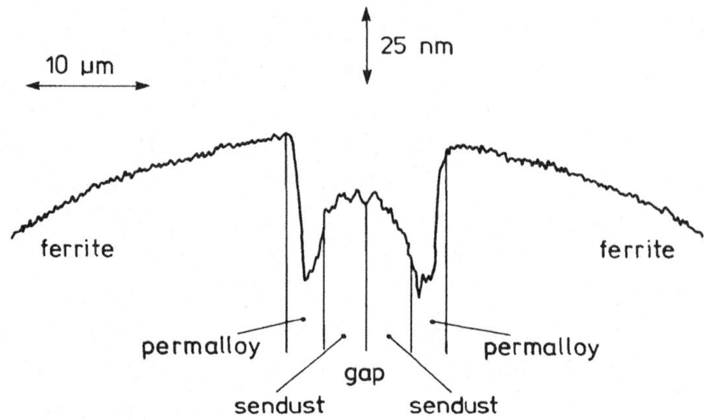

Fig. 12.18 Differential wear in the gap region of a permalloy/sendust MIG head measured after running against MP tape for 50 hours.

see section 8.2) is high (0.7) (pointing to small or zero stress in the ferrite and, so, in the gap cladding layers) as are the output and the signal-to-noise ratio.

12.2.7 Conclusion

In a permalloy/sendust MIG head, the differential wear in the vicinity of the gap is slight because of the presence of the sendust, whereas no pseudo gaps occur at the ferrite/gap cladding interface due to the presence of the permalloy. The SiO_2/Mo/Au gap which is applied in this type of head, can be prepared at relatively low temperatures, has a reproducible length and is magnetically sharp. The sendust properties satisfy the requirements for application in the MIG head.

As a consequence, the permalloy/sendust MIG head, which is technologically more simple and magnetically essentially better than other types of MIG head, can be used successfully for in-contact high-density recording in combination with MP tape.

References

[12.1]   H.N. Bertram and C.W. Steele, *Pole tip saturation in magnetic recording heads*, IEEE Trans. Magn., Mag-12, 702 (1976).

[12.2]   R.F.M. Thornley and H.N. Bertram, *The effect of pole tip saturation on the performance of a recording head*, IEEE Trans. Magn., Mag-14, 430 (1978).

[12.3]   F.J. Jeffers, R.J. McClure, W.W. French, N.J. Griffith, *Metal-In-Gap Record Head*, IEEE Trans. Magn., Mag-18, No. 6, 1146-1148 (1982).

[12.4]   W.K. Westmijze, *Field configuration around the gap and the gap-length formula*, Philips Research Reports 8, 161-183 (1953), equation 1.

[12.5]   H.N. Bertram and D.A. Lindholm, *Dependence of Reproducing Gap Null on Medium Permeability and Spacing*, IEEE Trans. Magn., Mag-18, 893 (1982).

[12.6]   Frederik W. Gorter, Jean-Paul Morel and Jacobus J.M. Ruigrok, *Magnetic write head with smooth frequency response*, United States Patent 4670807.

[12.7]   K. Matsuura, K. Oyamada, T. Yazaki, *Amorphous Video Head for High Coercive Tape*, IEEE Trans. Magn., Mag-19, No. 5, 1623-1625 (1983).

[12.8]   T. Kobayashi, M. Kubota, H. Satoh, T. Kumura, K. Yamauchi, S. Takahashi, *A Tilted Sendust Sputtered Ferrite Video Head*, IEEE Trans. Magn., Mag-21, No. 5, 1536-1538 (1985).

[12.9]   S. Otomo, T. Yamashita, S. Takayama, N. Saito and M. Kudo (Hitachi, Ltd.), *Composite type magnetic head and its manufacturing method*, European Patent Application, no. 0 125 891 A (1983).

[12.10]   Y. Ayabe (Nippon Victor Kabushiki Kaisha), *Composite type magnetic head and manufacturing method thereof*, UK Patent Application, no. GB 2 169 124 A (1984).

[12.11]   M. Ono, M. Ueda and O. Miyazaki (Matsushita Electrical Industrial Co. Ltd.), *Magnetic head and method of producing same*, European Patent Application, no. 0 201 255 (1986).

[12.12]   J.J.M. Ruigrok, *Cross measurements in magnetic recording*, IEEE Trans. Magn., Mag-20, No. 5, 875-877 (1984).

[12.13]   J.C. Mallinson, *Gap irregularity effects in tape recording*, IEEE Trans. Magn., Mag-4, 71 (1968).

[12.14]   H. Tomiyasu, K. Sato and K. Kanai, *Characteristics of Metal Magnetic Thin Film in MIG head and Head Performance*, J. Magnetics. Soc. Japan, Vol. 11, No. 2, 105-108 (1987).

[12.15]   J.J.M. Ruigrok, U.E. Enz and C.W.M.P. Sillen, *Magnetic transducing head having clad core faces*, European patent application 0246706.

Chapter 13

Bandpass heads

A very sensitive head is described for retaining information that is characterized by a restricted frequency band. The sensitive area of the head consists of a large number of small gaps at small distances from each other. The head has a high efficiency, a low inductance and a high quality factor and hence a high signal-to-noise ratio. These performances are all better than those reached by conventional heads designed for durable in-contact recording. By adjusting in particular the spacings between the gaps the head can be given any desired bandfilter action. This applies both to the amplitude and to the phase characteristic.

13.1 Introduction

The frequency characteristic of metal-in-gap (MIG) heads in chapter 12 pointed to the existence of non-magnetic or poor magnetic layers at the interfaces between the metal gap-cladding and the ferrite. We called these tiny layers residual gaps. Further considerations and model calculations showed that the sensitivity of this head in the peaks of the amplitude-versus-frequency characteristic is a little higher than without the residual gaps; see Fig. 12.2.

Analytical head-model considerations and calculations also showed a slightly decreased electric impedance.

The ideas of the model proposed in chapter 12 were used to design a multi-gap head with a well-defined bandpass characteristic, a very high sensitivity and a low impedance, and hence with a superior signal-to-noise ratio. This bandpass head (BPH) could be realized easily, using the same technology as for MIG heads.

Some embodiments of the BPH are shown schematically in Fig. 13.1, see also patent [13.1].

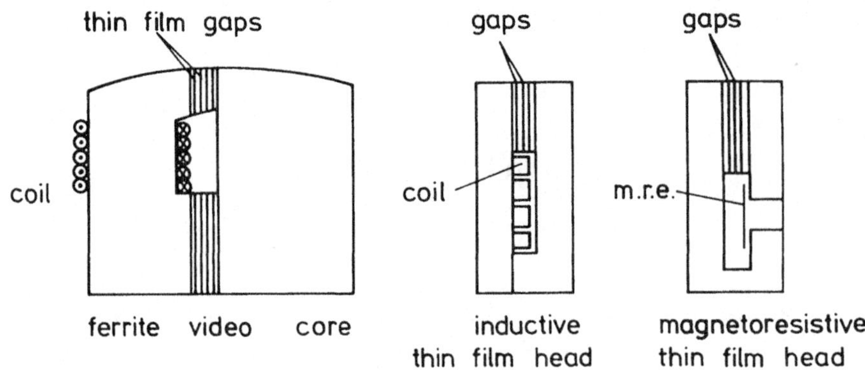

Fig. 13.1 Schematic representation of different embodiments of a bandpass head (BPH).

In this chapter we report on our initial experience with this head, after describing in detail the many theoretical aspects of the BPH.

13.2 Simple explanation of band filter action of bandpass head (BPH)

In the case of a *uniform* gap configuration the bandfilter action is easy to explain. Assume N_g ideal tiny gaps at equal distances a (between gap centres). A signal from the tape is sampled at the positions of the various gaps; see Fig. 13.2a. For the depicted sinusoidal component with wavelength λ, just $N_g a/\lambda$ periods are 'sampled' at N_g equidistant points. The number of 'effective periods' in the sampled signal (see dashed line in Fig. 13.2a) equals $|N_g a/\lambda - N_h N_g|$, where N_h is such a number (0, 1, 2 ...) that $|N_g a/\lambda - N_h N_g| < \frac{1}{2} N_g$. The total 'effective phase', φ, thus equals $2\pi |N_g a/\lambda - N_h N_g|$. The relative amplitude, i.e. compared with the situation where all samples are in phase ($N_g a/\lambda - N_h N_g = 0$, i.e. when $\lambda = a/N_h$), hence equals:

$$GLF = \frac{1}{\varphi} \int_{-\varphi/2}^{\varphi/2} \cos \varphi' \, d\varphi' = \mathrm{sinc} \, |N_g a/\lambda - N_h N_g|, \qquad (13.1)$$

where $\mathrm{sinc} \, x \equiv (\sin \pi x)/(\pi x)$.

The wavenumber $k_h = 2\pi/\lambda_h = 2\pi N_h/a$ at the maxima of the 'sensitivity function' (13.1) of the BPH increases proportionally to N_h.

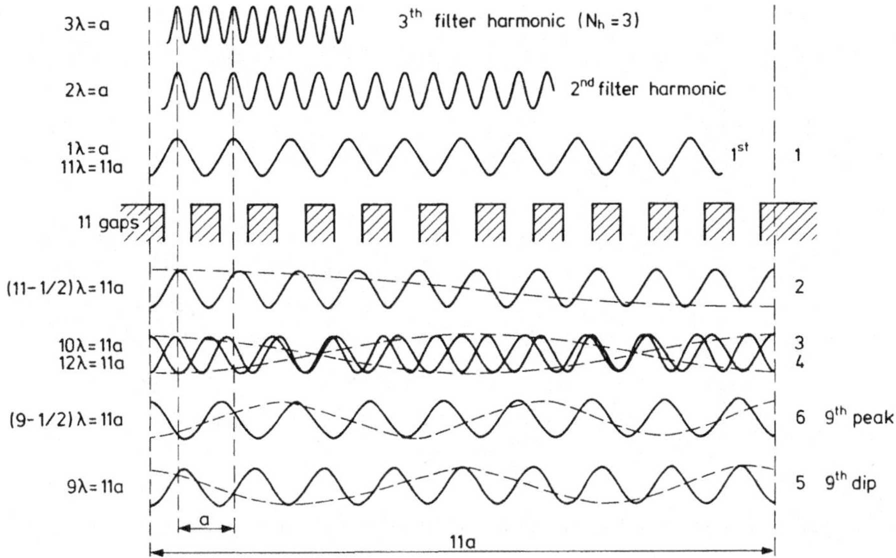

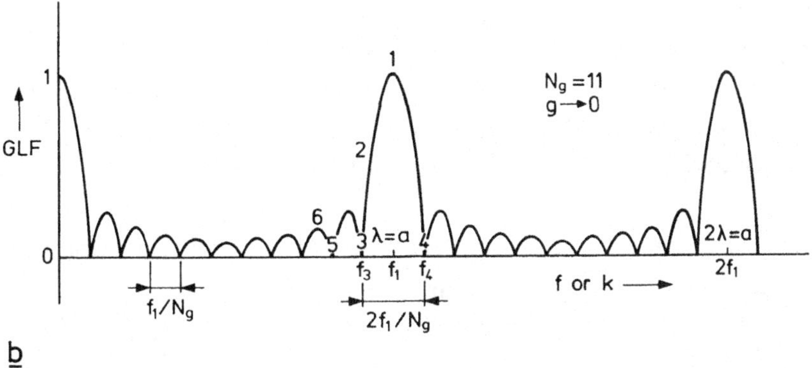

Fig. 13.2 The operation of a BPH
a) In the deep-gap field approximation each gap contributes a flux to the reading coil that is proportional to the integral of the magnetization in the tape over the gap length.

   Maxima in the frequency response therefore occur when $N_h\lambda = a$, since then there is no phase change, $\Delta\varphi$, over the series of gaps, i.e. $\Delta\varphi = 0$. Minima occur when $\Delta\varphi = 2\pi n$, see 3, 4 and 5.

b) The sketch of the major peaks ($N_h = 0$, 1 and 2) and the surrounding minor peaks in the frequency response of a BPH with an eleven-fold gap. The phase change over individual gaps is neglected ($g \rightarrow 0$).
   - In '1', $\Delta\varphi = 0$ (maximum).
   - In '3' and '4', $\Delta\varphi = 2\pi$ (minima).
   - In '5', $\Delta\varphi = 4\pi$ (minimum).
   - In '2', $\Delta\varphi = \pi$ (left flank of first major peak).
   - In '6', $\Delta\varphi = 5\pi$ (relative maximum).

The relative zero-bandwidth of the first major peak $(f_3 - f_4)/f_1 = (1/\lambda_4 - 1/\lambda_3)/(1/\lambda_1) = ((N_g + 1)/N_g - (N_g - 1)/N_g) = 2/N_g$. The higher-harmonic filter peaks are just as wide as the first major peak, all the minor peaks are half as wide.

Therefore N_h will be called the 'filter harmonic number'.

From (13.1) it follows immediately that the zero width (distance between two subsequent zeros) in the k domain of the minor peaks corresponds to one step in $N_g a/\lambda$, and that of major peaks (around filter harmonics $k_h = 2\pi N_h/a$) corresponds to two steps in $N_g a/\lambda$; see also Fig. 13.2b.

The samples are not infinitely short, but have a length equal to the gap length, g. The signal in the tape is thus averaged over the gap length. Hence another gap loss factor $\mathrm{sinc}(g/\lambda)$ (in the deep-gap field approximation) must be added to (13.1). This factor acts as a *hull curve* for the fine spectrum given by (13.1) since $g \ll N_g a$ and hence the first zero of $\mathrm{sinc}(g/\lambda)$ is at an $N_g a/g$ times larger wavenumber than the first zero of (13.1). At the major filter peaks this extra factor equals $\mathrm{sinc}(g/(a/N_h))$, which equals 0 when $g = N_h a$, 0.30 when $g = \frac{3}{4}N_h a$, 0.64 when $g = \frac{1}{2}N_h a$ and 0.90 when $g = \frac{1}{4}N_h a$. For the total GLF to approach 1 at a major peak it is therefore necessary to keep $g \lesssim \frac{1}{2} a/N_h$. It is convenient to choose $g = \frac{1}{4}a \sim \frac{1}{2}a$ and $N_h = 1$. When the BPH is tuned to a higher filter harmonic, as is necessary in some special BPH designs, it is desirable to choose g accordingly smaller. This is outlined in section 13.4.4.

For a uniform BPH with 19 gaps of finite length, instead of 11 gaps of vanishing length, the output spectrum is shown in Fig. 13.3. The assumptions concerning the write and read situations, necessary in the

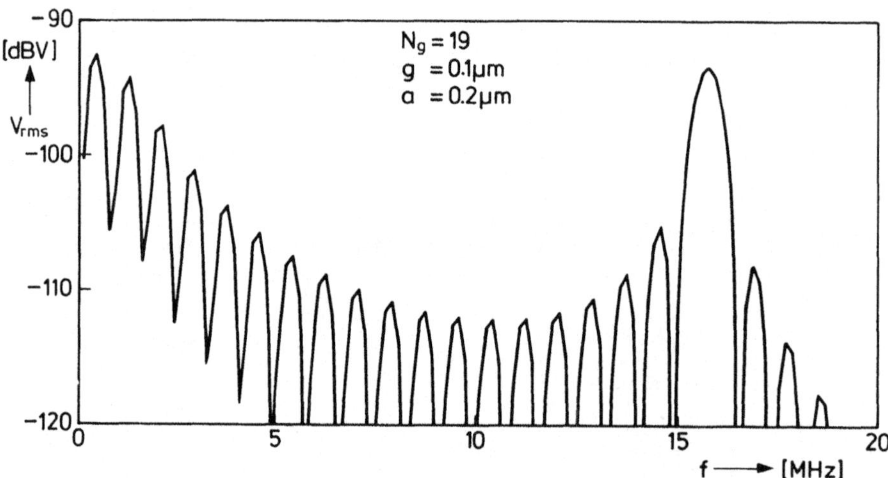

Fig. 13.3 Output of a BPH with 19 gaps of 0.1 µm on equal distances of 0.2 µm. $-80\ \mathrm{dB}V \equiv 100\ \mu V_{\mathrm{rms}}$.

calculation of this and other output spectra in this chapter, are given in Appendix 13.1. Now we observe that the first major peak is the 19th peak instead of the 11th. In addition the hull curve of the complete spectrum decreases with the frequency, due to the non-zero gap length. Finally the first large peak at $f = 0$ in the *GLF*, see Fig. 13.3, is not visible anymore, since $d\Phi/dt = 0$ and Φ is finite.

13.3 Theory of η, L and Q of BPH

In chapter 8 accurate expressions have been derived for the (field-) efficiency and electric impedance change of conventional heads (not valid for thin-film heads) when the gap length is varied; see section 8.1.6.

The mentioned expressions (8.29) and (8.30) read:

$$\frac{\eta(G)}{\eta_{\text{int}}} = \frac{\eta(G_0)/\eta_{\text{int}}}{1 - \dfrac{G - G_0}{G}\left(1 - \dfrac{\eta(G_0)}{\eta_{\text{int}}}\right)} \tag{13.2}$$

(see also Fig. 13.4) and

$$\frac{Z(G)}{Z(G_0)} = \frac{1 + \dfrac{G - G_0}{G_0}\left(1 - \dfrac{\eta_\Phi(G_0)}{\eta_{\Phi\text{int}}}\right)}{1 + \dfrac{G - G_0}{G_0}\dfrac{\eta(G_0)}{\eta_{\text{int}}}}, \tag{13.3}$$

(see also Fig. 13.5), where, according to section 5.7.2, the gap length G in its most general meaning is defined as

$$G = \int_{x_1}^{x_r} \frac{1}{\mu_r(x)}\,\mathrm{d}x. \tag{13.4}$$

G can here be interpreted as the 'sum' of all the 'gap lengths' in the BPH. Here $\eta(G_0)$ and $Z(G_0)$ are the efficiency and electric impedance of the same head but with gap length $G = G_0$. η_{int} and $\eta_{\Phi\text{int}}$ are the

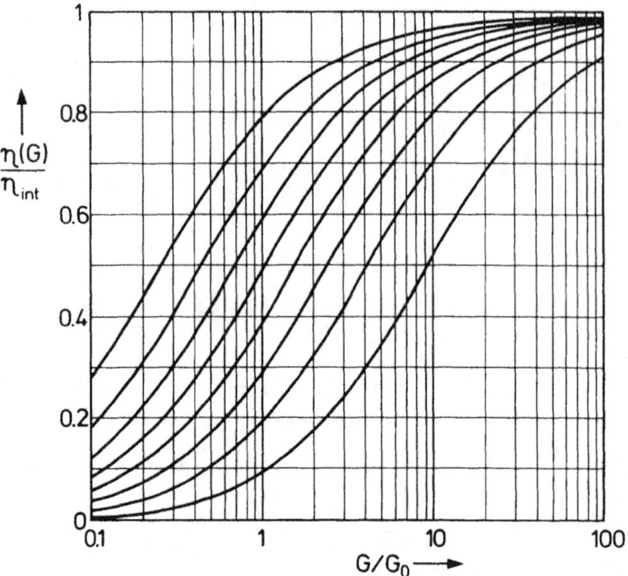

Fig. 13.4 The increase of the efficiency of (not thin-film) heads with the (generalized) gap length G. For simplicity, the permeability of the core is assumed to be real. Realize that the intersections with the $G/G_0 = 1$ line indicate that the curves have been drawn for $\eta(G_0)/\eta_{int} = 0.8, 0.7, 0.6 \ldots 0.1$.

(field) efficiency and the flux efficiency for very large and very small G respectively. In general these internal efficiencies do not reach 1, but absolute values of about 0.9 and 0.7 are reached at a few MHz for ordinary MnZn-ferrite videoheads (see table 8.8) and considerably lower values at frequencies above 10 MHz. If the head-to-tape velocity is not increased when going to these high frequencies, the gap length has to be reduced from 0.3 μm to say 0.1 μm. Otherwise the very small wavelength information cannot be read from the tape, which means that the efficiency will be reduced even further. Assume that the efficiency follows the curve for $\eta(G_0)/\eta_{int} = 0.6$, i.e. the third curve from the left in Fig. 13.4, and assume $\eta_{int} = 0.5$. Then the efficiency is only 0.17 when $g = 0.1$ μm. The inductive part ωL (see Fig. 13.5) and the ohmic part R of $Z \equiv R + j\omega L$ both increase when G decreases; see also section 8.15. The ratio $\omega L/R \equiv Q$ decreases too and may approach the 'Q factor of the material', μ', μ'', since the positive influence of the real gap reluctance on the Q factor of the head will vanish at vanishing G.

The change of the total noise voltage $\sqrt{N_{tot}}$ (electronics and tape-induced noise emf) with Q (at constant L and optimal noise match, see

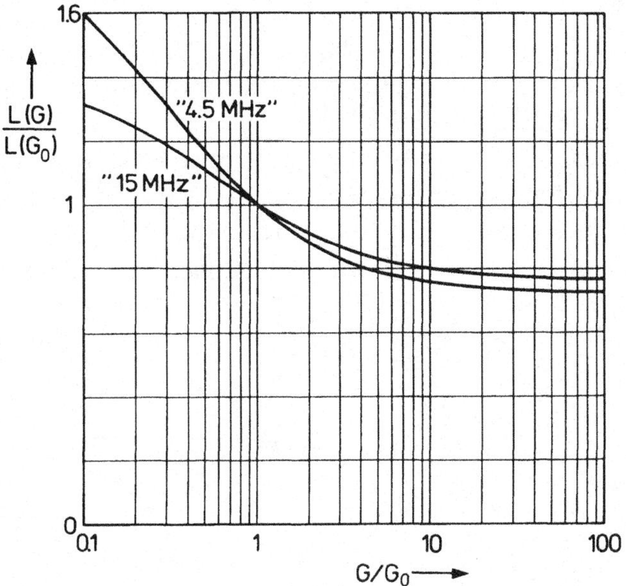

Fig. 13.5 The reduction of the inductance of (not thin-film) heads with (generalized) gap length G. The curves are given 'for an ordinary ferrite video head operating at 4.5 MHz and 15 MHz for $G_0 = 0.3$ μm.' (In the first case, see table 8.8, $|\eta_\Phi(G_0)|/|\eta_{\Phi\text{int}}| = 0.33/0.73 = 0.45$ and $|\eta(G_0)|/|\eta_{\text{int}}| = 0.67/0.88 = 0.76$. At the higher frequency when μ' is reduced from 550 to 150 and μ'' is reduced from 450 to 300, these two ratios read $0.31/0.61 = 0.51$ and $0.54/0.84 = 0.64$ respectively. In the application of equation (13.3) we neglected the complex character of the efficiencies for simplicity.)

appendix 13.2), is given in Fig. 13.6. Curves are given for different ratios between the noise power produced by the tape (e.g. particle and modulation noise from the tape and rubbing noise from the mechanical interaction between the head and the tape), N_{tin}, and the noise power from the head's resistance $R = \omega L/Q$ if Q is 10, called N_R ($Q = 10$). N_R is included in the noise of the electronics, N_{el}. When the tape-induced noise is negligible, the influence of Q on the total noise is most severe. The equivalent noise voltage source, V_r, and equivalent noise-current source, I_n, originating from various noise sources in the pre-amplifier but transformed to the input of the pre-amplifier, are assumed to be uncorrelated. They are represented by equivalent noise resistances R_r and R_n according to the definition $V_r^2/R_r \equiv 4\,kT$ and $I_n^2 R_n = 4\,kT$. Plots are given for two practical ratios of R_r/R_n in the 0.5-10 MHz range in the case of low-noise field-effect transistors (FETs). The plots in Fig. 13.6 show that the sensitivity to changes in Q is more pronounced at the smaller R_r/R_n ratio when the current noise is of less influence and at small values of the tape-induced noise.

Smaller R_r/R_n ratios are obtained at lower frequencies or with better FETs. For a systematic description of noise in high-performance amplifiers see E.H. Nordholt [13.2].

The decrease of the inductance is about equally important for a reduction of the noise as the decrease of R (or increase of Q) for the above case. L contributes to the noise voltage by the current of the input noise-current source flowing through L, see appendix 13.2. A

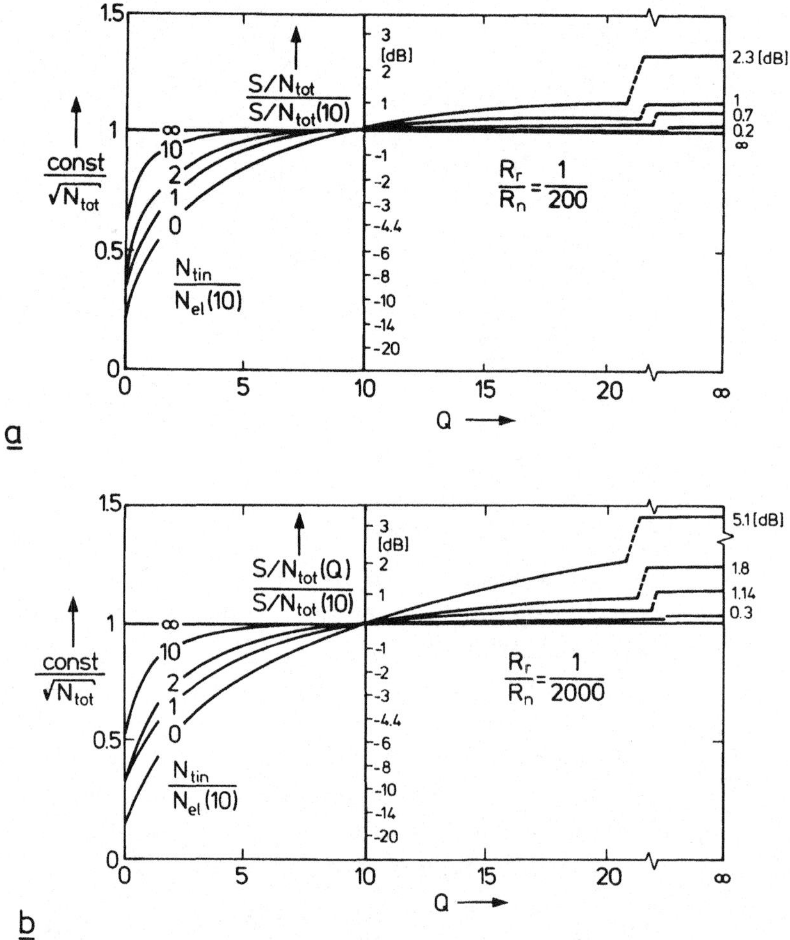

Fig. 13.6 The increase of the signal-to-(total) noise ratio versus Q with respect to the situation for $Q = 10$, at different ratios of the tape-induced noise to the electronics noise when $Q = 10$. Usually the tape-induced noise is of the order of the electronics noise (for Q = about 10). The left scale gives the factor by which the total noise amplitude decreases.
a) Low-noise amplifier at high frequencies (R_r/R_n = 1/200).
b) Low-noise amplifier at moderate frequencies (R_r/R_n = 1/2000).

small inductance also keeps the resonance frequency of the input circuit of the pre-amplifier above the frequency band that has to be transmitted.

At low permeabilities the electric impedance of the head will be small. The influence of the generalized gap length on the impedance, i.e. L and R, is then expected to be small, because then the 'interaction' between gap and coil is small and there is less to gain by increasing G. This is not directly visible from equation (13.3) but follows more directly from the equivalent equation (8.28) and immediately from the results in Fig. 13.5. On the other hand a low L is additionally important when a low $Z \equiv \omega L + R$ at a high radial frequency of operation, ω, and a resonant radial frequency $\omega_r \simeq 1/\sqrt{LC}$ above ω, are desired. A large increase of G may then be necessary to obtain the desired small L.

From the above considerations the advantages of a large generalized gap length G, realizable in a BPH and not in a conventional single-gap head without loosing much sensitivity for small wavelengths, is outlined. The *noise decreases, the efficiency increases and the resonance frequency increases.*

13.4 Theory of bandpass, (practical) limitations and model calculations

13.4.1 General

General expression for GLF(k)

In the most general case of a BPH the 'gaps' are located at arbitrary positions $x = p_i$ and all may have a different permeability profile. Without losing any generality about the total $1/\mu_r(x)$ profile, this profile can be expressed in the following summation:

$$\frac{1}{\mu_r(x)} = \sum_i \frac{1}{\mu_{ri}(x_i)} \equiv \sum_i A_i S_i(x_i), \tag{13.5a}$$

with $x_i \equiv x - p_i$ the new variable.

When pronounced peaks in $1/\mu_r(x)$ are present, i.e. gaps in $\mu_r(x)$, then A_i can be considered as the amplitude and $S_i(x_i)$ as the shape function of peak i in the $1/\mu_r(x)$ profile. For these cases it will be convenient to choose

$$\int S_i(x_i)\,dx_i = 1, \tag{13.5b}$$

such that:

$$\sum_i A_i = \int \frac{1}{\mu_r(x)}\,dx \equiv G, \tag{13.5c}$$

$$A_i = \int \frac{1}{\mu_{ri}(x_i)}\,dx_i \equiv g_i \tag{13.5d}$$

and

$$S_i(x_i) = \frac{1}{g_i\mu_{ri}(x_i)}. \tag{13.5e}$$

G is the (generalized) length of the collection of gaps and g_i is the (generalized) length of gap i.

With the above definitions, the deep-gap field approximation of the generalized gap-loss function of the BPH can be written as:

$$\boxed{GLF(k) = \frac{\displaystyle\int \frac{1}{\mu_r(x)}\,e^{-jkx}\,dx}{\displaystyle\int \frac{1}{\mu_r(x)}\,dx} = \frac{\displaystyle\sum_i g_i\,e^{-jkp_i}\,S_i(k)}{G}} \tag{13.6}$$

with

$$S_i(k) \equiv \int S_i(x_i)\,e^{-jkx_i}\,dx = \frac{\displaystyle\int \frac{1}{\mu_{ri}(x_i)}\,e^{-jkx_i}\,dx_i}{\displaystyle\int \frac{1}{\mu_{ri}(x_i)}\,dx_i}. \tag{13.7}$$

$GLF(k)$ and $S(k)$ are both normalized in such a way that they approach 1 when $k \to 0$.

For the wavelengths that are usually of interest it is sufficient to choose the lower and upper integral boundaries at locations x_l and x_r, left and right of the region of low permeabilities, where $1/\mu_r(x)$ vanishes. This is equivalent to assuming $1/\mu_r(x)$, or that the field is negligibly small outside this region.

Expression (13.6) is consistent with the definition of the $GLF(k)$ in (5.18). The deep-gap field, proportional to $1/\mu_r(x)$, is used here to represent the x-component of the field at the head's front plane $y = 0$. When the exact field is substituted in (13.6), then the accuracy of the sensitivity to very small and very large wavelengths would be improved. Usually we are not interested in sensitivities for these wavelengths.

The GLF of the BPH is described according to (13.6) by the one-dimensional Fourier transform. In the theory of Fourier images and in the formulation of the linear-exponent approximation of the Fraunhofer diffraction pattern in optics, the two-dimensional Fourier transform appears; see e.g. reference [13.3]. In this respect the BPH can be described more easily.

Because of the generality of the assumed $\mu_r(x)$ dependence, however, less can be said about the $S_i(k)$ or $GLF(k)$ in general, although it is not difficult to calculate (13.6) numerically. Of course, foreknowledge when designing BPHs is desirable, and this will be facilitated by reduction of the generality in the following sections.

Relations between widths in the x and k domains

For the present general case there is no 'rule of thumb' that relates the width of the function $GLF(k)$ in the k-domain with the width of the inverse Fourier transform $GLF(x)$ in the x-domain. There is not even any generally satisfactory definition of a width. Our problem is in fact similar to the problem in theoretical physics, namely 'to find useful concepts of width to express the uncertainty principle in quantum mechanics' for general wavefunctions; see e.g. Uffink and Hilgevoort [13.4].

The uncertainty principle reads $\sigma_x \sigma_{p'} > \hbar/2$, where p' is the momentum and $\hbar$ is the reduced form of Planck's constant, h. In our case this relation reads

$$\sigma_x \sigma_k > \tfrac{1}{2}, \tag{13.8}$$

where σ_x and σ_k are standard deviations ($\sigma_x \equiv \sqrt{\langle x^2 - \langle x \rangle^2 \rangle}$, $\langle\ \rangle$ is the weighted average defined as $\langle x \rangle \equiv \int xp(x)\,dx$, with $p(x) \equiv |GLF(x)|^2/\int |GLF(x)|^2 dx$ being the definition of the probability density function).

This 'uncertainty' relation only gives a lower bound and no upper bound. The drawback of using standard deviations where widths are desired is that standard deviations and width may be of a quite different order. For example $\sigma_k \to \infty$ for $\sin^2(kg/2)/(kg/2)^2$, while for instance the halfwidth (in the k domain) $W_{\frac{1}{2}}$ is of the order of $2\pi/g$. When two or more widths can be defined, for instance in our BPH a width related to the length of individual gaps and the width related to the overall length of the gap configuration, then the first width will be much smaller than the second. Usually the latter will correspond to the standard deviation, but this also depends on the chosen definition of width. Uffink and Hilgevoort [13.4] defined an *overall width*, W_x, of a wavefunction in quantum mechanics, which, 'translated' to our $GLF(x)$, reads:
W_x is the smallest number for which

$$\frac{\int_{\langle x \rangle - \frac{1}{2}W_x}^{\langle x \rangle + \frac{1}{2}W_x} |GLF(x)|^2 dx}{\int |GLF(x)|^2 dx} = N, \tag{13.9}$$

with N chosen close to 1.

They defined a *mean peak width* as the *smallest* number, w_x, for which the square of the autocorrelation function $g(w_x)$ of the (normalized) $GLF(x)$,

$$g(w_x) = \frac{\left| \int_{-\infty}^{\infty} GLF^*(x) \cdot GLF(x - w_x)\,dx \right|^2}{\int |GLF(x)|^2 dx} = M^2, \tag{13.10}$$

where $*$ means complex conjugate and M is chosen close to 1. They arrived at the following relations:

$$W_x \leqslant 4\sigma_x \quad \text{and} \quad w_x > \frac{1}{2\sigma_k} \tag{13.11}$$

and analogously for the overall width and mean peak width in the k-domain:

$$W_k \leqslant 4\sigma_k \text{ and } w_k > \frac{1}{2\sigma_x} \tag{13.12}$$

while

$$W \geqslant w \text{ if } M^2 + N^2 \geqslant 1 \tag{13.13}$$

in the x domain as well as the k domain.

For functions GLF that are well localized or functions that possess a well-defined peak structure, W and w are of the right order, i.e. of the order of the width that one intuitively would assign to the *overall* width and the *'internal'* mean peak width. However the inequalities still do not lead to upper bounds of w_x and W_x, given σ_k, w_k and W_k and vice versa, or still better to inverse proportionalities between the widths w_x and W_k etc. themselves. This is not a problem to be solved; it is in fact fundamentally impossible to derive such relations, even in an approximate way, for general functions. This will become clearer in the following sections, especially in section 13.4.4.

Practical limitations

Not all mathematically possible $1/\mu_r(x)$ profiles are practically possible. Hence not all $GLF(k)$ can be realized with BPHs. For instance the Fourier-transform pair,

$$GLF(x) = \frac{1}{G \cdot \mu_r(x)} = \frac{1}{G} \frac{\sin(xb_k/2)}{xb_k/2} \cos(k_0 x)$$

and

$$GLF(k) = \frac{\pi}{Gb_k} \left\{ \Pi\!\left(\frac{k + k_0}{b_k}\right) + \Pi\!\left(\frac{k - k_0}{b_k}\right) \right\} \tag{13.14}$$

where $\Pi((k - k_0)/b_k)$ is the rectangle function that is 1 when $k - k_0 > \frac{1}{2}b_k$ and 0 when $k - k_0 < \frac{1}{2}b_k)$, is not practically realizable. The reason is that the sign of $1/\mu_r(x)$ changes, which never can be realized within a single BPH. Manufacturing a specific gradual permeability, even when $\mu_r(x)$ is real as in the above case and without negative values would technologically already be a tremendous task. The above ideal rectangle bandpass with width b_k ($\approx w_k$, the internal width) and central wavenumber k_0 cannot therefore be realized with a BPH.

Hence it is useful to start the design of a BPH with the following possible limitations and consequences in mind.

1) $1/\mu_r(x)$ will be almost real, except when $\mu_r(x) \ggg 1$. This is because we assume that a controlled low permeability is obtained with a strong anisotropy in the y direction such that the ferromagnetic resonance frequency is very high and hence the imaginary part of the permeability is negligible at the frequencies of operation. This resonance frequency may approach the frequency of operation when $\mu_r(x) \ggg 1$, but the contribution from regions where $\mu_r(x)$ is large to $GLF(k)$ (and to G) is negligible since $1/\mu_r(x)$ is negligible there.

   Hence $1/\mu_r(x)$ will be assumed *real* and of course *positive*. As a consequence of $GLF(x)$ being real, the real part of $GLF(k)$ is an even function and the imaginary part an odd function (as easily can be proved).

2) Technologically it will be difficult to manufacture a head with a specific gradual permeability profile $\mu_r(x)$. It is most practical to assume *sharp gaps* with $\mu_r = 1$ inside the gap and $\mu_r \to \infty$ outside the gap. When the gap lengths are very small they will automatically become more gradual, but for our purposes they can still be approximated reasonably well by the ideal gap with length $g = \int (1/\mu_r(x))\,dx$,
   where the integral is carried out over a little larger area than in between the gap edges.

   Hence $1/\mu_r(x)$ will often be assumed 1 in the gap and 0 outside the gap.

3) For most applications a *linear phase characteristic* is desired, such that no phase distortion occurs. This means that it must be possible to write the gap loss function as $GLF(k) = |GLF(k)|\,\exp(jkx_0)$, with x_0 a constant. Hence we have the following Fourier transform pair:

$$GLF(x - x_0) \leftrightarrow GLF(k) \cdot e^{-jkx_0} = |GLF(k)|. \qquad (13.15)$$

Since $|GLF(k)|$ is real and $GLF(x)$ is also real in practice, it follows as in point 1 that $GLF(x - x_0)$ must be an even function. In other words, $GLF(x)$ must have a symmetrical shape. In practice a small deviation from this is allowed when the (internal) bandwidth is relatively small.

Hence $1/\mu_r(x)$ will often be chosen *symmetric* or almost symmetric.

In spite of the above practical limitations, the ideal rectangle bandpass of the example can be approximated as will be shown in section 13.4.4.

13.4.2 Gaps with equal shape

Unequal distances

When the N_g gaps of a BPH, located at arbitrary positions p_m, have the same permeability profile, such that $S_m(x_m) = S(x_m)$ with $x_m \equiv x - p_m$, but all gaps have a different 'amplitude', g_m, then (13.6) reduces to:

$$GLF(k) = S(k) \cdot g(k), \tag{13.16}$$

where

$$g(k) \equiv \frac{1}{G} \sum_{m=0}^{N_g - 1} g_m \, e^{-jk p_m}$$

and $S(k)$ is the Fourier transform of $S(x_m)$ from the x_m-domain to the k-domain.

The above procedure can be repeated and so on. For example, assume that each of the above 'gaps' is composed of a number of N_m' tiny gaps and that their positions p_{ml} relative to p_m, their amplitudes g_{ml} and the number of gaps N_m' are all independent of m (which is a rather hypothetical example). The following definitions can then be introduced: $N_g' \equiv N_m'$, $p_l' \equiv p_{ml} - p_m$ and $g_l' \equiv g_{ml}$. The shape function of each tiny gap is thus assumed to be independent of m and l and so can be written as $S_{ml}(x_{ml}) = S'(x - p_m - p_l)$ with $x_{ml} \equiv x - p_m - p_l$. Hence

$S(x_m) = \Sigma g_l' S'(x_{ml})$, and leads, analogously to (13.5a), to a Fourier series like (13.16):

$$GLF(k) = S'(k) \cdot g'(k) \cdot g(k), \qquad (13.17)$$

where

$$g'(k) \equiv \frac{1}{\Sigma g_l'} \sum_{l=0}^{N_g'-1} g_l' \, e^{-jkp_l'}$$

and $S'(k)$ is the Fourier transform of $S'(x_{ml})$ from the x_{ml}-domain to the k-domain.

$S'(k)$, $g'(k)$ and $g(k)$ are associated with the coarse, the fine and the hyperfine spectrum respectively.

This hypothetical example clearifies the *generality of the inverse relationships in x and k domain in the case of regular structures.*

Equal distances

Assume that the distances between the gaps and the distances between the possible tiny subgaps are constant, say 'a' and 'a'' respectively. When the origins of x_m and x_{ml} are appropriately shifted, the relative positions of the gaps and subgaps can be written as $p_m = ma$ and $p_l' = la'$. Then the above series $g(k)$ and $g'(k)$ are *Fourier series*. Hence they represent periodic functions with periods $\Delta k = 2\pi/a$ and $\Delta k' = 2\pi/a'$ in the k-domain. (Without the shifts of the origins, a linear phase factor $e^{-j \cdot k \cdot const}$ remains in the expressions for $g'(k)$ and $g(k)$.) Consequently $g'(x)$ and $g(x)$ in the inverse Fourier transform $GLF(x) = S'(x) * g'(x) * g(x)$, where $*$ means convolution, then represent series of N_g and N_g' delta functions at equidistant positions p_l' and p_m in the x-domain with amplitudes $g_l'/\Sigma g_l'$ and $g_m/\Sigma g_m$ respectively. The 'double' convolution restores the original 'permeability profile' $GLF(x) = 1/(G \cdot \mu_r(x))$.

At $k = 0$, all four functions in (13.17) are 1, because of the proper normalization conditions introduced earlier. At $k_{nn'} = 2\pi n/a + 2\pi n'/a$ this maximum of 1 is reached again. Hence $S'(k)$ is just the hull curve (envelope) of the fine spectrum $(g'(k))$, while in its turn $g'(k)$ is just the hull curve of the hyperfine spectrum $g(k)$! All three functions may be complex for $k \neq k_{nn'}$. The situation is sketched in Fig. 13.7, where all functions in the x- and k-domains are assumed for simplicity to be *real*. (Hence all functions are *even* functions as well, i.e. they have a symmetric shape and the origins are chosen properly in the middle.)

There are three methods to calculate or estimate the shape of the periodic functions.

Method 1. Calculate the Fourier series for different values of k numerically. For some cases analytical solutions are known.

Method 2. Make use of the discrete Fourier transform

$$\hat{g}_n = \frac{1}{\sqrt{N_g}} \sum_{m=0}^{N_g - 1} g_m e^{-j\frac{2\pi mn}{N}} \qquad (n = 0, 1, 2, ..., N - 1), \qquad (13.18)$$

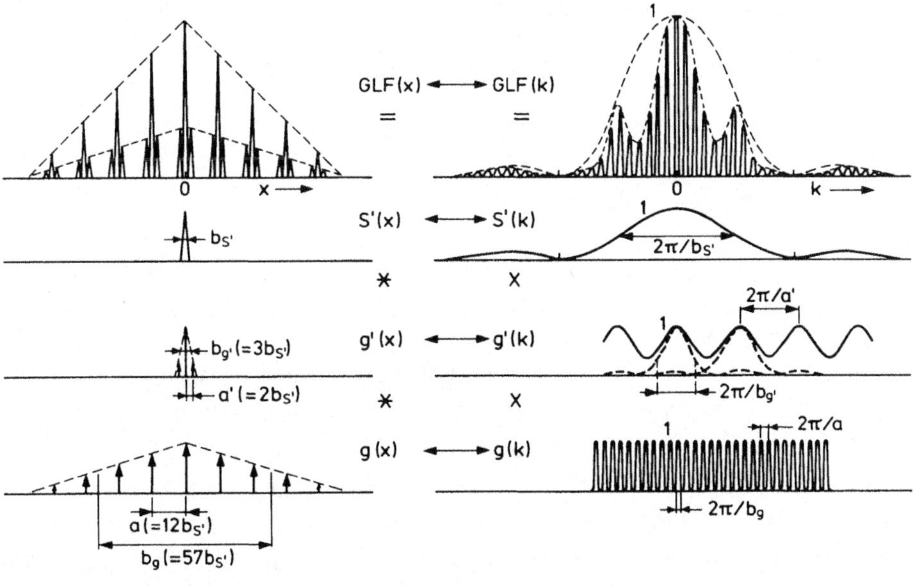

Fig. 13.7 Sketch of a symmetric $GLF(x)$ (proportional to the permeability profile $1/\mu_r(x)$) and the Fourier transform $GLF(k)$.

The sketch visualizes the relations between widths (and shapes) in the x-domain and k-domain in the case of (sub) gaps of equal shape (triangles in this example) at equal distances a', forming gap clusters (with a hull curve of triangular shape) at distances a. The hull curve of the series of clusters (dashed lines in upper and lower left figures) is also assumed to be triangular.

The dashed hull curves of $g'(x)$ and $g(x)$ in the two lower-left figures are associated with the dashed functions on the right-hand side by the Fourier transform, just as the relation between the delta functions etc. on the left-hand side and the periodic functions etc. on the right-hand side is given by the Fourier transform ($\leftrightarrow$). The few (approximately $b_g/a \approx N_g/2$) small peaks in between the major peaks in $g(k)$ and in $GLF(k)$ are not drawn in the sketch. Because of the equal shapes and equal distances, the permeability profile could be built up from two subsequent convolutions ($*$), see the three lower-left figures. Consequently the gap-loss function is found by two subsequent multiplications ($\times$), as shown in the three lower-right figures.

as defined in the subroutine C06EAF of the NAG Fortran library for real g_m.

The value of our $g(k)$ at the wavenumbers $k_n \equiv n2\pi/(Na)$ then follows immediately from the relation:

$$g(k_n) = a\sqrt{N_g}\,\hat{c}_n. \tag{13.19}$$

Method 3. Calculate the Fourier transform of the 'maximally flat' hull curve of the series of g_m values. (Maximally flat means here that the hull curve does not contain wavenumber components with a wavelength smaller than $2a$.) This (properly-normalized) hull-curve, $g^h(x)$, equals:

$$g^h(x) = \frac{1}{a \cdot G} \sum g_m \, \mathrm{sinc}\left(\frac{x - p_m}{a}\right) \tag{13.20}$$

with $p_m = ma$.

The Fourier transform $g^h(k)$ is easily rewritten into:

$$g^h(k) = \frac{1}{a \cdot G} \sum g_m \mathrm{e}^{-jkp_m} \int \mathrm{e}^{-jkx'} \mathrm{sinc}\,\frac{x'}{a}\, \mathrm{d}x' = \Pi\left(\frac{k}{2\pi/a}\right) \cdot \frac{1}{G} \sum g_m \, \mathrm{e}^{-jkp_m},$$

$$\tag{13.21}$$

where $\Pi\big(k/(2\pi/a)\big)$ is the rectangle function, which is 1 when $|k| < \pi/a$ and 0 when $|k| > \pi/a$ (and the wavelength would be smaller than $2a$). Hence $g^h(k)$ *exactly represents one period of the periodic function* $g(k)$ in (13.16) for 'gaps' at equal distances. The same applies to $g'(k)$. It is noted that the triangular hull curves chosen in Fig. 13.7 and denoted there by the dashed curves in the lower left figures do contain higher harmonics than necessary. Hence there is overlap of the corresponding dashed functions on the right-hand side.

Further Fig. 13.7 visualizes the *approximately inverse relationships between widths (and periods) in the x- and k-domains.* They may serve as *reliable 'rules of thumb'* only for the present case: BPHs with gaps of *equal shape at equal distances.* When for instance the gaps are not located at equal distances, then the widths in the k-domain may be much larger, as will be clear in section 13.4.4.

13.4.3 Ideal gaps

Unequal distances

It is most practical to make the gaps from layers of a non-magnetic material and the layers in between the gaps of a soft-magnetic material. As a result, the deep-gap field will be constant within the gaps ($\mu_r = 1$) and will approach zero in the core ($\mu_r \gg 1$). Therefore $S_i(x_i) = (1/g_i)\,\Pi(x_i/g_i)$ according to (13.5e), such that (13.6) reduces to:

$$GLF(k) = \frac{1}{G} \sum_i e^{-jkp_i} g_i \operatorname{sinc}(kg_i/2\pi) \; . \tag{13.22}$$

$GLF(x)$ cannot be expressed by a convolution of a series of delta functions, $g(x)$, with *one* shape function $S(x)$. Consequently $GLF(k)$ in (13.22) is not exactly expressed by the product of a fine-spectrum function $g(k)$ with an overall curve $S(k)$.

Equal distances

When the distance between subsequent gap centres is constant, the last remark translates into: 'the $GLF(k)$ is not exactly expressed by the product of a *periodic* fine-spectrum function $g(k)$ and an overall curve $S(k)$'.

This makes it more difficult to estimate the characteristics of the bandpass beforehand. Or, by approximating g_i by the average gap length $\langle g \rangle$, it makes the estimation less accurate.

In this *approximation*:

$$GLF(k) = S(k) \cdot g(k) \; , \tag{13.23}$$

where

$$S(k) = \operatorname{sinc}(k\langle g \rangle/2\pi)$$

and

$$g(k) = \frac{1}{N_g} \sum e^{-jkp_i} \; .$$

Now all results derived in section 13.4.2 for gaps at equal distances apply; see also Fig. 13.7 and the second of the following two examples: ideal gaps with equal length at equal distances.

Examples

example 1: ideal gaps with different lengths at equal distances
We calculated with the aid of expression (13.22) the characteristics of the bandpass of BPHs with different numbers N_g of ideal gaps at

Table 13.1 Definition of the locations, p_i, and gap lengths, g_i, of five BPHs with different numbers of gaps, N_g.
The calculated improvements in $|\eta|$, L and Q are given for the ordinary MnZn ferrite head defined in table 8.8, but operating at the frequency of the first major peak in the frequency response of the BPH, $f_1 = 15.6$ MHz, where $\mu' = 150$ and $\mu'' = 300$. Between brackets the improvements of the head's characteristics are visible at 4.5 MHz where $\mu' = 550$ and $\mu'' = 450$.

$\{p_i\}$	[μm]	$\{g_i\}$				[μm]
−1.8						0.01
−1.6						0.02
−1.4						0.03
−1.2						0.04
−1.0						0.05
−0.8					0.02	0.06
−0.6					0.04	0.07
−0.4				0.033	0.06	0.08
−0.2			0.05	0.066	0.08	0.09
0.0	0.08		0.10	0.10	0.10	0.10
0.2			0.05	0.066	0.08	0.09
0.4				0.33	0.06	0.08
0.6					0.04	0.07
0.8					0.02	0.06
1.0						0.05
1.2						0.04
1.4						0.03
1.6						0.02
1.8						0.01
G	[μm]	0.08	0.20	0.30	0.50	1.00
N_g		1	3	5	9	19
$|\eta|$		0.24 (0.39)	0.45 (0.60)	0.54 (0.67)	0.65 (0.75)	0.74 (0.81)
L	[μH]	1.52 (2.45)	1.58 (2.10)	1.58 (1.93)	1.54 (1.76)	1.46 (1.59)
Q		1.6 (3.2)	2.2 (5.3)	2.6 (6.7)	3.2 (8.7)	4.1 (11.6)

equal distances $a = 0.2$ μm and with a hull curve $g^h(x)$ of triangular shape; see the g_i values in Table 13.1. The first major peak is at $f = 15.6$ MHz, since $v = 3.14$ m/s and $a = 0.2$ μm; see the normalized output versus frequency curves in Fig. 13.8. According as the number of gaps increases, the efficiency and the Q factor increase and the inductance (if μ' is not too small with respect to μ'') and the bandwidth of the first major peak decrease. The variations of η, L and Q are calculated for the MnZn-ferrite video head given in table 8.8 and operating at about 15 MHz (i.e. $\mu' \approx 150$ and $\mu'' \approx 300$); see full curves in the plots of Fig. 13.9, by way of the complete analytical head model described in chapter 8. Application of the expressions in section 13.3 gives almost the same results. For comparison we have calculated the improvement that would be obtained if the same head were to operate at 4.5 MHz (i.e. $\mu' \approx 550$ and $\mu'' \approx 450$), and the results are given by the dashed curves in Fig. 13.9. In the table the improvements are indicated for $N_g = 1, 3, 5, 9$ and 19, corresponding to the increase in the length of the collection of gaps, $G = 0.08, 0.20, 0.30, 0.50$ and 1.00 μm respectively. The improvement in the efficiency is dramatic! Improvements in a thin-film head with the same gap surface, Wh, as in the above ferrite head are expected to be even greater.

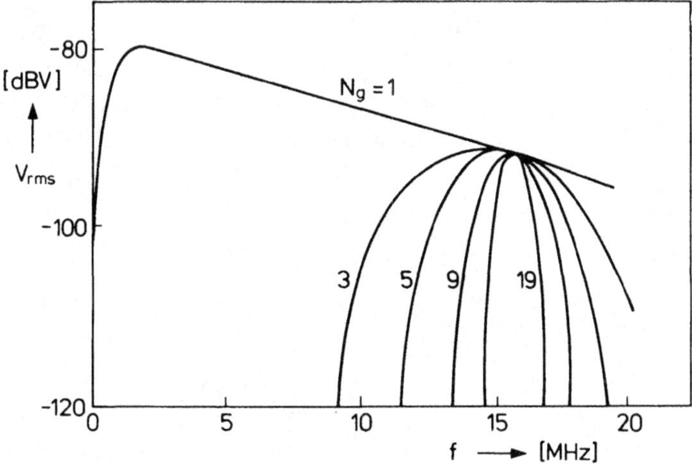

Fig. 13.8 The approximately inverse proportionality between the width of the first major peak in the output voltage spectrum of the BPHs defined in table 13.1 and the number of gaps, N_g. (For details concerning the write-head etc. see appendix 13.1.) For a reliable comparison of the output voltages of the various BPHs, the voltage must be scaled with the efficiency values given in table 13.1 at the frequency of interest, here about 15 MHz.

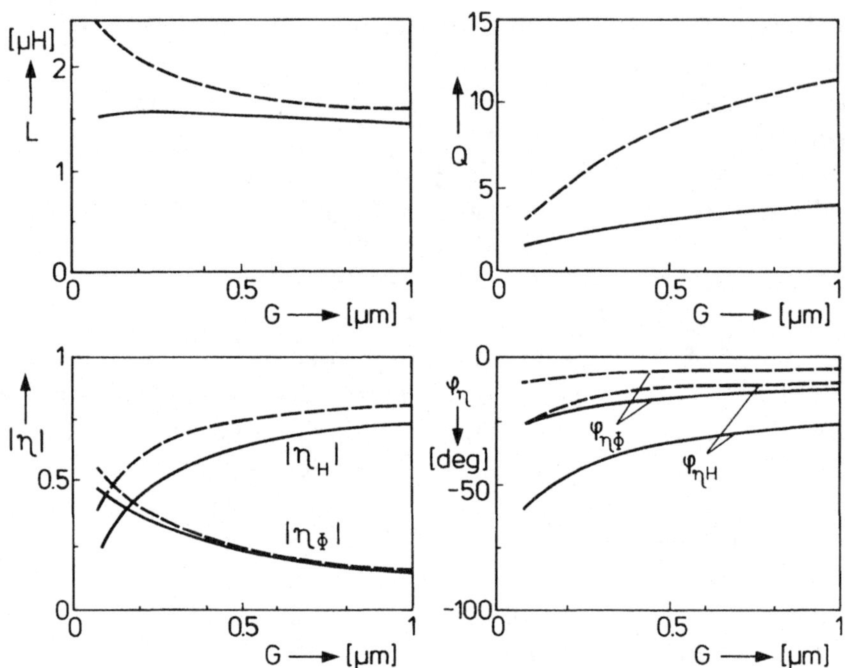

Fig. 13.9 Improvement of η, L and Q with G, calculated according to the analytical model described in chapter 8.
——— MnZn-ferrite head of table 8.8 operating at 15 MHz ($\mu' = 150$, $\mu'' = 300$).
---- MnZn-ferrite head of table 8.8 operating at 4.5 MHz ($\mu' = 550$, $\mu'' = 450$).

example 2: ideal gaps with equal lengths at equal distances

Assume that the BPH consists of N_g ideal gaps with length g and distance a from gap centre to gap centre. This situation is the easiest to implement, especially when N_g is odd, as it is in the experimental BPH that will be described in section 5. It is convenient to choose $x \equiv 0$ at the centre of the gap configuration, such that altogether expression (13.22) reduces to

$$GLF(k) = S(k) \cdot g(k) \qquad (13.24)$$

where

$$g(k) = \frac{1}{N_g} \sum_{i = -\frac{N_g - 1}{2}}^{i = \frac{N_g - 1}{2}} e^{-jkp_i}$$

and

$$S(k) = \mathrm{sinc}(kg/2\pi),$$

in analogy with (13.16); see also (13.23).

$S(k)$ is the overall hull curve and $g(k)$ the fine spectrum in the k-domain. The central period $(|k| < \pi/a)$ of the periodic function $g(k)$ is exactly the Fourier transform of the maximally flat hull curve defined by (13.20), $g^h(x) = (1/N_g a) \sum_i \mathrm{sinc}((x - p_i)/a)$. This hull curve, for the present case of constant $g_i \doteq g$, approximates the rectangle function $(1/N_g a)\,\Pi(x/N_g a)$, whose Fourier transform equals $\mathrm{sinc}(kN_g a/2\pi)$. Hence

$$\boxed{GLF(k) \approx \mathrm{sinc}\left(\frac{kg}{2\pi}\right) \cdot \mathrm{sinc}\left(\frac{kN_g a}{2\pi}\right)} \quad |k| \leq \frac{\pi}{a}. \qquad (13.25)$$

For large N_g the second factor becomes exact, while the first sinc function will always remain an approximation (the deep-gap field approximation) of the gap-loss function of an ideal gap and so of $S(k)$.

The output voltage spectrum of a BPH with 9 gaps, with distances of 0.4 µm between the gap centres and gap lengths of 0.2 µm, is plotted in Fig. 13.10. For this purpose the more accurate expression (13.22) was used for the $GLF(k)$ in the model calculations (for further details see the subsection 'output voltage' in section 13.2). The hull curve in Fig. 13.10 is obtained in the same way by substituting $S(k)$ for the $GLF(k)$, i.e. by assuming only one gap, located at the position $x = 0$

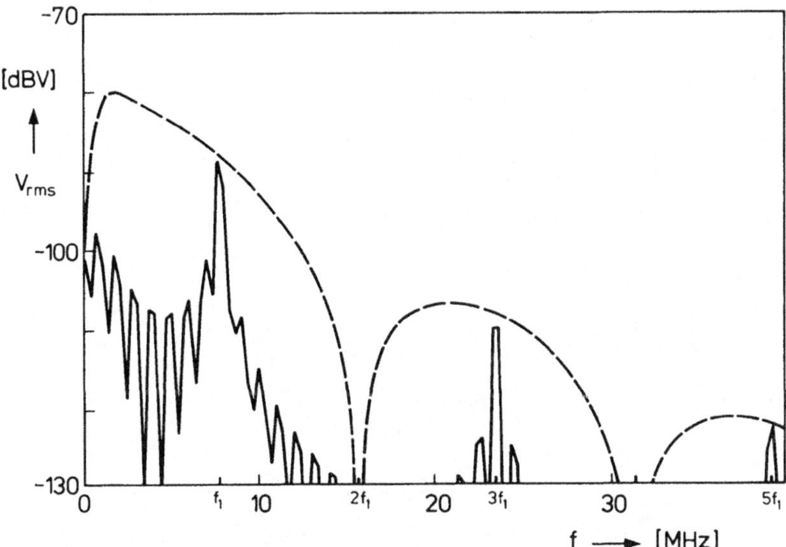

Fig. 13.10 Output spectrum of a BPH with nine ideal gaps, $a = 0.4$ µm and $g = 0.2$ µm. The dashed curve is the hull curve of the spectrum.

and with a gap length g in (13.22). Since $a = 2g$, the even filter harmonics coincide with the zeros of the overall hull curve and so no major peaks at $2f_1$ and $4f_1$ are visible in the figure.

Whereas in this example the frequency of operation is preferably chosen at the frequency of the first filter harmonic, f_1, in the following section BPHs will be described that preferably operate at frequencies near the frequencies of higher filter harmonics.

13.4.4 BPHs composed of several filters

The title of this section indicates that the gap configuration can be thought of as built up from several filters. We will call this number of filters N_f. Each filter has a different distance $a_j = a_1$ to a_{N_f} between its gaps. As an example the gap configuration of a BPH that consists of 3 filters with 6 or 7 gaps each, at distances of $a_1 = 0.60$ μm, $a_2 = 0.64$ μm and $a_3 = 0.68$ μm, is given by the 3 bar diagrams in Fig. 13.11. The

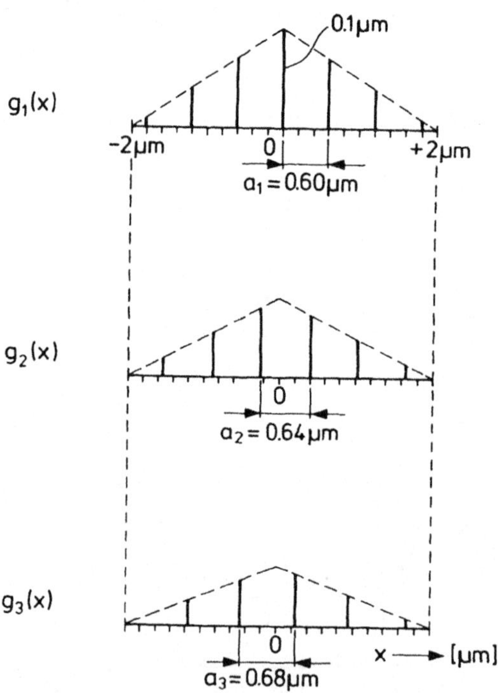

Fig. 13.11 Bar diagram of the gap length versus the gap position of the three sets of uniformly distributed gaps, (denoted by $g_1(x)$, $g_2(x)$ and $g_3(x)$) that can be distinguished in table 13.2. The (not maximally flat) triangular hull curves are represented by the dashed lines.

Table 13.2 Definition of the four heads in Fig. 13.12 and the calculated magnetic and electric characteristics of the heads at about 15 MHz, assuming that the core of the heads is made of MnZn ferrite and the shape of the core is as defined in table 8.8. p_i, g_i, a_j and G are in μm and L is given in μH. $|\eta|$ approaches $|\eta_{int}|(= 0.83)$ for heads 2, 3 and 4.

	head 1		head 2		head 3						head 4					
	p_i	g_i	$\{p_i\}$	$\{g_i\}$	$\{p_i\}$			$\{g_i\}$			$\{p_i\}$			$\{g_i\}$		
			−1.8	0.01	−1.800			0.01			−1.800			0.010		
			−1.6	0.02			−1.813			0.02			−1.813			0.012
			−1.4	0.03		−1.493			0.03			−1.493			0.027	
			−1.2	0.04	−1.200			0.04			−1.200			0.040		
			−1.0	0.05			−1.133			0.05			−1.133			0.030
			−0.8	0.06		−0.853			0.06			−0.853			0.054	
			−0.6	0.07	−0.600			0.07			−0.600			0.070		
			−0.4	0.08			−0.453			0.08			−0.453			0.048
			−0.2	0.09		−0.213			0.09			−0.213			0.081	
	0.0	0.08	0.0	0.10	0.000			0.10			0.000			0.100		
			0.2	0.09			0.227			0.09			0.227			0.054
			0.4	0.08		0.427			0.08			0.427			0.072	
			0.6	0.07	0.600			0.07			0.600			0.070		
			0.8	0.06			0.907			0.06			0.907			0.036
			1.0	0.05		1.067			0.05			1.067			0.045	
			1.2	0.04	1.200			0.04			1.200			0.040		
			1.4	0.03			1.587			0.03			1.587			0.018
			1.6	0.02		1.707			0.02			1.707			0.018	
			1.8	0.01	1.800			0.01			1.800			0.010		
G_j								0.34	0.33	0.33				0.340	0.297	0.198
a_j			0.2		0.60	0.64	0.68				0.60	0.64	0.68			

	head 1	head 2	head 3	head 4
G	0.08	1.00	1.00	0.835
N_g	1	19	19	19
$|\eta|$	0.24	0.74	0.74	0.72
L	1.52	1.46	1.46	1.48
Q	1.6	4.1	4.1	3.9

length of the gaps is proportional to the length of the bars. The positions and lengths of the gaps are also listed in Table 13.2 (head 4). Lengths and positions of the gaps are such that there is no overlap of gaps, which would be physically impossible. The output spectrum of this BPH is calculated with the aid of expression (13.22) and is plotted in Fig. 13.12 (curve 4). The third harmonics of the three filters are for $v = 3.14$ m/s at the frequencies $3f_1 = 3 \times 3.14/0.6 = 15.71$, $3 \times 3.14/0.64 = 14.73$ and $3 \times 3.14/0.68 = 13.86$ MHz respectively. The (zero) bandwith of a single filter is independent of the number of the filter harmonic, N_h (see also Fig. 13.2b) and equals about $2v/b_g = 3.14$ MHz since $b_g = 2$ μm for the triangular hull curves (it would be twice as large or $N_g a$ for a rectangular hull curve). Each filter is approximately sym-

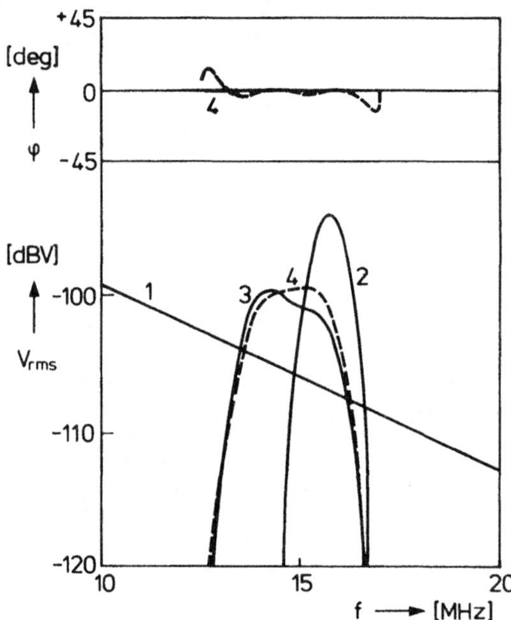

Fig. 13.12 Output voltage spectrum of a single-gap head and three BPHs with 19 gaps.
  Only the major peak at about 15 MHz is plotted ($v = 3.14$ m/s and the average $\langle a \rangle \approx 0.2$ μm). The heads are defined in table 13.2 and are characterized as follows:
head 1) Single-gap head ($N_g = 1$); $g = 0.08$ μm, which is about optimal for the reproduction of 15 MHz signals at $v = \pi$ m/s.
head 2) BPH ($N_g = 19$); one filter ($N_f = 1$), i.e. equal distances between subsequent gaps ($a = 0.2$ μm) and the filter-harmonic number N_h equals 1 at about 15 MHz; triangular hull curve, $g_{max} = 0.1$ μm.
head 3) BPH ($N_g = 7 + 6 + 6 = 19$); three filters ($N_f = 3$), i.e. different distances between the subsequent gaps of different filters ($a_1 = 0.6$ μm, $a_2 = 0.64$ μm and $a_3 = 0.68$ μm) and N_h equals 3 at about 15 MHz; triangular hull curve, $g_{max} = 0.1$ μm, like head 2.
head 4) BPH; equal to head 3, except that the amplitude of the triangular hull curves of $g_1(x)$, $g_2(x)$ and $g_3(x)$ decreases in this order, to compensate for the slope in the curves of heads 1 and 3 due to distance losses etc.

metric and their centres coincide at $x = 0$, hence the associated bandpass functions are real functions and can simply be added. In this circumstance the *zero bandwidth* of the co-operating filters of the BPH can safely be approximated by

$$B_0 \simeq N_h \left(\frac{\nu}{a_{\min}} - \frac{\nu}{a_{\max}} \right) + 2 \frac{\nu}{b_g} . \tag{13.26}$$

B_0 equals $(15.71 - 13.86) + 3.14 = 5$ MHz for the heads 3 and 4 and equals 3.14 MHz for head 2 with $a_{\min} = a_{\max}$. These bandwidths agree fairly well with those in Fig. 13.12.

To each filter an efficiency of $\eta_j = (G_j/G)\, \eta$ must be attributed, where G_j is the sum of the gap lengths of the gaps of filter j. By increasing G_j when the frequency of the filter increases one can compensate for the slope in the bandpass curve, due to increasing distance, gap and efficiency losses. Another consequence of the fact that the efficiency values that must be attributed to the individual filters is about η/N_f, is a decreasing amplitude when the number of filters increases. The effect is stronger as the separation between the frequencies of the filters is larger, and there is no effect when all filters have the same frequency. The expression

$$\langle V \rangle\, B_0 \approx \text{const.} , \tag{13.27}$$

where $\langle V \rangle$ is the average output within B_0, includes the effects of larger separation between the filters and more filters.

Hence the output of heads 3 and 4 is several dBs lower than the output of head 2, although the efficiencies of the heads are approximately the same. The difference between the output of head 1 and head 2 is (mainly) of a quite different origin, namely the great difference in efficiency due to the considerable difference in G.

From the large amount of freedom concerning the possible values of $a_{\min}$ and $a_{\max}$ in expression (13.26) it becomes clear that it was impossible to derive in section 13.4.1 a generally valid expression for the upper bound of the internal bandwidth (here B_0) based on only the overall width (here the width of the hull curve, b_g) and possibly the internal widths (here the gap lengths, g_i, or the 'auto-correlation' length, w_x, as defined in (13.10)). Only a lower bound ($a_{\min} = a_{\max}$) that corresponds to the uncertainty principle ($\sigma_k\sigma_x > \frac{1}{2}$) can be formulated.

Finally we will argue that it is often convenient to operate the BPH at the frequency of the filter harmonic with harmonic number N_h equal to the number of filters N_f. This is simply because then $\lambda \approx \langle a \rangle$ and so g can be chosen rather close to a, say $g_{max} \lesssim 0.5a$, such that G and thus η are almost maximum, whereas there is still no excessive deterioration of the hull curve $S(k)$ at $\lambda \approx 2g_{max}$. (Note that $\langle a \rangle$ is meant to be the average distance between all the gaps of the BPH and not only between those of only one filter.) From another point of view, if g_{max} becomes too large and approaches $\langle a \rangle$, then a few or more gaps tend to operate like one large gap with large gap losses, and good filter action can no longer be expected.

More precisely g_{max} can be calculated for maximum output by calculating the g_{max} or the average $\langle g \rangle$ for which the product $|\eta|$ times $|S|$ is maximum. Then the relative increase of the efficiency just compensates the relative decrease of the hull curve with increasing $\langle g \rangle$. Hence $\langle g \rangle_{opt}$ follows from

$$\frac{1}{|\eta|} \frac{\partial |\eta(f, N\langle g \rangle)|}{\partial \langle g \rangle} + \frac{1}{|S|} \frac{\partial |S(k, \langle g \rangle)|}{\partial \langle g \rangle} = 0, \qquad (13.28)$$

where $f = v/\langle a \rangle$ and $k = 2\pi/\langle a \rangle$ are fixed.

In the average gap length approximation (13.23):

$$S = \frac{\sin(\pi\langle g \rangle/\langle a \rangle)}{\pi\langle g \rangle/\langle a \rangle}, \qquad (13.29)$$

and using the accurate expression for the efficiency in the Thevenin representation of a head (8.25):

$$\eta = \frac{\eta_{int}}{1 + \mu_0 Wh\, Z_{int}/N_g\langle g \rangle}, \qquad (13.30)$$

this maximization leads to the following expression for the *optimum ratio* $\langle g \rangle_{opt}/\langle a \rangle$:

$$\boxed{\cot(\pi y) - \frac{1}{\pi} \frac{y + \alpha x}{(y + \alpha x)^2 + (x\sqrt{1 - \alpha^2})^2} = 0}, \qquad (13.31)$$

where
$y \equiv \langle g \rangle_{\text{opt}}/\langle a \rangle$ is the 'optimum filling factor',
$x \equiv \mu_0 Wh|Z_{\text{int}}|/N_g\langle a \rangle$ is the ratio between 'an effective internal core
     length' and 'the overall length of the series of gaps'.
and $\alpha \equiv \text{Re}\{Z_{\text{int}}\}/|Z_{\text{int}}|$.

The efficiency at the optimum filling factor equals, according to (13.30):

$$\boxed{\eta_{\text{opt}} = \frac{\eta_{\text{int}}}{1 + xy}}. \qquad (13.32)$$

The product $|\eta||S|$ is calculated from (13.29) and (13.30) for the bandpass heads 2, 3 and 4 at 15 MHz, where $Z_{\text{int}} = 143 + j205$ [MA/Vs], and the result is given in Fig. 13.13 for a varying filling factor, $\langle g \rangle/\langle a \rangle$. In agreement with (13.31), the maximum at $\langle g \rangle/\langle a \rangle = 0.22$ is such that the twice as large $\langle g \rangle_{\text{max}}$ of the triangular hull curves of the gap configurations of the heads 2, 3 and 4 will be close to $0.5\langle a \rangle$.

The results of the implicit expression (13.31) for the optimal filling factor are plotted in Fig. 13.14. The shape of the curve of the optimum filling factor y versus the ratio x between the 'effective internal core length' and the 'overall length of the series of gaps' depends slightly on the ratio between the real and imaginary part of Z_{int}. All practical ratios $\eta_{\text{opt}}/\eta_{\text{int}}$ are included in the figure by the choice of the variable x between

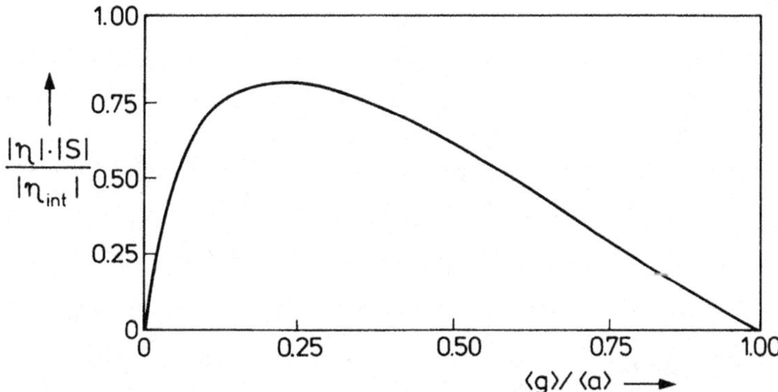

Fig. 13.13 The change of the output of the BPHs 2, 3 and 4 of table 13.2 and fig. 13.12 (described by the product $|\eta||S|$ divided by the constant $|\eta_{\text{int}}|$) versus the filling factor $\langle g \rangle/\langle a \rangle$.

According to the gap length data in table 13.2, $\langle g \rangle/\langle a \rangle = 0.22$ for head 4 and 0.26 for heads 2 and 3. This is close to the optimum ratio $\langle g \rangle_{\text{opt}}/\langle a \rangle = 0.22$ read from the plot.

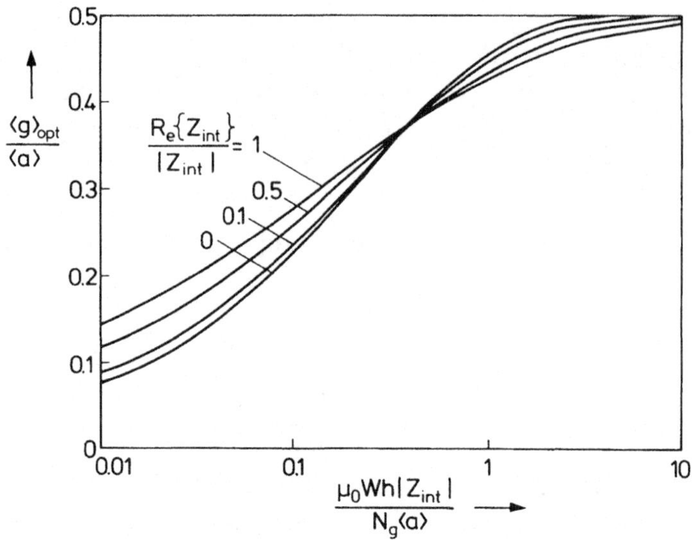

Fig. 13.14 The optimum filling factor, $\langle g\rangle_{\mathrm{opt}}/\langle a\rangle$, versus the ratio between the 'effective internal core length' and the 'overall length of the series of gaps', $\mu_0 Wh|Z_{\mathrm{int}}|/N_g\langle a\rangle$.

0.01 and 10, since according to expression (13.32) and the results in the figure for y, $|\eta_{\mathrm{opt}}|/|\eta_{\mathrm{int}}|$ ranges from 0.9 to 0.05.

It can easily be derived from (13.31) that:

$$\langle g\rangle_{\mathrm{opt}}/\langle a\rangle \; \genfrac{}{}{0pt}{}{\to 0}{\to \frac{1}{2}} \; \text{when } \left(\mu_0 Wh|Z_{\mathrm{int}}|/N\langle a\rangle\right) \genfrac{}{}{0pt}{}{\leqslant}{\geqslant} 1/\pi \;, \qquad (13.33)$$

see also Fig. 13.14.

This is as expected; when the efficiency increase with $\langle g\rangle/\langle a\rangle$ is small then of course $\langle g\rangle_{\mathrm{opt}}/\langle a\rangle$ is small, and when the efficiency increase is large then $\langle g\rangle_{\mathrm{opt}}/\langle a\rangle$ is large and approaches the maximum value $\frac{1}{2}$ when the efficiency is so small that the relative increase in η is constant up to $\langle g\rangle = \frac{1}{2}\langle a\rangle$.

As a rule of thumb, $N_h = N_f$ and $\langle g\rangle \approx \frac{1}{4}\langle a\rangle$ can be chosen.

13.4.5 Tolerances

From the manufacturing method described in the following section it will become clear that the length of the central gap is very likely to deviate from the intended value. The effect of this possible deviation on the location and shape and amplitude of the first major peak and surrounding minor peaks in the frequency spectrum of head 2 in table 13.2 has been calculated, and the results are shown in Fig. 13.15.

The main effect of the 50% too large or too small central gap is a shift of the major peak of the BPH to a 0.33 MHz, i.e. 2.1%, lower or higher frequency. This effect can be seen from the change of $\langle a \rangle$ by the factor $(19a \pm \tfrac{1}{2}g_c)/19a = 1 \pm 0.013$ for the present original distance of $a = 0.2$ μm between all the 19 gaps and original central gap length of $g_c = 0.10$ μm. The actual effect, 2.1%, is larger than this calculated 1.3% because the distance near the relatively large central gaps changed. (For the same reason the width of the major and minor peaks is about twice as large for this BPH with a triangular hull curve than for a BPH with a rectangular hull curve with the same amount of gaps and the same distances between the gaps. See also figure 13.7 in section

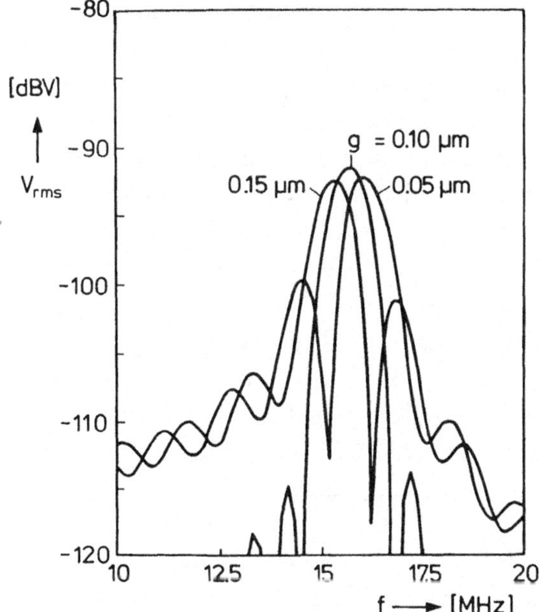

Fig. 13.15 Change of the frequency spectrum of head 2 due to an approximately 50 percent change of the length of the central gap.

(The distances between the subsequent gaps are kept constant in the calculation, except the distance between the centre of the central gap to the centres of the two neighbouring gaps. These two distances are changed by half the change of the length of the central gap. In this way just the change of the central gap length, which is possible in practice, is modelled. The efficiency is assumed constant; $\eta = 1$.)

13.4.2 and the remark concerning the bandwidth of an individual filter of the BPH described in section 13.4.4.)

Another effect is the considerable increase in the amplitude of the minor peaks at the cost of a slight reduction of the amplitude of the major peak. This was also observed in the spectrum of the 3-fold filter BPH in the previous section, where the effect was much more pronounced due to the much larger deviation in the series a_i values.

However no widening of the major peak nor any clear change of the shape of the (central part) of the major peak is observed when the central gap deviates from its nominal value of 0.1 μm.

What can be said in general about the permitted tolerance of the length of the central gap and the permitted tolerance of the layer thicknesses forming the other gaps and all the soft-magnetic layers between the gaps?

In practice the deviation of the thickness of those layers with respect to the nominal thicknesses will be approximately the same for all the magnetically-isolating layers forming the gaps (except the central gap) and will be about the same for all the intermediate layers. The deviation of the thickness of one pair of layers, Δa_i, is then roughly independent of i. Such a systematic error Δa may lead to a shift Δf_{N_h} of the frequency of the N_h^{th} filter-harmonic that exceeds the zero bandwidth of that filter, B_0, especially when the number of gaps of that particular filter is large. Therefore one must require that:

$$\Delta f_{N_h} \ll B_0 \tag{13.34}$$

Division by the frequency of the N_h^{th} filter harmonic $f_{N_h} = N_h f_1$ on both sides and choosing $N_h = N_f$ (while realizing that $\Delta f_{N_h} = N_h \Delta f_1$, $\Delta f_1/f_1 = \Delta a/a$ and $B_0/N_f f_1 \simeq 1/N_g$ for a rectangular hull curve and $2/N_g$ for a triangular hull curve) leads to:

$$\boxed{\frac{\Delta a}{a} \ll \frac{1}{N_g} \text{ to } \frac{2}{N_g}} \, , \tag{13.35}$$

for the assumed *systematic errors*.

The spread, Δa, for *non systematic* random errors is allowed to be $\sqrt{N_g}$ times larger, i.e.

$$\boxed{\frac{\Delta a}{a} \ll \frac{1}{\sqrt{N_g}} \text{ to } \frac{2}{N_g}} . \qquad\qquad (13.36)$$

When only one layer thickness deviates, e.g. the layer that determines the central gap length, then the error is even allowed to be about N_g times larger than in the case of the systematic error. The reason is that $\Delta\langle a\rangle = \Delta g_c/N_g$ then roughly replaces Δa in (13.35). The correctness of this replacement was shown by the results in Fig. 13.15, at the beginning of this section. Hence:

$$\boxed{\frac{\Delta g_c}{g_c} \ll 1} . \qquad\qquad (13.37)$$

In the above expression $\ll$ means approximately $< 1/5$.

13.5 A simple experimental BPH

Before designing a simple BPH to be manufactured, we will summarize some rules of thumb for BPHs with gaps at equal distances, $a_i = a$. The given values are valid for a rectangular hull curve $g^h(x)$. The values in the case of a triangular hull curve are given within brackets if they deviate.

13.5.1 Summary of design rules for simple BPHs

1) First major peak frequency: $f_1 = v/a$.
2) Relative zero and half-height bandwidth:

$$\frac{B_0}{f_1} = \frac{2}{N_g}\left(\approx \frac{4}{N_g}\right) \qquad \frac{B_{\frac12}}{f_1} = \frac{1}{N_g}\left(\approx \frac{2}{N_g}\right) .$$

3) First null of the hull curve S in the frequency domain:

$$f_0 = \frac{v}{g}\left(\approx \frac{v}{\langle g_i\rangle}\right) .$$

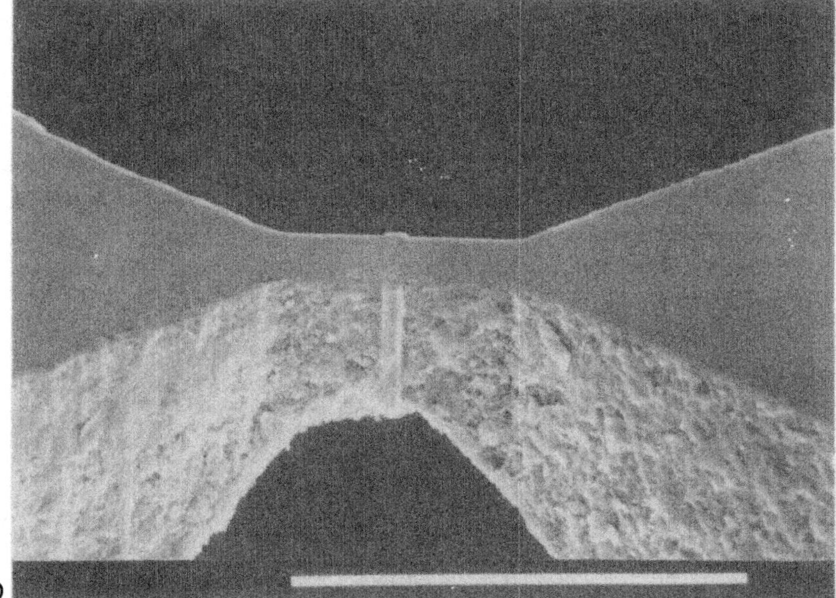

Fig. 13.16 SEM pictures of the BPH.
a) Side view of the BPH (1 mm/div.).
b) Top/side view of the bridge of the BPH (100 μm/div.).

c) Top view of the bridge (10 µm/div.).
d) Top view of the gaps in the bridge (1 µm/div.).

4) Permitted tolerances,
    if systematic: $\Delta a/a \lesssim 1/4N_g\ (1/2N_g)$,
    if non-systematic: $\Delta a/a \lesssim 1/4\sqrt{N_g}\ (1/2\sqrt{N_g})$,
    if only central gap: $\Delta g_c/g_c \lesssim \frac{1}{4}$.

13.5.2 Head technology

The BPH technology is very much like that of the MIG head, de-scribed in section 12.2.5. The substrate material is a [100] oriented polished MnZn ferrite. On this flat substrate layers of Mo and permal-loy (80/20 weight-percent NiFe) are alternately sputtered. On the last permalloy layer a Mo layer and then a Au layer are sputtered with a total thickness of $\frac{1}{2}g_c$. As in the MIG head, the Mo layer is used as an adhesive layer between the gold and the permalloy, for a good adhesion between the gold gap and ferrite core. Two of the above substrates are bonded by thermodiffusion of the Au layers at a temperature of 300°C and a pressure of 63 MPa for 10 minutes. The result is visible in the SEM picture of the cross-section of the series of layers in Fig. 13.16d. The bright line in the centre is the Au part of the central gap, g_c, which is surrounded by thin layers of Mo (hardly visible light-grey stripes). The other two times five light-grey stripes are the other Mo gaps and the ten dark-grey stripes are the soft-magnetic permalloy layers. Hence the number of gaps $N_g = 11$.

From the blocks of layers thus bonded head cores are sawn. The track width and coil chamber are shaped by laser-cutting; see the results in Fig. 13.16a, b and c. The orientation of the gap faces is now [100] while the large side faces and top and bottom faces of the BPH are [110] oriented, like the ferrite video heads in table 8.8 and the rings in Fig. 8.60.

Note that, when two identical substrates have been used to bond together at the Au, the number of gaps N_g is always odd.

13.5.3 Design

For $v = 3.14$ m/s, a bandpass at $f_1 = 7.85$ MHz with a bandwidth of $B_1 = 0.8$ MHz, we arive at the following BPH when choosing a rectan-gular hull curve:

$a = 0.4$ μm, $N_g = 11$ and $g = 0.2$ μm,
with the requirement that
$\Delta a/a \lesssim 2\%$ and $\Delta g_c/g_c \lesssim 25\%$.

So successively five layers of 0.2 μm Mo and five 0.2 μm permalloy layers are sputtered. Finally a 0.04 μm thick Mo and a 0.06 μm thick Au layer are sputtered to form half of the central gap.

The dimensions as well as the orientation of the core of the BPH are equal to those of the core of the MnZn ferrite head in table 8.8, except that the bridge length b_1 is 52 μm and the track width W is 21 μm. The number of windings N is 17.

13.5.4 Electrical and recording measurements and model calculation

The measured electrical characteristics L and Q of the BPH are accurately predicted by the analytical-head model calculation described in chapter 8; see the results in Fig. 13.17. L is lower and Q is considerably higher than in conventional ferrite video heads.

In contrast to the model calculation concerning Q until now, we took into account in this case the dc resistance of the windings, which is about 0.9 Ω for all heads in this book. The dc resistance causes the linear increase of Q with f up to a few megahertz, while the influence of μ'' of the core material is most pronounced above 3 MHz.

The calculated playback efficiency, $|\eta_H|$, is very high. It was not possible to measure this efficiency in the way described in section 8.2, since

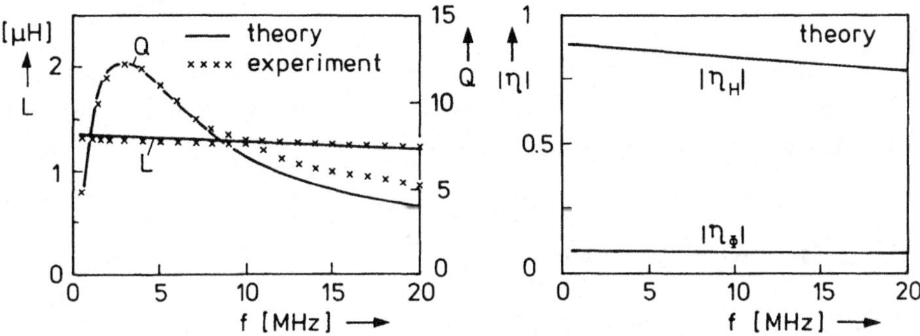

Fig. 13.17 Measured and calculated characteristics of the BPH with 11 gaps of 0.2 μm shown in Fig. 13.16. For the model calculations we took into account the μ', μ'' spectrum given in Fig. 8.60 and, as regards the Q factor, the 0.9 Ω dc resistance of the copper windings.

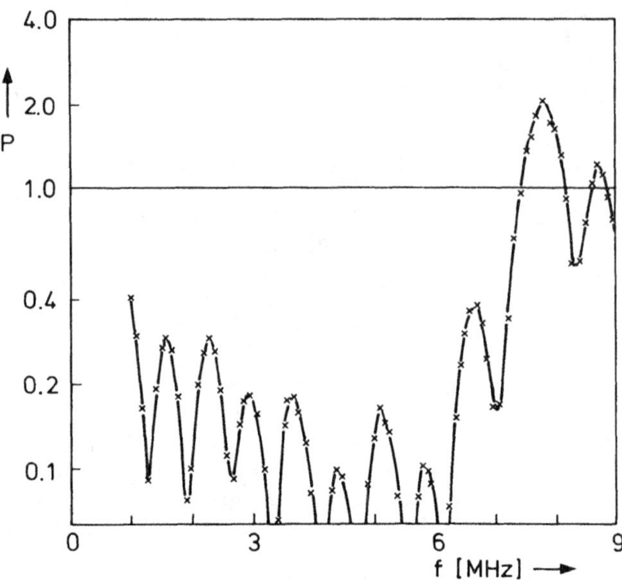

Fig. 13.18 Playback of the BPH, measured relative to that of a good 'reference' ferrite head (with an output of 60 mV per metre track width and per turn, at 5 MHz and against metal-powder tape).

the currents necessary to saturate the head are too large, due to the large total gap length G of 2.2 μm. So playback measurements must prove the high efficiency of the BPH by a superior read performance at 7.8 MHz. To check this, nothing is more suited than the playback figure, P, obtained from the cross measurements described in chapter 10. Fig. 13.18 shows the convincing results. At 7.8 MHz the output is two times higher than the output of a good ferrite head. The cross measurement was carried out on MP tape after running-in against CrO_2 tape.

After a longer time running-in against MP tape the output decreased, due to an increase of the hollow-out of the gaps region with respect to the surrounding ferrite.

13.5.5 Hollow-out measurements

Hollow-out is a result of differential wear. For the numerous metals and amorphous magnetic materials between single-crystalline MnZn ferrite, running against several tapes at a speed of a few metres per second, we always found hollow-out of the metal or amorphous material and never protrusion.

After running-in against the rather abrasive CrO_2 tape the hollow-out of the gap region of the BPH is only about 30 nm with respect to the ferrite level just beside this region.

This hollow-out measurement is carried out mechanically with the aid of a rebuilt Tencor α-step (see result in Fig. 13.19) or optically with an interference microscope (see result in Fig. 13.20). For a correct

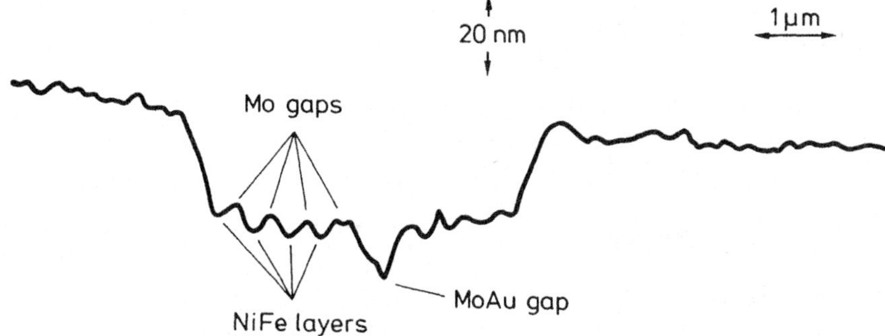

Fig. 13.19 Accurate mechanical measurement of hollow-out of the BPH's gap region after running-in against abrasive CrO_2 tape.

Even the enhanced hollow-out on the tiny NiFe layers and MoAu gap is visible. The hollow-out is about 30 nm on average.

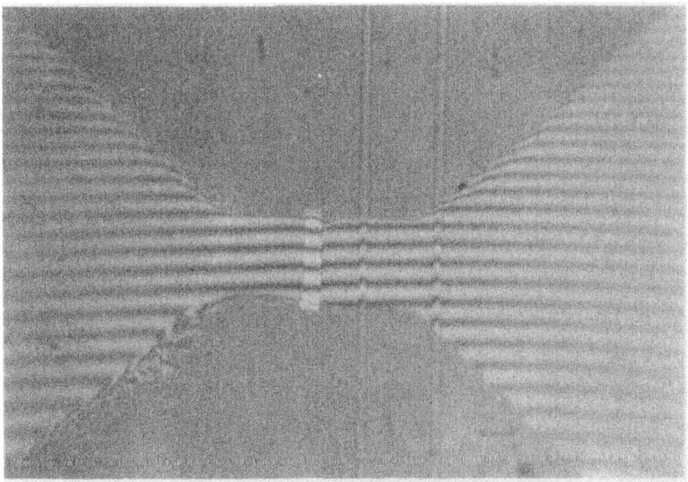

Fig. 13.20 Quick measurement of hollow-out of the BPH's gap region after running-in against CrO_2 tape.

Accurate measurement of the shift of the interference lines at the bright gap region points to a hollow-out of almost 40 nm when no correction is made for the difference in phase of the light reflected on the electrically-conducting metallic gap region and the light reflected from the electrically-insulating MnZn ferrite. With this correction the hollow-out value is about 30 nm, consistent with the mechanically-obtained result in Fig. 13.19.

interpretation of the interference picture in our case of hollow-out of one material with respect to another surrounding material, we refer to the following subsection.

After a longer time running against MP tape the hollow-out considerably increases and stabilizes at a value between 50 and 80 nm. The effects of hollow-out are pointed out in Section 12.2.3. For the present playback head the effect of the hollow-out, h_o, is approximately an extra playback distance-loss factor $\exp(-kh_o)$.

Interpretation of interference picture

The level difference is most quickly measured by way of an interference microscope. This microscope adds the spacial images of a flat mirror (with two vertical scratches, visible in Fig. 13.20) at approximately the same optical distance as the head's tape-facing surface to the about as bright image of the head's tape-facing surface. Hence the level difference between two subsequent bright or dark lines on the interference picture is half the wavelength of the about normal-incident light. Since we applied the green line of mercury light this is $\frac{1}{2} \times 546$ nm $= 273$ nm.

The shift of the lines from the ferrite to the Mo-NiFe gaps region is about $\frac{1}{7}^{\text{th}}$ of the distance between subsequent dark (or bright) lines. So one might assume a hollow-out of $\frac{1}{7} \times 273$ nm $\simeq 39$ nm. However we calculated, see Appendix 13.3, from data obtained by ellipsometry [13.5] that the phase of the wave reflected at metallic layers lags about $9°$ with respect to the phase of the wave reflected at ferrite, for the green mercury line. Hence the hollow-out is about $\frac{1}{2} \times (9°/360°) \times 546$ nm $= 7$ nm less than that above 39 nm. With this correction in mind the average results of the many interference measurements coincide with the mechanical hollow-out measurements on MIG heads. This is also observed in the case of the present BPH, as can be seen from the results of the accurate mechanical (tencor) measurement in Fig. 13.19.

So one has to be alert when measuring level differences between different materials with the aid of interference pictures; a correction may then be of importance when the level difference is small compared to the wavelength of the incident light and, as a rule of thumb, differences in the brightness of the two materials are high.

13.6 Discussion

In the BPH we used Mo as an adhesive layer in the central gap and as gap material in the remaining gaps. As was pointed out in section 12.2.6, this has the drawback of reducing the length and the sharpness of the gaps when no diffusion and electrochemical barrier like SiO_2 is applied between the originally non-magnetic Mo and the permalloy or sendust soft-magnetic layers. In principle one can compensate for the reduction of the gap length in the design, so as to obtain the optimum gap length for the highest product of efficiency times hull curve $S(k)$. The effect of the reduced sharpness of the gaps cannot be compensated, as appears from the description of the gradual gap in the Chapters 4 and 12. Another problem is the hollow-out of the gaps region due to the use of permalloy soft-magnetic layers. Both gradual gaps and hollow-out lead to a worsening of the playback behaviour at small wavelengths.

In a better design, which is absolutely necessary when going to an even smaller wavelength than 0.4 μm, gradual gaps and hollow-out have to be avoided. This can be done, for instance, by means of the measures that were successfully applied to our MIG heads, which originally faced the same problems:
– Use SiO_2 gaps and a SiO_2-Mo-Au central gap, to avoid diffusion of substances into the Mo and electrochemical reactions.
– Use *sendust* instead of permalloy in order to reduce the hollow-out.
The necessary small gap lengths (less than 0.1 μm) can readily be produced in thin-film techniques.

Because of the small bandwidth the application of a single BPH is restricted to the playback of information with a low data rate. For application in for instance a video recorder one has the chroma or the (hifi) audio signals to consider. Transformation of the small-bandwidth signal to the bandpass of the BPH is possible with the aid of a carrier having a frequency equal to the central frequency of the bandpass. Because of the superior sensitivity of the BPH this can be a frequency above the highest frequencies usually applied. If this is applied to the chroma signal, a wider frequency band becomes available for the luminance signal, such that the quality of the video television pictures will improve. The concept can be applied to digital information as well.

The simultaneous use of several BPHs, with partly overlapping bandpasses, opens the possibility of broadband application of the superior BPHs, but makes high demands upon technology and electronic timing. It is questionable whether this application will materialize, at least as far as consumer products are concerned.

The phase characteristic can just as well be adjusted as the amplitude characteristic of the BPH by choosing a particular non-symmetric gap configuration. It may be desirable to derive rules of thumb for the design of non-linear phase characteristics.

13.7 Conclusion

It has been shown, theoretically as well as experimentally, that BPHs may have better magnetic and electric characteristics than other types of heads. When the gap height is very small, as is allowed in heads that are not in contact with the medium and do not wear, those advantages vanish.

Some remaining problems, in the experimental head, gradual gaps and hollow-out, will have to be solved before going to smaller wavelengths and higher frequencies where the BPH will then be far superior to all other known head types for in-contact recording. There is experimental evidence from our experience with these problems in MIG heads that they can be solved by application of the up-to-date technology used in our present MIG heads.

It has been shown theoretically that the efficiency of the head increases and the impedance decreases as the bandwidth of the head becomes smaller.

Considerable attention has been paid to models for accurate designs, criteria for optimum designs and rules of thumb for rough designs of BPHs. Both the amplitude characteristic and the phase characteristic can be adjusted at will, within certain limitations. Subsequent equalization would then no longer be necessary.

Appendix 13.1 Assumptions concerning write and read situation

In order to calculate the output voltage, the write process for the writing head has to be modelled and the distance loss during reading with the BPH has to be incorporated. This has been done in chapter 7 and in the chapters 2 to 6 respectively. In Appendix 7.1 the final expression for the rms value of the output voltage has been given. For the *GLF* in that expression (a7.1) the *GLF* of the BPH must now be substituted and for the efficiency, η, the efficiency of the BPH. This efficiency still has to be calculated in section 13.3.

In all computed examples in this chapter the following data have been used:
- Writing head with a saturation magnetic flux density of $B_s = 1$ T and a gap length of $g_r = 0.3$ μm.
- Head-to-tape distance both for the writing head and the BPH $d_4 = 0.02$ μm. Head-to-tape velocity $v = 3.14$ m/s.
- Recording medium: vapour-deposited metal (ME) tape with $H_c = 66$ kA/m, $B_r = 0.38$ T, $\mu_r = 1.75$ and a coating thickness t_3 of 0.16 μm.
- BPH: number of windings $N = 18$, track width $W = 20$ μm and efficiency $\eta = 1$ unless otherwise mentioned.

Due to an inaccuracy in the early expression that described the permeability effects and is used for most calculations in this chapter, there may be slight differences with respect to the graphs obtained from the correct expression (a7.1).

Appendix 13.2 Noise

In Fig. a13.1 the equivalent circuit for playback with a head and an amplifier producing noise is given. Disregarding a possible small corre-

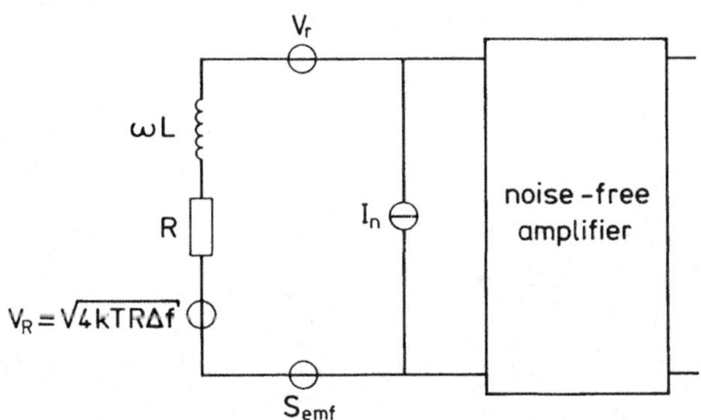

Fig. a13.1 Equivalent circuit of a pre-amplifier and head producing noise.
- U_R is the Johnson noise produced by the real part R of the head's impedance $Z \equiv R + j\omega L$, which automatically includes Johnson noise produced in the dissipative core of the head.
- S_{emf} is the induction voltage and includes signal and noise from the tape and noise from irregular domain-wall movements and so on in the core.
- V_r and I_n are the equivalent input noise sources of the amplifier actually producing noise.

lation between the equivalent noise-current and noise-voltage sources, I_n and V_r, and disregarding the input capacitance, the noise figure F defined as the ratio between the total noise power (spectral density) and the noise power caused by the real part of the head's impedance then equals:

$$F \equiv \frac{N_\mathrm{tot}}{N_\mathrm{R}} = \frac{4kTR + I_\mathrm{n}^2|Z|^2 + V_\mathrm{r}^2}{4kTR}. \tag{a13.1}$$

This ratio is minimal with respect to $|Z|$, while Q and hence $|Z|/R$ will remain constant, when:

$$|Z| = |Z|_\mathrm{opt} = \frac{V_\mathrm{r}}{I_\mathrm{n}} = \sqrt{R_\mathrm{r}R_\mathrm{n}} \tag{a13.2}$$

where $R_\mathrm{r} \equiv V_\mathrm{r}^2/4kT$ and $R_\mathrm{n} \equiv 4kT/I_\mathrm{n}^2$ have been introduced. Hence:

$$F_\mathrm{opt} = 1 + 2\,\frac{|Z|_\mathrm{opt}}{R_\mathrm{opt}}\,\sqrt{R_\mathrm{r}/R_\mathrm{n}}. \tag{a13.3}$$

In Fig. 13.6 it is assumed that the above condition is fulfilled by transforming the actual head impedance Z by way of an ideal transformer to the desired value $(N_2/N_1)^2|Z| = |Z|_\mathrm{opt}$, where N_1 is the number of transformer windings on the head side and N_2 the number of windings on the amplifier side. Hence

$$F_\mathrm{opt} = 1 + 2\,\frac{|Z|}{R}\,\sqrt{R_\mathrm{r}/R_\mathrm{n}}. \tag{a13.4}$$

For ordinary video heads, $|Z|/R$ is 5 to 10 at 4.5 MHz; see also table 8.8. Practical values for the equivalent-noise resistances in the MHz range for a pre-amplifier with a field-effect transistor and some feedback in the first stage, are $R_\mathrm{r} \simeq 100\ \Omega$ and $R_\mathrm{n} \simeq 2.5/f^2$ MΩ with f in MHz. Hence $R_\mathrm{r}/R_\mathrm{n} \simeq 1/1000$ and $|Z|_\mathrm{opt} \simeq 3$ kΩ at 5 MHz. Our video heads from table 8.7 have $L \simeq 1 \sim 2\ \mu$H or $|Z| \simeq 50\ \Omega$ at 5 MHz, such that a transformer ratio of $N_2/N_1 = 7$, which we use, is almost optimal at 5 MHz. If Q is 10, then $F_\mathrm{opt} \simeq 1.3$. According to (a13.1), most of this noise is caused by $R \neq 0$. Consequently a considerable decrease of the system noise is possible when Q increases.

The total noise power for a given quality factor Q with respect to the total noise in a situation *with the same L* but with a quality factor Q_{ref} equals:

$$\frac{N_{tot}(Q)}{N_{tot}(Q = Q_{ref})} = \frac{R_{el}(Q) + R_{tin}}{R_{el}(Q_{ref}) + R_{tin}} \qquad (a13.5)$$

with, according to (a13.4):

$$R_{el} = R + 2|Z|\sqrt{R_r/R_n} = \omega L/Q + 2\sqrt{(\omega L)^2 + (\omega L/Q)^2}\ \sqrt{R_r/R_n}.$$

The inverse of this ratio represents the signal-to-noise ratio. For different ratios of the tape-induced noise (characterized by the equivalent noise resistance $R_{tin} \equiv V_{tin}^2/4kT$) to the noise of the electronics when Q_{ref} is chosen to be 10, $R_{tin}/R_{el}(10) \equiv N_{tin}/N_{el}(10)$, results are given in Fig. 13.6a ($R_r/R_n = 1/200$) and in Fig. 13.6b ($R_r/R_n = 1/2000$).

The sensitivity of the signal-to-noise ratio to changes in Q is large when the noise of the amplifier and the tape-induced noise are small.

Appendix 13.3 Correction for hollow-out values determined from interference patterns

From Maxwell's equation it follows easily for one-dimensional waves that:

$$\frac{\partial E_z}{\partial y} = -j\omega\mu H_x \qquad\qquad -\frac{\partial H_x}{\partial y} = (\sigma + j\omega\varepsilon)E_z \qquad (a13.6a)$$

and

$$\frac{\partial E_x}{\partial y} = j\omega\mu H_z \qquad\qquad \frac{\partial H_z}{\partial y} = (\sigma + j\omega\varepsilon)E_x. \qquad (a13.6b)$$

Comparison with equation (11.2 + 3) in section 11.3.1 shows that R_f is replaced by $j\omega\mu$ and R_{gf} by $(\sigma + j\omega\varepsilon)^{-1}$ for both sets of equations, while U is replaced by E_z or E_x and Φ is replaced by $-H_x$ or H_z. With these replacements in mind, the results in section 11.3.1 directly apply to E_z and $-H_x$ and to E_x and H_z. For instance the real characteristic length λ is replaced by the following complex λ_c:

$$\lambda_c = \frac{1}{\sqrt{(\sigma + j\omega\varepsilon)j\omega\mu}}, \tag{a13.7}$$

which describes now both the phase change and the fall-off of the waves $\sim \exp(\mp y/\lambda_c)$ in the $\pm y$ direction by $\mathrm{Im}\{1/\lambda_c\}$ and $\mathrm{Re}\{1/\lambda_c\}$ respectively. The reflection factor $\Gamma \equiv H_{x-}/H_{x+} = H_{z-}/H_{z+} = E_{x-}/E_{x+} = E_{z-}/E_{z+}$ for a wave in the $-y$ direction becomes:

$$\Gamma = \frac{Z_L - Z_0}{Z_L + Z_0} \tag{a13.8}$$

with $Z = \sqrt{j\omega\mu/(\sigma + j\omega\varepsilon)}$. No subscript denotes the material (medium) to be characterized and the subscript 0 denotes free space. With the aid of the complex refraction index, n, defined as the ratio between the 'complex characteristic length' in free space and that in the surface of the medium:

$$n \equiv \frac{\sqrt{(\sigma + j\omega\varepsilon)j\omega\mu}}{j\omega\sqrt{\varepsilon_0\mu_0}}, \tag{a13.9}$$

the reflection factor can be expressed in the more common form for calculations in optics:

$$\Gamma = \frac{1 - n/\mu_r}{1 + n/\mu_r}. \tag{a13.10}$$

The relative permeability μ_r can always be taken 1 at light frequencies. Usually n is written as $(1 - j\varkappa)\mathrm{Re}\{n\}$, where $\varkappa$ is the so-called absorption index. With this definition the wave (in the $-y$ direction) is proportional to

$$e^{y/\lambda_c} = e^{j2\pi y/\lambda_m} \cdot e^{y/(\lambda_m/2\pi\varkappa)} \tag{a13.11}$$

with $\lambda_m \equiv v_m/f$ being the wavelength in the medium and $v_m = 1/(\sqrt{\varepsilon_0\mu_0}\,\mathrm{Re}\{n\})$ the velocity of light in the medium. The first factor in (a13.11) shows that the wavelength and velocity in the medium are reduced by the factor $\mathrm{Re}\{n\}$ compared to those in vacuum. The second factor in (a13.11) describes the loss in the medium due to the conductivity and amounts to a factor $e \simeq 2.71$ for every distance $\lambda_m/2\pi\varkappa$ in the $-y$ direction.

The complex refraction index is determined from the change of the ellipticity of the reflected light with respect to that of the incident light at a non-normal angle of incidence, called reflection (or surface) ellipsometry [13.5].

For metals one finds in the literature values for $\mathrm{Re}\{n\} \simeq 2$ and for the absorption index $\varkappa \simeq 1.7$. Due to this strong absorption caused by the high surface conductivity, the reflection is rather high, as will be clear from substitution of the data in (a13.10). For the electrically insulating single-crystalline MnZn ferrite, the following data are measured: $\mathrm{Re}\{n\} = 2.36$ and $\varkappa = 0.31$ (by courtesy of J.W.D. Martens). The low value of the absorption index due to the small (surface) conductivity explains the rather low reflectivity of the MnZn ferrite in the interference picture of Fig. 13.20. The ellipsometry measurements were carried out at the green mercury line ($\lambda = 546$ nm) that was also used for the interference picture.

Substitution of the above data with $\mu_r = 1$ into (a13.10) leads to a phase lag of $25°$ for the light reflected from the metal gap-cladding, whereas this is only $16°$ for the light reflected from the ferrite. Hence it seems from the interference pattern that the gap-cladding material is $((25° - 16°)/360°) \times \frac{1}{2}\lambda = 7$ nm further away. In the case of hollow-out of the metal with respect to the ferrite, one has to reduce the hollow-out value by this amount.

References

[13.1] J.J.M. Ruigrok, *Multigap magnetic reading head*, US Patent 4669015.
[13.2] Ernst H. Nordholt, *Design of high-performance negative feedback amplifiers*, Elsevier Scientific Publishing Company, Amsterdam, Oxford, New York, 1983.
[13.3] See for example: E.U. Condon and Hugh Odishaw (editors), *Handbook of Physics*, McGraw-Hill Book Company, chapter 5, 1967.
[13.4] J.B.M. Uffink and J. Hilgevoord, *Uncertainty principle and uncertainty relations*, Foundations of Physics, Vol. 15, 925-944 (1985).
[13.5] R.M.A. Azzam and N.M. Bashara, *Ellipsometry and polarized light*, North-Holland Publishing Company, Amsterdam, Oxford, New York, Tokyo, 1987.

Author index

Subject index

Printed in Dunstable, United Kingdom